Formation of the microcolonies on surfaces is an important bacterial survival strategy. These biofilms occur on both inert and living systems, making them important to a wide range of scientific disciplines.

This book first provides an analysis of the chemical, ecological and physical processes involved in the development of biofilms and their interactions with surfaces.

The next section deals with biofilms on non-living surfaces. Biofilms have important engineering implications, such as in mining industries, the corrosion of pipelines and pure and waste water industries. They also have medical significance when associated with the mouth, urinary tract and urogenital tract. In addition, they form in plant root systems and in animals, for example in the ruminant digestive tract, and so are agriculturally important. The final section examines these interactions with living surfaces.

PLANT AND MICROBIAL BIOTECHNOLOGY RESEARCH SERIES: 5
Series Editor: James Lynch

Microbial Biofilms

PLANT AND MICROBIAL BIOTECHNOLOGY RESEARCH SERIES
Series Editor: James Lynch

Microbial Biofilms

Edited by

Hilary M. Lappin-Scott
University of Exeter

and

J. William Costerton
Montana State University

CAMBRIDGE
UNIVERSITY PRESS

Published by the Press Syndicate of the University of Cambridge
The Pitt Building, Trumpington Street, Cambridge CB2 1RP
40 West 20th Street, New York, NY 10011–4211, USA
10 Stamford Road, Oakleigh, Melbourne 3166, Australia

First published 1995

Printed in Great Britain at the University Press, Cambridge

A catalogue record for this book is available from the British Library

Library of Congress cataloguing in publication data

Microbial biofilms/edited by Hilary M. Lappin-Scott and J. William Costerton.
   p. cm. – (Plant and microbial biotechnology research series; 5)
Includes bibliographical references and index.
ISBN 0 521 45412 3
1. Biofilms.   I. Lappin-Scott, Hilary M.   II. Costerton, J. W. III. Series.
QR100.8.B55M525   1995
576′.15–dc20   94–39377   CIP

ISBN 0 521 45412 3   hardback

WS

Contents

Part III Biofilms on the Surfaces of Living Cells

Contributors

M. J. Bazin
Division of Life Sciences
King's College
Campden Hill Road
Kensington
London W8 7AH

T. J. Beveridge
Department of Microbiology
College of Biological Science
University of Guelph
Guelph
Ontario N1G 2W1
Canada

M. G. Brading
Department of Biological Sciences
University of Exeter
Exeter EX4 4PS

M. R. W. Brown
Department of Pharmaceutical Sciences
Aston University
Birmingham B4 7ET

A. W. Bruce
Division of Urology
Department of Surgery
University of Toronto
Toronto
Ontario M5G 2C4
Canada

D. E. Caldwell
Department of Applied Microbiology and Food
Science
University of Saskatchewan
Saskatoon
Saskatchewan S7N 0W0
Canada

K.-J. Cheng
Research Station
Agriculture Canada
Lethbridge
Alberta T1J 4B1
Canada

J. W. Costerton
Center for Biofilm Engineering
Montana State University
Bozeman
Montana 59717
USA

F. G. Ferris
Department of Geology
Earth Sciences Centre
University of Toronto
22 Russel Street
Toronto
Ontario M5S 3B1
Canada

A. Fomsgaard
Department of Clinical Microbiology 7806
Rigshospitalet
Tagensvej 20
DK-2200 Copenhagen N
Denmark

P. Gilbert
Department of Pharmacy
University of Manchester
Manchester M13 9PL

A. E. Goodman
School of Microbiology and Immunology
University of New South Wales
PO Box 1
Kensington
NSW 2033
Australia

W. A. Hamilton
Department of Molecular and Cell Biology
Marischal College
University of Aberdeen
Aberdeen AB9 1AS

N. Høiby
Department of Clinical Microbiology 7806
Rigshospitalet
Tagensvej 20
DK-2200 Copenhagen N
Denmark

J. Jass
Department of Biological Sciences
University of Exeter
Exeter EX4 4PS

E. T. Jensen
Department of Clinical Microbiology 7806
Rigshospitalet
Tagensvej 20
DK-2200 Copenhagen N
Denmark

H. K. Johansen
Department of Clinical Microbiology 7806
Rigshospitalet
Tagensvej 20
DK-2200 Copenhagen N
Denmark

C. W. Keevil
Centre for Applied Microbiology and Research
Porton Down
Salisbury SP4 0JG

K. J. Kennedy
Department of Civil Engineering
University of Ottawa
Ottawa
Ontario K1N 6N5
Canada

A. Kharazmi
Department of Clinical Microbiology 7806
Rigshospitalet
Tagensvej 20
DK-2200 Copenhagen N
Denmark

S. Kinniment
School of Pure and Applied Biology
University of Wales
PO Box 915
Cardiff CF1 3TL

D. R. Korber
Department of Applied Microbiology and Food Science
University of Saskatchewan
Saskatoon
Saskatchewan S7N 0W0
Canada

G. Kronborg
Department of Clinical Microbiology 7806
Rigshospitalet
Tagensvej 20
DK-2200 Copenhagen N
Denmark

H. M. Lappin-Scott
Department of Biological Sciences
University of Exeter
Exeter EX4 4PS

J. R. Lawrence
National Hydrology Research Institute
11 Innovation Boulevard
Saskatoon
Saskatchewan S7N 3H5
Canada

J. W. C. Leung
Division of Gastroenterology
UC Davis Medical Center
45th & X Street, FOLB II, Building D
Sacramento CA 95817
USA

J. M. Lynch
School of Biological Sciences
University of Surrey
Guildford GU2 5XH

T. A. McAllister
Research Station,
Agriculture Canada
Lethbridge
Alberta T1J 4B1
Canada

R. J. C. McLean
Department of Biology
Southwest Texas State University
San Marcos
Texas 78666
USA

C. W. Mackerness
Centre for Applied Microbiology and Research
Porton Down
Salisbury
Wiltshire SP4 0JG

P. D. Marsh
Pathology Division
PHLS
Centre for Applied Microbiology and Research
Porton Down
Salisbury
Wiltshire SP4 0JG

K. C. Marshall
School of Microbiology and Immunology
University of New South Wales
PO Box 1
Kensington
NSW 2033
Australia

M. W. Mittelman
Center for Infection and Biomaterials
Toronto Hospital Bell Wing
200 Elizabeth Street
Toronto
Ontario M5G 2C4
Canada

J. C. Nickel
Department of Urology
Queen's University
Kingston
Ontario K7L 3N6
Canada

M. E. Olson
Health Sciences Centre
The University of Calgary
Calgary
Alberta T2N 4N1
Canada

D. Pearce
Division of Life Sciences
King's College
Campden Hill Road
Kensington
London W8 7AH

S. S. Pedersen
Department of Clinical Microbiology 7806
Rigshospitalet
Tagensvej 20
DK-2200 Copenhagen N
Denmark

T. Pressler
Danish Cystic Fibrosis Centre
Department of Pediatrics G 5002
Rigshopitalet
Blegdamsvej 9
DK-2100 Copenhagen Ø
Denmark

G. Reid
Department of Microbiology and Immunology
University of Western Ontario
London
Ontario N6A 5B8
Canada

J. Rogers
Centre for Applied Microbiology and Research
Porton Down
Salisbury
Wiltshire SP4 0JG

G. Southam
Department of Biological Sciences
Northern Arizona University
Flagstaff
Arizona 86011–5640
USA

J. J. Y. Sung
Department of Medicine
Prince of Wales Hospital
Chinese University of Hong Kong
Shatin
Hong Kong

J. T. Walker
Centre for Applied Microbiology and Research
Porton Down
Salisbury
Wiltshire SP4 0JG

J. W. T. Wimpenny
School of Pure and Applied Biology
University of Wales
PO Box 915
Cardiff CF1 3TL

R. C. Wyndham
Institute of Biology
Carleton University
Ottawa
Ontario K1S 5B6
Canada

Series Preface
Plant and Microbial Biotechnology

The primary concept of this series of books is to produce volumes covering the integration of plant and microbial biology in modern biotechnological science. Illustrations abound: for example, the development of plant molecular biology has been heavily dependent on the use of microbial vectors, and the growth of plant cells in culture has largely dawn on microbial fermentation technology. In both of these cases the understanding of microbial processes is now benefitting from the enormous investments made in plant biotechnology. It is interesting to note that many educational institutions are also beginning to see things in this way and are integrating departments previously separated by artificial boundaries.

Many definitions have been proposed for biotechnology but the only one which has specifically defined *environmental biotechnology* is that of the European Federation of Biotechnology as *The specific application of biotechnology to the management of environmental problems, including waste treatment, pollution control and integration with non-biological technologies.* The study of microbial biofilms is clearly an excellent illustration of environmental biotechnology. The manipulation and control of biofilms is of great interest to industries, including agriculture, chemicals and healthcare.

One of the leaders in the study of biofilms has been Bill Costerton, especially in his early studies when he produced superb electron micrographs to demonstrate the fascinating microbial assemblages which developed in biofilms. However, he rapidly proceeded to demonstrate important physiological functions which occured in these interesting layers. In 1986, Hilary Lappin-Scott joined him to work partly in Cambridge and partly in Calgary on the biofilms associated with oil wells, so starting a long and productive association. Hilary went on to the University of Exeter in 1990 to create a research group on biofilms which is proving to have substantive inputs in a range of environmental and industrial fields. I had known Hilary since her days as a research student at the University of Warwick and found myself talking to her about biofilms on more than one occasion when we were both warming-up at the start of London marathons! I was delighted when Hilary said that she would be prepared to contribute a volume to the series with Bill Costerton. They have produced a textbook which covers not only the fundamentals of this important subject, but also provides a range of diverse applications.

Jim Lynch

Introduction to Microbial Biofilms

J. William Costerton and Hilary M. Lappin-Scott

In any scientific examination that addresses a subject as basic as the mode of growth of bacteria it is prudent to begin by considering the successful prokaryotic communities that clearly predated the development of the eukaryotic cell. During the millions of years in which bacteria constituted the only life form on Earth, we visualize an extremely oligotrophic aquatic environment in which specific ecosystems were impacted by many factors (e.g. heat, acid) hostile to their survival. It is the nature of aquatic systems to flow from one ecosystem to another and we can imagine a primitive stream connecting permissive and non-permissive bacterial habitats in the nascent Earth. Once bacterial cells had evolved, the planktonic (floating) mode of growth would deliver them from one habitat to another until they perished in the first non-permissive locus. The sessile mode of growth as attached bacteria would allow these primitive organisms to colonize a permissive habitat and persist therein. Biofilm formation would allow these sessile organisms to trap and retain scarce organic compounds and to develop a focused attack on complex or refractory nutrients whose processing required time and/or the cooperation of one or more bacterial species. Biofilm formation would also change the microenvironment at the colonized surface in a colonized habitat and render its inhabitants less susceptible to hostile chemical, physical, or even biological (e.g. bacteriophage) factors. Each colonized habitat would become a stable crucible of genetic adaption and physiological cooperativity that would flourish in its own location but would also shed its component organisms as planktonic cells so that, if they sur-vived, they could establish a similar integrated biofilm community in any permissive habitat downstream.

Our image of aquatic systems in the nascent Earth militates against the survival of planktonic bacteria, and leads us to suggest that the sessile mode of growth and biofilm formation may have been the *sine qua non* of survival of newly evolved bacteria in this hostile environment. It is therefore germane to examine hostile oligotrophic environments on the modern planet to determine which mode of growth of prokaryotic cells is most successful. The ubiquity and predominance of bacterial biofilms was first noted in very oligotrophic high altitude alpine streams in Canada (Geesey *et al.* 1977) and subsequent detailed examinations of these systems clearly show that bacterial populations can only be maintained in their turbulent waters if these organisms live in biofilms adherent to available surfaces. In the equally hostile and oligotrophic Antarctic desert environment bacteria and algae can invade the exposed surfaces of rocks to produce complex biofilms or 'varnishes' whose matrices trap scarce rainwater and permit growth and primary production based on photosynthesis.

All modern bacteria are obviously descendant from the primitive forms that successfully colonized the planet Earth early in its biological history and their basic strategies of colonization and survival depend on patterns of phenotypic expression of their genetic material that made them successful in that primitive milieu. These patterns affect many modern processes because heat exchangers are fouled, pipelines are corroded, and medical devices are infected by recalcitrant slimy bacteria, because bacteria have long ago evolved a set of basic strategies to colonize and persist and to survive in permissive habitats.

Laboratory cultures represent the planktonic mode of growth

The phenomenon of bacterial adhesion to surfaces is clearly visible in routine light microscopic examinations of natural populations and it was elegantly described (ZoBell 1943) long before its relationship to ubiquitous biofilm formation was recognized. Later, descriptions of this process emphasized its initial reversibility (Marshall *et al.* 1971) and its putative mechanisms (Fletcher & Loeb 1979) but what has emerged is a whole spectrum of adhesion phenomena, that range from the very specific pilus-mediated adhesion of bacteria to specific tissues to totally non-specific exopolysaccharide-mediated adhesion of natural wild bacteria to all surfaces within a stream (Geesey *et al.* 1977). What is perhaps most important in this ongoing area of research, which is mired in detail but driven by the search for colonization-resistant materials, is that genetic examinations of the best known adhesion mechanisms show that they are highly conserved during evolution. It has long been recognized that simple animal or natural ecosystem passage of a bacterial strain that has lost surface structures and adhesion capability during repeated subcultures as a planktonic single species culture restores these structures (pili, exopolysaccharides) and this capability. In some instances these surface structures and the adhesion capability can be restored by culture in menstrua that contain surfactants or antibiotics at concentrations that kill planktonic cells totally lacking in protective surface structures (Govan 1975) but allow the survival of glycocalyx enclosed wild type cells. These simple observations, some of which date back to the 1930s, probably should have alerted us to the fact that the planktonic single species laboratory culture exerts a powerful selective pressure on a bacterial genome that eventually produces a 'stripped down' cell lacking in protective and adhesive surface structures that simply cannot survive in natural environments where adhesion and protection are of paramount importance.

It is very sobering to realize, over a century after the development of the planktonic single species laboratory culture (Koch 1881), that the cells we have been studying so assiduously are phenotypically locked in a planktonic mode of growth. This is at the opposite end of a phenotypic spectrum from the sessile mode of growth clearly seen to predominate in most natural environments. The classic laboratory culture has been extremely useful for the exploration of the genome-driven activities of bacterial species, but bacteria are protean creatures whose survival depends on their phenotypic responses to environmental factors and we have generally studied cells locked by their test tube environment into the planktonic mode of growth. Decades of productive research have yielded dividends in the control of planktonic diseases and in modern genetic engineering but have been less successful in the control of biofilm diseases and industrial and environmental microbiology. Now that we realize that new culture methods, several of which are described by Caldwell in Chapter 3, can mimic the biofilm mode of growth that predominates in nature, and in many heretofore recalcitrant bacterial diseases, we can look forward to a new and equally exciting explosion of practical sequelae of modern microbiological biofilm research.

Phenotypic responses to adhesion

Modern research using reporter genes has clearly shown that the adhesion event triggers the expression of genes controlling the production of bacterial components (for example, the alginate of *Pseudomonas aeruginosa*) necessary for continued adhesion and biofilm formation. Reporter gene systems constructed by Chakrabarty and by Deretic have been used by Geesey's group (Davies *et al.* 1993) and by Costerton's group (Hoyle *et al.* 1993) to show that adhesion triggers the expression of the alg C and other genes that control the production of phosphomannomutase and of other enzymes in the alginate synthesis pathway.

Parallel work with Gram positive pathogens, notably *Staphylococcus epidermidis*, has shown that adhesion triggers the expression of enzymes which produce exopolysaccharides that are pivotal in continued adhesion and biofilm formation and in the aetiology of device related bacterial infections (Costerton *et al.* 1987). These complex and focused reporter gene techniques have shown that adhesion triggers the rapid and specific phenotypic expression of several specific genes whose products are concerned with adhesion and biofilm formation. Parallel, general examinations, comparing the proteinaceous gene products made by sessile bacteria with those made by planktonic cells of the same species have shown (H. Yu and J. W. Costerton, unpublished observations) that adhesion changes the phenotypic expression of at least 30% of the proteins detectable in cell extracts by gel chromatography. Recent studies in Deretic's laboratory (Martin *et al.* 1993) indicate that a sigma factor similar to that involved in sporulation, and in the reversible rough–smooth lipopolysaccharide transformation in Gram negative bacteria, may be involved in the adhesion-mediated change between planktonic cells and sessile biofilm cells of the same bacterial species. If this fascinating hypothesis stands up under current intense scrutiny, the battery of phenotypic changes that occur as cells of bacterial species alternate between planktonic and sessile modes of growth will come to be regarded as a phase change mediated by a sigma factor that controls a whole cassette of genes related to adhesion and to biofilm formation. If biofilm bacteria do, in fact, constitute a different phase of phenotypic expression of the bacterial genome many of their observed characteristics, such as their almost complete resistance to antibiotics that are effective against planktonic cells of the same species (Nickel *et al.* 1985), will be partially explained.

We are presently studying the rate at which bacteria revert to the planktonic phase of phenotypic expression after they have become detached from established biofilms, by active shedding mechanisms or by simple fragmentation. These studies will provide insights into the nature of a phase change that may enable bacteria to control their cell surface components and to alternate between sessile and planktonic modes of growth to facilitate their colonization and survival within permissive habitats.

Biofilm structure

The confocal scanning laser (CSL) microscope has enabled us to examine living, fully hydrated,

biofilms and the use of this microscope has provided structural information that is especially valuable because it is direct and consequently unequivocal (Lawrence *et al.* 1991). The examination of hundreds of biofilms formed by dozens of different pure cultures and by several natural bacterial populations has clearly shown that biofilm bacteria grow predominantly in microcolonies of similar morphotypes (Fig. 0.1) interspersed between water channels that contain few bacterial cells and appear to contain a more permeable matrix material. These clearly heterogeneous surface-associated bacterial populations can now be examined, living and fully functional, by CSL microscopy and by the use of non-intrusive chemical probes and of physical microprobes (5–10 μm tip diameter) that can be positioned at any location within the biofilm and visualized by the CSL microscope. The concerted use of these complementary analytical tools has produced a conceptual image of bacterial biofilms that is truly amazing in its complexity and sophistication.

The structural heterogeneity of biofilms

Soon after the initial adhesion of bacteria to a surface, in either a single species culture or a mixed natural population, certain adherent cells proliferate and elaborate exopolysaccharides until they produce a microcolony in which morphologically similar 'sister' cells are embedded in a thick polysaccharide matrix (Fig. 0.1). As the biofilm thickens and matures individual microcolonies may lose their associations with the colonized surface and, in multispecies populations, cells of several species may come together to produce functional consortia (Kudo *et al.* 1987a, b) that carry out complex physiologically cooperative processes such as methane production (MacLeod *et al.* 1990). The microcolony is the basic growth unit of the biofilm and we consider bacterial growth to be sessile in nature if these microcolonies are produced, even if their final geometric configuration differs from that depicted in Fig. 0.1. Specific data attest to the limited permeability of the thick matrices surrounding individual microcolonies, in that CSL microscopy has shown that fluorescein-conjugated dextrans and other permeability probes penetrate the water channels but fail to penetrate the microcolonies within biofilms. Very recent work by

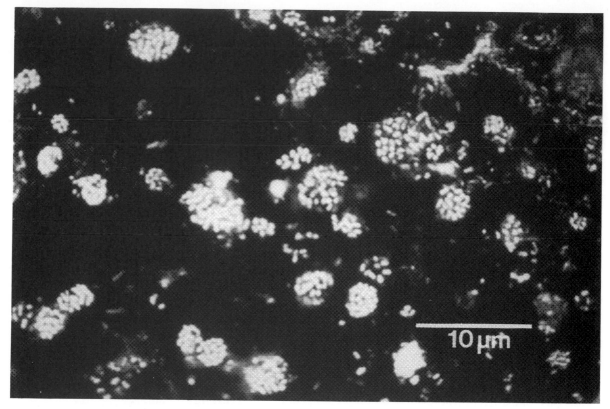

Fig. 0.1.   Confocal scanning laser micrograph of an optical section of a mixed species natural biofilm parallel to a colonized surface. Note the occurrence of discrete bacterial microcolonies interspersed with broad and relatively unpopulated water channels within this living, fully hydrated, biofilm that developed in the Bow River, Alberta, Canada. (Photograph courtesy of Garth James.)

Dr Lewandowski's group at the Center for Biofilm Engineering has indicated that the water channels that lie between and sometimes below these microcolonies are actually sufficiently permeable to allow convective fluid flow, and the same group has obtained NMR data to confirm flow within these channels. These data, obtained in direct examinations of living biofilms, combine to present a concept of biofilm structure that is revolutionary in its complexity and sophistication. Biofilm bacteria clearly live in dense matrix-enclosed microcolonies, where they are exposed to a bathing flow of modified bulk fluid through the less dense water channels that anastomose throughout even the thickest and most mature biofilms. These morphological data suggest a biofilm within which bacteria live in specialized microniches that are served by a primitive circulatory system within a stationary matrix-protected population adherent to surfaces within a flowing system.

The chemical and physical heterogeneity of biofilms

The planktonic mode of growth affords each individual bacterial cell an almost identical ecological niche, in that all cells communicate almost directly with the bulk fluid by simple diffusion. The simple immobilization of a bacterial cell within an anionic matrix introduces heterogeneity because these cells carry out many chemical functions, such as proton extrusion and oxygen consumption, and the matrix areas near the cells must, necessarily, differ from these further from the cells. If we then visualize different microcolonies, containing one or more physiological types of bacteria, within a biofilm, we must expect that the metabolic activities of these clusters of cells would produce loci with sharply different chemical environments. If we consider the simple cases of acid generation and oxygen consumption, we can state that adjacent areas of the biofilm will be different at a given moment in time, depending on the extent to which acid generation or oxygen consumption exceed the diffusion of protons or of dissolved oxygen through the biofilm matrix.

Because of the development of the CSL microscope we can now introduce both chemical (Lawrence *et al.* 1991) and physical probes (Lewandowski *et al.* 1993) into living biofilms and record such parameters as pH and dissolved oxygen concentrations at particular loci. Chemical probes are difficult to calibrate and physical probes may be somewhat intrusive but both serve to show local differences very accurately. Early work with pH sensitive chemical probes clearly showed that some bacterial microcolonies in both pure culture and mixed natural biofilms operated at pH values significantly lower than the water channels (Lawrence *et al.* 1991) and that individual cells within microcolonies were surrounded by an acid zone that may be produced by proton extrusion. Recent work with dissolved oxygen microelectrodes has produced equally unequivocal data to indicate heterogeneity within biofilms. When the microelectrode (tip diameter 5–10 μm) is advanced from the bulk fluid through the biofilm interface and into a bacterial microcolony (Fig. 0.2a) the dissolved oxygen concentration is seen to decrease at the interface and to reach truly anaerobic levels within the microcolony (Fig. 0.2b). When the same microelectrode is traversed only 100 μm laterally and advanced from the bulk fluid through the biofilm interface and into a water channel (Fig. 0.2c) much higher levels of dissolved oxygen are recorded (Fig. 0.2d). These simple and direct measurements of pH and of dissolved oxygen concentration are made in living biofilms and they provide unequivocal evidence of the basic chemical heterogeneity of these structurally complex adherent populations.

If we grasp these basic concepts of the structural and chemical heterogeneity of biofilms and begin to apply them to natural biofilm populations that have been described and defined during the past two decades, a fascinating picture of the sessile mode of growth begins to emerge. Ultrastructural observations of cellulose digestion by some cellulolytic bacteria (Cheng *et al.* 1984) showed that these organisms adhere to this insoluble substrate and produce deep pits into which they and their progeny eventually penetrate. We can infer a local concentration of cellulolytic enzymes, within a biofilm, that mediate a focused attack on a surface that is typical of many instances of biodegradation. Acid generation by specific microcolonies within a biofilm could mediate local focused attack on surfaces ranging from dental enamel to stainless steel. In instances in which the concerted metabolic activities of several bacterial species are necessary to biodegrade a complex substrate (e.g. bitumen) cells of these species would form a microcolony within a biofilm and that microcolony would mediate a local attack on the substratum. One of the most important inherent characteristics of bacterial biofilms is their capability of focused and cooperative biodegradation, and this characteristic depends entirely on the sustained juxtaposition of cells with each other and with surfaces that is a feature of the biofilm mode of growth.

If we re-examine the structure and activity of bacterial consortia (MacLeod *et al.* 1990) in the light of these recent revelations of biofilm heterogeneity a similarly gratifying concept emerges. The biofilm mode of growth positions a wide variety of bacterial cells at a surface and the individual cells replicate to initiate microcolony formation at a rate that depends on how well their particular microniche suits their physiological requirements. If a particular cell is unable to replicate it may simply persist, entrapped in the biofilm, until suitable conditions develop. If a particular cell requires acetate it will replicate if this substrate is supplied by a neighbouring cell, and this type of metabolic cooperativity often produces structural consortia of considerable complexity and metabolic efficiency (MacLeod *et al.* 1990). The chemistry of a particular microniche within a biofilm depends on both the delivery of bulk fluid components through the water channels and the metabolic activity of neighbouring cells. The rapid asexual reproduction of bacteria enables them to react very quickly to favourable chemical changes within a specific microniche and their starvation survival strategies (Kjellberg *et al.* 1987) enable them to persist for very long periods of time in nonpermissive conditions. Spatial juxtaposition within biofilms is essential to the development of

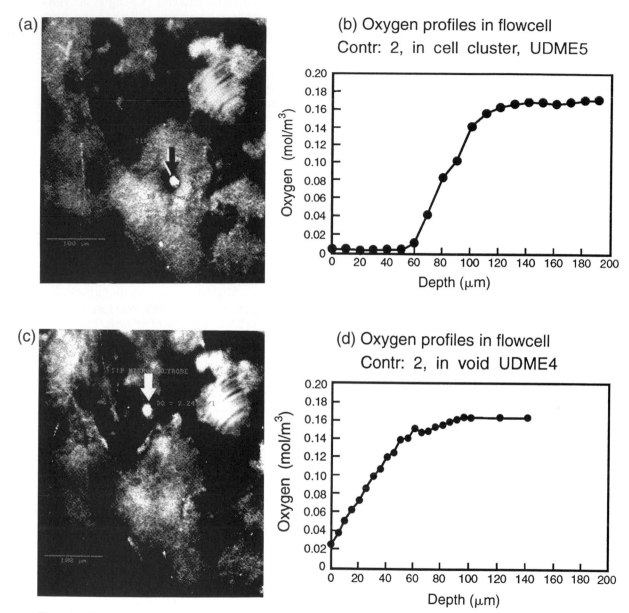

Fig. 0.2.   Confocal scanning laser micrographs (a), (c) of a single species biofilm showing the location (arrows) of a dissolved oxygen electrode within a microcolony (a) and within an adjoining water channel (c). (b) and (d) show dissolved oxygen levels as the probe is advanced from the bulk fluid into the biofilm 80–100 μm from the surface and thereafter deeper into the biofilm towards the colonized surface. (b) Shows dissolved oxygen levels as the probe is advanced through the microcolony in (a); the centre of the microcolony is essentially anaerobic. (d) Shows dissolved oxygen levels in the water channel in (c); this water channel contains dissolved oxygen at all depths. (Photograph courtesy of Z. Lewandowski and colleagues.)

the very efficient bacterial consortia seen in natural bacterial populations, and the physical retention of non-dividing cells within biofilms is the basis of their ability to react to changing conditions by developing new and equally efficient consortia to process new substrates. The sustained juxtaposition of metabolically cooperative cells is impossible in truly planktonic populations and non-replicating species are rapidly displaced under laboratory growth conditions.

If we assume the basic structural and chemical heterogeneity of bacterial biofilms, and if we extrapolate to even greater levels of structural complexity and chemical heterogeneity in natural adherent populations, we must expect that these biofilms will also be heterogeneous in several important physical parameters. The differences in electrical potential between adjacent loci within a biofilm combine to produce a measurable corrosion potential (Little *et al.* 1987) if the colonized surface is conductive. These corrosion potentials set up classic electrical 'corrosion cells' on biofilm-colonized surfaces, and the further reinforcement of effective cathodes within the biofilm produces metal loss at functional anodes and initiates the microbially influenced corrosion that causes huge losses in industry. Electrical potential differences exist within living biofilms, whether or not the colonized surface is conductive, and we must assume that the movement of ions and charged molecules will be influenced by these electrical gradients. This raises the fascinating possibility that charged antibiotic and biocide molecules that readily penetrate the water channels of a biofilm may be excluded from the dense matrix-enclosed microcolonies by both diffusion limitation and electrical gradients. We therefore predict that electrical heterogeneities will eventually be seen to be as important as structural and chemical heterogeneities in rationalizing the unique characteristics of bacterial biofilms. Measurements of AC impedance and open circuit potentials in living biofilms are currently being made by Dr Lewandowski and his colleagues, at the Center for Biofilm Engineering, and it may soon be possible to relate these electrical parameters to the structure and to the metabolic activity of biofilms.

It focuses the mind, wonderfully, to consider the essential differences between planktonic bacterial cells and sessile cells living in microcolonies within a biofilm. Much of the emphasis of the first six chapters of this book is centred on the complex heterogeneities of the biofilm mode of growth, and much of their detail documents these essential differences between planktonic and sessile cells. All of these chapters are based on data directly derived from living biofilms and we really cannot escape the contemplation of these complex communities by spurious arguments about extrapolation because, here, there is no extrapolation. We see living cells within microcolonies in fully hydrated biofilms and we see that these sessile populations predominate in almost all natural ecosystems. We can conclude that cells thus immobilized in exopolysaccharide matrices, many of whose chemical structures we understand, are highly conditioned by these matrices and are affected by molecules that diffuse from adjacent cells or from the bulk fluid via the water channels. Thus each biofilm cell lives in a microniche and its response to the special conditions of that microniche dictate its physiological activities, including its reproduction. Complex consortia of biofilm bacteria stand as a proof of widespread biofilm heterogeneity and we are only now beginning to suggest mechanisms that may have produced what we must accept because we can visualize it directly. Many early microbiologists valued direct observation above almost all else and the field is perhaps now just beginning to recover from its fascination with pure single species cultures and the perilous process of extrapolation.

Biofilm formation on inert surfaces

Chapters 7 to 11 deal with the formation of biofilms on inert surfaces. The consequences of this non-specific bacterial accretion onto inert surfaces range from simple fouling to the complex microbial aetiologies of microbially influenced corrosion and of recalcitrant device related infections.

Inert surfaces in aqueous environments rapidly accrete organic molecules and inorganic ions to form a layer called, in its most elaborate form, a 'conditioning film'. Therefore, in all but the most oligotrophic ecosystems, planktonic bacteria actually adhere to the surfaces of surface films that may or may not vary with the chemical nature of the inert surface being colonized. The use of planktonic cells that have lost important adhesion determinants during extended subculture in single species laboratory cultures has generated a large amount of contradictory data, many of which cannot be extrapolated to real ecosystems. The surfaces of these planktonic cells have often been altered by the loss of surface structures (pili, capsules) and cells with different degrees of surface modification may exhibit different adhesion behaviour on inert surfaces. We submit that meaningful data concerning the formation of bacterial biofilms on inert surfaces in aqueous systems are best obtained in studies in which these surfaces are presented to wild mixed planktonic bacterial populations in a menstruum in which realistic conditioning films will develop.

When biofilm development studies are conducted in this realistic mode, inert surfaces in aqueous systems are usually colonized rapidly and non-specifically, and surface topography and surface chemistry appear to be much less important parameters than was indicated in laboratory studies. The usefulness of these realistic studies is especially obvious in the examination of the putative colonization resistance of new materials that have been developed for use in medical devices. Literally hundreds of materials that have appeared to resist colonization by laboratory strains have formed luxuriant biofilms in a few hours when challenged with wild strains of the same species in real or simulated body fluids. The readily accessible aqueous systems that illustrate this principle most graphically may be rivers and streams, in which very similar mature biofilms are found on inert surfaces with a bewildering variety of topography and surface chemistry ranging from rock to wood to discarded plastic trash.

These general observations of non-specificity do not, of course, preclude instances of very specific adhesion and biofilm formation in specialized ecosystems such as the bovine rumen, in which cellulolytic bacteria adhere avidly and very specifically to cellulose (Minato & Suto 1978). It has been suggested (Kudo *et al.* 1987a, b) that this specific adhesion system may depend on the specific affinity of cell-associated enzymes for the substrate (cellulose) of those enzymes. Similarly, the well documented affinity of bacterial exopolysaccharides for specific metals may mediate specific adhesion and subsequent biofilm formation on the inert surfaces of ores that may be leached by this process. Objectivity is best served by the examination of a wide variety of inert surfaces in a natural ecosystem to detect the consequences of specific colonization and, in instances where species specificity is seen, to attribute this specificity clearly to the adhesion event rather than to the special suitability of the organism to the microniche that it has colonized.

If, as we contend, bacterial adhesion to inert surfaces in aqueous systems is a function of the association of bacterial surface components (notably exopolysaccharides) with components of the conditioning film that covers these surfaces (also notably polysaccharides) the adhesion event may directly involve neither the inert surface nor the bacterial cell wall. This perception may explain the avid colonization of silver and copper surfaces by bacterial species whose planktonic cells are exquisitely sensitive to ionic forms of these metals.

Biofilm formation on the surfaces of living cells

Chapters 12 to 18 deal with the formation of bacterial biofilms on the surfaces of living cells. Bacteria show an enormous range of avidity for these very specifically structured surfaces, some of which are readily colonized (e.g. buccal and vaginal epithelia) while others (e.g. kidney epithelia) are colonized only by bacteria with very specific pathogenic ligands.

If we use the direct analysis of the adherent bacterial populations actually seen on the surfaces of living cells and tissues as our criterion it is clear that living cells vary between widely separated extremes in their tendency to serve as suitable substrata for bacterial adhesion and subsequent biofilm formation. Dry tissue surfaces that slough external cells in a regular pattern (the epithelia of mammalian skin and of plant root hairs) are seen to be colonized by a wide variety of bacteria and fungi some of which invade the deeper epithelia and develop a somewhat commensal relationship with the colonized tissues. The moist epithelial tissues of organs that are heavily colonized by bacteria (mouth, gut, vagina) are often covered by very thick (>400 μm) mucus layers (Rozee *et al.* 1982) that provide a viscous environment whose peculiar chemical characteristics tend to select its primary microbial inhabitants. Special adhesion mechanisms are clearly required (Cheng *et al.* 1979) by organisms that adhere and form *de facto* biofilms in these viscous systems, and many species that predominate (e.g. *Lactobacillus*) in these ecosystems also produce chemical antagonists to the growth of competing microorganisms. Even in these mucus-lined organs inert surfaces (e.g. teeth) differ from healthy living tissue surfaces (e.g. gums) in their tendency to accrete microbial biofilms.

Even though heavily colonized mammalian organs are often directly connected to other organs, the internal organs can often maintain tissue surface sterility by the exercise of effective means of limiting bacterial adhesion and biofilm formation. The healthy respiratory tree is essentially sterile while the oropharynx is heavily colonized; the biliary system is sterile while the gut is colonized; and the uterus and bladder are essentially sterile while the vagina and distal urethra are very heavily colonized. The surfaces of epithelial cells in these non-colonized tissues are covered with an expolysaccharide glycocalyx composed largely of hyaluronic acid, as are the epithelial cells of all other normally sterile internal organ systems. It is obvious that the surfaces of these epithelial cells are challenged by planktonic bacterial cells, for example from adjacent organs, but that bacterial adhesion and biofilm formation occur only very rarely on these healthy tissues. Inert surfaces introduced into these normally sterile organ systems fare much less well than natural tissue surfaces and are often colonized by these same planktonic bacteria to produce biofilms that contribute notably refractory characteristics to device related bacterial infections (Costerton *et al.* 1987).

The contemplation of the bacterial colonization of tissue surfaces in living organ systems is really an exercise in microbial ecology. These tissue surfaces comprise a series of ecological niches whose conditions are entirely permissive for bacterial adhesion, growth, and biofilm formation, if the natural defences of the organ system are compromised as by instrumentation or system failure. However, in normal circumstances, planktonic bacteria do not adhere and survive to initiate the colonization process even when the neighbouring inert surfaces of medical devices are heavily colonized. Living tissue surfaces are actually comprised of the uronic acid containing polysaccharides of the glycocalyx that overlies the plasma membrane and, in mammals, this surface is covered with plasma proteins (fibronectin, lamellin) and richly supplied with surfactants and tissue associated antibodies (IgM). In one of the simplest cases, that of the eye, the environment of the tissue surface and its associated fluids is sufficiently hostile to bacteria that the insertion of planktonic bacteria and a biofilm colonized inert foreign body (the contact lens) from a biofilm infested storage case usually fail to initiate bacterial colonization and infection of the optical epithelia. In a more complex case the female bladder and uterus usually remain uncolonized even though the introitus, vagina, and cervix are heavily colonized by autochthonous and by potentially pathogenic bacteria. The bladder appears to employ mucus and tissue sloughing as a defensive measure, and the uterus appears to employ phagocytic cell activity and periodic tissue sloughing to prevent bacterial colonization. The structure of the sphincter of Oddi and the generally bactericidal characteristics of bile appear to prevent colonization of the common bile duct by intestinal bacteria and surfactants, phagocytic cells, ciliary activity, and the 'mucous escalator' usually keep the deep airways of the pulmonary system free of adherent bacteria.

Colonization resistant surfaces

For the past 15 years the corridors of industry and the groves of Academe have echoed to the exuberance of scientists who are confident that they have discovered a material whose surface is inherently resistant to bacterial colonization. Laboratory tests have been uniformly encouraging and practical tests have been equally uniformly disappointing and we believe that we can now conclude that bacteria will eventually adhere to and colonize the surface of any man-made material. However, we must consider the inherent colonization resistance of the surfaces of tissues that are not normally colonized even though they are exposed to planktonic bacteria in large numbers. We must ask the question – is it the detailed chemical structure and peculiar topography of these tissue surfaces that prevents bacterial colonization, or is it the activity of surface located defence mechanisms such as surfactants, antibodies, and phagocytic cells? If the former is correct then simple mimicry of tissue surface characteristics will suffice to produce a colonization-resistant material for use in medical devices. If the latter is correct then this mimicry of the tissue surface must be sufficiently effective to accrete surfactants, antibodies, and phagocytic cells and to allow their unimpeded antibacterial activity. We expect that the latter is correct because most tissue surfaces are readily colonized by bacteria when the surface environment is disturbed by such processes as general infection or instrumentation.

This book attempts to capture the revelations concerning the structure and activity of bacterial biofilms that now occur at very regular intervals. These data are perhaps most readily understood by comparing the microenvironment of a bacterial cell within a biofilm with that of a planktonic cell of the same species. These microenvironments are profoundly different and the phenotypic reaction of bacteria to these environmental differences provides the beginnings of an explanation of the singular characteristics of biofilm organisms (for example, resistance to antibacterial agents).

Because of our general fascination with lock-and-key molecular mechanisms we seized on a few instances in which specific pili mediate the attachment of certain pathogenic bacteria to specific ligands on tissue surfaces and we anticipated similar specificity in the ubiquitous bacterial colonization of surfaces in aquatic environments. Generally, the past two decades have taught us that bacteria adhere to a very wide variety of surfaces with remarkable avidity, especially if we examine wild bacteria in natural ecosystems. Ecological factors appear to exert a more profound effect on the bacterial colonization of surfaces than actual surface chemistry and topography. We note that the only well documented surfaces that consistently resist bacterial colonization because of their inherent characteristics are those of some living tissues. We suggest that bacterial biofilms will form readily on all surfaces except those that resemble living tissues in combining structural and environmental factors that prevent bacterial adhesion and colonization.

References

Cheng, K.-J., McCowan, R. P. & Costerton, J. W. (1979). Adherent epithelial bacteria in ruminants and their roles in digestive tract function. *American Journal of Clinical Nutrition*, **32**, 139–48.

Cheng, K.-J., Stewart, C. S., Dinsdale, D. & Costerton, J. W. (1984). Electron microscopy of bacteria involved in the digestion of plant cell walls. *Animal Feed Science and Technology*, **10**, 93–120.

Costerton, J. W., Cheng, K.-J., Geesey, G. G. *et al.* (1987). Bacterial biofilms in nature and disease. *Annual Reviews of Microbiology*, **41**, 435–64.

Davis, D. G., Chakrabarty, A. M. & Geesey, G. G. (1993). Exopolysaccharide production in biofilms: substratum activation of alginate gene expression by *Pseudomonas aeruginosa*. *Applied and Environmental Microbiology*, **59**, 1181–6.

Fletcher, M. & Loeb, G. I. (1979). Influence of substratum characteristics on the attachment of a marine pseudomonad to solid surfaces. *Applied and Environmental Microbiology*, **37**, 67–72.

Geesey, G. G., Richardson, W. T., Yeomans, H. G., Irvin, R. T. & Costerton, J. W. (1977). Microscopic examination of natural sessile bacterial populations from an alpine stream. *Canadian Journal of Microbiology*, **23**, 1733–6.

Govan, J. R. W. (1975). Mucoid strains of *Pseudomonas aeruginosa*: the influence of culture medium on the stability of mucus production. *Journal of Medical Microbiology*, **8**, 513–22.

Hoyle, B. D., Williams, L. J. & Costerton, J. W. (1993). Production of mucoid exopolysaccharide during development of *Pseudomonas aeruginosa* biofilms. *Infection and Immunity*, **61**, 777–80.

Kjelleberg, S., Hermansson, M., Marden, P. & Jones, G. W. (1987). The transient phase between growth and non-growth of heterotrophic bacteria, with emphasis on the marine environment. *Annual Reviews of Microbiology*, **41**, 25–49.

Koch, R. (1881). Zur untersuchung von pathogenen organismen. *Mittheilungen aus dem Kaiserlichen Gesundheitsamte*, **1**, 1–48.

Kudo, H., Cheng, K.-J. & Costerton, J. W. (1987a). Electron microscopic study of the methylcellulose-mediated detachment of cellulolytic rumen bacteria from cellulose fibers. *Canadian Journal of Microbiology*, **33**, 267–72.

Kudo, H., Cheng, K.-J. & Costerton, J. W. (1987b). Interactions between *Treponema bryantii* and cellulolytic bacteria: *in vitro* degradation of straw cellulose. *Canadian Journal of Microbiology*, **33**, 244–8.

Lawrence, J. R., Korber, D. R., Hoyle, B. D., Costerton, J. W. & Caldwell, D. E. (1991). Optical sectioning of microbial biofilms. *Journal of Bacteriology*, **173**, 6558–67.

Lewandowski, Z., Altobelli, S. A. & Fukushima, E. (1993). NMR and microelectrode studies of hydrodynamics and kinetics in biofilms. *Biotechnology Progress*, **9**, 40–5.

Little, B. J., Wagner, P. A., Gerchakov, S. M., Walsh, M. & Mitchell, R. (1987). Involvement of a thermophilic bacterium in corrosion processes. *Corrosion*, **42**, 533–6.

MacLeod, F. A., Guiot, S. R. & Costerton, J. W. (1990). Layered structure of bacterial aggregates produced in an upflow anaerobic sludge bed and filter reactor. *Applied and Environmental Microbiology*, **56**, 1598–1607.

Marshall, K. C., Stout, R. & Mitchell, R. (1971). Mechanisms of the initial events in the sorption of marine bacteria to surfaces. *Journal of General Microbiology*, **68**, 337–48.

Martin, B. W., Holloway, B. W. & Deretic, V. (1993). Characterization of a locus determining the mucus status of *Pseudomonas aeruginosa*: Alg U shows sequence similarities with a bacillus sigma factor. *Journal of Bacteriology*, **175**, 1153–64.

Minato, H. & Suto, T. (1978). Technique for fractionation of bacteria in rumen microbial ecosystem. II. Attachment of bacteria isolated from bovine rumen to cellulose powder *in vitro* and elution of attached bacteria therefrom. *Journal of General and Applied Microbiology*, **24**, 1–16.

Nickel, J. C., Ruseska, I., Wright, J. B. & Costerton, J. W. (1985). Tobramycin resistance of *Pseudomonas aeruginosa* cells growing as a biofilm on urinary catheter material. *Antimicrobial Agents and Chemotherapy*, **27**, 619–24.

Rozee, K. R., Cooper, D., Lam, K. & Costerton, J. W. (1982). Microbial flora of the mouse ileum mucous layer and epithelial surface. *Applied and Environmental Microbiology*, **43**, 1451–63.

ZoBell, C. E. (1943). The effect of solid surfaces on bacterial activity. *Journal of Bacteriology*, **46**, 39–56.

Part I STRUCTURE, PHYSIOLOGY AND ECOLOGY OF BIOFILMS

1

Growth of Microorganisms on Surfaces

Darren R. Korber, John R. Lawrence, Hilary M. Lappin-Scott and
J. William Costerton

Introduction

Traditional microbiological investigations have focused on the culture and analysis of pure cell lines of bacteria, in either batch or chemostat culture. However, it has been clearly established that in nature, disease and industry, the majority of bacteria exist attached to surfaces within biofilms (Costerton *et al.* 1978, 1987; Lappin-Scott & Costerton 1989; Characklis *et al.* 1990a). Furthermore, it has also been established that the bacteria which exist in biofilms, termed sessile bacteria, are inherently different from bacteria existing in the planktonic state. In the sessile state, bacteria may express different genes, alter their morphologies, grow at different rates, or produce extracellular polymers in large amounts (Costerton *et al.* 1978; Wright *et al.* 1988; Gilbert *et al.* 1990; Dagostino *et al.* 1991; McCarter *et al.* 1992). One significant consequence of sessile growth is that biofilm bacteria are more resistant to medical and industrial control strategies than their planktonic counterparts (Brown *et al.* 1988; Nichols 1989; Eng *et al.* 1991; Blenkinsopp *et al.* 1992).

The development of complex attached and aggregated communities is also important for the survival and reproductive success of microorganisms. These communities have been considered to act as reservoirs for diverse species, sites of specific limited niches, and protective refuges from competition, predation or harsh environmental conditions, allowing otherwise poor competitors to survive. Integration into a biofilm or bioaggregate may be regarded as a survival strategy beyond that of maximizing or increasing the growth rate. Placed in the context of survival and integration into a community, the evolutionary and ecological rationale behind the enormous range of diversity in cell surface chemistry, cell morphology, development of life cycles, complex pathways of microcolony development, and formation of syntrophic associations or consortia can be more easily rationalized.

In part, these adaptive strategies are related to the physical nature and scale of the microbial environment. An individual adherent bacterial cell is unable to alter significantly its surrounding chemical and physical environment (Koch 1991). In addition, many of the strategies are dictated by requirements for the growth of bacteria in close associations or groupings represented by biofilms or other bioaggregations. This is meaningful because in pure cultures, the requirements for growth of some groups of bacteria, such as sulphate reducing bacteria (SRB), methanogens, and syntrophic organisms, may only be met in specialized environments. The maintenance of a stable attached bacterial population is dependent on a constant flux of nutrients and wastes to and from the cells, therefore a sessile existence may not be sufficient to ensure bacterial reproductive success if conditions become unfavourable.

Wimpenny (1992) considered that the concept of an activity domain (that is, zones under the metabolic influence of one bacterium) is fundamentally linked to biofilm growth; when substrate/product sources and sinks associated with two physiologically distinct activity domains spatially overlap, the potential for beneficial or detrimental interaction exists. Functional consortia and microcolonies within quasi-organized biofilms provide a wide range of growth requirements, for within these systems temporal and spatial development of chemical microzones (for example, E_h and pH), positioning of syntrophic partners, and establishment of complementary metabolic pathways may all occur. Thus, while a single bacterium may live in a hostile, diffusion dominated environment, sessile bacteria overcome these limitations by modifying their surroundings through association and diversity.

While adaptations to surface colonization are widely varied, there are some traits common to all sessile bacteria. For example, all must attach to surfaces or other bacteria, utilize available resources for growth and reproduction, then redistribute over surfaces and to new locations when conditions become less favourable. Such behaviour has been reported by a number of researchers describing stages of surface colonization and biofilm formation (Lawrence *et al.* 1987, 1992; Power & Marshall 1988; Korber *et al.* 1989a). Successful growth of sessile bacteria does not necessarily mean rapid growth. Rather, sessile bacteria must be able to survive and adapt over time. They do so by remaining simple, diverse, minimizing the constraints which a large gene load would place upon them, but remaining metabolically adaptable as a group. The requirements for the formation of favourable chemical and physical microenvironments make pure culture biofilms in nature a rarity, although within a biofilm distinct microcolonies may remain identifiable.

The array of tools available for the analysis of biofilms is impressive. These include high resolution video microscopy (Kjelleberg *et al.* 1982; Power & Marshall 1988; Shapiro & Hsu 1989), environmental scanning electron microscopy (Little *et al.* 1991), microelectrode analysis (Lens *et al.* 1993; Lewandowski *et al.* 1993), microbalance applications (Nivens *et al.* 1993) and scanning confocal laser microscopy (SCLM). Of these new methods, SCLM and microelectrode analyses are widely accepted for microbiological studies (Lawrence *et al.* 1991; Caldwell *et al.* 1992a, b; Korber *et al.* 1992; Lens *et al.* 1993; Lewandowski *et al.* 1993). The data generated from these techniques have altered our understanding of biofilms. Both pure culture and mixed species biofilms are now accepted as being spatially and temporally heterogenous systems containing microscale variations in biofilm architecture and chemistry. The extent of analyses possible using SCLM is broad, because visual information may be correlated with fluorescent signals. SCLM is amenable to *in situ* analyses on a variety of surfaces. When used in conjunction with appropriate fluorophores, SCLM has the potential to define biofilm architecture, estimate cellular viability, identify microenvironmental chemical gradients, quantify molecular transport and molecular interactions, or phylogenetically analyse population architecture (Amann *et al.* 1992).

The nature of the surface microenvironment

Organisms move from the bulk aqueous phase to microenvironments on a surface where foci of growth have established. A number of investigations have examined surface physicochemistry or the effect of adsorbed molecular films on microbial attachment success (Fletcher 1976; Fletcher & Loeb 1979). Such models of microbe–surface interactions are valuable because, in nature, newly exposed surfaces become coated with organic material present within the aqueous solution, often masking the original chemistry of the surface. These coatings may then stimulate chemotactic bacteria to colonize these surfaces, or act as a potential nutrient source.

Surface conditioning films

Surface conditioning, or molecular, films are rapidly adsorbed whenever clean surfaces are exposed to natural and *in vitro* solutions containing organic molecules. The movement of organics from the bulk phase to the surface is primarily the result of molecular diffusion, described in quiescent or laminar systems by Fick's law of diffusion (Koch 1991). Theoretical and experimental studies both indicate that molecular diffusion occurs rapidly, resulting in significant organic deposits after only 15 min (Bryers 1987; Characklis *et al.* 1990a). For example, molecular coatings have been shown to result in 0.8–15 mg organic matter per m^2 of exposed surface material, equating to film thicknesses ranging from 30 to 80 nm (Loeb & Neihof 1975). Experimental results obtained using Fourier-Transformed Infrared spectroscopy (FTIR), Multiple

Attenuated Internal Reflectance Infrared spectroscopy (MAIR-IR), and Infrared spectroscopy (IR) have indicated that polymers of glycoproteins, proteins, and possibly humic acids, each of which may contain as many as 100 000 molecular units per chain, may be involved in the formation of organic films (Marshall *et al.* 1971; Baier 1980; Rittle *et al.* 1990). Thus, an abundance of reactive binding sites would be available for further reaction with either additional solute molecules or with chemical groups expressed on bacterial surfaces.

Exposure of surfaces to organic compounds has resulted in various, measurable effects on the original surface properties. For example, both negatively and positively charged surfaces acquired net negative surface charges, and zeta potentials, surface contact angles and free energy vary as a function of their original surface energy (Fletcher 1976; Fletcher & Marshall 1982; Absolom *et al.* 1983; van Loosdrecht *et al.* 1987; Sjollema *et al.* 1988; Busscher *et al.* 1990b; Vanhaecke *et al.* 1990). The manner by which molecular films influence the adsorption of bacteria to surfaces still remains unclear except for specific cases; however, the general viewpoint is that chemical groups of the organic coating interact with chemical groups on bacterial appendages, such as pili, flagella or bacterial exopolysaccharides (EPS) (Paerl 1975; Dazzo *et al.* 1984; Vesper & Bauer 1986; Sjollema *et al.* 1990b). Such appendages extend through the free energy barrier and make contact with the organic matrix where the structure may form reversible short range electrostatic, covalent, or hydrogen bonds. The presence of organic molecules on surfaces also provides an explanation for the chemotactic movement and preferential positioning of many bacterial species on solid substrata (Berg & Brown 1972; Adler 1975; Kjelleberg *et al.* 1982; Block *et al.* 1983; Berg, 1985; Hermansson & Marshall 1985).

Surface related physicochemical forces

All surfaces (coated or non-coated) possess physicochemical characteristics which may influence the adsorption of bacteria and other molecules. At relatively short distances from the surface (≤ 1–10 nm), attractive and repulsive forces may influence molecular sorption (Bowen & Epstein 1979), defined by the DLVO theory.

However, the critical surface tension of the surface may also dictate whether a system containing bacteria will undergo a reduction in free energy: that is, whether bacteria adsorb or stay in suspension (Absolom *et al.* 1983). These physicochemical theories have been exhaustively tested as predictive models for bacterial–surface interactions.

DLVO theory

The DLVO, or double-layer theory, was originally formulated from the works of Derjaguin & Landau (1941) and Verwey & Overbeek (1948), and equates electrostatic forces and London–van der Waals forces active at the surface with the propensity for colloidal attraction and adhesion, generally given by Equation 1:

$$V_T(l) = V_A(l) + V_R(l) \qquad (1)$$

The total interaction energy (V_T) of a particle as a function of its separation distance (l) from a solid surface is the sum of the van der Waals attraction (V_A) and the electrostatic interaction (V_R) (van Loosdrecht *et al.* 1987). The theory predicts that there are two regions where particle attraction may occur, that is, regions where small distances (≤ 1 nm) separate the substratum and approaching surface, and regions where larger distances (5–10 nm) separate the two surfaces (Bowen & Epstein, 1979; Characklis *et al.* 1990b). These locations are known as the primary minimum and secondary minimum, respectively, and are quite small in scale when compared with the thickness of the hydrodynamic boundary layer (see Brading *et al.* Chapter 2). Between these two positions is located an energy level where the surfaces experience maximum repulsion (an electrostatic repulsion occurs because the cell and substratum surfaces both carry a net negative charge), the magnitude of which is dependent on the surface potential of the particle and the substratum, the separation distance, and the electrolytic strength of the aqueous medium. Furthermore, because these interactions occur within the quiescent boundary layer, convective forces and attractive events may be mathematically isolated (Adamczyk & van de Ven 1981). More detailed DLVO based mathematical expressions incorporate the radius of the approaching particle (Sjollema *et al.* 1990a). As the radius decreases, so the repulsive energy barrier decreases.

A number of problems must be addressed if the electrical double-layer theory is to be adapted for the prediction of microbial adsorption tendencies. The original theorem was developed for shear free systems (that is, within the hydrodynamic boundary layer), whereas the majority of natural systems have dynamic fluid movement applying shear force to attaching cells approaching surfaces (Christersson *et al.* 1988; Characklis *et al.* 1990a; Mittelman *et al.* 1990). Also, a significant size differential exists between colloidal particles and bacterial cells, as well as the biogenic chemical differences caused by the ion and pH gradients at the boundaries of the cell (Boone *et al.* 1989). Lastly, geometrical considerations must be taken into account (Bowen & Epstein 1979), as cellular appendages (Isaacson *et al.* 1977; Vesper & Bauer 1986) alter the cell effective diameter near the surface and hence alter the diameter dependent repulsive effects experienced within regions of maximal repulsion between the primary and secondary minimum (Sjollema *et al.* 1990a). Overall, the application of DLVO based theory remains popular during formulation of theoretical cell adhesion models, despite discrepancies between observed and predicted results.

Surface free energy/hydrophobicity theory

The initial adsorption of bacteria may also be explained in terms of system free energy. If the total free energy of a system is reduced by cell contact with a surface, then adsorption of the cell to the substratum will occur (Absolom *et al.* 1983). Thermodynamically, a free energy balance may be developed on the basis of the interfacial free energies of the surfaces involved, as shown in Equation 2:

$$\Delta F_{adh} = \gamma_{bs} - \gamma_{sl} - \gamma_{bl} \qquad (2)$$

where ΔF_{adh} is the free energy of adhesion, γ_{bs} is the interfacial free energy for the bacterial–substratum interface, γ_{sl} is the interfacial free energy for the substratum-liquid interface, and γ_{bl} is the interfacial free energy for the bacterial–liquid interface. The free energy, or wetting theory, does not depend on chemical measurements, but rather relies on an appropriate method for determining critical surface tension (σ_c) of the bacteria and substratum. This is commonly performed microscopically by measuring contact angles between defined liquids and surfaces or surface preparations of microbial cells (Dexter *et al.* 1975; Fletcher & Loeb 1979; Fletcher & Marshall 1982; Fletcher 1988; Rittle *et al.* 1990). The Young equation describes the relationship between contact angle and the interfacial energies of the involved phases and their contact angles (Equation 3):

$$\gamma_{sv} - \gamma_{sl} = \gamma_{lv} \cos\theta_e \qquad (3)$$

where γ_{sv} represents the solid/vapour phase, γ_{sl} equals the solid/liquid phase, and γ_{lv} the liquid/vapour phase.

In theory, a liquid which provides a contact angle of 0° should have a surface tension equivalent to the surface free energy of the substratum, where an increase in the measured contact angle corresponds to a increase in the surface hydrophobicity and a decrease in surface free energy. The use of dried bacterial lawns for the calculation of bacterial contact angles has been disputed, with supporters of alternative techniques maintaining that irreversible changes in the adsorbed macromolecules occur which influence the determination. Fletcher & Marshall (1982) suggested that bacterial surface free energies should be measured in a more natural state (i.e. within a liquid) using the bubble contact angle method, where the effect of adsorbed macromolecules or bacteria can be conducted in aqueous solution. Both methods use the contact angle of the liquid or bubble with the substratum to determine the critical surface tension for the test surface after conditioning. Studies utilizing these criteria to explain bacterial adhesion events have demonstrated a range of results using various substrata, bacterial species, and assay methodology (Dexter 1979; Absolom *et al.* 1983; van Pelt *et al.* 1985; Burchard *et al.* 1990; Rittle *et al.* 1990; Sorongon *et al.* 1991). This variability of experimental results is due largely to the lack of standard methods as well as the heterogenous, physicochemical nature of the bacterial cell.

Overall, physicochemical models of surface interactions assume that the surfaces are small, smooth and energetically homogenous; however, this is rarely true for bacteria (Costerton *et al.* 1978; Busscher *et al.* 1990a). As a consequence, simple physicochemical approaches which fail to incorporate the microscopic condition of the cell outer surface or adaptive microbial behaviour will never fully explain all aspects of bacterial adhesion.

System hydrodynamics

Bacterial surface colonization success is dependent on a flux of cells from the bulk phase to surface microenvironments where attachment may occur. Bacterial flux is largely governed by the hydrology of the fluid system (such as static, turbulent, or laminar), the concentration gradient, the nature of the organism (such as size or surface characteristics) and behaviour (Powell & Slater, 1983; Sjollema *et al.* 1988, 1989a; van Loosdrecht *et al.* 1989; Korber *et al.* 1990). Fluid hydrodynamics are important and often paramount considerations in industrial, medical and natural systems (Costerton *et al.* 1978; Lappin-Scott & Costerton 1989; Characklis *et al.* 1990a; Christensen & Characklis 1990), where cell deposition precedes metal corrosion and microbial infection, as well as a number of other deleterious and beneficial processes. In addition, the flow regime also influences the nature of the physical and chemical surface microenvironment after a biofilm has been established. It is therefore important that the dimensions of flow systems and fluid velocity be carefully controlled or defined (for example, Reynolds number) to reflect the system being modelled. Further details of these parameters, including the hydrodynamic boundary layer, are covered by Brading *et al.* (Chapter 2).

Static systems

Within quiescent environments, factors such as Brownian diffusion, gravity and motility all assist the movement of small particles, including bacteria, to surfaces (Bryers 1987). However, the relative importance of each of these mechanisms varies with specific conditions. Based on the size of most bacterial cells, Brownian motion or diffusion may contribute to the overall attachment of non-motile cells; however, the extent of these contributions is reportedly low (Characklis *et al.* 1990b). It is generally concluded that motility greatly increases the chance that a bacterium may contact a potential attachment surface (Fletcher 1977; Marmur & Ruckenstein 1986) with enough potential energy to overcome the electrical repulsive forces which exist between the bacterium and substratum. During studies where the motility of bacteria was observed to decrease with culture age, concurrent decreases in cellular

adsorption were also noted (Fletcher 1977). Similarly, cell transport resulting from gravitational cell sedimentation in static systems has also been documented as an effective surface colonization mechanism (Walt *et al.* 1985).

In the absence of flow, the thickness of the hydrodynamic boundary layer (Brading *et al.* Chapter 2) can be effectively large, consequently limiting the success of organisms established in these zones by restricting nutritional transport to diffusion and convection. In nature, static aqueous systems are rare as some mechanism for mixing usually exists. However, studies applying static regimes may be useful as the conditions within biofilms are largely insulated from flow, therefore applying a static model to biofilms formed under flowing conditions (for example, removing flow after a growth period) may provide insight into nutrient depletion responses of sessile bacteria, as well as the formation or dissipation of chemical microzones.

Overall, the use of batch static systems for the study of bacterial deposition has been described by some as being poorly controlled and random (Sjollema *et al.* 1989a), dependent to a large extent on the suspension volume to total substratum area ratio, and not an appropriate model of most environments. However, the general trends observed during such studies remain valuable provided that the testable hypothesis incorporates an appropriate control where relative differences in adhesion success may be measured.

Diffusion

An understanding of diffusion is fundamental to biofilm ecology. In general terms diffusion is a process which drives motion in biology (Koch 1991). Diffusion is caused by the random, thermally energetic movement of particles from regions of high concentration to low concentration, as described by Fick's first law. This flux (dq/dt) is dependent on the solute concentration gradient (dC/dX), the area of the material through which the solute is diffusing (A), and the diffusion constant for the solute (D) (Equation 4).

$$dq/dt = -D\,A\,dC/dX \qquad (4)$$

During the early stages of biofilm development in flowing microenvironments, attached organisms may be influenced by advection and flow. However, as growth proceeds, the hydrodynamic

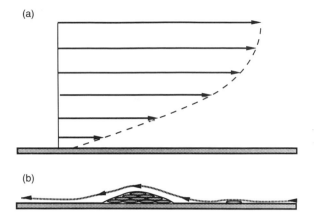

(a)

(b)

Fig. 1.1. (a) Illustration showing the distance-dependent nature of flow associated with surfaces, where bulk phase flow velocities are experienced at large distances from the surface and a no-slip condition is found at very small distances from the surface. (b) The effect of microcolony/biofilm formation on the nature of the hydrodynamic boundary layer (i.e. moves it away from the surface). Even in systems dominated by laminar flow, turbulent vortices may form in the regions of obstructions such as microcolonies or biofilm biomass.

boundary layer moves upward with the biofilm–liquid interface as the biofilm increases in depth (Fig. 1.1). Within the EPS matrix, rate limitations to growth and mass transport are now theoretically governed solely by diffusion, and as source and sink concentrations of various compounds invariably approach steady state, solute flux becomes a constant proportional to the distance of migration.

The thickness of microbial biofilms is known to influence a number of biotic and abiotic processes in flowing systems. Heat transfer efficiency, fluid frictional resistance, and chemical transformations are industrially significant examples of detrimental biofilm effects (Trulear & Characklis 1982; Characklis *et al.* 1990b). Biofilm thickness may also affect the transfer of essential nutrients such as oxygen and other metabolites into growing biofilms by increasing the diffusional length where physiological substrate consumption may limit the concentration of the metabolite before penetration to the base of the film (Sanders 1966; Howell & Atkinson 1976; Cox *et al.* 1980; Christensen & Characklis 1990). Hindered diffusion may also play a role within the biofilm cell–EPS matrix. Hindered or restricted diffusion occurs when the effective

diffusivity of a solute within a pore of comparable size is less than its value in bulk solution (Bohrer *et al.* 1984). Within aqueous solution, the diffusion of many biologically significant molecules are adequately described by Einstein's solution in Equation 5:

$$D_\infty = kT/f_\infty \qquad (5)$$

where D_∞ is the bulk solution diffusion coefficient, k is Boltzmann's constant, T is temperature, and f_∞ is the molecular friction coefficient in bulk solution. Assuming that the diffusing molecule is spherical, f_∞ may be determined using $f_\infty = 6\pi\mu r_s$, where r_s is the solute radius and μ the solvent kinematic viscosity. This conceptual model of molecular diffusion through barriers with a defined mesh size has been confirmed during biofilm and membrane diffusion studies, with reduced, or hindered diffusion being measured in the biofilm or membrane system relative to the theoretical rates of diffusion in the aqueous phase (Acker & Steere 1962; Bohrer *et al.* 1984). Hindered diffusion thus has relevance to biofilm systems with regard to molecular sieving and size dependent exclusion of molecules from the EPS matrix (Lawrence *et al.* 1994).

Fluid flow has a profound effect on diffusion even though the effective diffusion coefficient (D_e) remains constant for a given biofilm matrix (Freeze & Cherry 1979; Koch 1991). However, maintenance of a concentration gradient by flow replenishment of, for example, substrate molecules in the hydrodynamic boundary layer can lead to large increases in diffusion limited biofilm events such as growth (Characklis *et al.* 1990a). In addition to maintaining a steeper concentration gradient, high flows tend to reduce the thickness of the hydrodynamic boundary layer, decreasing the distance between molecular sources and sinks. The importance of flow to aqueous systems is much greater than within the atmosphere where much slower rates of diffusion for similar molecules in water have been reported (Vogel 1983). It thus becomes apparent that within aqueous systems of increasing viscosity, diffusion becomes a less effective mechanism of molecular transport.

Many articles examining diffusion or relevant to diffusion in biological systems have been published, the reader is directed to several of the excellent review articles on this topic (Freeze & Cherry 1979; Characklis *et al.* 1990b; Koch

1991). In addition to providing useful insight into molecular transport and nutrition, it is significant that diffusion governs the movement of bacteria to surface sites, demonstrated by predictive mathematical models of non-motile and motile bacterial transport based on Fick's law of diffusion (Jang & Yen 1985). Modifications of these formulae have also been used to predict directed chemotactic migration of bacteria (Rivero & Lauffenburger 1986; Kelly *et al.* 1988).

Quantifying behaviour and kinetics of surface colonization

The events which occur following initial attachment, including microcolony formation, aggregation, and development of syntrophic interactions or consortia, are increasingly recognized as being extremely important for understanding the adaptive significance of surface associated microbial growth. Thus, a large number of techniques can be and have been applied to these studies. The most common approach has been the use of a variety of transparent and opaque substrates for culture and subsequent observation by direct microscopic or other non-conventional techniques.

Cultural methods

Cultural methods include the use of glass slide incubation (Cholodny 1930; Henrici 1933; Bott & Brock 1970; Staley 1971), incubation of electron microscope grids (Drucker & Whittaker 1971; Hirsch & Pankratz 1971), capillary techniques (Perfil'ev & Gabe 1969), and agar block microculture (Torrella & Morita 1981). A variety of flow cells and microperfusion chambers have also been used to study attached growth by microorganisms (Duxbury 1977; Caldwell & Lawrence 1986).

Observation and analysis

Bacterial behaviour and the development of colonies and biofilms may be followed using cinemicrographic techniques (Wilkins *et al.* 1972), video, time lapse video recording, or digitization (Kjelleberg *et al.* 1982; Caldwell 1985; Power & Marshall 1988). Typically, images have been obtained using transmitted, incident or polarizing light microscopy techniques, as outlined by Marshall (1986). These images are all amenable to digital image analysis for extraction of data on growth rates, cell size, rates of motility and a variety of other parameters (Lawrence *et al.* 1989a, b; Caldwell *et al.* 1992a, 1993). Over the past decade, computer enhanced microscopy and image analysis have been utilized to study the positioning of bacteria within surface microenvironments, permitting rapid, precise analysis and quantitative data collection. While the majority of these studies have been directed at the enumeration of cells present or being deposited on the surface of glass slides (Sieracki *et al.* 1985; Sjollema *et al.* 1988, 1989b), computer enhanced microscopy (CEM) has also been applied to the analysis of surface growth, positioning, and locomotion of cells colonizing surface environments (Gualtieri *et al.* 1985, 1988; Caldwell & Lawrence 1986; Lawrence & Caldwell 1987; Lawrence *et al.* 1987; Korber *et al.* 1989a, 1990). Other established techniques include the application of transmission electron microscopy (TEM) and scanning electron microscopy (SEM) for observation of microcolonies on a variety of surfaces (Costerton *et al.* 1981; Handley *et al.* 1985) and studies of bioaggregates (Bochem *et al.* 1982; MacLeod *et al.* 1990).

Recent advances in techniques

More recently the application of direct microscopy utilizing scanning confocal laser microscopy (SCLM) with molecular probes and image analysis has proven an effective method for analysis of microcolonies and biofilms (Lawrence *et al.* 1991; Caldwell *et al.* 1992a, 1993). Application of SCLM technology has led to rapid developments in the study of thick biofilms without the need for disruptive or sacrificial preparation procedures (Lawrence *et al.* 1991; Caldwell *et al.* 1992a, b; Korber *et al.* 1992, 1993; Wolfaardt *et al.* 1994). Use of defined fluorescent molecular probes such as viability indicators, 16S rRNA oligonucleotides, and chemically sensitive molecules in conjunction with SCLM may help elucidate the underlying mechanisms and environmental conditions responsible for behavioural variability of sessile bacteria (Amann *et al.* 1992; Caldwell *et al.* 1992a; Rodriguez *et al.* 1992). Methods such as X-ray microanalysis, FTIR and NMR (nuclear magnetic resonance) have also provided valuable insights into events within biofilms and microcolonies (Geesey & White 1990). Use of microelectrode techniques to study bacteria on surfaces and in aggregates also provides a wide range of data on activities within bacterial associations (Alldredge & Cohen 1987; Lewandowski *et al.* 1989, 1993; Revsbech 1989; Lens *et al.* 1993).

The techniques mentioned above have been extensively reviewed (Marshall 1986; Revsbech & Jorgenson 1986; Shotton 1989; Geesey & White 1990; Caldwell *et al.* 1992a, 1993).

Microcolony formation strategies

Bacterial behaviour and life cycles at surfaces

The behaviour of microorganisms includes interactions with their environment, such as the solutions surrounding the bacterium, the physical conditions to which the bacterium is exposed, and other bacteria inhabiting similar habitats (Lauffenburger *et al.* 1984; Lawrence & Caldwell 1987; Arnold & Shimkets 1988; Shapiro & Hsu 1989; Shimkets 1990; Lawrence *et al.* 1992). Where the conditions of study are defined and constant, any pattern evident in the growth, development and translocation of cells or colonies may be deemed as a behavioural trait characteristic of that particular bacterial strain (Lawrence & Caldwell 1987; Lawrence *et al.* 1987).

The formation of microcolonies or cellular groups is one of what may be a series of states comprising a life cycle. Life cycles of surface colonizing bacteria range from that of alternation between stalked mother cell and swarmer cell in *Caulobacter* spp. or an attached cell giving rise to free living progeny (e.g. *Rhizobium* spp.), to the complex multifaceted life cycles of myxobacteria (Arnold & Shimkets 1988; Shimkets 1990). The simplest form of the life cycle is that which separates growth and reproduction from a migratory or dispersal phase, a pattern which is repeated throughout many genera of sessile bacteria. During the attached phase of the life cycle, growth and microcolony formation proceed through a range of patterns, including the well documented behaviour of the prosthecate bacteria, as well as many not as extensively studied (Hirsch 1984; Lawrence & Caldwell 1987;

Lawrence *et al.* 1987, 1992; Power & Marshall 1988). Such behaviour represents the diversity of approaches to colonization and occupation of the substratum, redistribution on the substratum, and dispersal of daughter cells to the aqueous phase.

Motile attachment behaviour

The motile attachment hypothesis is based on behavioural and CEM analyses of cells moving along surfaces (Lawrence *et al.* 1987). During motile attachment, bacteria may travel upstream or cross stream against flow velocities exceeding the cells' maximum rate of motility or simply move along the surface while remaining in a semi-attached condition (Lawrence *et al.* 1987). Additional support for the existence of motile attachment behaviour was provided by Korber *et al.* (1989b) following observations of flow independent behaviour of *Pseudomonas fluorescens* on surfaces where the laminar flow velocity was $200 \, \mu m \, s^{-1}$, and also by Malone (1987), who reported that various *Rhizobium* spp. travelled upstream against flow rates exceeding their maximum reported rates of motility. However, the mechanisms responsible for flow independent cell movement are not known. Consequences of motile attachment include lateral colonization (also termed backgrowth) against the direction of flow. For example, motility facilitated the backgrowth of a mot$^+$ *P. fluorescens* strain in continuous flow slide culture against microenvironmental flow rates greater than the maximum rate of motility (Korber *et al.* 1989b).

Gliding motility is well documented for *Myxococcus xanthus*, where locomotion is achieved by the smooth movement of cells in the direction of their long axis without the use of flagella; however, the mechanism of cell translocation is poorly understood. Overall, reports of movement of surface-associated cells are common, although the significance of such behaviour remains unclear. Explanations for motile attachment include bacterial positioning, nutrient capture strategies, and cellular migration within rapidly flowing systems. Perhaps the most obvious microbial behaviour in this context is the chemosensory and motile responses which they exhibit in natural and experimental aqueous environments (Smith & Doetsch 1969; Pilgram & Williams 1976; De Weger *et al.* 1987; Willey &

Waterbury 1989). Less apparent are the cell–cell and cell–surface interactions exhibited by motile microorganisms during colonization and growth on surfaces (Shapiro & Hsu 1989; Shimkets 1990). However, the analysis of interactive and surface associated microbial behaviour represents a relatively new area of research and further investigations are needed in this topic.

Reversible and irreversible attachment

The initial attachment of bacteria to surfaces is considered to occur in two stages: the reversible attachment phase and the irreversible attachment phase (Marshall *et al.* 1971; Lawrence *et al.* 1987; Marshall 1988). These initial attachment events frequently involve the attachment of bacteria to surfaces by a portion of the cell or flagellum while the cell continues to revolve (Meadows 1971; Lawrence *et al.* 1987; Malone 1987; Marshall 1988). Spinning behaviour may then be followed by either irreversible cell attachment or cell detachment. Powell & Slater (1983) concluded that if it was assumed that all cells which contacted the surface became irreversibly adsorbed, the resultant population of bound cells would be grossly overestimated, and that reversible attachment was predominant in nature. Marshall (1988) reported that cells became irreversibly attached to surfaces only after a period of unstable attachment, during which time cells revolved around the axis of attachment and frequently emigrated from the attachment site rather than becoming irreversibly attached. Others have made comparable observations during continuous flow slide culture of monoculture strains (Lawrence *et al.* 1987; Malone 1987) and also by surface colonizing bacteria observed in a marine microcosm (J. R. Lawrence, D. R. Korber and D. E. Caldwell, unpublished data).

Spinning behaviour by cells reversibly bound to surfaces has also been exploited to examine the mechanisms of rotation. Silverman & Simon (1974) studied rotation of motile *Escherichia coli* artificially tethered to surfaces by antiflagellum antibodies. One explanation for the existence of spinning behaviour in nature might be the function of a chemosensory mechanism whereby the cells determine, in some manner, the suitability of potential colonization sites by binding, rotating, and sensing the ambient conditions through chemoreceptors (Lawrence *et al.* 1987).

It has been considered that the absence of a functional flagellum should limit the reversible attachment phase as well as the overall attachment, recolonization, and distribution success of non-motile organisms (Lawrence *et al.* 1987). These hypotheses have been confirmed by the studies of Korber *et al.* (1989b, 1990, 1993), during examinations of the adaptive significance of motility with respect to cell positioning, recolonization, and structural rearrangements within biofilms.

Bacterial surface colonization strategies

Strategies involving solitary cells

Some bacteria are adapted to grow at interfaces and readily attach; however, the progeny may return to the planktonic state. Lawrence & Caldwell (1987) termed this phenomenon shedding behaviour, where undifferentiated cells adhere and grow in a perpendicular manner, releasing daughter cells successively after discrete growth intervals (Lawrence & Caldwell 1987; Lawrence *et al.* 1989a). Similar behaviour has been described for attached cells by Henrici & Johnson (1935), Helmsetter (1969), Bott & Brock (1970), and Marshall & Cruickshank (1973). Bacteria utilizing this mode of attachment behaviour appear to have a some specialized polar structure, that is, the adhesins of *Caulobacter* spp. or the hydrophobic/adhesive polar regions described for *Flexibacter* sp. (Marshall & Cruickshank 1973).

In some instances this surface colonization pattern has been studied in more detail. Kjelleberg *et al.* (1982) described the attachment of starved *Vibrio* DW1 adhering to a surface in a perpendicular position and following growth of the attached cells; motile daughter cells were released from the mother cells at regular intervals. In the case of *Rhizobium* spp., motile daughter cells were released which subsequently reattached at vacant surface sites and underwent growth and release of their own daughter cells. Additional variations on this behavioural theme have been reported by other authors. Power & Marshall (1988) described *Vibrio* MH3 which grows associated with the surface but completes cell division in the planktonic state.

Attached bacteria may grow and divide, with progeny gradually but continuously moving away from other members of the spreading micro-colony. Power & Marshall (1988) described this type of behaviour in *Pseudomonas* JD8. After each growth and division cycle the daughter cells of JD8 migrated away from each other while still adhering to the substratum. The mechanism proposed was related to utilization of the surface bound substrate stearic acid. When the nutrient concentration declined in the microenvironment surrounding the cell it became reversibly attached and moved until encountering more substrate. This cycle was repeated until all substrate was utilized, at which time the cells returned to the planktonic state. Organisms which exhibit such spreading surface colonization behaviour have also been observed in natural stream communities (Lawrence & Caldwell 1987). This type of behaviour can only be detected by continuous observation and thus its significance may be underestimated.

Other organisms grow and divide on surfaces while in a semi-adherent or motile state. Gliding bacteria are the most common bacteria that exhibit this type of surface association. A type of motile attachment, or rolling behaviour, has been documented by Lawrence & Caldwell (1987), where organisms were followed for extended periods with the growth, division, and release of daughter cells all conducted without the parent cell ever becoming fixed to the substratum. It is thought that the microorganisms moved in contact with the surface within the secondary minimum of the diffuse double layer, but were not immobilized.

Studies comparing the growth rates of attached bacteria indicated that solitary cells growing in the same environment exhibited higher specific growth rates than bacteria forming extensive microcolonies or groups (Lawrence & Caldwell 1987). This suggested that these pathways functioned as positioning mechanisms optimizing intercellular distances, as well as balancing substrate removal and waste accumulation (Lawrence & Caldwell 1987). However, detailed study of the growth rates of shedding cells has also indicated that the attached mother cell exhibited a declining growth rate with each subsequent generation. Observations performed on soil and marine biofilms have indicated that the presence of predation pressure virtually eliminated bacteria exhibiting these solitary behavioural patterns with the apparent exception of prosthecate bacteria (J. R. Lawrence & D. R.

Korber, unpublished data). Thus, these strategies may be those of pioneer colonizing organisms whose significance in mature biofilm communities is not known.

The budding and prosthecate bacteria

These bacteria represent a specialized group of bacteria with unique cell division and cell association strategies. However, other than the obvious division between planktonic swarmer and attached mother cell in the life cycle, and possible advantages in niche and resource exploitation, no clear understanding of the particular advantages of the strategies of these bacteria exists. Their patterns of growth have been extensively described (Hirsch 1974). For example, attached *Pedomicrobium* spp. initiated hyphal elongation with 1–4 branching hyphae arising from the central mother cell. Hyphal elongation continued with the initiation of bud formation and subsequent bud growth prior to the release of daughter cells.

Lawrence *et al.* (1989b) used computer enhanced image analysis to quantitate the surface growth and behaviour of *Pedomicrobium* spp. present within a marine microcosm and cultured using continuous flow slide culture. Behaviour similar to that reported by Hirsch (1974) was observed for *Pedomicrobium* spp., with cells attaching and developing hyphae, and later bearing daughter cells on their apical tips. The hyphae extended along the surface and also into the surrounding medium away from the attachment surface. Image analysis was used to increase the effective depth of field using images obtained from different focal planes. These allowed the determination of cell area, hyphal length, and the growth rates of mother and daughter cells. Hyphae doubled in length every 2.6 h until the onset of bud formation, during which time only slight extension of hyphae occurred. Growth rate differences between mother and daughter cells on surfaces were also documented (that is, the time to double for mother cell, 24 h; time to double for daughter cell, 1.2 h). The daughter cells were released into the bulk phase and eventually recolonized vacant sites within the flow cell system, resulting in the formation of a network of hyphae and cells. Rosette formation in these bacteria has also been extensively described and studied (Stove-Poindexter 1964; Hirsch 1984).

Colonial strategies

Hirsch (1984) documented many forms of microcolonies including coenobia, nets, mycelia, multiple parallel cell arrangements, and rosette formation. In most cases, it is not established how these various groupings of cells arise or what their adaptive significance may be. However, with the increasing interest in this aspect of microbial behaviour, some insight into pathways and even genetic and environmental control of these pathways has been obtained. Unique paths of microcolony development have been documented for a number of genera including *Pseudomonas* spp., *Rhizobium* spp., *Vibrio* spp. and *Caulobacter* spp. (Lawrence *et al.* 1989b, 1992; Caldwell *et al.* 1992a) and various unidentified bacteria in environmental samples (Torrella & Morita 1981; Lawrence & Caldwell 1987). Each of these pathways of microcolony development contains a variety of strategies for releasing daughter cells and returning them to the planktonic or migratory state.

The arrangement of cells within a microcolony is a function of the cells' shape, size, mode of cell wall growth, polymer production, and poorly understood events occurring at the moment of cell division. For example, a snapping cell division in *Corynebacterium* spp. gave rise to V-shaped daughter pairs (Hirsch 1984). Similar movements at the time of separation of daughter cells have been described for other bacteria such as *P. fluorescens* and *E. coli* (Lawrence *et al.* 1987; Shapiro & Hsu 1989). The specific alignments of cells that occur during microcolony formation have been interpreted to indicate that a variety of complex cell–cell interactions do occur within the developing microcolonies (Shapiro & Hsu, 1989). Extracellular polymers also play a critical role in microcolony development. For example, Allison & Sutherland (1987) showed that a freshwater bacterium and a non-EPS producing mutant colonized surfaces equally, but only the former could produce microcolonies.

What has been termed 'packing' colony formation behaviour is commonly observed during surface growth in natural communities by pseudomonads and vibrios (Torrella & Morita, 1981; Lawrence & Caldwell, 1987; Lawrence *et al.* 1987, 1992; Korber *et al.* 1989a). In this case, the irreversibly attached cells grow and divide while forming a tightly packed group of cells at a discrete surface location. P. *fluorescens* strain CC840406-E followed a developmental pathway that proceeded from single cells to a monolayer microcolony containing between 8–16 cells (after 4–5 h surface growth: Lawrence *et al.* 1987). Following the 8–16 cell stage, additional growth resulted in the release of some colony members into the aqueous phase, those detached bacteria being termed 'recolonizing' or migratory cells. Cells which successfully recolonized were observed to reattach at vacant surface sites, thereby creating new focal points of growth and aiding in the formation of a confluent biofilm.

Other *Pseudomonas* spp. exhibited paths of microcolony development and timing of recolonization different to that of *P. fluorescens*. Delaquis (1990) observed that *P. fragi* microcolonies displayed a high degree of structural organization. Colonies developed via a packing manoeuvre, resulting in a monolayer of regular, circular colonies. After 7 h, detachment and recolonization of motile cells occurred, marked by an increase in the number of microcolonies per field and also by an increase in the number of motile cells present in the surface boundary layer. After approximately 8 h surface growth, it was apparent that a secondary structure of polarly attached cells had developed on microcolonies, which subsequently matured into a palisade layer. The palisade layer may remain intact and contribute to the development of a confluent biofilm structure.

Vibrio parahaemolyticus is an autochthonous estuarine organism isolated from water sediment, marine organisms and inanimate surfaces, and is also associated with gastroenteritis in humans. As such, it exhibits a broad array of adaptive flexibility. Microcolonies of *V. parahaemolyticus* formed by two pathways which were variations of the packing behaviour described above. Behavioural analysis of *V. parahaemolyticus* strains in CFSC showed that during growth on surfaces, microcolonies consisted of a variety of functional types (Lawrence *et al.* 1992). These included laterally flagellated progeny adapted for attachment in low viscosity environments or migration in high viscosity environments. In addition, polarly flagellated progeny adapted for detachment and dispersal in low viscosity environments were also produced. The mechanisms controlling the formation of these types included induction of the lateral flagella gene system (McCarter *et al.* 1992). In addition, phenotypic switching between a transluscent variant and an opaque variant occurred, the latter only having polar flagella (Lawrence *et al.* 1992). Thus, *V. parahaemolyticus* microcolonies could form opaque or translucent variants, or more commonly as mixed populations. This system provided laterally flagellated bacteria adapted for attachment to the surface in low viscosity environments and motility in high viscosity environments, or polarly flagellated progeny adapted for detachment and dispersion. The opaque variant provided a continuous source of variation, creating a population suited for planktonic growth (Lawrence *et al.* 1992). During the migration or recolonization phase which occurred after 5 h of surface growth, *V. parahaemolyticus* cells detached from microcolonies and either remained planktonic, or reattached. The sites of reattachment indicated that cell–cell interactions were also important in *V. parahaemolyticus* behaviour, because cells aggregated at the initial sites of colonization. In addition, extensive interactions were observed between laterally flagellated cells and microcolonies in high viscosity microenvironments (Lawrence *et al.* 1992).

Dispersal

In each of the cases of microcolony development described above, after a species specific period of growth on a surface, attached bacteria underwent a redistribution phase. Lawrence *et al.* (1987) termed this event 'recolonization', a phase which for *P. fluorescens* included the emigration of attached cells from the surface to the suspended phase and the subsequent reattachment of those cells to vacant surface sites. The timing of the onset of this phase varied between species as well as with the conditions under which the cells were growing. For example, attached *P. fluorescens* growing in continuous flow slide culture recolonized after approximately 5 h, whereas *V. parahaemolyticus* grown in 75% 2216 medium recolonized after approximately 4 h. A typical data set for *V. harveyi* illustrating the growth phase, increase in number of motile cells and increase in microcolony number is shown in Fig. 1.2. A series of darkfield images showing the actual growth and development of the microcolonies is shown in Fig. 1.3. In this instance, redistribution occurred following only 2 h of surface associated growth. These variations in the

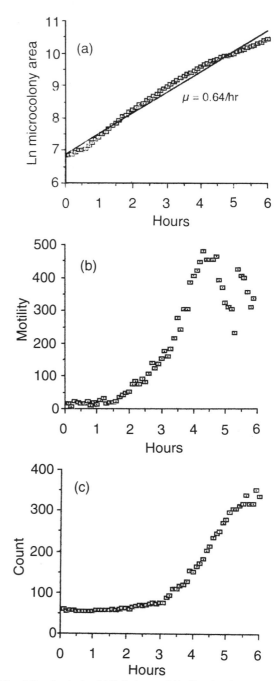

Fig. 1.2. Analysis of *Vibrio harveyii* biofilm development showing (a) biofilm development rate determined using darkfield microscopy and digital image analysis, (b) increase in the number of motile cells in the hydrodynamic boundary layer, and (c) increase in number of attached cells and microcolonies with time. Note the increase in number of motile cells at approximately 2 h and the subsequent increase in number of attached cells and microcolonies.

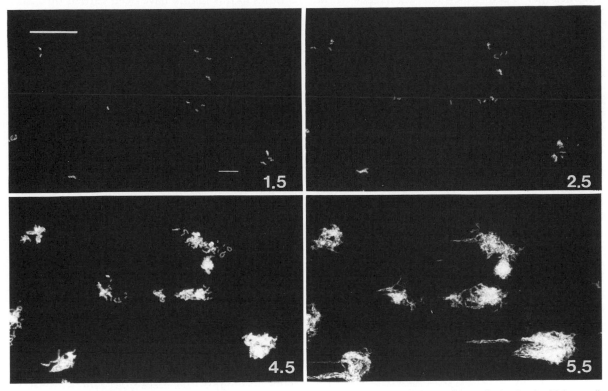

Fig. 1.3. Darkfield photomicrographs showing growth and microcolony formation during surface colonization by *V. harveyii*. Times indicated are 1. 5, 2. 5, 4. 5 and 5. 5 h after inoculation of the flow cell. The phenomena of formation of elongated cells and of aggregation around initial microcolonies are also evident in these photomicrographs. Bar = 20 µm.

timing of cell redistribution represent important behavioural characteristics which appear to be species specific. This event provided a mechanism whereby cells emigrated from heavily colonized regions, where growth factors may be depleted, to regions more amenable to rapid growth. However, in detailed studies of individual cells within microcolonies, no evidence for a change in growth rate as a trigger has been detected (J. R. Lawrence & D. R. Korber, unpublished data). It is more likely that this phenomenon represents variable expression of traits, spontaneous variants, or other mechanisms which generate diversity allowing organisms to adapt to changing environments and providing at least two morphogenic states optimal for different circumstances.

Aggregation and coaggregation

Behaviour such as aggregation appears to play an important role in the development of micro-colonies, and subsequently, in biofilm architecture. Kolenbrander (1989) noted that nearly all human oral bacteria participate in intergeneric coaggregation. Sjollema *et al.* (1990a) used the term cooperative effects among oral bacteria to describe a microbial process by which adhering cells can modify their surrounding environment into a more favourable one for further attachment. In addition, a unanimous conclusion from surveys on more than 700 strains of oral bacteria was that partner recognition during bacterial coaggregation is very specific (Kolenbrander 1989). Coaggregation may occur intergenerically, intragenerically and multigenerically, and these interactions are extremely important in the colonization and occupation of substrata. These phenomena have been reviewed in detail by Kolenbrander (1989) and Kolenbrander & London (1992).

The existence of extensive aggregation during the development of *V. parahaemolyticus* microcolonies has been described (Lawrence *et al.* 1992). In addition, cell aggregation was still evident in the form of distinct microcolonies attached to the surface at the base of the biofilms after 12 and 24 h growth. However, the architecture of the biofilms was also influenced by migration within the biofilm, a phenomenon apparent in the development of a diffuse basal layer with numerous channels in *V. parahaemolyticus* biofilms after 24 h growth (Lawrence *et al.* 1991). Similarly, observations of the growth and development of *P. fragi* microcolonies indicated that after onset of migration an extensive layer of vertically attached cells developed on the surface of the initial microcolonies. This layer has also been observed in biofilms of *P. fluorescens*, suggesting that some cells may aggregate to form specialized structural elements during biofilm development (Korber *et al.* 1993; Lawrence & Korber 1993). The nature of the cellular interactions in biofilm formation may be further investigated through the use of non-motile mutants.

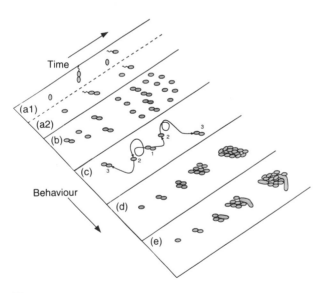

Fig. 1.4. Illustration showing the strategies exhibited by various surface-colonizing bacteria. (a1) and (a2), the attachment of a mother cell, in polar and horizontal orientations, with growth and subsequent release of the daughter cell to the planktonic state. (b) the attachment, growth, and slow migration of daughter cells (spreading). (c), growth and division during motile attachment. (d), typical microcolony development (packing) exhibited by Pseudomonas and other species. (e), microcolony development with variant production and aggregation exhibited by *Vibrio* spp.

Studies of this nature have been carried out using wild type and mot⁻ mutants of *P. fluorescens* (Korber *et al.* 1989b, 1992, 1993). These latter studies have concluded that many of the structures which develop in microcolonies and within biofilms require active, motility dependent, cell–cell interactions. The range of surface colonization behaviour exhibited by various bacteria is summarized in Fig. 1.4.

Control of microcolony formation

Surfaces may stimulate, inhibit, or have no effect on the activity of microorganisms at interfaces (van Loosdrecht *et al.* 1990). However, microscopy has shown that in many cases surface associated growth results in the formation of microcolonies, following developmental pathways that are species specific and modulated by a number of factors such as flow rates, viscosity, nutrient status, starvation and light intensity (Hirsch 1968, 1974; Kjelleberg *et al.* 1982; Marshall 1988; Power & Marshall 1988; Lawrence *et al.* 1989a, 1992). In contrast, *P. fluorescens* strain CC840406-E, when grown in either minimal or complex medium, followed the same developmental pathway. This pattern of surface growth by *P. fluorescens* has also been shown to be unaffected by factors such as initial cell density, nutrition, and laminar flow velocity (Korber *et al.* 1989b).

However, behavioural sequences have been shown to vary with respect to environmental conditions or induction of specific genes. Nutritional status affects the sequence of events which occur during surface associated growth by *P. syringae*, where the formation of microcolonies in CFSC was inhibited by growth in defined, glucose amended minimal medium. Under these conditions, *P. syringae* initially adhered to the surface and then detached without a growth phase. In contrast, *P. syringae* cells cultured using complex media attached, grew to form microcolonies, and exhibited a recolonization phase after four generations. With *Vibrio* DW1, overcrowding on the surface resulted in nutrient depletion, cessation of growth, and induction of a starvation survival phase (Kjelleberg *et al.* 1982; Marshall 1989). In the case of a natural stream community (Lawrence & Caldwell 1987), one morphological type responded to the cessation of flow by immediately emigrating from the surface. The microcolony in question had developed over a period of 10–12 h without any previous dispersal of daughter cells, indicating a possible response to changes in the microenvironment as a consequence of loss of flow.

Studies of *V. parahaemolyticus* represent the most comprehensive investigations of surface induced gene expression (McCarter *et al.* 1992). The genetic control of *V. parahaemolyticus* lateral flagellum production has been shown to be physical, that is, increasing the viscosity of the medium was sufficient to induce the laterally flagellated phenotype. In this instance, the polar flagellum has been shown to act as a tactile sensory device informing the bacterium about its ability to move. Additional studies have shown that iron deficiency is also required to initiate production of swarmer cells (McCarter *et al.* 1988; McCarter & Silverman 1989). Dagostino *et al.* (1991) reported that genes related to other specific functions may only be switched on at surfaces and not while the bacteria are growing planktonically. All of the above observations suggest that some surface colonizing bacteria respond to a broad set of environmental factors, including concentrations of nutrients, micronutrients, O_2, CO_2, etc. It is also possible that substances such as pheromones or signal proteins similar to those active in myxobacteria (Nealson 1977; Kim & Kaiser 1990) may be involved in controlling these surface growth related phenomena.

Thus, bacteria have a variety of strategies for adapting to life in a surface habitat and also for releasing progeny from the attached to the planktonic state. Behavioural adaptations are an integral part of surface colonization and represent: (i) adaptations for colonization and occupation of the substratum, (ii) optimization of substrate availability, (iii) strategies for reproductive success, (iv) balancing of intra- and interspecific competition, and (v) responses to environmental stress. However, more information is required regarding the quantitative relationships between growth rate, reproductive success, resistance to stress, and the relative merits of specific behavioural patterns. In addition, it is clear that cell–cell interactions play a role in the establishment of the structural matrix of microcolonies and biofilms.

Advantages of sessile growth

It is considered that surface colonization provides an advantage for bacteria that exist in oligotrophic environments (Kjelleberg *et al.* 1982; Kjelleberg & Hermansson 1984). The open ocean is representative of such an environment as the concentration of dissolved organic matter in these habitats is commonly less than 10 mg L^{-1} (ZoBell 1943; Wahl 1989). Surface conditioning films adsorb to surfaces and tend to concentrate scarce nutrients, making attached growth by bacteria nutritionally advantageous.

However, eutrophic conditions do not necessarily result in the exclusive existence of planktonic bacteria, suggesting that the role of bacterial attachment cannot be solely based on nutrient capture (Paerl 1975; Robinson *et al.* 1984; MacLeod *et al.* 1990). Alternative explanations for bacterial surface colonization include resistance from washout from the surface, predator avoidance behaviour, heterotrophic relationships, co-metabolism of recalcitrant compounds, positioning behaviour, syntrophic dependencies and optimization of solution chemistry (Caldwell & Caldwell 1978; Boone & Bryant 1980; McInerney *et al.* 1981; Brannan & Caldwell 1983; Lawrence *et al.* 1987; Thiele *et al.* 1988; Korber *et al.* 1989a; Decho 1990; Wolfaardt *et al.* 1994). In many cases, attachment behaviour may not be easily explained in terms of a single advantage, but rather by the combined effect of a number of factors.

Nutrient scavenging

One explanation for the attachment of bacteria existing in nutrient limiting environments is to gain access to the film of organic molecules adsorbed to the substratum (ZoBell 1943; Kjelleberg 1980; Kefford *et al.* 1982; Kjelleberg *et al.* 1982; Hermansson & Marshall 1985). Heukelekian & Heller (1940) observed that *E. coli* failed to grow in suspension where the concentration of organic nutrients was less than 0.5–2.5 mg L^{-1} except in the proximity of solid surfaces. At higher nutrient concentrations ($>25 \text{ mg L}^{-1}$), no surface dependent growth effects were observed. Fletcher (1986) measured the uptake of radiolabelled glucose by attached and free living cells, determining that the attached populations assimilated 2–5 times more glucose and also that the respiration of surface associated cells occurred at greater rates than for suspended cells. This confirmed earlier observations made by Bright & Fletcher (1983), and led to explanations for increased activity of cells on surfaces which included altered cell physiological responses triggered by: (i) induction of surface induced pathways (e.g. increased substrate transport), or (ii) increased flux of nutrients at the surface through formation of organic films or improved nutrient movement at the surface as a result of fluid hydrodynamics.

Results have been more conclusive within flowing environments. Caldwell & Lawrence (1986) demonstrated flow dependent growth kinetics of a *P. fluorescens* strain in continuous flow slide culture. At high nutrient concentrations (1 g L^{-1} glucose), it was determined that laminar flow could be halted without affecting the rate of microcolony development. At lower nutrient concentrations (100 mg L^{-1} glucose), microcolony growth ceased when flow was halted, suggesting that glucose depletion within the hydrodynamic boundary layer rather than oxygen depletion was responsible for growth limitation. Similarly, Confer & Logan (1991) provided evidence that shear strength resulted in significant increases in the utilization of radiolabelled macromolecules, suggesting that diffusion limitation through the zone surrounding the cell (termed the dead zone), in combination with enzymatic hydrolysis, decreased the rate of uptake by up to 12.5 times. In contrast, low molecular weight compounds were demonstrated to show little shear dependency on uptake. Thus, constant replenishment of the surface microenvironment surrounding attached bacteria may provide a mechanism for maintaining a steep concentration gradient (such as of nutrients and wastes) required for rapid diffusion and growth.

An alternative mechanism for nutrient scavenging on surfaces is based on the non-adhesive association of bacteria with the surface. Hermansson & Marshall (1985) reported the surface associated behaviour of marine isolate *Vibrio* MH3 which reversibly adhered to surfaces coated with stearic acid without becoming permanently attached. It was determined that *Vibrio* MH3 had to be in close proximity with the surface to take up the fatty acid, even though only 6% of the measured cell respiration was due to irreversibly

bound MH3 cells. This provided evidence that irreversible binding was not a prerequisite for the uptake of surface bound nutrients.

Microbially produced EPS is considered to play a significant role in the nutrition of sessile bacteria. The sorptive capacity of EPS for dissolved organic matter and metals has been widely demonstrated for sewage and marine systems (Brown & Lester 1982; Lion *et al.* 1988; Ferris *et al.* 1989). Thus, organisms can concentrate scarce nutrients extracellularly and utilize these compounds via the activity of exoenzymes. As a consequence, the uptake of dissolved organic matter from the suspending solution has been shown to be up to 60% greater for sessile biofilms than for planktonic bacteria. The storage of nutrients within EPS has also been suggested; however, there is little direct evidence that bacteria are capable of utilizing either their EPS or sorbed nutrients on their EPS when conditions for growth become inadequate. In addition to protecting cells from surface grazing predators, EPS, which are rich in sugars, may also constitute an important energy source for protozoa (Sibbald & Albright 1988). Examination of the amount of cellular biomass present within multiple trophic level systems has demonstrated that there is not enough bacterial biomass to support the measured biomass of grazing protozoans (for example, only <10–25%: Juniper 1987), and since there always remains a fairly large microbial population associated with surfaces even in the presence of extensive grazing, another mechanism of carbon transfer to higher trophic levels must exist. This supports epifluorescent microscopy observations made by J. R. Lawrence, D. R. Korber & D. E. Caldwell (unpublished data), where flagellate and ciliates actively feeding in a marine microcosm containing attached bacteria stained with fucose specific, FITC (fluorescein isothiocyanate) conjugated lectin were observed to have large amounts of fluorescent material in their food vacuoles. It was considered that the predators in this system were removing the abundant bacterial EPS but not the bacteria, thereby transferring nutrients up the food chain without detrimentally affecting populations of primary producers. The reactivity of microbial exopolymeric substances in a broad range of microbially associated events has been described by Decho (1990).

Positioning and resistance to washout

Bacteria persist in locations conducive to their growth and development. Motile responses to variations in light and solution chemistry, such as photo- and chemotaxis (Willey & Waterbury 1989) provide well documented examples of position maintenance by bacteria, as does the formation of intracellular gas vacuoles (Caldwell & Tiedje 1975). A wide range of indigenous bacteria also rely on adherent behaviour to position themselves where the supply of organic nutrients is optimal, the pH allows utilization of specific growth factors, oxygen requirements are met, or where temperature ranges are favourable (Caldwell & Caldwell 1978; Caldwell *et al.* 1983). The colonization of sulphur springs by *Thermothrix thiopara* provides an example of bacterial positioning as a result of cellular attachment (Brannan & Caldwell 1983; Caldwell *et al.* 1983). *T. thiopara* is a sulphur oxidizing bacterium which lives and grows in sulphur springs at temperatures of 74 °C. The maintenance of *T. thiopara* filaments at sulphide–oxygen interfaces permits the use of reduced sulphur as a source of energy. Cells growing beyond the interface into oxidizing conditions are continuously shed, reattaching where favourable conditions are encountered.

Adherent positioning mechanisms utilized by bacteria may involve either attachment to substrata or coaggregation, adaptations responsible for the formation of marine snow and sewage sludge granules (Conrad *et al.* 1985; LeChevallier *et al.* 1987; Alldredge & Silver 1988; Merker & Smit 1988; Thiele *et al.* 1988; MacLeod *et al.* 1990). In high rate, anaerobic waste treatment reactors, sludge must be immobilized or retained by some mechanism in order to decrease the dependence of sludge retention on the rate of flow within the system. It has been shown that upflow reactors may bear several times the loading factors utilized in systems lacking the upflow design, and that the pivotal factor responsible for this increased loading is the formation and persistence of large (diameter *c.* 3 mm) bioactive granules. The factors contributing to the formation of such granules have been a topic for conjecture: however, flow resistance, microcolony positioning, and syntrophic cross-feeding are all potential advantages for degradative bacteria which exist in a granular consortium. Adherent behaviour is

also observed within the bovine rumen (Russel *et al.* 1981; Rasmussen *et al.* 1989), where bacteria with cellulolytic capabilities adhere to cellulose fibres, causing the release of glucose monomers for growth by the attached organism and preventing the rapid loss of bacterial cells to bovine digestive processes. Similarly, the adherence of heterotrophic bacteria to cyanobacterial filaments may optimize the uptake of O_2 by the bacteria while simultaneously producing CO_2 for cyanobacterial photosynthesis (Caldwell & Caldwell 1978). Similar observations have been made by Lupton & Marshall (1981), where EPS produced by cyanobacterial heterocysts were observed to be populated by bacteria. Positioning behaviour is also common in streams, at aerobic–anaerobic interfaces in iron rich waters, on the surfaces of pipes and water supply equipment, in biological reactors, and in model flow systems (Kinner *et al.* 1983; Robinson *et al.* 1984; Lawrence & Caldwell 1987; Lawrence *et al.* 1987; Characklis *et al.* 1990a). Thus, adherent colonization strategies are indirectly responsible for the formation of diverse and unique microhabitats.

Resistance to predation

Bacterial attachment also improves the survival of microorganisms which might otherwise be preyed upon by protozoa feeding on plankton or bacteria on the surface of the biofilm (Porter 1976; Decho & Castenholtz 1986; Decho 1990). Cell aggregation constitutes another form of cellular adherence which may similarly improve species survival. Caron (1987) observed that attached or aggregated bacteria were grazed to a lesser extent by microflagellates than planktonic cells. Heijnen *et al.* (1991) investigated the feeding success of a flagellate, *Bodo saltans,* on *Rhizobium leguminosarum* in liquid suspension amended with bentonite clay. The authors found that addition of the clay strongly reduced the predation success of *B. saltans*, and ascribed these effects to coating of either the rhizobia or flagellates with clay particles. Thus, by association with surfaces or other cells, bacteria may become less susceptible to the action of predators. However, there have been reports of protozoan feeding behaviour adapted to surfaces exhibiting selective grazing on aggregated bacteria.

The formation of a complex EPS coating by

the bacterium provides strong adhesive forces which physically inhibit removal of the cell from the attachment surface (Porter 1976; Costerton *et al.* 1985, 1987; Decho & Castenholtz 1986), and may also slow the penetration of digestive enzymes secreted by surface predators.

Development of heterogenous biofilm communities

Concept of the microenvironment

The ambient conditions associated with clean surfaces are temporally and spatially modified by the colonization, growth and activity of bacteria. Microbial consumption of nutrients, production of wastes, and synthesis of cellular and extracellular materials all act in concert to define physicochemically what is known as the microbial microenvironment (Hamilton 1987). These microenvironments persist over time due to diffusion limitations in combination with sustained metabolic activity. The presence of concentration gradients can influence bacterial population diversity and spatial distribution as well as microbial metabolic activity within biofilms, allowing the survival of other organisms with specific, and often stringent, growth requirements. The microenvironment is therefore believed to be the foun-

dation for much of the observed microbial diversity in nature. Although this is a fundamental concept, its presence has proven difficult to directly verify. However, application of pH sensitive probes has provided some information on microenvironments, indicating 1–3 μm zones of reduced fluorescence surrounding cells and

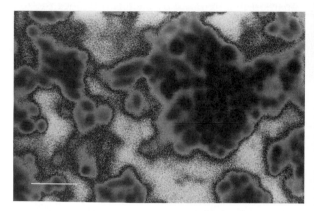

Fig. 1.5. Scanning confocal laser image of fluorescence intensity of the pH sensitive probe 5 and 6 carboxyfluorescein in the voids and spaces surrounding cells and microcolonies of V. parahaemolyticus in a 48 h biofilm. The image shows a dark boundary layer approximately 1–3 μm thick surrounding individual cells and microcolonies. This zone represents the approximate dead volume surrounding bacterial cells and may represent the scale of microzonal influence for these cells. Bar = 5 μm.

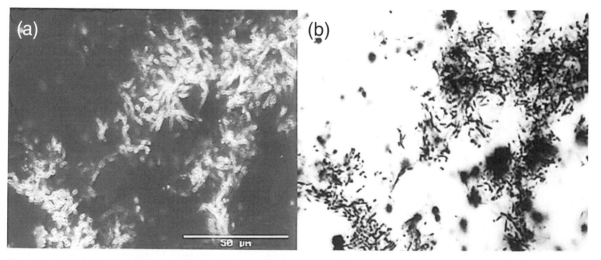

Fig. 1.6. (a), SCLM horizontal optical thin section showing the position of exopolysaccharide material produced by a microcolony within a sludge degrading biofilm. A polyanionically charged fluorescently labelled dextran (70 kD.) was found to bind this colony type preferentially when compared with other consortium members, indicating spatial variability in the polymer chemistry of the biofilm. (b) is a negatively stained image of the same region, showing the position of the microcolony bacteria which produced the EPS. Bar = 50 μm.

microcolonies within biofilms as shown in the SCLM micrograph (Fig. 1.5). Distribution of unique extracellular polymers surrounding different members of biofilm consortia may also contribute to the development of chemical microzones (Fig. 1.6a, b).

One of the best examples of temporal/chemical modification of biofilms results from the activity of aerobic heterotrophic bacteria. Within a newly formed biofilm, rapid utilization of available nutrients lead to an increase in microbial biomass with a concurrent increase in the demand for oxygen. Thus, anaerobic zones form due to metabolic oxygen depletion and diffusion limitation, potentially allowing proliferation of anaerobic microorganisms while limiting the success of aerobes. In pure laboratory cultures, the cultivation of strict anaerobes involves reducing the growth medium and eliminating any exposure to oxidized equivalents during transfer, growth or subculture. Within the biofilm microenvironment, these same organisms flourish because the conditions are met within microniches created chemically or biologically.

Development of biofilm architecture – spatial heterogeneity

Microbial biofilms consist of a spatially heterogeneous arrangement of bacteria, biogenic extracellular material, and void space (McFeters *et al.* 1984; Gujer 1987; Lawrence *et al.* 1991; Keevil & Walker 1992; Stewart *et al.* 1993). Repeating patterns of cells and EPS in biofilms formed by pure bacterial cultures indicate that this spatially defined architecture is often species specific, and that distinct microcolony morphology evident at early stages of development is sometimes detectable following prolonged biofilm growth (Lawrence *et al.* 1991). Biofilm biomass distribution can exhibit species specific trends (Lawrence *et al.* 1991; Korber *et al.* 1993). For example, significant differences in the distribution of biomass within biofilms of *P. fluorescens*, *P. aeruginosa* and *V. parahaemolyticus* have been determined using SCLM and image analysis (Lawrence *et al.* 1991). These trends in biomass distribution were likely to be a reflection of the different surface colonization strategies used by these different bacteria (Lawrence & Caldwell 1987; Korber *et al.* 1989a; Lawrence *et al.* 1992).

Motility also affects the distribution and arrangement of biomass within biofilms. For example, the amount of cell biomass present at different depths in motile and non-motile strains of 24 h *P. fluorescens* biofilms differed (Korber *et al.* 1993). At the base of the mot$^-$ biofilm (depth 0 μm) the measured mean cell area value was 4.2 ± 1.5 μm^2 per 50 μm^2, at the 8 μm depth, 2.6 ± 2.0 μm^2 per 50 μm^2, and at the 16 μm depth, 1.8 ± 2.1 μm^2 per 50 μm^2. This arrangement of non-motile biofilm cells culminated in formation of typical pyramidal biofilm colonies, with the highest measured cell areas present at the base. In contrast, the mot$^+$ strain formed thicker biofilms (depth = 19 ± 10 μm vs 42 ± 19 μm for the mot$^-$ and mot$^+$ strains, respectively), they were relatively less variable, and possessed greater amounts of cellular biomass at the attachment surface and the biofilm–liquid interface. Interestingly, the 8 μm section depth of the mot$^+$ *P. fluorescens* biofilms was diffuse, relative to over- and underlying regions, and may have played a role in facilitating diffusion of nutrients to the base of the biofilm.

Analyses have also shown mixed species biofilms to be quite variable (Robinson *et al.* 1984; Caldwell *et al.* 1992a; Wolfaardt *et al.* 1994). For example, Keevil & Walker (1992) reported that natural biofilms, observed with DIC and epifluorescence microscopy, consisted of a basal cell layer approximately 5–10 μm thick, with stacks of polymer associated cells extending from the surface into the bulk aqueous phase at regular intervals. These structures were presumed to have facilitated the flux of nutrients to the basal biofilm layers, thereby supporting the growth of aerobic heterotrophs, while creating chemical microenvironments which would support the growth of anaerobic organisms as well.

This general theme of cell aggregates interspersed by channels has been confirmed during other pure culture and mixed species investigations, with biofilm structure varying for different hydrodynamic regimes, incubation times, species composition, or nutrient status (Santos *et al.* 1991; Korber *et al.* 1994b; Stewart *et al.* 1993). For example, Bakke & Olsson (1986) detected channels within *P. aeruginosa* biofilms using low magnification microscopy. Depth measurements indicated that variability increased with age, although the average depth of the biofilm did not. The channels were considered by the authors to have increased surface roughness and decreased diffusional length, thus facilitating mixing and nutrient exchange. An SCLM study of flow velocity effects (flow range, 0.01–0.21 cm s^{-1}) on the architecture of *P. fluorescens* biofilms determined that the extent of channel formation was flow velocity dependent, with increased flow velocities causing an increase in the average depth of the biofilms and a concurrent decrease in the variability of depth measurements (Korber *et al.* 1994b). This study, performed over large biofilm regions (>1 000 000 μm^2) and at low magnification, demonstrated that at low flow rates biofilms were less deep and contained relatively more channels (0.01 cm s^{-1}; mean depth, 12.3 ± 6.8 μm) than at high flow rates (0.21 cm s^{-1}; mean depth, 83.7 ± 10.2 μm). The increase in depth and decrease in the total area of channels suggested that limitations to growth present at low flow may be offset by increased exchange of nutrients and/or oxygen across the hydrodynamic boundary layer at high flow rates. Figure 1.7 shows the presence of channels and cell aggregates within a typical *P. fluorescens* biofilm using SCLM at low magnification in the *xy* plane (showing the pattern of cell aggregates and channels), as well as at high magnification (Fig. 1.8) in the *xz* plane (showing in sagittal profile, individual channels and cell aggregates). In addition to pores and channels, other unique architectural features have also been observed. For example, bridging structures, formed as part of biofilms developed under flowing conditions, have been observed (Korber *et al.* 1993b). The basal layer of a biofilm (Fig. 1.9a) and the bridge structure extending across the channel can also be observed (Fig. 1.9b). Pore development also constitutes a situation where the biofilm has overgrown a small channel, resulting in a void space not connected to the bulk aqueous phase. Similar structures have been described for many biofilms including those on rotating biological contactors and other systems (Kinner *et al.* 1983). The significance of such structures is as yet not clear.

The presence of channels within biofilms (Figs. 1.7, 1.8) may be analogous to the growth habit of barnacle colonies, which form relatively high hummocks about 5 cm across and are also interspersed by water channels at regular intervals (Barnes & Powell 1950; Knight-Jones & Moyse 1961). The largest barnacle individuals are found at the peak of the hummocks, where more nutrients are available for growth as a consequence of more rapidly flowing water. Hummocks would also cause turbulent mixing thereby preventing buildup of a nutrient depleted boundary layer, with associated benefits for other members of the barnacle community. For biofilm bacteria, channels and cell aggregates may function similarly, with flow and advection replenishing nutrients and facilitating diffusion in regions not directly exposed to the bulk phase. Depth dependent variability in growth rates for biofilm bacteria has previously been inferred by Korber *et al.* (1994a), where treatment with fleroxacin, a DNA gyrase inhibiting fluoroquinolone, caused greater cell elongation at the bulk–liquid interface than at the base of the biofilm, even though visual determinations of fleroxacin penetration indicated that migration of this compound was not significantly impeded by the bacterial–EPS matrix. Similar observations have been made for penetration of antimicrobial agents (Nichols 1989). Consequently, the increase in cell length following application of fleroxacin should be proportional to the amount of cell growth. These observations agree with the premise that biofilms grow mainly at the biofilm–liquid interface and not from the biofilm base, and have implications not only for understanding the ecology of biofilm systems, but also for interpreting the effects of antimicrobial agents on biofilms (Gilbert *et al.* 1990; Eng *et al.* 1991).

Population architecture of biofilms

While studies of the physical architecture of microbial biofilms are now routinely conducted, population architecture may also be directly studied using fluorescent conjugates of oligonucleotide probes, using phylogenetic probes and PCR (polymerase chain reaction) methods (Stahl *et al.* 1988; Amann *et al.* 1992). If microchemical gradients are also examined, valuable information describing chemical and spatial requirements of highly structured biofilm communities may be obtained. Recently, Amann *et al.* (1992) conducted studies using natural populations of SRB within multispecies biofilms, demonstrating that two types of cell distributions were observed when biofilms were stained with SRB specific oligonucleotides. One probe distribution pattern indicated a microcolony growth habit for *D. vulgaris*-like sequences, whereas *D. desulfuromonas* was more randomly distributed. Time course

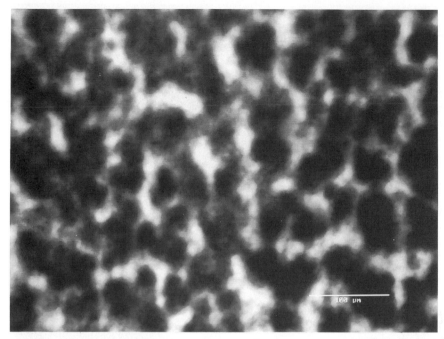

Fig. 1.7. Low magnification SCLM optical thin section (*c*. 80 000 μm^2) clearly showing a spatial pattern of channels and cell aggregates. Bar = 100 μm.

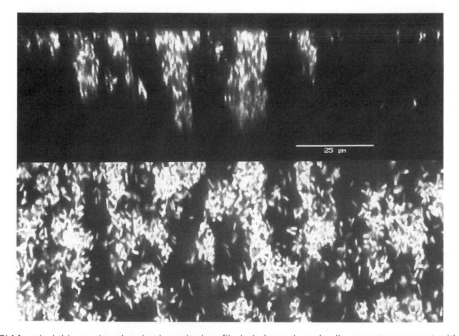

Fig. 1.8. SCLM optical thin section showing in sagittal profile (*xz*) channels and cell aggregates present within a positively-stained 24 h *P. fluorescens* biofilm (top). The bottom portion of the micrograph is a horizontal optical thin section, the top portion of which corresponds with the position of *xz* sectioning. Bar = 25 μm.

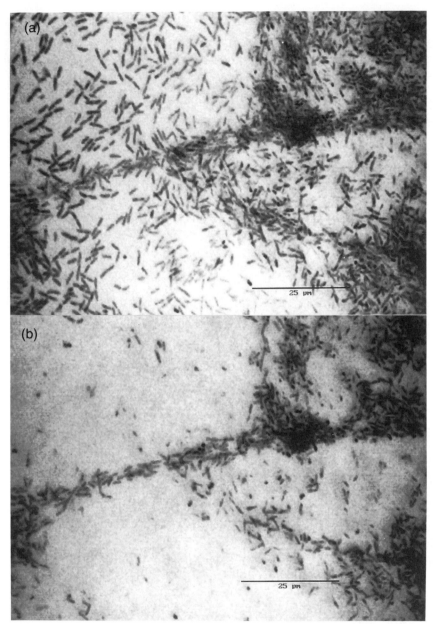

Fig. 1.9.   A projection of four SCLM optical thin sections obtained from the basal region of a negatively stained *P. fluorescens* biofilm (a), showing the distribution of cells near the attachment surface, as well as a bridge structure evident near the 6 μm sectioning depth. (b) is a single optical thin section from the 6 μm section depth, showing the bridge structure alone. Bar = 25 μm.

investigations using these techniques may provide important information regarding the temporal success of organisms which have specific chemical requirements. For example, development of anaerobic zones may be found to corre- late with the occurrence of SRB microcolonies.

Investigations using 16S, 5S and 23S rRNA probes are now being used to study a wide range of natural systems, allowing investigations into community structure without the need for cultivation (Stahl *et al.* 1988; Amann *et al.* 1990, 1992). This is important because members of a biofilm community isolated from the environment may not be easily culturable. While still undergoing rapid development, these methods, used in conjunction with either SCLM or epifluorescence microscopy, will provide valuable insight into the spatial contiguity of obligate or syntrophic cell associations.

Competition and succession in biofilms

In natural systems, there exists a high degree of cellular interaction and competitive behaviour (Connell & Slatyer 1977; Fredrickson 1977). Much of this interactive behaviour results from the competition of cells for available resources, whether those resources be attachment sites or growth substrate. Higher organisms may also influence the outcome of microbial surface colonization, and events such as grazing by predators have the potential to limit the success of one strain while having relatively little effect on other consortium members. Few studies have focused on the effect of competitive pressure during the development of microbial biofilms. Cowan *et al.* (1991) utilized image analysis to measure the extent of surface coverage during colonization of pure cultures and mixed species over a 27 day period; in the presence of *Xanthomonas maltophilia,* greater surface colonization was observed during coculture experiments than where *X. maltophilia* was absent. The researchers also noted limitations of their system in that three dimensional cell aggregates could not effectively be discriminated from two dimensional growth and development using phase contrast image analysis. Also noted was the inability to discriminate between bacterial species in order to determine dominance of any particular colonizing species within the system.

Radiochemical labelling of bacteria has also been used to evaluate competitive behaviour in biofilm cocultures (Banks & Bryers 1991). Discrimination between the two strains used during this study was possible because a *Pseudomonas putida* strain metabolized glucose exclusively whereas *Hyphomicrobium* sp. metabolized only methanol. Replicate culture systems were used to cultivate either mixed species biofilms grown in the presence of ^{14}C-glucose and unlabelled methanol, or ^{14}C-methanol and unlabelled glucose. Success of coculture members was deter-

mined by comparison of radioactivity accumulated over time, sacrificing flow cells for analytical purposes. When both strains were simultaneously added to flow cells, the *P. putida* isolate outnumbered the *Hyphomicrobium* sp. after 24 and 48 h. Similarly, when *P. putida* was added to an established *Hyphomicrobium* sp. biofilm, the *P. putida* cells became the dominant biofilm species after only 48 h. In both cases, the *P. putida* strain outnumbered the *Hyphomicrobium* sp. by approximately 5 to 1 after 50 h of biofilm growth. The authors attributed these results to the differences in growth rates of the two species, whereas the effect of microbial colonization behaviour was not addressed. Current limitations to the study of complex microbial communities lie in the inability to discriminate visually between consortium members that have similar shapes. Use of SCLM in conjunction with fluorescently labelled poly/monoclonal antibodies has potential in this area of study, as laser optical thin sectioning may then be used to help elucidate the position of fluorescent antibody stained bacteria present within the biofilm matrix (James *et al.* 1993).

Observations of adherent populations in natural habitats provides evidence that these systems are in constant flux (Connell & Slatyer 1977; Baier 1984; Wahl 1989). A sequence of events occurs during surface fouling, resulting in a succession of surface colonizing species which dominate at various times following immersion of clean surfaces in natural waters. In a process termed epibiosis (Wahl 1989), the initial colonizing species are subsequently colonized by organisms before others, with defined requirements, become involved with the surface consortium. The succession is based on a sequence of physical and biological events, initiated by adsorption of organic films and closely followed by surface colonization by bacterial species (Fletcher & Loeb 1979; Characklis 1981; Baier 1984). The time frames for these two initial phases of surface colonization are relatively short, with the molecular film forming over a period of minutes, and with significant bacterial colonization occurring within 24 h. J. R. Lawrence, D. R. Korber & D. E. Caldwell (unpublished data) noted that only brief periods were required for the onset of bacterial deposition from a simulated marine microcosm, and that common marine species (*Hyphomicrobium, Pedomicrobium, Pseudomonas, Caulobacter* spp.) were primarily involved.

While bacteria continue to adsorb to exposed surfaces, larger fouling organisms (including amoeba, flagellates, ciliates, diatoms and larvae) subsequently colonize the surface over a period of days for predatory eukaryotes and diatoms, or weeks for larval and spore deposition (Wahl 1989; Decho 1990). This successional pattern of surface colonization has been frequently observed during long term observation of immersed surfaces in flowing or static systems. Researchers have reported that rod-shaped bacteria are frequently the primary colonizers following exposure of clean surfaces to natural aqueous systems (Marshall *et al.* 1971; Marszalek *et al.* 1979), followed by stalked bacteria such as *Caulobacter* spp. (Dempsey 1981). Subsequent colonization by filamentous algae, diatoms, larvae, etc. has reportedly been observed in stream, river and oceanic environments, with eventual attraction of predators which feed on the biofilm (Marszalek *et al.* 1979; Baier 1985; Wahl 1989; Rittle *et al.* 1990).

The recent developments in studying biofilms described in this chapter have allowed us to describe them as both variable and heterogeneous (Keevil & Walker 1992; Stewart *et al.* 1993; Korber *et al.* 1993). In addition, biofilms form through a variety of strategies based on cell–cell interactions that function as multicellular behaviour of microorganisms. The techniques are now available to study the microenvironment in real time using non-destructive methods. They also allow analysis of the spatial arrangements of cells within the context of their natural surroundings. SCLM permits optical sectioning of thick biological specimens without interference from light scattering (Heertje *et al.* 1987; White *et al.* 1987) and has recently been utilized to examine the internal structure and architecture of mature biofilms without diffraction based limitations in microscopy resolution (Lawrence *et al.* 1991; Caldwell *et al.* 1992a, b; Korber *et al.* 1992, 1993; Wolfaardt *et al.* 1994). This, combined with the wide range of environmentally sensitive reporter molecules, means that the microenvironment may be defined and quantitatively analysed. The application of molecular techniques provides the ability to analyse natural communities (Amann *et al.* 1992). These techniques allow the study of biofilms and help to relate the occurrence of bacterial populations to the measurement of specific processes.

Application of these tools should provide new understanding of the functioning of biofilms in the diverse industrial, agricultural, medical and natural environments where they are of major importance as discussed in later chapters.

Acknowledgements

We acknowledge the Natural Sciences and Engineering Research Council, National Hydrology Research Institute and PestPlan for their financial support.

References

Absolom, D. R., Lamberti, F. V., Policova, Z., Zingg, W., van Oss, C. J. & Neumann, A. W. (1983). Surface thermodynamics of bacterial adhesion. *Applied and Environmental Microbiology*, **46**, 90–7.

Acker, G. K. & Steere, R. L. (1962). Restricted diffusion of macromolecules through agar-gel membranes. *Biochimica et Biophysica Acta*, **59**, 137–49.

Adamczyk, Z. & van de Ven, T. G. M. (1981). Deposition of particles under external forces in laminar flow through parallel-plate and cylindrical channels. *Journal of Colloid and Interface Science*, **80**, 340–57.

Adler, J. (1975). Chemotaxis in bacteria. *Annual Reviews of Biochemistry*, **44**, 341–55.

Alldredge, A. L. & Cohen, Y. (1987). Can microscale chemical patches persist at sea? Microelectrode study of marine snow, fecal pellets. *Science*, **235**, 689–91.

Alldredge, A. L. & Silver, M. W. (1988). Characteristics, dynamics, and significance of marine snow. *Progress in Oceanography*, **20**, 41–82.

Allison, D. G. & Sutherland, I. W. (1987). The role of exopolysaccharides in adhesion of freshwater bacteria. *Journal of General Microbiology*, **133**, 1319–27.

Amann, R. I., Krumholz, L. & Stahl, D. A. (1990). Fluorescent oligonucleotide probing of whole cells for determinative phylogenetic, and environmental studies in microbiology. *Journal of Bacteriology*, **172**, 762–70.

Amann, R. I., Stromley, J., Devereux, R., Key, R. & Stahl, D. A. (1992). Molecular and microscopic identification of sulfate-reducing bacteria in multispecies biofilms. *Applied and Environmental Microbiology*, **48**, 614–23.

Arnold, J. W. & Shimkets, L. J. (1988). Inhibition of cell–cell interactions in *Myxococcus xanthus. Journal of Bacteriology*, **170**, 5765–70.

Baier, R. E. (1980). Substrate influence on adhesion of microorganisms and their resultant new surface properties, In *Adsorption of microorganisms to surfaces*, ed. G. Bitton & K. C. Marshall, pp. 59–104. New York: Wiley.

Baier, R. E. (1984). Initial events in microbial film formation, In *Marine biodetermination: an interdisciplinary approach*, ed. J. D. Costlow & R. C. Tipper, pp. 57–62. London: E. & F. N. Spon Ltd.

Baier, R. E. (1985). Adhesion in the biologic environment. *Biomaterials Medical Devices and Artificial Organs*, **12**, 133–59.

Bakke, R. & Olsson, P. Q. (1986). Biofilm thickness measurements by light microscopy. *Journal of Microbiological Methods*, **5**, 93–8.

Banks, M. K. & Bryers, J. D. (1991). Bacterial species dominance within a binary culture biofilm. *Applied and Environmental Microbiology*, **57**, 1874–9.

Barnes, H. & Powell, H. T. (1950). The development, general morphology, and subsequent elimination of barnacle populations, *Balanus crenatus* and *B. balanoides* after a heavy initial settlement. *Journal of Animal Ecology*, **19**, 175–9.

Berg, H. C. (1985). Physics of bacterial chemotaxis, In *Sensory perception and transduction in aneural organisms*, ed. G. Colombetti, F. Linci & P.-S. Song, pp. 19–30. London: Plenum Press.

Berg, H. C. & Brown, D. A. (1972). Chemotaxis in *Escherichia coli* analyzed by three-dimensional tracking. *Nature*, **239**, 500–4.

Blenkinsopp, S. A., Khoury, A. E. & Costerton, J. W. (1992). Electrical enhancement of biocide efficacy against *Pseudomonas aeruginosa* biofilms. *Applied and Environmental Microbiology*, **58**, 3770–3.

Block, S. M., Segall, J. E. & Berg, H. C. (1983). Adaptation kinetics in bacterial chemotaxis. *Journal of Bacteriology*, **154**, 312–24.

Bochem, H. P., Schoberth, S. M., Sprey, B. & Wengler, P. (1982). Thermophilic biomethanation of acetic acid: morphology and ultrastructure of a granular consortium. *Canadian Journal of Microbiology*, **28**, 500–10.

Bohrer, M. P., Patterson, G. D. & Carroll, P. J. (1984). Hindered diffusion of dextran and ficoll in microporous membranes. *Macromolecules*, 1170–3.

Boone, D. R. & Bryant, M. P. (1980). Propionate-degrading bacterium, *Syntrophobacter wolinii* sp. nov. gen. nov., from methanogenic ecosystems. *Applied and Environmental Microbiology*, **40**, 626–32.

Boone, D. R., Johnson, R. L. & Liu, Y. (1989). Diffusion of the interspecies electron carriers H_2 and formate in methanogenic ecosystems and its implications in the measurement of K_m for H_2 or formate uptake. *Applied and Environmental Microbiology*, **55**, 1735–41.

Bott, T. L. & Brock, T. D. (1970). Growth and metabolism of periphytic bacteria: methodology. *Limnology and Oceanography*, **20**, 191–7.

Bowen, B. D. & Epstein, N. (1979). Fine particle deposition in smooth parallel-plate channels. *Journal of Colloid and Interface Science*, **72**, 81–97.

Brannan, D. K. & Caldwell, D. E. (1983). Growth kinetics and yield coefficients of the extreme thermophile *Thermothrix thiopara* in continuous culture. *Applied and Environmental Microbiology*, **45**, 169–73.

Bright, J. J. & Fletcher, M. (1983). Amino acid assimilation and respiration by attached and free-living populations of a marine *Pseudomonas* sp. *Microbial Ecology*, **9**, 215–26.

Brown, M. R. W., Allison, D. G. & Gilbert, G. (1988). Resistance of bacterial biofilms to antibiotics: a growth-rate related effect? *Journal of Antimicrobial Chemotherapy*, **22**, 777–80.

Brown, M. J. & Lester, J. N. (1982). Role of bacterial extracellular polymers in metal uptake in pure bacterial culture and activated sludge–II: Effects of mean cell retention time. *Water Research*, **16**, 1549–60.

Bryers, J. D. (1987). Biologically active surfaces: processes governing the formation and persistence of biofilms. *Biotechnology Progress*, **3**, 57–68.

Burchard, R. P., Rittschof, D. & Bonaventura, J. (1990). Adhesion and motility of gliding bacteria on substrata with different surface free energies. *Applied and Environmental Microbiology*, **56**, 2529–34.

Busscher, H. J., Bellon-Fontaine, M.-N., Mozes, N. *et al.* (1990a). Deposition of *Leuconostoc mesenteroides* and *Streptococcus thermophilus* to solid substrata in a parallel plate flow cell. *Biofouling*, **2**, 55–63.

Busscher, H. J., Bellon-Fontaine, M.-N., Sjollema, J. & van Der Mei, H. C. (1990b). Relative importance of surface free energy as a measure of hydrophobicity in bacterial adhesion to solid surfaces. In *Microbial cell surface hydrophobicity*, ed. R. J. Doyle & M. Rosenberg, pp. 335–59. Washington, DC: American Society for Microbiology.

Caldwell, D. E. (1985). New developments in computer-enhanced microscopy. *Journal of Microbiological Methods*, **4**, 117–25.

Caldwell, D. E., Brannan, D. K. & Kieft, T. L. (1983). *Thermothrix thiopara*: Selection and adaptation of a filamentous sulfur-oxidizing bacterium colonizing hot spring tufa at pH 7. 0 and 74 °C. *Environmental Biogeochemical Ecology Bulletin*, **35**, 129–34.

Caldwell, D. E. & Caldwell, S. J. (1978). A *Zoogloea* sp. associated with blooms of *Anabaena flos-aquae*. *Canadian Journal of Microbiology*, **24**, 922–31.

Caldwell, D. E., Korber, D. R. & Lawrence, J. R. (1992a). Confocal laser microscopy and digital image analysis in microbial ecology. *Advances in Microbial Ecology*, **12**, 1–67.

Caldwell, D. E., Korber, D. R. & Lawrence, J. R. (1992b). Imaging of bacterial cells by fluorescence exclusion using scanning confocal laser microscopy. *Journal of Microbiological Methods*, **15**, 249–61.

Caldwell, D. E., Korber, D. R. & Lawrence, J. R. (1993). Analysis of biofilm formation using 2-D versus 3-D digital imaging. *Journal of Applied Bacteriology* (Symposium Supplement), **74**, 52S–66S.

Caldwell, D. E. & Lawrence, J. R. (1986). Growth kinetics of *Pseudomonas fluorescens* microcolonies within the hydrodynamic boundary layers of surface microenvironments. *Microbial Ecology*, **12**, 299–312.

Caldwell, D. E. & Tiedje, J. M. (1975). The structire of anaerobic bacterial communities in the hypolimnia of several lakes in Michigan. *Canadian Journal of Microbiology*, **21**, 377–85.

Caron, D. A. (1987). Grazing of attached bacteria by heterotrophic microflagellates. *Microbial Ecology*, **13**, 203–18.

Characklis, W. G. (1981). Bioengineering report. Fouling biofilm development: a process analysis. *Biotechnology and Bioengineering*, **23**, 1923–60.

Characklis, W. G., McFeters, G. A. & Marshall, K. C. (1990a). Physiological ecology in biofilm systems. In *Biofilms*, ed. W. G. Characklis & K. C. Marshall, pp. 341–93. New York: Wiley.

Characklis, W. G., Turakhia, M. H. & Zelver, N. (1990b). Transfer and interfacial transport phenomena. In *Biofilms*, ed. W. G. Characklis & K. C. Marshall, pp. 265–40. New York: Wiley.

Cholodny, N. (1930). Über eine neue Method zur Untersuchung der Bodenflora. *Archiv für Mikrobiologie*, **1**, 620–52.

Christensen, B. E. & Characklis, W. G. (1990). Physical and chemical properties of biofilms. In *Biofilms*, ed. W. G. Characklis & K. C. Marshall, pp. 93–130. New York: Wiley.

Christersson, C. E., Glantz, P.-O. J. & Baier, R. E. (1988). Role of temperature and shear forces on microbial detachment. *Scandinavian Journal of Dental Research*, **96**, 91–8.

Confer, D. R. & Logan, B. E. (1991). Increased bacterial uptake of macromolecular substrates with fluid shear. *Applied and Environmental Microbiology*, **57**, 3093–100.

Connell, J. H. & Slatyer, R. O. (1977). Mechanisms of succession in natural communities and their role in community stability and organization. *American Naturalist*, **111**, 1119–44.

Conrad, R., Phelps, T. J. & Zeikus, J. G. (1985). Gas metabolism evidence in support of the juxtaposition of hydrogen-producing and methanogenic bacteria in sewage sludge and lake sediments. *Applied and Environmental Microbiology*, **50**, 595–601.

Costerton, J. W., Cheng, K.-J., Geesey, G. G. *et al.* (1987). Bacterial biofilms in nature and disease. *Annual Reviews of Microbiology*, **41**, 435–64.

Costerton, J. W., Geesey, G. G. & Cheng, K.-J. (1978). How bacteria stick. *Scientific American*, **238**, 86–95.

Costerton, J. W., Irvin, R. T. & Cheng, K.-J. (1981). The bacterial glycocalyx in nature and disease. *Annual Reviews of Microbiology*, **35**, 299–324.

Costerton, J. W., Marrie, T. J. & Cheng, K.-J. (1985). Phenomena of bacterial adhesion. In *Bacterial adhesion: mechanisms and physiological significance*, ed. D. C. Savage & M. Fletcher, pp. 3–43. New York: Plenum Press.

Cowan, M. M., Warren, T. M. & Fletcher, M. (1991). Mixed-species colonization of solid surfaces in laboratory biofilms. *Biofouling*, **3**, 23–34.

Cox, D. J., Bazin, M. J. & Gull, K. (1980). Distribution of bacteria in a continuous-flow nitrification column. *Soil Biology and Biochemistry*, **12**, 241–6.

Dagostino, L., Goodman, A. E. & Marshall, K. C. (1991). Physiological responses induced in bacteria adhering to surfaces. *Biofouling*, **4**, 113–19.

Dazzo, F. B. Truchet, G. L., Sherwood, F. E., Hrabak, E. M., Abe, M. & Pankratz, S. H. (1984). Specific phases of root hair attachment in the *Rhizobium trifolii*–clover symbiosis. *Applied and Environmental Microbiology*, **48**, 1140–50.

Decho, A. W. (1990). Microbial exopolymer secretions in ocean environments: their role(s) in food webs and marine processes. *Oceanography and Marine Biology Annual Review*, **28**, 73–153.

Decho, A. W. & Castenholtz, R. W. (1986). Spatial patterns & feeding of meiobenthic harpacticoid copepods in relation to resident microbial flora. *Hydrobiologica*, **131**, 87–96.

Delaquis, P. J. (1990). Colonization of model and meat surfaces by *Pseudomonas fragi* and *Pseudomonas fluorescens*. Ph.D. thesis, University of Saskatchewan, Saskatoon, Canada.

Dempsey, M. J. (1981). Marine bacterial fouling: A scanning electron microscope study. *Marine Biology*, **61**, 305–15.

Derjaguin, B. V. & Landau, L. (1941). Theory of the stability of strongly charged lyophobic sols and of the adhesion of strongly charged particles in solutions of electrolytes. *Acta Physicochimica URSS*, **14**, 633–62.

Dexter, S. C. (1979). Influence of substratum critical surface tension on bacterial adhesion – *in situ* studies. *Journal of Colloid and Interface Science*, **70**, 346–53.

Dexter, S. C., Sullivan, J. D. Jr, Williams, J. & Watson, S. W. (1975). Influence of substrate wetability on the attachment of marine bacteria to various surfaces. *Applied Microbiology*, **30**, 298–308.

De Weger, L. D., van Der Vlugt, C. I. M., Wijfjes, A. H. M., Bakker, P. A. H. M., Schippers, B. &

Lugtenberg, B. (1987). Flagella of a plant-growth-stimulating *Pseudomonas fluorescens* strain are required for colonization of potato roots. *Journal of Bacteriology*, **169**, 2769–73.

Drucker, D. B. & Whittaker, D. K. (1971). Microstructure of colonies of rod-shaped bacteria. *Journal of Bacteriology*, **108**, 515–25.

Duxbury, T. (1977). A microperfusion chamber for studying growth of bacterial cells. *Journal of Applied Bacteriology*, **42**, 247–51.

Eng, R. H. K., Padberg, F. T., Smith, S. M., Tan, E. N. & Cherubin, C. E. (1991). Bactericidal effects of antibiotics on slowly growing and nongrowing bacteria. *Antimicrobial Agents and Chemotherapy*, **35**, 1824–28.

Ferris, F. G., Schultze, S., Witten, T. C., Fyfe, W. S. & Beveridge, T. J. (1989). Metal interactions with microbial biofilms in acidic and neutral pH environments. *Applied and Environmental Microbiology*, **55**, 1249–57.

Fletcher, M. (1976). The effects of proteins on bacterial attachment to polystyrene. *Journal of General Microbiology*, **94**, 400–4.

Fletcher, M. (1977). The effects of culture concentration and age, time, and temperature on bacterial attachment to polystyrene. *Canadian Journal of Microbiology*, **23**, 1–6.

Fletcher, M. (1986). Measurement of glucose utilization by *Pseudomonas fluorescens* that are free-living and that are attached to surfaces. *Applied and Environmental Microbiology*, **52**, 672–76.

Fletcher, M. (1988). Attachment of *Pseudomonas fluorescens* to glass and influence of electrolytes on bacterium–substratum separation distance. *Abstracts of the 88th Annual Meeting of the American Society for Microbiology*, Washington DC, USA.

Fletcher M. & Loeb, G. I. (1979). Influence of substratum characteristics on the attachment of a marine pseudomonad to solid surfaces. *Applied and Environmental Microbiology*, **37**, 67–72.

Fletcher, M. & Marshall, K. C. (1982). Bubble contact angle method for evaluating substratum interfacial characteristics and its relevance to bacterial attachment. *Applied and Environmental Microbiology*, **44**, 184–192.

Fredrickson, A. G. (1977). Behaviour of mixed cultures of microorganisms. *Annual Reviews of Microbiology*, **33**, 63–87.

Freeze, R. A. & Cherry, J. A. (1979). Groundwater. Englewood Cliffs, NJ: Prentice Hall.

Geesey, G. G. & White, D. C. (1990). Determination of bacterial growth and activity at solid–liquid interfaces. *Annual Review of Microbiology*, **44**, 579–602.

Gilbert, P., Collier, P. J. & Brown, M. R. W. (1990). Influence of growth rate on susceptibility to antimicrobial agents: biofilms, cell cycle, dormancy, and stringent response. *Antimicrobial Agents and Chemotherapy*, **34**, 1856–68.

Gualtieri, P., Colombetti, G. & Lenci, F. (1985). Automatic analysis of the motion of microorganisms. *Journal of Microscopy*, **139**, 57–62.

Gualtieri, P., Francesco, G., Passarelli, V. & Barsanti, L. (1988). Microorganism track reconstruction: an image processing approach. *Computers in Biology and Medicine*, **18**, 57–63.

Gujer, W. (1987). The significance of segregation of biomass in biofilms. *Water Science Technology*, **19**, 495–503.

Hamilton, W. A. (1987). Biofilms: microbial interactions and metabolic activities. In *Ecology of microbial communities*, ed. M. Fletcher, T. R. G. Gray & J. G. Jones, Symposium of the Society for General Microbiology 41. Cambridge: Cambridge University Press.

Handley, P. S., Carter, P. L., Wyatt, J. E. & Hesketh, L. M. (1985). Surface structures (peritrichous fibrils and tufts of fibrils) found on *Streptococcus sanguis* strains may be related to their ability to coaggregate with other oral genera. *Infections and Immunity*, **47**, 217–27.

Heertje, I., van der Vlist, P., Blonk, J. C. G., Hendrickx, H. A. C. M. & Brakenhoff, G. J. (1987). Confocal scanning laser microscopy in food research: some observations. *Food Microstructure*, **6**, 115–20.

Heijnen, C. E., Hok-A-Hin, C. H. & van Veen, J. A. (1991). Protection of rhizobium by bentonite clay against predations by flagellates in liquid culture. *FEMS Microbial Ecology*, **85**, 65–72.

Helmsetter, C. E. (1969). Sequence of bacterial reproduction. *Annual Reviews of Microbiology*, **23**, 223–38.

Henrici, A. T. (1933). Studies of freshwater bacteria. I. A direct microscopic technique. *Journal of Bacteriology*, **25**, 277–86.

Henrici, A. T. & Johnson, D. E. (1935). Studies of freshwater bacteria. II. Stalked bacteria, a new order of Schizomycetes. *Journal of Bacteriology*, **30**, 61–93.

Hermansson, M. & Marshall, K. C. (1985). Utilization of surface localized substrate by non-adhesive marine bacteria. *Microbial Ecology*, **11**, 91–105.

Heukelekian, H. & Heller, A. (1940). Relationship between food concentration and surface for bacterial growth. *Journal of Bacteriology*, **40**, 547–58.

Hirsch, P. (1968). Biology of budding bacteria. *Archiv für Mikrobiologie*, **60**, 201–16.

Hirsch, P. (1974). Budding bacteria. *Annual Reviews of Microbiology*, **28**, 391–433.

Hirsch, P. (1984). Microcolony formation and consortia. In *Microbial adhesion and aggregation*, ed. K. C. Marshall, pp. 373–93. New York: Springer-Verlag.

Hirsch, P. & Pankratz, S. H. (1971). Studies on

bacteria populations in natural environments by use of submerged electron microscope grids. *Zeitschrift für Allgemeine Mikrobiologie*, **10**, 589–605.

Howell, J. A. & Atkinson, B. (1976). Sloughing of microbial film in trickling filters. *Water Research*, **10**, 307–15.

Isaacson, R. E., Nagy, B. & Moon, H. W. (1977). Colonization of porcine small intestine by *Escherichia coli*: colonization and adhesion factors of pig enteropathogens that lack K88. *Journal of Infectious Diseases*, **135**, 531–38.

James, G. A., Caldwell, D. E. & Costerton, J. W. (1993). Spatial relationships between bacterial species within biofilms. Abstract, Canadian Society of Microbiologists/Society for Industrial Microbiology annual meeting, Toronto, Canada.

Jang, L. K. & Yen, T. F. (1985). A theoretical model of diffusion of motile and nonmotile bacteria toward solid surfaces, In *Microbes and oil recovery*, 1, ed. J. E. Zajic & E. C. Donaldson, pp. 226–46. International Bioresources Journal.

Juniper, S. K. (1987). Deposit-feeding ecology of *Amphibola crenata*. I. Long-term effects of deposit feeding on sediment microorganisms. *New Zealand Journal of Marine and Freshwater Research*, **21**, 235–46.

Keevil, C. W. & Walker, J. T. (1992). Nomarski DIC microscopy and image analysis of biofilms. *Binary: Computing in Microbiology*, **4**, 93–5.

Kefford, B., Kjelleberg, S. & Marshall, K. C. (1982). Bacterial scavenging: utilization of fatty acids localized at a solid–liquid interface. *Archives of Microbiology*, **133**, 257–60.

Kelly, F. X., Dapsis, K. J. & Lauffenburger, D. A. (1988). Effect of bacterial chemotaxis on dynamics of microbial competition. *Microbial Ecology*, **16**, 115–31.

Kim, S. K. & Kaiser, D. (1990). C-factor; a cell–cell signalling protein required for fruiting body morphogenesis of *M. xanthus*. *Cell*, **61**, 19–26.

Kinner, N. E., Balkwill, D. L. & Bishop, P. L. (1983). Light and electron microscopic studies of microorganisms growing in rotating biological contactor biofilms. *Applied and Environmental Microbiology*, **45**, 1659–69.

Kjelleberg, S. (1980). Effects of interfaces on survival mechanisms of copiotrophic bacteria in low-nutrient environments. In *Microbial adhesion to surfaces*, ed. R. C. W. Berkeley, J. M. Lynch, J. Melling, P. R. Rutter & B. Vincent, pp. 151–9. Chichester: Ellis Horwood.

Kjelleberg, S. & Hermansson, N. (1984). The effect of interfaces on small starved marine bacteria. *Applied and Environmental Microbiology*, **48**, 497–503.

Kjelleberg, S., Humphrey, B. A. & Marshall, K. C. (1982). The effect of interfaces on small, starved marine bacteria. *Applied and Environmental Microbiology*, **43**, 1166–72.

Knight-Jones, E. W. & Moyse, E. (1961). Intraspecific competition in sedentary marine animals. *Symposium of the Society for Experimental Biology*, **15**, 72–95.

Koch, A. L. (1991) Diffusion: the crucial process in many aspects of the biology of bacteria. In *Advances in microbial ecology*, ed. K. C. Marshall, pp. 37–70. New York: Plenum Press.

Kolenbrander, P. E. (1989). Surface recognition among oral bacteria: multigeneric coaggregations and their mediators. *Critical Reviews in Microbiology*, **17**, 137–59.

Kolenbrander, P. E. & London, J. (1992). Ecological significance of coaggregation among oral bacteria. In *Advances in Microbial Ecology*, **12**, 183–217.

Korber, D. R., James, G. A., Costerton, J. W. (1994a). Evaluation of flevoxacin efficacy against established *Pseudomonas fluorescens* biofilms. *Applied and Environmental Microbiology*, **60**, 1663–9.

Korber, D. R., Lawrence, J. R., Cooksey, K. E., Cooksey, B. & Caldwell, D. E. (1989a). Computer image analysis of diatom chemotaxis. *Binary*. **1**, 155–68.

Korber, D. R., Lawrence, J. R., Hanson, K. G., Caldwell, D. E. & Costerton, J. W. (1994b). The effect of environmental laminar flow velocities on the architecture of *Pseudomonas fluorescens* biofilms. Abstract, American Society of Microbiology 94th annual meeting, Las Vegas.

Korber, D. R., Lawrence, J. R., Hendry, M. J. & Caldwell, D. E. (1992). Programs for determining representative areas of microbial biofilms. *Binary*. **4**, 204–10.

Korber, D. R., Lawrence, J. R., Hendry, M. J. & Caldwell, D. E. (1993). Analysis of spatial variability within mot$^+$ and mot$^-$ *Pseudomonas fluorescens* biofilms using representative elements. *Biofouling*, **7**, 339–58.

Korber, D. R., Lawrence, J. R., Sutton, B. & Caldwell, D. E. (1989b). The effect of laminar flow on the kinetics of surface recolonization by mot$^+$ and mot$^-$ *Pseudomonas fluorescens*. *Microbial Ecology*, **18**, 1–19.

Korber, D. R., Lawrence, J. R., Zhang, L. & Caldwell, D. E. (1990). Effect of gravity on bacterial deposition and orientation in laminar flow environments. *Biofouling*, **2**, 335–50.

Lappin-Scott, H. M. & Costerton, J. W. (1989). Bacterial biofilms and surface fouling. *Biofouling*, **1**, 323–42.

Lauffenburger, D., Grady, M. & Keller, K. H. (1984). An hypothesis for approaching swarms of myxobacteria. *Journal of Theoretical Biology*, **110**, 257–74.

Lawrence, J. R. & Caldwell, D. E. (1987). Behaviour of bacterial stream populations within the hydrodynamic boundary layers of surface microenvironments. *Microbial Ecology*, **14**, 15–27.

Lawrence, J. R., Delaquis, P. J. Korber, D. R. & Caldwell, D. E. (1987). Behaviour of *Pseudomonas fluorescens* within the hydrodynamic boundary layers of surface microenvironments. *Microbial Ecology*, **14**, 1–14.

Lawrence, J. R., Korber, D. R. & Caldwell, D. E. (1992). Behavioural analysis of *Vibrio parahaemolyticus* variants in high- and low-viscosity microenvironments by use of digital image processing. *Journal of Bacteriology*, **174**, 5732–9.

Lawrence, J. R., Korber, D. R. & Caldwell, D. E. (1989a). Computer-enhanced darkfield microscopy for the quantitative analysis of bacterial growth and behavior on surfaces. *Journal of Microbiological Methods*, **10**, 123–38.

Lawrence, J. R., Korber, D. R., Hoyle, B. D., Costerton, J. W. & Caldwell, D. E. (1991). Optical sectioning of microbial biofilms. *Journal of Bacteriology*, **173**, 6558–67.

Lawrence, J. R., Malone, J. A., Korber, D. R. & Caldwell, D. E. (1989b). Computer image enhancement to increase depth of field in phase contrast microscopy. *Binary*, **1**, 181–85.

Lawrence, J. R., Wolfaardt, G. M. & Korber, D. R. (1994). Monitoring diffusion in biofilm matrices using confocal laser microscopy. *Applied and Environmental Microbiology*, **60**, 1166–73.

LeChevallier, M. W., Babcock, T. M. & Lee, R. G. (1987). Examination and characterization of distribution system biofilms. *Applied and Environmental Microbiology*, **53**, 2714–24.

Lens, P. N. L., De Beer, D., Cronenberg, C. C. H., Houwen, F. P., Ottengraf, S. P. P. & Verstraete, W. H. (1993). Heterogeneous distribution of microbial activity in methanogenic aggregates: pH and glucose microprofiles. *Applied and Environmental Microbiology*, **59**, 3803–15.

Lewandowski, Z., Altobelli, S. A. & Fukushima, E. (1993). NMR and microelectrode studies of hydrodynamics and kinetics in biofilms. *Biotechology Progress*, **9**, 40–5.

Lewandowski, Z., Lee, W. C., Characklis, W. G., & Little, B. (1989). Dissolved oxygen and pH microelectrode measurements at water immersed metal surfaces. *Corrosion*, **45**, 92–8.

Lion, L. W., Shuler, M. L., Hsieh, K. M. & Costerton, W. C. (1988). Trace metal interactions with microbial biofilms in natural and engineered systems. *CRC Critical Reviews in Environmental Control*, **17**, 273–305.

Little, B., Wagner, P., Ray, R., Pope, R. & Scheetz, R. (1991). Biofilms: an ESEM evaluation of artifacts introduced during SEM preparation. *Journal of Industrial Microbiology*, **8**, 213–22.

Loeb, G. I. & Neihof, R. A. (1975). Marine conditioning films. *Advances in Chemistry Series*, **145**, 319–35.

Lupton, F. S. & Marshall, K. C. (1981). Specific adhesion of bacteria to heterocysts of *Anabaena*

spp. and its ecological significance. *Applied and Environmental Microbiology*, **42**, 1085–92.

MacLeod, F. A., Guiot, S. R. & Costerton, J. W. (1990). Layered structure of bacterial aggregates produced in an upflow anaerobic sludge bed reactor. *Applied and Environmental Microbiology*, **56**, 1598–607.

Malone, J. A. (1987). Colonization of surface microenvironments by *Rhizobium* spp. M.Sc. thesis, University of Saskatchewan, Saskatoon, Canada.

Marmur, A. & Ruckenstein, E. (1986). Gravity and cell adhesion. *Journal of Colloid and Interface Science*, **114**, 261–66.

Marshall, K. C. (1986) Microscopic methods for the study of bacterial behaviour at inert surfaces. *Journal of Microbiological Methods*, **4**, 217–27.

Marshall, K. C. (1988). Adhesion and growth of bacteria at surfaces in oligotrophic habitats. *Canadian Journal of Microbiology*, **34**, 503–6.

Marshall, K. C. (1989). Growth of bacteria on surface-bound substrates: significance in biofilm development. In *Recent advances in microbial ecology*, ed. T. Hattori, Y. Ishida, Y. Maruyama, R. Y. Morita & A. Uchida, pp. 146–50. Tokyo: Japan Scientific Societies Press.

Marshall, K. C. & Cruickshank, R. H. (1973). Cell surface hydrophobicity and the orientation of certain bacteria at interfaces. *Archiv für Mikrobiologie*, **91**, 29–40.

Marshall, K. C., Stout, R. & Mitchell, R. (1971). Mechanisms of the initial events in the sorption of marine bacteria to solid surfaces. *Journal of General Microbiology*, **68**, 337–48.

Marszalek, D. S., Gerchakov, S. M. & Udey, L. R. (1979). Influence of substrate composition on marine microfouling. *Applied and Environmental Microbiology*, **38**, 987–95.

McCarter, L., Hilmen, M. & Silverman, M. (1988). Flagellar dynamometer controls swarmer cell differentiation of *Vibrio parahaemolyticus*. *Cell*, **54**, 345–51.

McCarter, L. L., Showalter, R. E. & Silverman, M. R. (1992). Genetic analysis of surface sensing in *Vibrio parahaemolyticus*. *Biofouling*, **5**, 163–75.

McCarter, L. L. & Silverman, M. (1989). Iron regulation of swarmer cell differentiation of *Vibrio parahaemolyticus*. *Journal of Bacteriology*, **171**, 731–6.

McFeters, G. A., Bazin, M. J., Bryers, J. D. *et al.* (1984). Biofilm development and its consequences: group report. In *Microbial adhesion and aggregation*, ed. K. C. Marshall, pp. 109–24. Berlin: Springer-Verlag.

McInerney, M. J., Bryant, M. P., Hespell, R. B. & Costerton, J. W. (1981). *Syntrophomonas wolfei* gen. nov. sp. nov., an anaerobic, syntrophic, fatty acid-oxidizing bacterium. *Applied and Environmental Microbiology*, **41**, 1029–39.

Meadows, P. S. (1971). The attachment of bacteria to solid surfaces. *Archiv für Mikrobiologie*, **75**, 374–81.

Merker, R. I. & Smit, J. (1988). Characterization of the adhesive holdfast of marine and freshwater Caulobacters. *Applied and Environmental Microbiology*, **54**, 2078–85.

Mittelman, M. W., Nivens, D. E., Low, C. & White, D. C. (1990). Differential adhesion, activity, and carbohydrate:protein ratios of *Pseudomonas atlantica* monocultures attaching to stainless steel in a linear shear gradient. *Microbial Ecology*, **19**, 269–78.

Nealson, K. H. (1977). Autoinduction of bacterial luciferase: occurrence, mechanism and significance. *Archives of Microbiology*, **112**, 73–79.

Nichols, W. W. (1989). Susceptibility of biofilms to toxic compounds. In *Structure and function of biofilms*, ed. W. G. Characklis & P. A. Wilderer, pp. 321–31. New York: Wiley.

Nivens, D. E., Chambers, J. Q., Anderson, T. R. & White, D. C. (1993). Long-term, on-line monitoring of microbial biofilms using a quartz crystal microbalance. *Analytical Chemistry*, **65**, 65–9.

Paerl, H. W. (1975). Microbial attachment to particles in marine and freshwater ecosystems. *Microbial Ecology*, **2**, 73–83.

Perfil'ev, B. V. & Gabe, D. R. (1969). *Capillary methods of investigating micro-organisms* (translated by J. M. Shewan). Toronto: University of Toronto Press.

Pilgram, W. K. & Williams, F. D. (1976). Survival value of chemotaxis in mixed cultures. *Canadian Journal of Microbiology*, **22**, 1771–3.

Porter, K. G. (1976). Enhancement of algal growth and productivity by grazing zooplankton. *Science*, **192**, 1332–4.

Powell, M. S. & Slater, N. K. H. (1983). The deposition of bacterial cells from laminar flows onto solid surfaces. *Biotechnology and Bioengineering*, **25**, 891–900.

Power, K. & Marshall, K. C. (1988). Cellular growth and reproduction of marine bacteria on surface-bound substrates. *Biofouling*, **1**, 163–74.

Rasmussen, M. A., White, B. A. & Hespell, R. B. (1989). Improved assay for quantitating adherence of ruminal bacteria to cellulose. *Applied and Environmental Microbiology*, **55**, 2089–91.

Revsbech, N. P. (1989). Diffusion characteristics of microbial communities determined by use of oxygen microsensors. *Journal of Microbiological Methods*, **49**, 111–22.

Revesbech, N. P. & Jørgenson, B. B. (1986). Microelectrodes: their use in microbial ecology. *Advances in Microbial Ecology*, **9**, 252–93.

Rittle, K. H., Helmstetter, C. E., Meyer, A. E. & Baier, R. E. (1990). *Escherichia coli* retention on solid surfaces as functions of substratum surface energy and cell growth phase. *Biofouling*, **2**, 121–30.

Rivero, M. & Lauffenburger, D. A. (1986). Analysis of the capillary assay for determination of bacteria motility and chemotaxis parameters. *Biotechnology and Bioengineering*, **28**, 1178–83.

Robinson, R. W., Akin, D. E., Nordstedt, R. A., Thomas, M. V. & Aldrich, H. C. (1984). Light and electron microscopic examinations of methane-producing biofilms from anaerobic fixed-bed reactors. *Applied and Environmental Microbiology*, **48**, 127–36.

Rodriguez, G. G., Phipps, D., Ishiguro, K & Ridgway, H. F. (1992). Use of a fluorescent redox probe for direct visualization of actively respiring bacteria. *Applied and Environmental Microbiology*, **58**, 1801–8.

Russel, J. B., Cotta, M. A. & Dombrowski, D. B. (1981). Rumen bacterial competition in continuous culture: *Streptococcus bovis* versus *Megasphaera elsdenii*. *Applied and Environmental Microbiology*, **41**, 1394–9.

Sanders, W. M. (1966). Oxygen utilization by slime organisms in continuous-culture. *Journal of Air and Water Pollution*, **10**, 253–76.

Santos, R., Callow, M. E. & Bott, T. R. (1991). The structure of *Pseudomonas fluorescens* biofilms in contact with flowing systems. *Biofouling*, **4**, 319–36.

Shapiro, J. A. & Hsu, C. (1989). *Escherichia coli* K-12 cell–cell interactions seen by time-lapse video. *Journal of Bacteriology*, **171**, 5963–74.

Shimkets, L. J. (1990). Social and developmental biology of the myxobacteria. *Microbiological Reviews*, **54**, 473–501.

Shotton, D. M. (1989). Confocal scanning optical microscopy and its applications for biological specimens. *Journal of Cell Science*, **94**, 175–206.

Sibbald, M. J. & Albright, L. J. (1988). Aggregated and free bacteria as food sources for heterotrophic microflagellates. *Applied and Environmental Microbiology*, **54**, 613–16.

Sieracki, M. E., Johnson, P. W. & Sieburth, J. M. (1985). Detection, enumeration, and sizing of planktonic bacteria by image-analyzed epifluorescence microscopy. *Applied and Environmental Microbiology*, **49**, 799–810.

Silverman, M. & Simon, M. (1974). Flagellar rotation and the mechanism of bacterial motility. *Nature*, **249**, 73–4.

Sjollema, J., Busscher, H. J. & Weerkamp, A. H. (1988). Deposition of oral streptococci and polystyrene latices onto glass in a parallel plate flow cell. *Biofouling*, **1**, 101–12.

Sjollema, J., Busscher, H. J. & Weerkamp, A. H. (1989a). Experimental approaches for studying adhesion of microorganisms to solid substrata: applications and mass transport. *Journal of Microbiological Methods*, **9**, 79–90.

Sjollema, J., Busscher, H. J. & Weerkamp, A. H. (1989b). Real-time enumeration of adhering microorganisms in a parallel plate flow cell using automated image analysis. *Journal of Microbiological Methods*, **9**, 73–8.

Sjollema, J., van der Mei, H. C., Uyen, H. M. & Busscher, H. J. (1990a). Direct observations of cooperative effects in oral streptococcal adhesion to glass by analysis of the spatial arrangement of adhering bacteria. *FEMS Microbiology Letters*, **69**, 263–70.

Sjollema, J., van der Mei, H. C., Uyen, H. M. W. & Busscher, H. J. (1990b). The influence of collector and bacterial cell surface properties on the deposition of oral streptococci in a parallel plate flow cell. *Journal of Adhesion Science and Technology*, **4**, 765–77.

Smith, J. L. & Doetsch, R. N. (1969). Studies on negative chemotaxis and the survival value of motility in *Pseudomonas fluorescens*. *Journal of General Microbiology*, **55**, 379–91.

Sorongon, M. L., Bloodgood, R. A. & Burchard, R. P. (1991). Hydrophobicity, adhesion, and surface-exposed proteins of gliding bacteria. *Applied and Environmental Microbiology*, **57**, 3193–9.

Stahl, D. A., Flesher, B., Mansfield, H. R. & Montgomery, L. (1988). Use of phylogenetically based hybridization probes for studies of ruminal microbial ecology. *Applied and Environmental Microbiology*, **54**, 1079–84.

Staley, J. T. (1971). Growth rates of algae determined *in situ* using an immersed microscope. *Journal of Phycology*, 13–17.

Stewart, P. S., Peyton, B. M., Drury, W. J. & Murga, R. (1993). Quantitative observations of heterogeneities in *Pseudomonas aeruginosa* biofilms. *Applied and Environmental Microbiology*, **59**, 327–9.

Stove-Poindexter, J. (1964). Biological properties and classification of the caulobacter group. *Bacteriological Reviews*, **28**, 231–95.

Thiele, J. H., Chartrain, M. & Zeikus, J. G. (1988). Control of interspecies electron flow during anaerobic digestion: role of floc formation in syntrophic methanogenesis. *Applied and Environmental Microbiology*, **54**, 10–19.

Torrella, F. & Morita, R. Y. (1981). Microcultural study of bacterial size changes and microcolony and ultramicrocolony formation by heterotrophic bacteria in seawater. *Applied and Environmental Microbiology*, **41**, 518–27.

Trulear, M. G. & Characklis, W. G. (1982). Dynamics of biofilm processes. *Journal of the Water Pollution Control Federation*, **54**, 1288–301.

Vanhaecke, E., Remon, J.-P., Moors, M., Raes, F., de Rudder, D. & van Peteghem, A. (1990). Kinetics of *Pseudomonas aeruginosa* adhesion to 304 and 316-L stainless steel, role of cell surface hydrophobicity. *Applied and Environmental Microbiology*, **56**, 788–95.

van Loosdrecht, M. C. M., Lyklema, J., Norde, W. & Zehnder, A. J. B. (1989). Bacterial adhesion: a physicochemical approach. *Microbial Ecology*, **17**, 1–15.

van Loosdrecht, M. C. W., Lyklema, J., Norde, W. & Zehnder, A. J. B. (1990). Influence of interfaces on microbial activity. *Microbiological Reviews*, **54**, 75–87.

van Loosdrecht, M. C. M., Norde, W. & Zehnder, A. J. B. (1987). Influence of cell surface characteristics on bacterial adhesion to solid surfaces. *Proceedings of the 4th European Congress on Biotechnology*, pp. 575–80.

van Pelt, A. W. J., Weerkamp, A. H., Uyen, M. H. W. J. C., Busscher, H. J., de Jong, H. P. & Arends, J. (1985). Adhesion of *Streptococcus sanguis* CH3 to polymers with different surface free energies. *Applied and Environmental Microbiology*, **49**, 1270–5.

Verwey, E. J. W. & Overbeek, J. T. G. (1948). *Theory of the stability of lyophobic colloids*. Amsterdam: Elsevier.

Vesper, S. J. & Bauer, W. D. (1986). Role of pili (fimbriae) in attachment of *Bradyrhizobium japonicum* to soybean roots. *Applied and Environmental Microbiology*, **52**, 134–41.

Vogel, S. (1983). *Life in moving fluids: the physical biology of flow*. Princeton, NJ: Princeton University Press.

Wahl, M. (1989). Marine epibiosis. I. Fouling and antifouling: some basic aspects. *Marine Ecology Progress Series*, **58**, 175–89.

Walt, D. R., Smulow, J. B., Turesky, S. S. & Hill, R. G. (1985). The effect of gravity on initial microbial adhesion. *Journal of Colloid and Interface Science*, **107**, 334–6.

White, J. G., Amos, W. B. & Fordham, M. (1987). An evaluation of confocal versus conventional imaging of biological structure by fluorescence light microscopy. Journal of Cell Biology, **105**, 41–8.

Wilkins, J. R., Darnell, W. L. & Boykin, E. H. (1972). Cinemicrographic study of the development of subsurface colonies of *Staphylococcus aureus* in soft agar. *Applied Microbiology*, **24**, 786–97.

Willey, J. M. & Waterbury, J. B. (1989). Chemotaxis toward nitrogenous compounds by swimming strains of marine *Synechococcus* spp. *Applied and Environmental Microbiology*, **55**, 1888–94.

Wimpenny, J. W. T. (1992). Microbial systems: patterns in time and space, In *Advances in microbial ecology*, ed. K. C. Marshall, pp. 469–522. New York: Plenum Press.

Wolfaardt, G. M., Lawrence, J. R. Robarts, R. D., Caldwell, S. J. & Caldwell, D. E. (1994). Multicellular organization in a degradative biofilm community. *Applied and Environmental Microbiology*, **60**, 434–46.

Wright, J. B., Costerton, J. W. & McCoy, W. F. (1988). Filamentous growth of *Pseudomonas aeruginosa*. *Journal of Industrial Microbiology*, **3**, 139–46.

ZoBell, C. E. (1943). The effect of solid surfaces upon bacterial activity. *Journal of Bacteriology*, **46**, 39–56.

2

Dynamics of Bacterial Biofilm Formation

Melanie G. Brading, Jana Jass and Hilary M. Lappin-Scott

Introduction

Biofilm formation is important in a wide variety of situations: for instance, colonization of pipe surfaces in the food and water industries, metal corrosion due to sulphate reducing bacteria in the shipping and oil industries, and in medicine associated with infections of various tissues (osteomeylitis and endocarditis), dental decay (Addy *et al.* 1992) and prosthetic implants (Dougherty 1988). Whereas biofilm formation in a chemostat is considered merely an operating nuisance (Bryers 1984), in industrial fermentors such fouling can cause physical damage by the production of metabolites at points on the surface. Biofilms may lead to reduced heat efficiency transfer and reduction in flow rates, and can also act as a resevoir for potential pathogens (Lappin-Scott & Costerton 1989).

Although biofilm formation is frequently associated with being harmful and detrimental, in many instances it can also be beneficial. Biofilms are used in wastewater treatment for the degradation of soluble organic or nitrogenous waste. In nature microbial decomposition of cellulose fibres requires prior attachment of cellulolytic bacteria and *Rhizobium* cells form biofilms on the roots of leguminous plants where nodules are formed to fix atmospheric nitrogen. Bar-Or (1990) stated the importance of biofilms in stabilizing soil either by acting as cementing agents or flocculating soil particles, thereby improving aeration and water percolation and allowing further microbial growth.

Biofilm formation is difficult to control. A number of authors have reported that biofilm bacteria (sessile) are more resistant to antimicrobial agents than suspended bacteria (planktonic) of the same species (Brown *et al.* 1988; Anwar *et al.* 1989). Most commercial biocides and antibiotics were developed and tested for their ability to kill planktonic bacteria (Chopra 1986; Gilbert *et al.* 1987). This is important in medicine where biofilms may be associated with pulmonary and urogenital tract infections as well as being associated with medical implants and contact lenses. In industrial situations, Marshall (1992) stated that absolute control of biofilm formation on surfaces is not possible although it can be controlled with biocides or by incorporating toxic heavy metals into antifouling paint. Indeed, within industry biofouling is a huge problem, as described by Costerton and Lappin-Scott (1989). To control biofilm formation effectively where it is unwanted or to develop a biofilm where it is beneficial, it is important to understand the processes involved in bacterial adhesion.

Although scientists are becoming increasingly aware of the importance that biofilms play in medical infections and industrial processes, there is still only a limited understanding of the dynamic nature of biofilm development. The dynamic properties of biofilm formation, including the modes and rates of bacterial attachment, subsequent detachment from surfaces and the influences of fluid dynamics will be covered in this chapter.

Biofilm formation

In natural environments bacteria tend to exist as sessile populations as opposed to survival in the planktonic phase. Watkins & Costerton (1984) stated their belief that for every planktonic bacterium in an aqueous system there are between 1000 and 10000 bacteria actually attached to surfaces. Biofilm formation occurs on virtually any submerged surface in any environment in which bacteria are present. The principal reason for this is that biofilm bacteria are able to exploit essential nutrients which accumulate in the form of ions and macromolecules at the surface–water interface giving them a distinct ecological advantage in an otherwise nutritionally unfavourable environment (Brown *et al.* 1977; Dawson *et al.* 1981). Although bacteria require aqueous conditions for growth they will adhere to any solid surface, be it inorganic, living or dead materials (such as clays, plants, sand grains, animals), or organic remains (Marshall 1980). Bacteria behave as colloidal particles, according to Marshall (1980) as a result of their small size (1.0 µm in length), a density only slightly greater than that of water and their negatively charged surface. The soil environment was considered (Marshall 1980) to be the most complex of all microbial habitats for biofilm formation because within soils an almost infinitely variable array of pore spaces is represented, which in turn may or may not contain water. In marine habitats bacteria are found associated with surfaces of inorganic particulates and detritus as well as other microorganisms. In freshwater systems, which are often low in nutrients, both oligotrophic and copiotrophic bacteria can act as primary colonizers of surfaces (Marshall 1985) but biofilm formation can be particularly rapid if a regular nutrient supply is provided to the bacteria by flowing water systems (Marshall 1992).

Initial biofilm development

A biofilm is a biologically active matrix of cells and extracellular products attached to a solid surface (Bakke *et al.* 1984). It is difficult to age a biofilm because of the dynamic structure formed by the continuous combination of growth and sloughing processes (Santos *et al.* 1991). Characklis & Cooksey (1983) considered that there were five stages of biofilm development. At first, organic molecules and microbial cells are transported to the wetted surface where adsorption takes place resulting in a conditioned surface. This accumulation of nutrients at a surface

occurs because most solid surfaces assume a net negative charge when immersed in water and as a result cations and a variety of macromolecular and colloidal materials are attracted to the surface–water interfaces (Marshall 1980). It is not clear whether bacterial cells probe through this conditioning film to interact directly with the surface or whether macromolecular components of the bacterial cell wall interact directly with the film components (Marshall 1992), but the result is that bacteria commence adhesion. Characklis & Cooksey (1983) stated that adhesion may be reversible (cells can be removed by mild rinsing) or irreversible (after the production of exopolysaccharide). Metabolism by the microorganisms results in more attached cells and associated material before finally detachment occurs on initial portions of the biofilm. Examples of monolayers of Gram negative and Gram positive bacteria can be seen in Figs. 2.1 and 2.2.

Production of the glycocalyx

The microenvironment of the surface alters as the first colonizers grow and divide and produce exopolysaccharide (Lappin-Scott & Costerton 1989). The film traps more organic and inorganic matter (Bryers 1984; Marshall 1992), as well as microbial products and other microorganisms which join to form consortia protected within the glycocalyx. Often daughter cells may also become trapped in the matrix adding to the thickness of the biofilm. Heterogeneous systems are thus formed, bacteria forming specific interactions depending on their growth requirements. The bacterial glycocalyx itself is very important in biofilm formation and is critical for bacterial persistence and survival on a surface. Costerton *et al.* (1985) defined the glycocalyx as the polysaccharide containing structure of bacterial origin, lying outside the integral elements of the outer membrane of Gram negative cells and the peptidoglycan of Gram positive cells. Known as either the slime layer or the capsule, it is composed of either fibrous polysaccharides or globular glycoproteins (Costerton *et al.* 1985), and contains about 99% water in its hydrated state. Costerton *et al.* (1985) believed that within the biofilm a highly organized microbial community was formed within which substrate and hydrogen transfer was facilitated. The glycocalyx comprises polyanionic matrices that are expected to act as

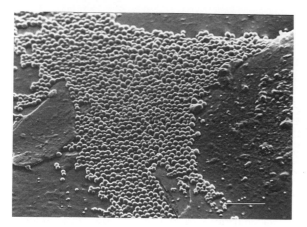

Fig. 2.1.   A scanning electron micrograph of a monolayer of *Enterococcus faecium* attached to a silastic rubber surface. Bar = 5 μm.

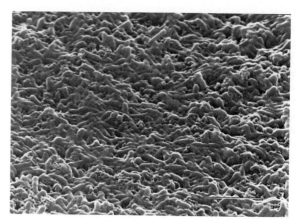

Fig. 2.3.   Scanning electron microscopy demonstrating a thick biofilm of *Pseudomonas fluorescens* covered with exopolysaccharide. The bacterium was colonized onto silastic rubber for 30 h. Bar = 5 μm.

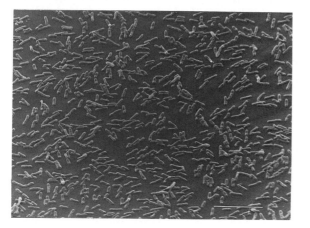

Fig. 2.2.   A scanning electron micrograph of *Pseudomonas fluorescens* attached sparsely on a silastic rubber surface. Bar = 10 μm.

ion exchange resins, attracting and building charged ions and molecules within the matrix that surrounds the bacterial cells, forming adherent microcolonies and eventually confluent biofilms (Fig. 2.3).

The maturing biofilm

With time a level of organization develops in which cells of different species achieve physiological cooperation (Blenkinsopp & Costerton 1991). As heterogeneity increases within the biofilm, chemical and physical microgradients develop including pH, oxygen and nutrient gradients. Oxygen is not usually a limiting factor

under initial stages of colonization (Marshall 1992) but, as explained by Anwar *et al.* (1992), in multiple layers of cells in thick biofilms the location of individual cells within the biofilm determines its physiological status. Cells in the upper regions have easy access to nutrients and oxygen and have few problems with discharging toxic waste products (Bar-Or 1990). Consequently these cells are similar to those grown planktonically. However, cells enmeshed in the glycocalyx are likely to lack nutrients and oxygen and to accumulate waste products. According to Marshall (1992), the formation of an anoxic zone means bacteria maintain respiratory activity by utilizing nitrate and other inorganic compounds as alternative electron acceptors. In complete anaerobic conditions near the substratum fermentative bacteria are particularly active.

The processes of attachment and detachment of bacteria from biofilms are closely related, detachment and desorption of bacteria and related biofilm material occurring continuously from the moment of initial attachment. Escher & Characklis (1990) described detachment as the movement of cells or other components from the biofilm into the bulk liquid. This is in contrast to desorption which may occur at the same time and is described as the loss of components from the substratum, often due to changes in cell surface properties or physicochemical conditions that can occur through chemical or physical treatment.

The literature includes a number of physical, chemical and biological processes that govern biofilm formation at a surface–liquid interface. Most microbiologists concentrate on the biological factors of adhesion but the chemical and physical components are as important in the final biofilm formation. The relative contribution of each process to net biofilm accumulation changes throughout the period of biofilm development (Applegate & Bryers 1991). They control the complete biofilm dynamics and thus are important in both attachment and detachment of bacteria near surfaces.

Biological processes

The biological processes involved in bacterial attachment, biofilm formation and detachment from surfaces are summarized in Fig. 2.4. The primary biological processes addressed in this section include species of organisms, nutrient status, the stage of the growth cycle, polysaccharide production and cell wall composition.

Attachment

Bacterial attachment occurs continuously on most surfaces. Meers (1973) stated that most natural environments are constantly subject to invasion by a wide variety of microorganisms. As already explained, most surfaces in nature are covered in a conditioning film prior to colonization by bacteria. These films are formed by adsorption of macromolecules and other low molecular weight hydrophobic molecules to the

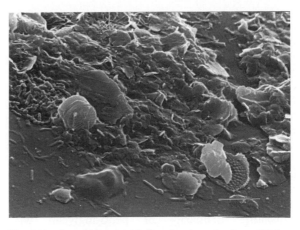

Fig. 2.5. A multispecies biofilm developed onto a silastic rubber surface after immersion in pondwater for 2 weeks. The presence of diatoms, different bacterial species and other particulate matter is clearly seen. Bar = 5 μm.

surface. Once the bacteria are at the surface they begin to secrete extracellular polysaccharide (EPS) or the glycocalyx, which binds both the bacteria and other extracellular material into a matrix. In nature, diverse multispecies biofilms develop on to surfaces (e.g. Fig. 2.5).

Passive versus active adhesion

The processes of bacterial attachment are, however, much more complicated than described above. In some bacteria adhesion is a passive process, the adhering bacteria possessing the necessary attachment structures prior to adhesion, such as holdfasts or pili which are usually 5–25 nm wide and 1–2 μm long (Fletcher & Pringle 1983; Kogure 1989). Santos *et al.* (1991) found that strains of *Pseudomonas fluorescens* possessed both flagella and pili to aid in attachment, and quoted (Brown *et al.* 1977) *Caulobacter* as an example of an organism that develops holdfasts to enable the bacteria to attach to glass or fibrous surfaces as well as to other microorganisms. Active adhesion, on the other hand, is when prolonged exposure is needed for the bacteria to attach firmly to a surface, often with the production of exopolysaccharide. Bacteria may use different techniques with different surfaces.

Cell wall characteristics

Fletcher & Pringle (1983) believed that the process of bacterial attachment to solid surfaces

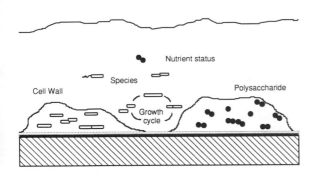

Fig. 2.4. A diagram showing the biological processes involved in bacterial attachment and detachment from surfaces and biofilm formation.

comprised three components: the bacterial face, the substratum and the liquid medium. Of these, the bacterial face is of direct biological significance as it can affect cell attachment, depending on whether the cell envelope is Gram negative or Gram positive in structure and composition. The Gram negative envelope structure contains protein, lipid and peptidoglycan with lipopolysaccharide on the outer surface. The lipopolysaccharide has varying polysaccharide chain lengths so that bacteria may vary accordingly in the degree of hydrophobicity of the cell surface and their consequent rates of attachment. The envelope structure of Gram positive bacteria is much simpler, consisting primarily of peptidoglycan with smaller amounts of teichoic and teichuronic acids, polysaccharides or proteins. The presence of such complex molecules increases the number of possible types of adhesion interaction for an organism.

Other authors have reported how the bacterial surface may affect adhesion. Martinez-Martinez *et al.* (1991) reported that surface hydrophobicity and net surface charge appeared to be important in influencing the rate of bacterial attachment. Stenström (1989) discovered that adhesion of *Salmonella typhimurium* as well as *Streptococcus faecalis* and *Escherichia coli* to the mineral particles of quartz, albite, feldspar and magnetite was related to the hydrophobicity of the cell surface, higher values giving enhanced adhesion to mineral particles. Busscher *et al.* (1986) reported that bacteria with a low surface free energy desorbed only from substrata with a high surface free energy and vice versa.

Cell morphology and structure

The rate of attachment is also affected by the type of adhering bacteria, rods and coccoids having a low surface roughness when compared with filamentous bacteria (Marshall 1992). When the colonization of glass slides and electron microscope grids immersed in seawater was investigated (Marshall *et al.* 1971) it was found that there was a selective irreversible sorption of small rods.

The differences in adhesion between gliding and non-gliding bacteria have also been investigated. McEldowney & Fletcher (1988b) looked at the permanent adhesion of the gliding *Flexibacter* sp. and the three non-gliding bacteria *P. fluorescens*, *Enterobacter cloacae* and a *Chromobacterium* sp. to polystyrene substrata and concluded that the differences seen could be due to different adaptations to different ecological niches. Permanent adhesion of the *Flexibacter* sp. appeared to be related to growth, levels of adhesion increasing rapidly with increased growth rate in continuous culture and declining rapidly with death phase in batch culture. Adhesion also decreased with increasing pH and temperature. However, with the three non-gliding bacteria there appeared to be no relationship between growth rate and the levels of permanent adhesion. Maximum adhesion was seen at a pH between 5.5 and 7 and between temperatures of 20 and 30 °C.

Whether the bacteria are mucoid or not is also important. Martinez-Martinez *et al.* (1991) investigated the adherence of six non-mucoid and three mucoid strains of *P. aeruginosa* to polyvinyl chloride, polyurethane and siliconized latex. Initially greater numbers of mucoid than non-mucoid strains adhered to all three materials, hydrophobic non-mucoid strains adhering more efficiently than hydrophilic strains. The two strains also showed differences in colonization. Non-mucoid strains increased with time, peaked and gradually decreased whereas mucoid strains increased and reached steady state.

The growth cycle

Adhesion may be affected by the growth cycle. This was investigated by Allison *et al.* (1990) on the dispersal of *E. coli* from biofilms. It was found that although there appeared to be no significant difference between surface hydrophobicity of both the sessile and planktonic cells isolated from a chemostat, the hydrophobicity decreasing with growth rate in both cases, daughter cells dislodged from the biofilm were significantly more hydrophilic than those remaining, indicating that hydrophobicity also changes with the division cycle. From this Allison put forward the hypothesis that dispersal of cells from adhesive biofilms and recolonization of new surfaces may reflect cell cycle mediated events. Gilbert *et al.* (1991) found that the proportion of cells able to adhere decreased significantly during active growth (5–8 h incubation). With *E. coli* and *Staphylococcus epidermidis* adhesiveness and surface hydrophobicity decreased in early to mid-exponential phase. Cell surface charge became more electronegative for *E. coli* as the cells proceeded to divide but electroneutral for *S. epidermidis*. Fletcher (1977) reported that one of the parameters affecting the attachment of a marine pseudomonad to polystyrene was the growth phase of the culture. The number of cells attaching and the rate of attachment was greatest with log phase cultures and progressively decreased with stationary and death phase cultures. Fletcher (1977) concluded that the influence of culture age on attachment was probably due to changes in cell motility and changes in the quantity or quality of the cell surface polymers. Indeed, she found the culture concentration to be directly related to attachment, the number of attached cells increasing with increasing culture concentration.

Starvation versus nutrient rich conditions
There are many reports of adhesion of bacteria being initiated by starvation conditions. This will be briefly covered here: further details are available in Korber *et al.*, Chapter 1. By so doing these bacteria are exploiting a source of essential nutrients which may be in short supply in the surrounding environment, for instance in oceans or low nutrient waters. Brown *et al.* (1977) believed that attachment occurred under starvation conditions because carbon source receptors are not saturated on the substratum so that bacteria can then interact and attach. However, Bar-Or (1990) explained that it is also vital for the bacteria to be able to detach and migrate when prevailing conditions become too harsh or in situations of population explosion. Kefford *et al.* (1986) found that in leptospires starvation resulted in greater adhesion and that the provision of an energy substrate to starved cells resulted in a decrease in adhesion. Dawson *et al.* (1981) discovered that adhesion was a tactic in the survival strategy of a marine *Vibrio*. Starvation induced dwarfs showed an enhanced rate of adhesion to siliconized glass surfaces. They also found that very small rod shaped bacteria were the primary colonizers on glass surfaces immersed into various seawater samples. These attached dwarfs had the capacity to grow into normal sized bacteria, a phenomenon also reported by Marshall (1992). Wrangstadh *et al.* (1986) studied starved marine *Pseudomonas* spp. and found that the presence of polysaccharide on the cell surface correlated with changes in the degree of adhesion to hydrophobic surfaces,

polysaccharide coated cells having a lower degree of adhesion than those devoid of polymer.

Detachment

Biological factors must also be considered in detachment of bacteria from biofilms. Release of bacteria from a biofilm may be due to production of unattached daughter cells through attached cell replication (Characklis 1990). Lack of oxygen and the accumulation of toxic waste products may affect biologically active cells (Bar-Or 1990). In a natural environment, biological grazing or predator harvesting, which results from protozoa feeding on the outer surface of a biofilm, is also a contributing factor (Rittmann 1989).

The growth cycle
Detachment may also increase at certain times of the cell cycle. *Acinetobacter calcoaceticus* produces polysaccharide to form a capsule around the cell at stationary growth phase, making the cells more hydrophilic and causing them to detach from the interface (Bar-Or 1990). Bakke *et al.* (1984) commented that biofilm detachment can be the rate limiting process in determining the average cellular growth rate. They observed the phenomenon of massive detachment when substrate loading to the biofilm was rapidly doubled. The specific substrate removal rate and product formation rate increased although biofilm numbers remained constant. A number of authors have reported similar results (Speitel & DiGiano 1986; Chang & Rittmann 1987, 1988), in particular, the very high detachment rates during periods of bioregeneration of the activated carbon in water treatment plants. This implies that very rapidly growing bacteria do not attach well in biofilms.

Bacterial species present
Variations in the bacterial species themselves may also be important, especially in mixed culture biofilms. Detachment losses can also affect the species distribution (Rittmann, 1989). This could explain why succession in species is seen as the biofilm develops. Mixed culture biofilms are susceptible to sloughing. Gantzer *et al.* (1989) suggested that this could occur for two reasons. First, although bacteria do not appear to degrade their own EPS, they may depolymerize those of another species, so making the biofilm less stable. Secondly, in addition to this, the EPS and micro-

bial surfaces may be incompatible, thus creating a flaw at which sloughing could start.

The adhesion of bacteria to solid surfaces from mixed cell suspension was investigated by McEldowney & Fletcher (1987). They examined the behaviour of *A. calcoaceticus*, *Staphylococcus aureus*, a second *Staphylococcus* sp. and a coryneform, isolated from continuous culture, on tin plate, glass and nylon. In their detachment experiments they found that detachment of bacteria was influenced by subsequent attachment of a second species in one of three ways. First, the preattached species increased in detachment; for instance, *S. aureus* did not detach from glass over a 2 h period in pure culture controls but did detach when the coryneform was subsequently allowed to attach. This could be due to the coryneform having the greater affinity of the two for attachment to glass as a surface. Secondly, the detachment of the preattached species appeared to be inhibited when the coryneform was the primary colonizer. The coryneform normally detached from glass but subsequent attachment by *S. aureus* appeared to stabilize attachment of the coryneform, thus preventing detachment. This was not due to physical contact between two types of cell as microscopic examination revealed that the two species were spatially separated on the surface. The third influence of attachment of a second species on a preattached species is that it had no effect on detachment. That is, the same proportion of attached cells detached both in buffer and when exposed to a second organism. McEldowney & Fletcher (1987) stated that in many experiments there was no effect at all, as nothing appeared consistently to influence attachment or detachment. Attachment by different bacterial species in mixed culture suspensions might be expected to be competitive, thus resulting in reduced attachment of the component species. However, this was clearly not the case: the results could not be accounted for, according to McEldowney & Fletcher (1987), by simple competition for attachment sites. They considered that both attachment and detachment could also be affected by the secretion of adsorbable polymers or metabolites by a secondary attaching organism.

Chemical processes

Although it is difficult to separate the chemical and biological processes involved in attachment and detachment, there are some distinct differences. Fig. 2.6 presents a simplified overview of the chemical processes involved in biofilm maintenance.

Attachment

Substratum characteristics
Fletcher & Pringle (1983) stated the importance of the substratum in bacterial attachment. Costerton *et al.* (1985) agreed that the rate of initial biofilm formation by bacterial adhesion is to a degree dependent on the chemical nature of the surface. Absolom *et al.* (1983) investigated the adhesion of five strains of bacteria, including two *E. coli* strains and a *S. aureus* strain to various polymeric surfaces such as Teflon and polystyrene. The authors found that adhesion is more extensive to hydrophilic substrata (high surface tension) than to hydrophobic substrata when the surface tension of the bacteria is larger than that of the suspending medium. When the surface tension of the suspending liquid is larger than that of the bacteria the opposite pattern of behaviour prevails. Dexter *et al.* (1975) investigated the attachment rate of marine bacteria to structural materials and found that substrate wettability also appeared to be important.

Surface bonding
The type of bonding that exists between the bacteria and the substratum must also be taken into consideration. Although the surface and the bac-

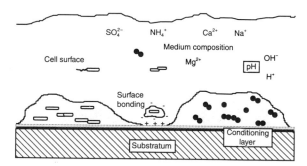

Fig. 2.6.   A diagram summarizing the chemical processes influencing bacterial attachment and detachment from surfaces and subsequent biofilm development.

teria may have a net negative charge van der Waals forces can exceed electrical repulsion at some distance from the surface (10–20 nm), so attracting the bacteria towards the surface (Marshall 1992). Satou *et al.* (1987) looked at the adherence of *Streptococcus* spp. to surface modified glass slides. Hydrophobic bonds and ionic interactions appeared to be important with hydrogen bonding contributing least to adherence, although Marshall (1992) found that when extracellular polysaccharide acted as the bridge between the bacterium and the substratum surface hydrogen bonding was important, as well as electrostatic and covalent bonds and dipole interactions. McEldowney & Fletcher (1986) believed that adhesion cannot be attributed to any one type of adhesive interaction and Marshall (1992) agreed, concluding that the same bacterial strain may adhere with different degrees of adhesive strength to substrata with different surface properties because of different types of short-range forces involved. Other types of bonding must also be taken into account such as the lectin–sugar interaction that occurs between *Rhizobium* cells and the roots of leguminous plants (Bar-Or 1990).

Surrounding medium

Fletcher & Pringle (1983) also believed that the liquid medium surrounding the substratum has an effect on the process of bacterial adhesion. The liquid medium controls the amount of dissolved nutrients in the surronding waters as well as having an effect on substratum interactions. Fletcher (1988) investigated the influence of the cations Na^+, Ca^{2+}, La^{3+} and Fe^{3+} on the adhesion of *P. fluorescens* and found that the addition of each cation caused a decrease in separation distances which was sometimes irreversible. Fletcher proposed that this was caused by the electrolytes neutralizing negative charges on bacterial surface polymers and that the different effects obtained with different strains were due to their different adhesion abilities. Gordon & Millero (1984) performed similar experiments to determine the effect of electrolyte concentration on attachment of *Vibrio alginolyticus* to hydroxyapatite. At concentrations below 0.1 M attachment of bacteria increased with increasing ionic strength but at higher concentrations bacterial affinity for the surface decreased with increasing concentrations of cations and was not related to

ionic strength changes in the medium. They therefore concluded that different mechanisms must be working.

The surrounding liquid medium is not only important in controlling the nutient load to the biofilm but may also have an effect on bacterial attachment, depending on the flow rate over the biofilm. Santos *et al.* (1991) grew *P. fluorescens* on the inside of glass tubes and found that films which developed under low water flow conditions (0.5 m s^{-1}) were less compact and thicker with a random distribution in attachment compared with those grown under higher flow rates (2.5 m s^{-1}) which were thinner, the cells being aligned in the direction of flow in order to offer the least resistance. The liquid medium may contain surface active agents as well which have an influence on surface tension and therefore bacterial attachment. Fletcher & Pringle (1983) found that dimethyl sulphoxide and a series of low molecular weight alcohols appeared to affect numbers of attached cells by altering the liquid surface tension and a number of other authors have reported substances which inhibit or significantly reduce the attachment of bacteria, such as metabolic inhibitors, antibiotics, inhibitors of protcin synthcsis and inhibitors of cnergy generation (Fletcher 1987). Ghannoum (1990) found adhesion of *Candida* spp. to buccal epithelial cells was significantly reduced after both long and short time exposure of yeast to aqueous garlic extracts and that preincubation with the extract prevented adhesion. Humphries *et al.* (1986) studied the effect of a range of biological polymers and synthetic surfactants on the adhesion of a *Pseudomonas* sp. to hydrophilic glass and hydrophobic polystyrene. They found that Brij 56 (polyethylene oxide cetyl ether) was the only compound that had a significant effect, almost totally inhibiting the adhesion of the pseudomonad to hydrophobic polystyrene (although it had little or no effect on hydrophilic glass). Carballo *et al.* (1991) have also looked at this phenomenon, determining the adherence of five coagulase negative bacterial strains onto different types of catheter after either or both the bacteria and catheters had been treated with citrated human plasma, human serum albumin or fibrinogen. Plasma and serum albumin were found to produce a marked inhibition of bacterial adherence to all types of catheters whereas fibrinogen enhanced the number of bound bacteria.

Detachment

Surrounding medium

Many chemical factors also affect detachment and desorption rates once the bacteria are attached to a surface. Turakhia et al. (1983), reported the dramatic increase in detachment on the addition of chelants such as EGTA. These remove calcium from the biofilm, which as previously explained leads to a decrease in the cohesiveness of the biofilm. Calcium in the bulk water, however, decreases erosion rates in P. aeruginosa biofilms. In addition, Characklis (1990) reported that chlorine has been used for specific detachment purposes, as has trypsin (Corpe 1974), 5% sodium hypochlorite (Costerton et al. 1985) and other detergent treatments (McEldowney & Fletcher 1986).

Nutrient availability

Nutrient availability is another important factor when considering the detachment of bacteria from biofilms. In a further study, Delaquis et al. (1989) studied the detachment of P. fluorescens from glass surfaces in response to nutrient stress, and found that depletion of either glucose or nitrogen led to the active detachment of cells from the biofilm. Under such limiting conditions attached cell biomass declined rapidly but was accompanied by a simultaneous increase in unattached cell biomass in the bulk phase. When glucose or nitrogen was available in excess, no decline in attached biomass was evident, providing further evidence that active emigration was induced by glucose or nitrogen respectively (Delaquis et al. 1989). Detachment in this case was a nutrient induced form of behaviour. This phenomenon had been reported previously by Yu et al. (1987), who observed that the detachment of Vibrio furnissii occurred on the omission of key nutrients such as lactate, nitrogen (in the form of ammonia) or phosphorus.

More recently, Applegate & Bryers (1991) have reported on the effects of carbon or oxygen limitations on the detachment rates of bacteria. They investigated biofilm removal processes in a turbulent flow system using P. putida ATCC 11172. They found that oxygen limited biofilms reached a higher steady state biofilm organic carbon level than carbon limited biofilms. In addition calcium concentrations were also found to be important, oxygen limited biofilms having a higher calcium content than carbon limited biofilms. Calcium (Ca^{2+}), and to a lesser extent magnesium (Mg^{2+}), accumulate within the biofilm matrix and act as cross-linking agents in exopolysaccharide production within the biofilm. Hence oxygen limited biofilms with higher exopolysaccharide and calcium content are more cohesive but were found to be more susceptible to catastrophic sloughing events, although shear removal rates were only 20–40% of those for carbon limited biofilms. The latter biofilms did not slough even when they were subjected to long term deprivation of all nutrients.

Other factors

Cation exchange, substrate concentration or ionic strenghs may also lead to destabilization of the matrix polymers and consequent detachment (Gantzer et al. 1989). Chemical changes in the substratum itself could lead to biofilm detachment, an important factor in situations where the substratum is also the major growth substrate, for instance in the degradation of cellulose. Charge accumulation in the biofilm must also be taken into account. Applegate & Bryers (1991) stated that if the electrochemical gradient is large enough to cause an extreme osmotic pressure difference, the biofilm could literally be pushed off the substrate. Alternatively, they suggested that rapid fluctuations in ambient ion or proton concentrations could create repeated swelling and contracting of the biofilm, eventually weakening the entire matrix and causing sloughing.

Physical processes

Physical processes are very important in attachment and detachment of bacteria from surfaces. The involvment of these physical mechanisms are frequently overlooked in many of the accounts of biofilm formation. Fig. 2.7. gives a summary of the physical mechanisms involved in attachment and detachment.

Surface texture

Pirbazari et al. (1990) found that surface texture of a solid appeared to affect biofilm formation. Chang & Rittmann (1988) discovered that a rough irregular carbon source provided better initial biofilm growth than a spherical medium, as the rough surface provided both attachment sites

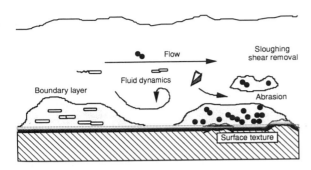

Fig. 2.7. A simple model describing the physical processes involved in bacterial attachment and detachment from surfaces and how these processes affect biofilm maintenance.

and a degree of protection to the developing biofilm. However, different carbon sources have different structures and a bacterial species may attach preferentially to one type of surface and not another (Fletcher 1987).

Environmental factors

A number of authors have discussed the effects that both temperature and pH can have on bacterial adhesion. Stenström (1989) noted that the alteration of pH between 4 and 9 did not significantly affect the adhesion process. Fletcher (1977), when investigating the attachment of a marine pseudomonad to polystyrene, reported that a temperature of 3 ± 1 °C noticably decreased the proportion of stationary phase cells which attached compared with cells at 20 ± 1 °C. Fletcher (1977) also found that the time allowed for attachment was important in the attachment process, the number of attached cells increasing with the time allowed for attachment until the attachment surface approached saturation.

Detachment

Shear removal and sloughing

Rittmann (1989) divided the three main physical processes into erosion or shear removal, sloughing and abrasion. Shearing, the continuous removal of small portions of biofilm, is highly dependent on fluid dynamic conditions, increasing with increased biofilm thickness and fluid shear stress at the biofilm–fluid interface. For this reason the process may not be evident in the initial stages of biofilm formation when the substratum is rough (Chang & Rittmann 1988). In contrast, sloughing is the rapid and massive loss from a biofilm, which may be occurring at the same time as erosion, but is more frequently witnessed with thicker biofilms developed in nutrient rich environments (Characklis & Cooksey 1983). Applegate & Bryers (1991) also stated that while shear removal is a continuous process of biomass erosion from the outer surfaces, sloughing appears to be a random and sporadic event with large amounts or entire biofilm being removed from the entire depth of the biofilm itself. Howell & Atkinson (1976) explained that sloughing is thought to be due to nutrient or oxygen depletion deep within the biofilm or to some dramatic change in the immediate environment of the biofilm. For instance, Applegate & Bryers (1991) suggested that in some cases, sloughing is caused by limitations of oxygen transfer in thick biofilms. As a result, facultative aerobic bacteria become anaerobic and convert organic substrates into volatile fatty acids (decreasing the pH in that area) and insoluble gases, both of which may serve to weaken the biofilm structure. Jansen & Kristensen (1980) also observed that the formation of nitrogen bubbles under denitrifying biofilms resulted in sloughing when pressure changes caused the bubbles to expand.

Abrasion

In addition to the two processes of shearing and sloughing, Rittmann (1989) also described abrasion as an important process in the detachment of bacteria from a biofilm. This is loss of biofilm due to repeated collisions between substratum particles, and is frequently witnessed in fluidized bed systems. He considered that the physiological changes caused by abrasion strongly affect the specific detachment rate coefficient, the value of which is lower in thin dense biofilms that have developed under high abrasion conditions compared with values obtained for biofilms grown with less abrasion. The former, denser biofilms, were therefore more resistant to abrasion.

Environmental parameters

Environmental parameters are also important, such as pH, temperature and the presence of organic macromolecules either absorbed on the substratum or dissolved in the liquid phase. These parameters were investigated by McEldowney & Fletcher (1988a) when investigating the desorption and detachment of three bacteria from food container and food processing surfaces. The bacteria investigated were *S. aureus*, *A. calcoaceticus* and a coryneform from glass, tin plate, stainless steel, nylon and polypropylene surfaces. When investigating pH, it was found that *S. aureus* did not detach from the substrata at any pH investigated (pH 5–9), although the other two sometimes did detach depending on pH and substratum composition. The degree of bacterial detachment from the substrata was not related to bacterial respiration at experimental pH values. It was also found that the rate of bacterial desorption was not affected by temperature (4–30 °C) nor by an absorbed layer of peptone and yeast extract on the substrata. McEldowney & Fletcher (1988a) concluded therefore that bacterial desorption and hence bacterial removal or transfer via liquids flowing over colonized surfaces is likely to vary with the surface composition and bacterial species colonizing the surfaces.

Fluid dynamics

The concept of fluid dynamics is important in many natural and man-made ecosystems and not just in the study of microbial ecology. The principles of fluid flow can be applied to a wide range of situations: from the movement of oil and water through industrial and domestic pipelines to the circulation of blood in the human body. It is an area that is fundamental in understanding biofilm formation, but as yet its implications have not been fully appreciated by many microbiologists. The basic principles of fluid mechanics will be considered and how they can be applied to both planktonic and biofilm bacteria.

Laminar flow and turbulent flow

Two contrasting extremes of flow exist, namely laminar flow and turbulent flow. Laminar flow is the smooth flow of water through a pipe or duct with no lateral mixing (Fletcher & Marshall 1982). Once the movement at any particular point in the flow becomes erratic and irregular the flow is defined as turbulent. This phenomenon was first reported in 1883 by a British engineer, Osbourne Reynolds. In an experiment that is still in use today, he monitored the flow of

water through a transparent pipe by injecting dye at one point in the flow and watching what happened as the dye moved along the length of the pipe. The velocity of the fluid in the pipe is controlled by a tap present at the end of the pipe which is opened or closed as necessary. When the tap is first switched on the flow is at a low velocity and the dye moves through the system in a continuous thread with no mixing (Fig. 2.8a). This is typical of laminar flow, when the fluid particles move in parallel lines maintaining the same relative positions in successive cross-sections (Douglas *et al.* 1984). The fluid is said to move in sheets or laminae (Characklis *et al.* 1990). Increasing the velocity of the flow will at first have no effect on the dye in the pipe but there will come a point when the velocity of the fluid is such that the dye thread will begin to oscillate and break up (Fig. 2.8b). If the velocity is increased further the path of dye through the pipe will be completely lost as it is broken up by increased mixing in the system due to the forma-

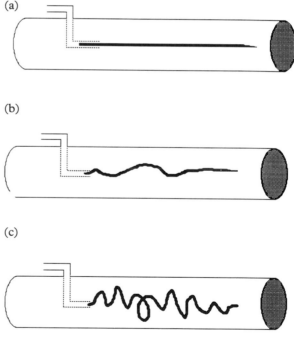

Fig. 2.8. The characteristic dye threads for three different flow regimes. (a) Laminar flow occurs at low velocity, producing no lateral mixing and is dominated by viscous forces. (b) The transitional flow is a condition which varies between laminar and turbulent flow. (c) Turbulent flow is a random and unpredictable flow with constant change in direction and velocity.

tion of eddy currents. The flow of fluid becomes random and unpredictable (Fig. 2.8c) and is termed turbulent flow. In this the flow is no longer orderly and fluid particles occupy different relative positions in successive cross-sections (Douglas *et al.* 1984).

Most flows in natural and engineered systems are turbulent. However, turbulent flow is very complex and difficult to predict. Although many engineering studies have concentrated on turbulent flow, most experimental work performed by microbiologists has been with laminar flow because it can be described analytically and because it is of interest in studying the movement of motile bacteria through slow moving or quiescent water (Characklis *et al.* 1990).

Reynolds number

It is almost impossible to predict exactly when laminar flow will cease and turbulent flow will occur; however, an estimation can be achieved using a dimensionless parameter called the Reynolds number (*Re*) (Munson *et al.* 1990). This describes the relative magnitude of inertia to viscous forces and is given as:

$$Re = \frac{pvd}{\mu} \qquad (1)$$

where *d* is a constant length in the system, such as the diameter of the pipe (in metres); *v* is the velocity of the fluid moving through the pipe (in m s^{-1}); *p* is the density of the fluid (in kg m^{-3}); and μ is the viscosity of the fluid (in N s m^{-2} or centipoise). Both *p* and μ are temperature dependent. It is generally accepted that a *Re* of about 2000 or less is representative of laminar flow, whereas turbulent flow is present when the *Re* is equal to or greater than 4000. If the *Re* is below 1 the flow is referred to as creeping flow (Munson *et al.* 1990). Flows with Reynolds numbers between 2000 and 4000 are known as transitional flows, where conditions can alternate between laminar and turbulent flow depending on environmental conditions such as the roughness of the pipe surface or the entrance conditions in the pipe (Characklis *et al.* 1990). Laminar flow can sometimes be maintained in flow up to *Re* = 50 000 given the right conditions (Douglas *et al.* 1984) but such a flow is unstable and will revert to turbulent flow with the slightest disturbance.

The Reynolds number described above is for the flow through a circular pipe of constant diameter. When the flow is through a rectangular cross-section, such as in the Robbins device then the hydraulic diameter *Dh* has to be substituted for *d* in the Reynolds equation, where *A* is the cross-sectional flow area and *P* is the wetted perimeter,

$$Dh = \frac{4A}{P} \qquad (2)$$

that is, four times the cross-sectional flow area divided by the wetted perimeter. (Munson *et al.* 1990).

When a fluid first enters a pipe it has an almost uniform velocity profile but as the fluid moves along the pipe, viscous effects cause it to stick to the pipe wall (Munson *et al.* 1990). Hence fluid moving near the centre is more rapid than fluid moving adjacent to the pipe walls due to the drag caused by this viscosity (Caldwell & Lawrence 1989). Characteristic velocity profiles exist, therefore, for both laminar and turbulent flow (Fig. 2.9a, b). Munson *et al.* (1990) explained that the differences observed between the two types of flow is due to shear stress. This is the stress caused by the shear force that occurs when two parts slide against each other. In laminar

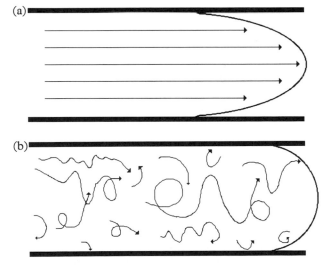

Fig. 2.9 Velocity profiles for laminar and turbulent flow. (a) In laminar flow the boundary layer increases in size until it is equal to the radius of the pipe. The central portion of the fluid moves much faster than the fluid near the sides of the pipe. (b) In turbulent flow a viscous sublayer exists at the surface where laminar flow predominates.

flow, it is the velocity gradient combined with the fluid viscosity that produces shear stress but in turbulent flow the shear stress is largely a result of momentum transfer among randomly moving, finite sized bundles of fluid particles.

The hydrodynamic boundary layer

In both laminar and turbulent flow the fluid next to the surface of the pipe walls begins to form a boundary layer in which the viscous forces are more important than the acceleration or inertia forces. This effectively separates the fluid in the boundary layer from the fluid outside the boundary layer where the viscous forces are negligible compared with other forces (Massey 1989). The boundary layer was described by Douglas *et al.* (1984) as the region of fluid close to the surface immersed in the flowing fluid. The fluid actually in contact with the pipe has zero velocity and a velocity gradient exists between the fluid in the free stream and the pipe surface. Laminar or turbulent flow may occur within the boundary layer. However, even when the boundary layer is turbulent the flow immediately next to the solid surface is not, due to the fact that fluid particles cannot pass through an impermeable solid surface and any perpendicular movement dies out (Massey 1989). In such an instance, therefore, an extremely thin layer exists adjacent to the solid surface (often less than 1 μm) in which the flow has negligible fluctuations of velocity. This is called the laminar sublayer or viscous sublayer and should not be confused with an entirely laminar boundary layer (Massey 1989). This sublayer has been quantified by Caldwell & Lawrence (1989) using continuous flow slide cultures.

Biofilm dynamics in laminar flow

The boundary layer increases in size as the flow develops until, once fully developed it actually takes up the whole of the pipe (Douglas *et al.* 1984). However, differences in fluid velocity still exist, the flow being much slower close to the surfaces, producing the characteristic parabolic velocity profile of laminar flow. The area next to the pipe surfaces is considered by some to be distinct from the main flow in the pipe. Caldwell & Lawrence (1989) referred to it as the surface microenvironment as opposed to the macroenvironment found in the bulk flow, or as Huang

et al. (1992) hypothesized for the trickling filter process, as the stagnant layer in a two layered liquid film (the other layer being the upper, moving liquid film) where mass transfer limitations occurred. Indeed, in laminar flow, the flow of fluid next to the surface can be negligible; this plays an important part in the limitation of biofilm growth, due to the consequent effects on the input of cells and nutrients.

In an environment where fluid moves in parallel laminae it is difficult to imagine how planktonic cells and nutrients in the bulk flow get to the pipe surfaces in the total absence of mixing, unless in the case of bacteria it is through their own motility. For non-motile cells in laminar flow, molecular diffusion is the important process by which planktonic cells reach the pipe surfaces (Characklis 1990). This is possible because of the ability of bacteria to act as colloidal particles (Marshall 1980). Gravity may also have an important effect. In contrast, motile cells have the ability to colonize as they choose, but also to detach and move into the bulk flow if necessary, for colonization further downstream. Caldwell (1987) suggested that flagellated bacteria, by moving perpendicular to the surface may move from relatively quiescent water layers into more rapidly moving flow lamina, which can transport them at rates of cm s^{-1}. Motility also allows bacteria to make use of the microenvironment next to the surface where the flow is much slower, in order to move against the main flow of fluid. Caldwell & Lawrence (1989) found that motile bacteria at the surface could move against a bulk flow of 10 cm s^{-1} or greater, contributing considerably to backgrowth in the system. Korber *et al.* (1989) investigated polarly flagellated, motile and non-motile strains of *P. fluorescens* under high and low flow (120 and 8 μm s^{-1} in the microenvironment within 0.2 μm of the surface, respectively) and found that under these conditions the bacteria actively moved upstream. Caldwell *et al.* (1992) found that the rate of backgrowth could be decreased by increasing the flow rate.

A number of authors have reported the effect flow can have on attachment and detachment of bacteria. Powell & Slater (1983) investigated attachment of *B. cereus* in laminar flows with Reynolds numbers between 0.4 and 16.0. They found that the deposition process can occur simultaneously with detachment, which is known to increase with fluid velocity. Fletcher &

Marshall (1982) found that the rates of attachment and detachment were dependent on the flow rates. Increasing the flow rates increased the attachment of slowly grown cells but decreased the atttachment of bacteria grown at faster rates. They also considered that attachment of fast growing bacteria may be principally governed by physiological processes whereas physicochemical adsorption is the dominant factor with slowly grown cells.

Under controlled laminar flow rates Lawrence & Caldwell (1987) found that in the actual process of attaching to a surface four different types of cell movement were seen in natural bacterial stream populations: that is, rolling, shedding (where daughter cells were released), packing and spreading. Attachment was also found to be different for motile and non-motile cells. The latter, due to their reliance on diffusion processes, are unable to control their final attachment orientation once irreversible attachment has occurred (Caldwell *et al.* 1992). They align themselves longitudinally with the direction of flow, presumably due to the presence of an adhesive region at one end. Motile cells, however, as one would expect attach much more randomly with respect to the direction of flow.

Once established, the growth of the biofilm in laminar flow may be limited, as already explained, especially at creeping velocities (*Re*<1). The lack of mixing and the slow velocity of fluid near to the surface means the biofilm rapidly depletes substrate adjacent to the pipe wall (Caldwell & Lawrence 1989). In addition, toxic metabolites and waste products will also build up. This leads to environmental stress on the biofilm. This will not only affect the growth of developing biofilms but may also cause the shearing of already developed, thick biofilms (Caldwell *et al.* 1992). The growth rate of the biofilm was found not only to be dependent on the laminar flow velocity of 10 cm s^{-1} (Caldwell & Lawrence 1986a) but also by the concentration of substrate. Caldwell & Lawrence (1986) found growth of *P. fluorescens* in biofilms to be dependent on laminar flow velocity with concentrations of glucose below 100ppm. Initially the growth was exponential but soon declined to zero in the absence of flow; growth resumed again when the flow was restarted, suggesting the biofilm to be diffusion limited. For concentrations above 100 ppm glucose, *P. fluorescens* grew at a maximum independent of flow rate, indicating that at this concentration of glucose, diffusion was not a limiting process. This is unusual for biofilms in their natural environments, mature biofilms usually being diffusion limited, even at high velocities. Caldwell & Lawrence (1986a) explained this as the disequilibrium that exists between the microenvironment at the surface and the remaining macroenvironment because of the high density of cells on the pipe walls compared to the fluid velocity in the pipe. An equilibrium will exist only if the laminar fluid velocity is high enough or the density of the cells is low enough, a situation that may only occur in developing biofilms at corresponding higher laminar flows.

Biofilm dynamics in turbulent flow

Turbulent flow is commonly found in most natural systems, as well as in pipelines. Many factors may affect turbulent flow, such as the presence of air bubbles or the roughness of the pipe surface. If the roughness of the tube wall becomes sufficiently coarse due to the build up of biofilm, eddies will develop even in laminar flow and will result in energy losses (Characklis *et al.* 1990). Hence, particles suspended within the fluid in the pipe are transported to the solid surface primarily by fluid dynamic forces (Characklis 1990) and not by diffusion as is the case in laminar flow. Turbulence affects rates of organism deposition, delivery of nutrients and increases transfer of heat and momentum between adjacent masses of water (Fletcher & Marshall 1982). Gravity appears to contribute little in turbulent flow. Stanley (1983) also found cell motility to be insignificant, in contrast to that reported for laminar flow. He found that although in laminar flow motile *P. aeruginosa* adsorbed more readily to stainless steel than non-motile bacteria, in turbulent flow the adsorption of the motile bacteria was decreased with the non-motile bacteria being unaffected.

The boundary layer is also very different in turbulent flow. It is considered to be the layer at the pipe surface in which laminar flow predominates and most of the resistance to mass transport occurs (Characklis *et al.* 1990). An important area for bacterial attachment and biofilm formation, the boundary layer remains close to the pipe surfaces and does not extend to fill the radius of the pipe as in laminar flow. There is no sharp limit to the boundary layer because of the continuous transition in fluid properties from the boundary layer region into the ideal fluid region (Silvester & Sleigh 1985). Due to the nature of turbulent flow this sublayer is constantly penetrated by turbulent fluctuations and bursts, causing local velocity pertubations in all directions, even opposite to the main direction of flow (Cleaver & Yates 1975, 1976). This is believed to be one way in which bacteria are transported to pipe surfaces in turbulent flow. Eddy currents cause upsweep and downsweep forces which either extend from the bulk flow of fluid and penetrate all the way to the pipe surface, or move away from the surface and into the main fluid flow, carrying colloidal particles such as bacteria with them in the process (Cleaver & Yates 1975; Escher & Characklis 1990). Alternatively, cells in turbulent flow may be transported to within short distances of the surface by eddy diffusion and propelled into the sublayer under their aquired momentum (Characklis *et al.* 1990). Viscous drag slows down the bacteria as they penetrate the sublayer because acceleration forces of microbial cells are very small due to their small size and density (in relation to water). The processes described above are also related to the transport of substrate and other nutrients into the biofilm.

Turbulence may also affect the structure of biofilms. Huang *et al.* (1992) stated that dense, thin biofilms were induced by high particle-to-particle contacts and liquid turbulence. Similarly, Santos *et al.* (1991) found that biofilms grown in a flow rate of 0.5 m s^{-1} were less compact but thicker than biofilms grown at 2.5 m s^{-1}. Turbulence may increase attachment in a biofilm but, if the biofilm is too thick, it may also cause detachment to increase. This will occur when the biofilm protrudes through the boundary layer, thereby increasing frictional resistance in the system. If the biofilm remains within the boundary layer there is no frictional resistance beyond that of the hydraulically smooth pipe value (Trulear & Characklis 1982). It can continue to grow, being affected little by the flow, except for the turbulent bursts which penetrate the sublayer and, as has been mentioned, bring nutrients and cells to the pipe surface (Cleaver & Yates 1975). It is not until the biofilm surface irregularities protrude through the sublayer that the frictional resistance increases, depending proportionally on the thickness of the biofilm (Applegate & Bryers 1991). At

this point the bacterial biofilm, by increasing the drag present in the system, has a direct effect on the flow, greatly decreasing the flow through the pipe (Watkins & Costerton 1984). Bryers (1987) believed that microbial films could increase fluid frictional resistance by *c.* 200–300%, causing both increased power consumption and maintenance costs due to periodic system shutdown, as well as causing costly fouling problems (Characklis *et al.* 1990). For the biofilm itself, protruding through the sublayer causes increased turbulence in the biofilm vicinity and increases the rate of detachment by erosion, sloughing and abrasion. These processes and also that of desorption are strongly influenced by the Reynolds number (Escher & Characklis 1990). The higher the Reynolds number, the higher the velocity in the pipe and the greater the detachment will be. Not only this but the thickness of the boundary layer decreases in size with higher Reynolds numbers, meaning that more of the biofilm will protrude beyond the sublayer with an increase in flow rate, allowing greater detachment to occur.

In the study of microbial ecology the fluid dynamics present in a system require a great deal of consideration. Although turbulent flows are more commonly found they are very complex and much harder to study with respect to the microbiological populations. Consequently most work in this field has been done with laminar flows.

Although engineers have investigated and appreciated the effects of turbulent flows, it is up to the microbiologist to utilize this information in understanding how fluid dynamics can affect the bacteria themselves, before further advances can be made in the control of detrimental biofilms in industrial systems.

Detachment and attachment are occurring simultaneously and continuously from a biofilm. It is almost impossible to prevent bacteria from attaching to surfaces in an aqueous environment. The ability to attach allows bacteria to exploit nutrients at the surface as well as providing a degree of protection from the surrounding environment. Detachment would therefore seem at first to disadvantage biofilm development, but it is an important process in biofilm formation and maintenance and may in certain cases be beneficial to the bacteria themselves. Biofilms with greater specific detachment rates have been found to have larger fractions of active bacteria, as the formation of a substantial inert zone near the substratum is precluded (Rittmann 1989). A number of authors have reported that detachment can occur in response to the need for dispersion under starvation conditions (Marshall 1985; van Loosdrecht *et al.* 1987). For instance, in soil, detachment could allow spreading either passively in pore water or by active means, to allow more nutritionally favourable locations to be exploited. Attachment and detachment are important procesess in biofilm dynamics but more understanding is needed before medicine and industry can control or exploit biofilm formation to their advantage.

Acknowledgements

We gratefully acknowledge the financial support of the University of Exeter in an award of a studentship to M. G. B. We thank John Boyle (Department of Engineering, Exeter University) for reading the manuscript.

References

Absolom, D. R., Lamberti, F. V., Policova, Z., Zingg, W., van Oss, C. J. & Neumann, A. W. (1983). Surface thermodynamics of bacterial adhesion. *Applied and Environmental Microbiology*, **46**, 90–7.

Addy, M., Slayne, M. A. & Wade, W. G. (1992). The formation and control of dental plaque – an overview. *Journal of Applied Bacteriology*, **73**, 269–78.

Allison, D. G., Evans, D. J., Brown, M. R. W. & Gilbert, P. (1990). Possible involvment of the division cycle in dispersal of *Escherichia coli* from biofilms. *Journal of Bacteriology*, **172**, 1667–9.

Anwar, H., Dasgupta, M., Lam, K. & Costerton, J. W. (1989). Tobramycin resistance of mucoid *Pseudomonas aeruginosa* biofilm grown under iron limitation. *Journal of Antimicrobial Chemotherapy*, **24**, 647–55.

Anwar, H., Strap, J. L. & Costerton, J. W. (1992). Establishment of aging biofilms: Possible mechanism of bacterial resistance to antimicrobial therapy. *Antimicrobial Agents and Chemotherapy*, **36**, 1347–51.

Applegate, D. H. & Bryers, J. D. (1991). Effects of carbon and oxygen limitations and calcium concentrations on biofilm removal processes. *Biotechnology and Bioengineering*, **37**, 17–25.

Bakke, R., Trulear, M. G., Robinson, J. A. & Characklis, W. G. (1984). Activity of *Pseudomonas aeruginosa* in biofilms: steady state. *Biotechnology and Bioengineering*, **26**, 1418–24.

Bar-Or, Y. (1990). The effect of adhesion on survival and growth of microorganisms. *Experientia*, **46**, 823–6.

Blenkinsopp, S. A. & Costerton, J. W. (1991). Understanding bacterial biofilms. *Trends in Biotechnology*, **9**, 138–43.

Brown, C. M., Ellwood, D. C. & Hunter, J. R. (1977). Growth of bacteria at surfaces: influence of nutrient limitation. *FEMS Microbiology Letters*, **1**, 163–6.

Brown, M. R. W., Allison, D. G. & Gilbert, P. (1988). Resistance of bacterial biofilms to antibiotics: a growth-rate related effect? *Journal of Antimicrobial Chemotherapy*, **22**, 777–83.

Bryers, J. D. (1984). Biofilm formation and chemostat dynamics: pure and mixed culture considerations. *Biotechnology and Bioengineering*, **26**, 948–58.

Bryers, J. D. (1987). Biologically active surfaces: processes governing the formation and persistence of biofilms. *Biotechnology Progress*, **3**, 57–68.

Busscher, H. J., Uyen, M. H. M. J. C., Weerkamp, A. H., Postma, W. J. & Arends, J. (1986). Reversibility of adhesion of oral streptococci to solids. *FEMS Microbiology Letters*, **35**, 303–6.

Caldwell, D. E. (1987). Microbial colonization of solid–liquid interfaces. *Biochemical Engineering V*, **506**, 274–80.

Caldwell, D. E. & Lawrence, J. R. (1986). Growth kinetics of *Pseudomonas fluorescens* microcolonies within the hydrodynamic boundary layers of surface microenvironments. *Microbial Ecology*, **12**, 299–312.

Caldwell, D. E. & Lawrence, J. R. (1989). Study of attached cells in continuous flow slide culture. In *Handbook of laboratory model systems for microbial ecosystem research*, ed. J. W. T. Wimpenny, pp. 117–38. Boca Raton, Fla: CRC Press.

Caldwell, D. E., Korber, D. R. & Lawrence, J. R. (1992). Confocal laser microscopy and digital image analysis in microbial ecology. In *Advances in Microbial Ecology*, **12**, 1–67.

Carballo, J., Ferreiros, C. M. & Criado, M. T. (1991). Importance of experimental design in the influence of proteins in bacterial adherence to polymers. *Medical Microbioloy and Immunology*, **180**, 149–55.

Chang, H. T. & Rittmann, B. E. (1987). Verification of the model of biofilm on activated carbon. *Environmental Science and Technology*, **21**, 280–8.

Chang, H. T. & Rittmann, B. E. (1988). Comparative study of biofilm shear loss on different adsorptive media. *Journal of the Water pollution and Control Federation*, **60**, 362–8.

Characklis, W. G. (1990). Biofilm processes. In *Biofilms*, ed. W. G. Characklis & K. C. Marshall, pp. 195–231. New York: Wiley.

Characklis, W. G. & Cooksey, K. E. (1983). Biofilms and microbial fouling. *Advances in Applied Microbiology*, **29**, 93–138.

Characklis, W. G., Turakhia, M. H. & Zelver, N. (1990). Transport and interfacial transport phenomena. In *Biofilms*, ed. W. G. Characklis & K. C. Marshall, pp. 265–340. New York: Wiley.

Chopra, I. (1986). Antibiotics and bacterial adhesion. *Journal of Antimicrobial Chemotherapy*, **18**, 553–6.

Cleaver, J. W. & Yates, B. (1975). A sub layer model for the deposition of particles from a turbulent flow. *Chemical Engineering Science*, **30**, 983–92.

Cleaver, J. W. & Yates, B. (1976). The effect of re-entrainment on particle deposition. *Chemical Engineering Science*, **31**, 147–53.

Corpe, W. A. (1974). Detachment of marine periphytic bacteria from surfaces of glass slides. *Developments in Industrial Microbiology*, **15**, 281–7.

Costerton, J. W. & Lappin-Scott, H. M. (1989). Behavior of bacteria in biofilms. *American Society for Microbiology News*, **55**, 650–4.

Costerton, J. W., Marrie, T. J. & Cheng, K.-J. (1985). Phenomena of bacterial adhesion. In *Bacterial adhesion*, ed. D. C. Savage & M. Fletcher, pp. 3–43. New York: Plenum Press.

Dawson, M. P., Humphrey, B. A. & Marshall, K. C. (1981). Adhesion: a tactic in the survival strategy of a marine vibrio during starvation. *Current Microbiology*, **6**, 195–9.

Delaquis, P. J., Caldwell, D. E., Lawrence, J. R. & McCurdy, A. R. (1989). Detachment of *Pseudomonas fluorescens* from biofilms on glass surfaces in response to nutrient stress. *Microbial Ecology*, **18**, 199–210.

Dexter, S. C., Sullivan, J. D. Jr, Williams, J. III & Watson, S. W. (1975). Influence of substrate wettability on the attachment of marine bacteria to various surfaces. *Applied Microbiology*, **30**, 298–308.

Dougherty, S. H. (1988). Pathobiology of infection in prosthetic devices. *Reviews of Infectious Diseases*, **10**, 1102–17.

Douglas, J. F., Gasiorek, J. M. & Swaffield, J. A. (1984). Motion of fluid particles and streams. In *Fluid mechanics*, pp. 97–117. London: Pitman Publishing.

Escher, A. & Characklis, W. G. (1990). Modeling the initial events in biofilm accumulation. In *Biofilms*, ed. W. G. Characklis & K. C. Marshall, pp. 445–86. New York: Wiley.

Fletcher, M. (1977). The effects of culture concentration and age, time and temperature on bacterial attachment to polystyrene. *Canadian Journal of Microbiology*, **23**, 1–6.

Fletcher, M. (1987). How do bacteria attach to solid surfaces? *Microbiological Sciences*, **4**, 133–6.

Fletcher, M. (1988). Attachment of *Pseudomonas fluorescens* to glass and influence of electrolytes on bacterium–substratum separation distance. *Journal of Bacteriology*, **170**, 2027–30.

Fletcher, M. & Marshall, K. C. (1982). Are solid surfaces of ecological significance to aquatic bacteria? In *Advances in Microbial Ecology*, **12**, 199–236.

Fletcher, M. & Pringle, J. H. (1983). The effect of surface free energy and medium surface tension on bacterial attachment to solid surfaces. *Journal of Colloid and Interface Science*, **104**, 5–14.

Gantzer, C. J., Cunningham, A. B., *et al.* (1989). Group report. Exchange process at the fluid–biofilm interface. In *Structure and function of biofilms*, ed. W. G. Characklis & P. A. Wilderer, pp. 73–89. New York: Wiley.

Ghannoum, M. A. (1990). Inhibition of *Candida* adhesion to buccal epithelial cells by an aqueous extract of *Allium sativum* (Garlic). *Journal of Applied Bacteriology*, **68**, 163–9.

Gilbert, P., Brown, M. R. W. & Costerton, J. W. (1987). Inocula for antimicrobial sensitivity testing: a critical review. *Journal of Antimicrobial Chemotherapy*, **20**, 147–54.

Gilbert, P., Evans, D. J., Evans, E., Duguid, I. G. & Brown, M. R. W. (1991). Surface characteristics and adhesion of *Escherichia coli* and *Staphylococcus epidermidis*. *Journal of Applied Bacteriology*, **71**, 72–7.

Gordon, A. S. & Millero, F. J. (1984). Electrolyte effects on attachment of an estuarine bacterium. *Applied and Environmental Microbiology*, **47**, 495–9.

Howell, J. A. & Atkinson, B. (1976). Sloughing of microbial film in trickling filters. *Water Research*, **10**, 307–15.

Huang, J., Hao, O. J., Al-Ghusain, I. A. *et al.* (1992). Biological fixed film systems. *Water Environmental Research*, **64**, 359–66.

Humphries, M., Jaworzyn, J. F. & Cantwell, J. B. (1986). The effect of a range of biological polymers and synthetic surfactants on the adhesion of a marine *Pseudomonas* sp. strain NCMB 2021 to hydrophilic and hydrophobic surfaces. *FEMS Microbiology Ecology*, **38**, 299–308.

Jansen, J. & Kristensen, G. H. (1980). Fixed film kinetics: denitrification in fixed films. In *Report 80-59, Department of Sanitary Engineering*, Technical University of Denmark.

Kefford, B., Humphrey, B. A. & Marshall, K. C. (1986). Adhesion: a possible survival strategy for Leptospires under starvation conditions. *Current Microbiology*, **13**, 247–50.

Kogure, K. (1989). Attachment of aquatic bacteria – overview. In *Recent advances in microbial ecology*, ed. T. Hattori, Y. Ishida, Y. Maruyama R. Y. Morita, & A. Uchida, pp. 131–4. Tokyo: Japan Scientific Societies Press.

Korber, D. R., Lawrence, J. R., Sutton, B. & Caldwell, D. E. (1989). Effect of laminar flow velocity on the kinetics of surface recolonization by mot[+] and mot[−] *P. fluorescens*. *Microbial Ecology*, **18**, 1–19.

Lappin-Scott, H. M. & Costerton, J. W. (1989).

Bacterial biofilms and surface fouling. *Biofouling*, **1**, 323–42.

Lawrence, J. R. & Caldwell, D. E. (1987). Behavior of bacterial stream populations within the hydrodynamic boundary layers of surface microenvironments. *Microbial Ecology*, **14**, 15–27.

Marshall, K.C (1980). Bacterial adhesion in natural environments. In *Microbial adhesion to surfaces*, ed. R. C. W. Berkeley, J. M. Lynch, J. Melling, P. R. Rutter & B. Vincent, pp. 187–96. Chichester: Ellis Horwood.

Marshall, K. C. (1985). Bacterial adhesion in oligotrophic habitats. *Microbiological Sciences*, **2**, 323–6.

Marshall, K. C. (1992). Biofilms: an overview of bacterial adhesion, activity, and control at surfaces. *American Society for Microbiology News*, **58**, 202–7.

Marshall, K. C., Stout, R. & Mitchell, R. (1971). Selective sorption of bacteria from seawater. *Canadian Journal of Microbiology*, **17**, 1413–16.

Martinez-Martinez, L., Pascual, A. & Perea, E. J. (1991). Kinetics of adherence of mucoid and non-mucoid *Pseudomonas aeruginosa* to plastic catheters. *Journal of Medical Microbiology*, **34**, 7–12.

Massey, B. S. (1989). *Mechanics of fluids*, 6th edition, pp. 148–50. London: Chapman & Hall.

McEldowney, S. & Fletcher, M. (1986).Variability of the influence of physicochemical factors affecting bacterial adhesion to polystyrene substrata. *Applied and Environmental Microbiology*, **52**, 460–5.

McEldowney, S. & Fletcher, M. (1987). Adhesion of bacteria from mixed cell suspension to solid surfaces. *Archives of Microbiology*, **148**, 57–62.

McEldowney, S. & Fletcher, M. (1988a). Bacterial desorption from food container and food processing surfaces. *Microbial Ecology*, **15**, 229–37.

McEldowney, S. & Fletcher, M. (1988b). Effect of pH, temperature, and growth conditions on the adhesion of a gliding bacterium and three nongliding bacteria to polystyrene. *Microbial Ecology*, **16**, 183–95.

Meers, J. L. (1973). Growth of bacteria in mixed cultures. *CRC Critical Reviews in Microbiology* (January), 139–79.

Munson, B. R., Young, D. F. & Okiishi, T. H. (1990). Viscous flow in pipes. In *Fundamentals of fluid mechanics*, ed. B. R. Munson, pp. 465–559. New York: Wiley.

Pirbazari, M., Voice, T. C. & Weber, W. J. Jr (1990). Evaluation of biofilm development on various natural and synthetic media. *Hazardous Waste and Hazardous Materials*, **7**, 239–50.

Powell, M. S. & Slater, N. K. H. (1983). Deposition of bacterial cells from laminar flows onto solid surfaces. *Biotechnology and Bioengineering*, **25**, 891–900.

Rittmann, B. E. (1989). Detachment from biofilms. In *Structure and function of biofilms*, ed. W. G. Characklis & P. A. Wilderer, pp. 49–58. New York: Wiley.

Santos, R., Callow, M. E. & Bott, T. R. (1991). The structure of *Pseudomonas fluorescens* biofilms in contact with flowing systems. *Biofouling*, **4**, 319–36.

Satou, N., Satou, J., Shintani, H. & Okuda, K. (1987). Adherence of Streptococci to surface-modified glass. *Journal of General Microbiology*, **134**, 1299–1305.

Silvester, N. R. & Sleigh, M. A. (1985). The forces on microorganisms at surfaces in flowing waters. *Freshwater Biology*, **15**, 433–48.

Speitel, G. E. Jr & DiGiano, F. A. (1986). Biofilm shearing under dynamic conditions. *Journal of Environmental Engineering*, **113**, 464–75.

Stanley, P. M. (1983). Factors affecting the irreversible attachment of *Pseudomonas aeruginosa* to stainless steel. *Canadian Journal of Microbiology*, **29**, 1493–9.

Stenström, T. A. (1989). Bacterial hydrophobicity, an overall parameter for the measurment of adhesion potential to soil particles. *Applied and Environmental Microbiology*, **55**, 142–7.

Turakhia, M. H., Cooksey, K. E. & Characklis, W. G. (1983). Influence of a calcium-specific chelant on biofilm removal. *Applied and Environmental Microbiology*, **46**, 1236–8.

Trulear, M. G. & Characklis, W. G. (1982). Dynamics of biofilm processes. *Journal of the Water Pollution and Control Federation*, **54**, 1288–1301.

van Loosdrecht, M. C. M., Lyklema, J., Norde, W., Schraa, G. & Zehnder, A. J. B. (1987). Electrophoretic mobility and hydrophobicity as a measure to predict the initial steps of bacterial adhesion. *Applied and Environmental Microbiology*, **53**, 1898–1901.

Watkins, L. & Costerton, J. W. (1984). Growth and biocide resistance of bacterial biofilms in industrial systems. *Chemical Times and Trends* (October), 35–40.

Wrangstadh, M., Conway, P. L. & Kjelleberg, S. (1986). The production and release of an extracellular polysaccharide during starvation of a marine *Pseudomonas* sp. and the effect thereof on adhesion. *Archives of Microbiology*, **145**, 220–7.

Yu, C., Lee, A. M. & Roseman, S. (1987). The sugar-specific adhesion/deadhesion apparatus of the marine bacterium *Vibrio furnissii* is a sensorium that continuously monitors nutrient levels in the environment. *Biochemical and Biophysical Research Communications*, **149**, 86–92.

3

Cultivation and Study of Biofilm Communities

Douglas E. Caldwell

The need for laboratory studies of biofilm communities

If microbial ecology is to move forward, it must go beyond the reduction of the complexities of bacteria into merely isolated cell lines, enzymes, and genetic sequences. A century of pure culture studies has provided extremely detailed information on the biochemistry, physiology and genetics of bacteria. What remains to be determined is how they function as successful members of interacting communities in biofilms and how microbial communities function as components of the environment.

Filling this gap in knowledge involves more than the *in situ* enumeration of cells, molecules, and genetic sequences. It requires that microbial communities be considered as functional units of ecological activity. Individual microorganisms are often tightly coupled with other community members through a complex network of interactions. The genetic programming of each species may be considered to be a reproductive strategy formulated over 2.5 billion years of natural selection, and intricately intertwined with the survival of other organisms (Margulis 1981). Consequently, the most rigorous measure of understanding must involve not only the cultivation of isolated cell lines, but also the successful cultivation and characterization of dynamic microbial communities, complete with their predators and parasites.

End of the pure culture era

The primary axiom of bacteriology is that organisms must be isolated prior to their identification and study, and prior to the description of new species (Koch 1881, 1884). This axiom is so pervasive that it impacts protozoology, mycology and algology as well as bacteriology. In the words of Pringsheim,

Progress in microbiology, as a result of the adoption of pure culture methods, has been so striking that in the opinion of some biologists a culture not free of bacteria is not a real culture at all.

(Pringsheim 1946)

In the words of Oscar Brefeld,

Work with impure cultures yields nothing but nonsense.

(Brefeld 1881)

However, on-line image analysis, scanning confocal laser microscopy (SCLM), and fluorescent molecular probes provide a non-destructive means of analysis that can be applied to individual cells and their associated microenvironments without isolation (Caldwell *et al.* 1992a, b). Consequently, it is no longer a necessity to isolate and clone bacteria to obtain enough for biochemical, genetic, and other analyses. In many cases isolation and cloning are not desirable. Isolation destroys spatial relationships between organisms and their microenvironment. Cloning constrains laboratory studies to the descendants of native organisms and selects genetic variants. The advent of non-destructive methods offers an alternative to these approaches, expands the domain of laboratory studies from pure cultures to complex natural communities, and represents the end of an era lasting more than a century in which knowledge of bacteriology came almost exclusively from pure cultures and plate counts.

Extension of habitat range through communal associations

The range of conditions under which a species can survive in nature (its habitat range) is limited by genetic and physiological capabilities. Thus its reproductive success is often dependent upon the development of close associations with other species. These interdependencies are most important in bacteria, the smallest free living microorganisms, possessing approximately 0.1% of the DNA contained within the cells of vertebrate animals. The demands of bacteria for genetic information are perhaps even greater than those of vertebrates. However, due to their small size and limited genetic resources, they cannot afford the luxury of redundant or unnecessary genetic coding. Consequently, they frequently adapt to environmental stress by forming close associations with other species or by transferring genetic information (Sonea & Panisset 1983).

Two of the more common examples of bacterial associations are biodegradative consortia and interspecies hydrogen transfer. However, the most successful communal associations may be the lichens. They are consortia consisting of fungi, algae, and in many cases a cyanobacterium. The members of the consortium are so highly interdependent that each lichen has been given a unique genus and species designation. Lichen associations extend the habitat range of their individual members from near zero to a vast assortment of terrestrial environments in which none of the individual members could achieve reproductive success alone. Although it is a simple matter to isolate and cultivate the individual members of a lichen, it is difficult or impossible to cultivate the intact lichen association. Thus the isolation of a cell line is often only the first step toward understanding the complex web of ecological interactions essential for the success of each species in nature.

Problems arising from the application of Koch's postulates to bacterial communities

The germ theory of disease and Koch's postulates (Koch 1881, 1884) require that microorganisms must be isolated as pure cultures and kept in isolation while demonstrating cause and effect relationships. This works well in medical situations where a single cell may slip past the host defences and cause a disease. However, it does not apply to situations where a community of organisms is the causative agent and individual cell lines are ineffective or less effective. Examples include anaerobic digestor granules, biofouling, degradative consortia, microbially influenced corrosion, spoilage biofilms, mixed bacterial infections, and dental plaque. Consequently, it may seem difficult to isolate and describe the causative agent for a medical, industrial, agricultural or environmental process if the mechanism involves an interacting community rather than single pure cultures. In this situation it is the community which must ultimately be studied and understood under defined laboratory conditions (Caldwell 1993). Studies of pure cultures are still necessary, but insufficient to obtain an accurate and predictive understanding of the mechanism involved.

The process of obtaining a pure culture can result in the creation of artefacts. Unlike atoms, bacteria are not immutable and cannot be purified in the same sense that an element such as gold or silver might be purified through separation and concentration. The process of bacterial purification is more accurately referred to as cloning, and the product of cloning is an isolated cell line or clone as opposed to, in the words of Koch 'a perfect pure culture'[1] (Koch 1881). A minimum of 20 generations is normally required to produce a visible cell clone on agar (10^6 cells). If the clone is subcultured again, an additional 20 generations are required. Assuming a spontaneous mutation rate of approximately one in 10^6 cell divisions implies that there are as many mutants in a clone as there are genes in the organism even before the first subculture. It is also logical to assume that an even larger number of cells may be spontaneous genetic variants or have lost a plasmid. As bacteria are passed through numerous subcultures there is a strong selection pressure in favour of those cells that grow rapidly by either deleting or repressing adaptations essential for reproductive success *in situ*. Similarly, there is a strong selection pressure in favour of cells that are easily diluted. Isolation

1 Translation by Tom Brock. 1961. Milestones in Microbiology. Published by the American Society of Microbiology. 273 pp.

is achieved through a dilution process that involves serial streaks on agar, dilution and plating, or dilution tubes alone. Consequently, non-adherent cells are selected and may not adequately represent adherent biofilm populations which are not readily amenable to the dilution process.

Communities, consortia, and microecosystems

Communities are groups of species populations occupying the same physical habitat. Although biofilm communities are among the most amenable to laboratory study, there are other examples. These include neuston, bioaggregates, anaerobic consortia, degradative consortia, lichens, aufwuchs, sulphureta, marine snow, soil aggregates, and numerous other discrete communities. They often have well defined boundaries, internal and external nutrient exchange, multiple trophic levels, spatial organization, internal microenvironments, and a dynamic quasi-steady state condition. They can thus be considered experimentally and functionally as microecosystems that are in many ways analogous to macroscopic ecosystems. However, the spatial scale is much smaller, a square kilometre of forest being the equivalent of a bacterial ecosystem occupying a square millimetre (Caldwell *et al.* 1992a). When two or more organisms enter into an association involving physical contact, the association is normally referred to as a consortium (Hirsch 1980, 1984) as opposed to a community or a microecosystem.

Biofilm communities have both a structure and an architecture. The population structure of a community refers to its species composition. This is the percentage of each species population within the community. For example, a soil biofilm might be comprised of 20% *Bacillus subtilus*, 13% *B. megaterium*, etc. The term architecture has been used to refer to the spatial relationships between the organisms, exopolymers, and microenvironments within biofilms (Lawrence *et al.* 1991). Thus community structure refers to species composition and biofilm architecture refers to spatial heterogeneity. Both the community structure and the physicochemical architecture must be determined regardless of whether the functional unit being considered is a community, consortium or microecosystem.

Community level methods of cultivation

The objective of community cultivation is the establishment, subculture and study of a well defined and reproducible climax community through colonization and succession (Caldwell *et al.* 1992a, b). In the case of biofilm communities, the surface to be colonized is irrigated with a solution containing native populations (Caldwell & Lawrence 1990). Typical solutions include surface water, groundwater, saliva, cooling water, cutting fluid, meat processing dips, urine, etc. When subculturing, the effluent from the irrigation of cultured biofilms is used as the inoculum. The cells within these solutions have undergone emigration from a pre-existing biofilm and are ready to begin the immigration (colonization) process within the new surface microenvironment. The colonizing organisms are not positioned on the new surface using an inoculating loop, as in the case of spread plates or streak plates. They position themselves, forming microcolonies and aggregates which eventually converge to form a confluent bacterial biofilm (Lawrence *et al.* 1987, 1992; Korber *et al.* 1989, 1991; see also Korber *et al.* Chapter 1). During surface colonization bacteria continually reposition themselves to maintain an optimal growth rate while simultaneously remaining attached. There are several distinct bacterial positioning manoeuvres that occur during this process. For example, some bacteria form microcolonies by congregating (Lawrence *et al.* 1992) while others slowly glide away from one another by 10 or 20 μm following each cell division (Lawrence & Caldwell 1987). To a casual observer the latter appear to be cells that have adsorbed and not grown. In some cases this colonization process involves the attachment of heterogeneous cell aggregates as well as planktonic unicells. This is particularly true of strict anaerobes which are protected from oxygen by forming aggregates.

Well defined environments yield well defined climax communities

Defined steady state environments produce defined steady state communities through natural selection (Fig. 3.1). Natural selection refers to the reproductive success of environmentally adapted species populations or communities when in competition with a wide variety of other

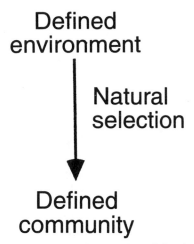

Defined environment

Natural selection

Defined community

Fig. 3.1. Defined environments produce defined communities through natural selection. This normally requires that the environment be defined in terms of both the concentration and the flux of substrates and other solutes. The formation of a biofilm community within a defined environment usually involves primary colonization, a gradual succession of populations, and the subsequent formation of a dynamic climax community in a quasi-steady state condition.

organisms. It involves reproductive strategies that go beyond rapid growth rates and which include mechanisms that increase the probability of survival for a species, and occasionally its symbionts or its entire community. One example is the aggregation of strict anaerobes from the hypolimnion of eutrophic lakes when exposed to oxygen (Caldwell & Tiedje 1975; Caldwell *et al.* 1975; Caldwell 1977). This adaptation protects cells within the centre of the aggregate from oxidation. Protection of interior cells is at the expense of those at the exterior, which serve as protective reductants. All cells trapped within the aggregate are protected regardless of whether they are capable of oxygen dependent aggregation or of whether they are in the same species. Aggregation mechanisms are also involved in the formation of plaque (Kolenbrander & London 1992) and many other biofilm communities. Synthesis of the required chemoreceptors may decrease the maximum specific growth rate of a species but it improves the probability of survival. Adaptations such as these optimize both reproductive success and the effective utilization of environmental resources.

If an environment is well defined and natural selection occurs, a reproducible climax community will be established through an orderly process of microbial succession. There are several microbiological culture systems that can be used to define adequately the environments used to support reproducible community cultures. These include continuous culture, continuous flow slide culture, the microstat, and colonization tracks. Each of these culture devices can be used to control not only the concentration of solute molecules but also their flux. This is essential to sustain and subculture dynamic microbial communities in a quasi-steady state. In some cases the development of a specific community may require periodic drying, transitions between static and dynamic (steady state) environmental conditions, etc. In these situations the cultivation system must be modified to provide the conditions required.

The chemostat

The chemostat provides a steady state environment in which both the flux and concentration of a limiting nutrient eventually reach a steady state (Monod 1949; Herbert 1956; Jannasch 1974; Veldkamp 1977) and in which the dilution rate (flow rate/volume) sets and is equal to the specific growth rate (the instantaneous fractional rate of increase in population size). Although the flux of substrate in the vessel can be controlled, the concentration of the limiting nutrient cannot. Consequently, it must be determined empirically once steady state has been obtained. The time required is normally at least three times the turnover time (three times the reciprocal of either the dilution rate or the specific growth rate). An axiom of chemostat kinetics is that cells do not adhere to the walls of the culture vessel. This limits their utility in studies of adherent biofilm communities unless the kinetics of surface attachment and growth are taken into consideration (Caldwell *et al.* 1981, 1983; Kieft & Caldwell 1983, 1984; Malone & Caldwell 1983).

Continuous-flow slide cultures

Continuous flow slide cultures (Fig. 3.2) contain microorganisms that have either colonized or adsorbed to the walls of a capillary sized flow cell (Caldwell & Lawrence 1988). An inoculated flow cell can be irrigated with a cell suspension or a sterile solution. When irrigating, both the flux

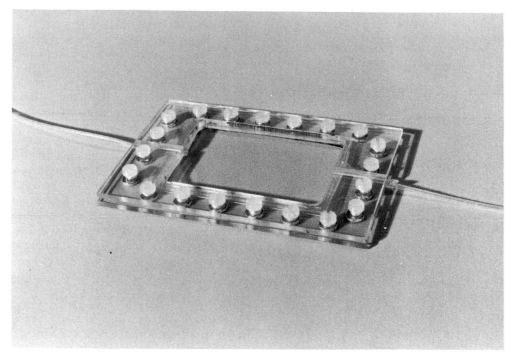

Fig. 3.2. A continuous flow slide culture used for confocal laser microscopy. Continuous flow slide cultures provide a constant flux and constant concentration of substrates and other solutes for studying the development and maturation of bacterial biofilms. The irrigation medium can be changed almost instantaneously, allowing studies of the effects of periodic environmental changes on biofilm development. The interior dimensions are 35 × 45 mm with a thickness of 1.0 mm. The total thickness of a flow cell should not exceed 1.4 mm, the maximum distance between the objective and the condensor for Kohler illumination using a 1.4 n.a., × 100 objective lens.

and concentration of solutes are well defined and can be changed almost instantaneously. The laminar flow velocity (cm s^{-1}) for the bulk phase is calculated by dividing the bulk flow rate (cm^3 s^{-1}) by the cross-sectional area (cm^2) of the flow cell. The flux of substrate (g cm^{-2} s^{-1}) and other solutes is then calculated by multiplying the laminar flow velocity by the concentration (g cm^{-3}). However, the laminar flow velocity within the microenvironment of attached cells is normally at least an order of magnitude less than that in the bulk phase of the flow cell (Lawrence *et al.* 1987). Consequently, the laminar flow velocity within the microenvironment must be determined empirically using 0.2 μm latex spheres and this value used in calculations. The continuous flow slide culture is one of the simplest and most effective methods of studying biofilm communities non-destructively during the time course of colonization and development.

Microstats

The microstat is sometimes referred to as a solid state chemostat (Caldwell 1993). It provides a constant steady state microenvironment for cells adsorbed to a solid–liquid interface. It also provides a two dimensional steady state concentration gradient across the surface being colonized. This permits study of relationships between environmental variation and the development of biofilm communities. The gradients are produced by diffusion through a gel located beneath the flow cell. Originally, only biofilms developing on the surface of the gel could be studied and only exponential gradients could be generated. More recently, linear gradients have been developed. While earlier computer simulations of the microstat predicted only concentrations within the gel, current simulations predict concentrations and fluxes within the gel and within the irrigation solution above the gel (Wolfaardt *et al.* 1993). Consequently, the development of communities on the glass surface above the flow cell can also be studied. These improvements make the microstat more amenable to analysis using scanning confocal laser microscopy (SCLM) by reducing the distance between the biofilm and the objective lens, thus allowing continuous *in vivo* microscopy at high resolution without the need to dismantle the chamber periodically. They also make it possible to expose a mature biofilm, grown without test gradients, to a two dimensional gradient periodically at various points in its development.

The microstat contains an agarose or acrylamide gel 25×25 mm in size and 3 mm thick (Fig. 3.3a) with chamber components, diffusion and flow paths through each component (Fig. 3.3b). There are two reservoirs used to supply test factors to two edges of the gel. These supply the compounds for which two dimensional steady state diffusion gradients are to be produced. The upper surface of the gel is irrigated with a cell suspension containing the native community. The microstat can be modified for laser microscopy without disassembling the chamber (Fig. 3.4). The methods used to analyse the growth of biofilm communities on two dimensional microstat gradients are discussed later in this chapter.

An alternative to traditional microstat gradients is the gel free microstat. This consists of a continuous flow slide culture with two 2 mm dialysis tubes. Each tube carries a flowing solution containing one of the two solutes used to generate the two dimensional gradient. One tube is fitted on the left side of the flow channel and the other is fitted on the right. Each of these tubes generates a diffusion gradient perpendicular to the direction of flow, originating from the sides of the flow channel and diffusing toward the centre. As the irrigation solution moves down-

Fig. 3.3. (a) A photograph of the microstat diffusion chamber used to produce two dimensional steady state diffusion gradients. It provides a constant steady state microenvironment for cells adsorbed to a solid–liquid interface.

(b)

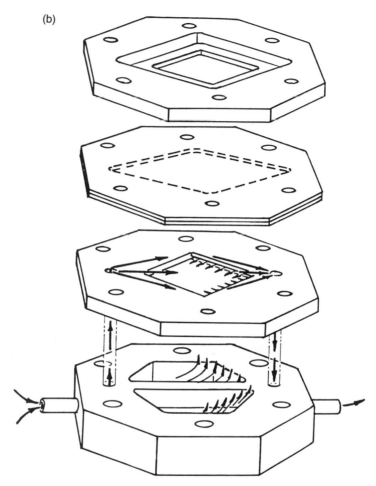

Fig. 3.3 *(cont.)*   (b) A diagram of the microstat chamber in (a) showing the component plates and the direction of flow or diffusion through each plate. The flow of the irrigation solution must always be against the direction of diffusion. (Diagram courtesy of E. W. Caldwell.)

stream through the flow cell, the diffusion fronts originating from the dialysis tubing move past one another, perpendicular to the flow path. Unlike the gel based microstats, the two dimensional gradients generated are rectangular rather than square. The advantage of the gel free microstat is that it eliminates problems with gel stability, allows exposure of precolonized surfaces, allows construction of thin microstats amenable to Kohler illumination at high magnification, greatly reduces the time to reach steady state by reducing the distance across which diffusion must occur, and simplifies the kinetics used to predict concentrations and fluxes at various points within the two dimensional gradient.

Colonization tracks (spoilage biofilms)

Biofilms develop most rapidly in flow situations as opposed to stagnant, static situations, probably because a flowing solution continuously supplies substrate and removes wastes. Flow also increases the effective concentration of substrates by providing high fluxes at low concentrations. This allows biofilm populations to obtain substrate from solution through sorption to exopolymers and subsequent degradation (Wolfaardt *et al.* 1994b).

There are also conditions under which biofilms develop in the absence of flow. One example is the development of spoilage biofilms on food surfaces. These can be studied by using colonization tracks. Using meat as an example, a large cube is surface sterilized and cut into strips using a sterile spatula. The strip is laid into a sterile Petri dish and inoculated by placing a spoiled cube of meat at one end. The spoilage biofilm then develops by colonizing from one end of the strip to the other. It is then subcultured by removing a cube from the distal end and placing it on a new strip or track. Through minor modification the colonization strip (or track) can be adapted to study other solid substrata upon which degradative biofilms develop. This method provides a colonization front which continuously moves along the colonization track. All stages of community development and succession can be seen at various points along this front and the community is continuously sustained in a dynamic steady state. However, the mechanism is

very different from that used in a chemostat, flow cell, or microstat.

Steady state community cultures versus batch enrichment cultures

The objective of community cultivation is to study communities as functional units of ecological activity. Community cultures are sustained in a quasi-steady state condition to understand better the dynamics of interrelationships among community members and to observe their collective response to spatial and temporal environmental variations. The objective of batch enrichment culture is to provide highly selective environmental conditions, making one species numerically dominant over all others and thus enabling isolation of highly adapted cell lines by subsequent dilution and plating (Beijerinck

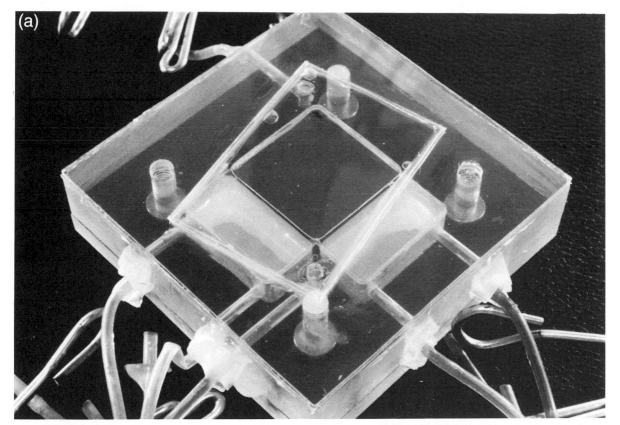

Fig. 3.4. (a) A modified microstat intended for scanning confocal laser microscopy, without the need to disassemble the culture chamber. Instead of a series of thick plates and gaskets, as shown in Fig. 3.3(a), the coverslip is held in place using a silicone adhesive to minimize the distance between the agar and the objective lens. The development of biofilm communities can be studied either on the surface of the gel or on the surface of the coverglass.

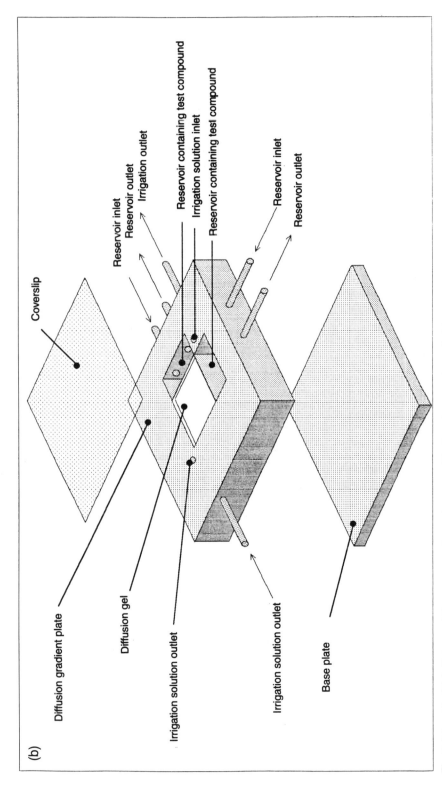

(b)

Coverslip

Reservoir inlet
Reservoir outlet
Irrigation outlet

Reservoir containing test compound
Irrigation solution inlet
Reservoir containing test compound

Reservoir inlet
Reservoir outlet

Diffusion gradient plate

Diffusion gel

Irrigation solution outlet

Irrigation solution outlet

Base plate

Fig. 3.4 *(cont.)* (b) A diagram of the components used to assemble the microstat shown in (a).

1901). The environment provided in batch enrichments is highly selective but continually changes during cultivation. Chemostat enrichments (Jannasch 1967; Schlegel & Jannasch 1967) provide better environmental control.

The philosophy of enrichment culture is highlighted by the following from Beijerinck (1901):

Because of our very imperfect understanding of the environmental requirements of the majority of microbes, it is impossible in most enrichment culture experiments to go further than to bring about a relative increase in the numbers of a desired form without leading to a complete disappearance of the other species present. Often this partial enrichment only occurs at a particular stage of the experiment, whereas earlier and later other forms predominate. Because of this, enrichment culture experiments can be called perfect or imperfect. In a perfect experiment a single species is isolated in all its varieties.

The assumption of enrichment culture is thus that the product of environmental selection in a precisely controlled environment is a single species population (Beijerinck 1901). It is based upon the tacit assumption that the basis of selection for bacterial enrichment is the maximum growth rate of the organism. This is often the situation during r selection (reproductive success is limited primarily by an organism's intrinsic rate of increase), which occurs during the exponential phase of batch culture. Under this condition the cell line(s) that grows most rapidly is selected and other factors can normally be ignored.

The rationale of community cultivation is that the ultimate result of selection is a steady state community, consisting of several interacting species populations rather than an individual species. These cultured communities consist of interdependent populations which utilize environmental resources more effectively than any individual population could in isolation. This is most likely during K selection (reproductive success is limited primarily by environmental resources) and is the normal situation in natural environments and in community cultures. However, periods of r selection may occur temporarily, as in the case of surface colonization prior to the formation of a mature biofilm (climax community).

Community culture versus mixed culture

Mixed cultures are often used to simulate interactions in natural microbial communities. However, the process of isolation results in the creation of genetic variants that may no longer be capable of interacting with other community members as their ancestors did. Consequently, it is preferable to cultivate and study natural communities prior to the isolation of pure cultures if possible. It is also informative to combine isolated cell lines and produce a mixed culture for comparison with the original community. In the case of degradative biofilms, mixed cultures seldom regain the degradative capacity of the original community or consortium (Wolfaardt *et al.* 1994a).

Community level methods of analysis

Scanning confocal laser microscopy

When laser microscopy is combined with fluorescent molecular probes, the identity, chemistry, metabolism, and gene expression of organisms within biofilms can be transformed into optical signals and represented as pseudocolour maps. These maps overlay the image of cells within biofilm communities to allow non-destructive real time analysis. Such methods can be used to study the organisms themselves, their chemistry and the chemistry of their ambient microenvironment, as well as their identity, metabolic rate and genetic control. Thus it is no longer necessary to disrupt biofilm communities to study the molecular, behavioural, metabolic or genetic aspects of their ecology. These developments represent the beginning of a new era of non-destructive analysis in microbial ecology.

The scanning confocal laser microscope (SCLM) transforms the optical microscope into an analytical spectrofluorimeter (Caldwell *et al.* 1992b). Each point of light (volume element or voxel) within an SCLM image represents the light emission from a defined volume element in the specimen. Scattered light from above and below this volume element is removed by the confocality of the imaging system and by the use of a scanning laser beam to image each element of the image separately. Consequently, each voxel is functionally equivalent to an optodigital spectrofluorimeter cell. The information from a series of optodigital thin sections is used to create a three dimensional digital reconstruction of the cells and their associated chemistry (Caldwell *et al.* 1993). This can be used to analyse spatial relationships between organisms and their microenvironments by rotating the three dimensional digital model.

Specific proteins are normally imaged using fluorescein or rhodamine conjugated antibodies. This approach can also be used to identify specific microbial populations (Schmidt 1972) and was first applied to microbial ecology by Hobson & Mann (1957). Polysaccharides are imaged using fluorescein isothiocyanate (FITC) and other conjugated lectins (Caldwell et al. 1992a, b). Nucleic acids (particularly specific genetic sequences) were imaged within in situ microbial communities by DeLong et al. using 16S rRNA probes (Delong et al. 1990; Ward et al. 1992). Growth rates were visualized by Caldwell & Germida using difference imaging (Caldwell & Germida 1985). Cell respiration was visualized directly using a redox probe (resorufin) to image intracellular oxygen depletion, thus effectively using individual cells as respirometers (Caldwell & Lawrence 1990). Many of these and other fluorescent molecular probes are variations of fluor-

Resorufin ß-D-galactopyranoside

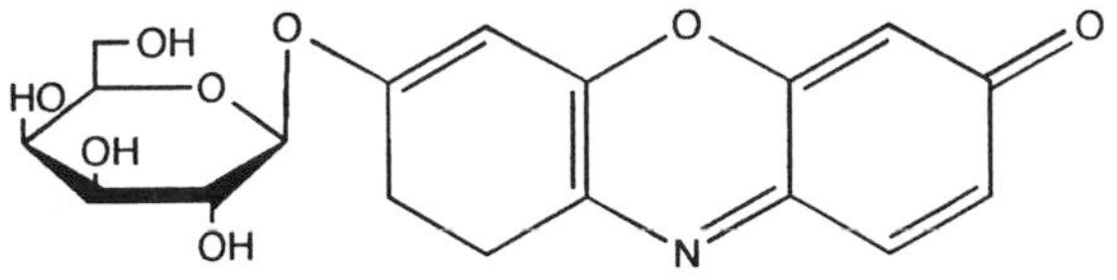

Resorufin acetate

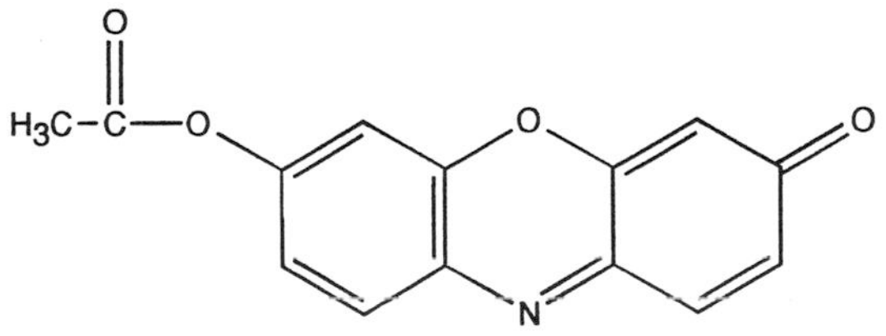

Fig. 3.6. The structures of resorufin ß-D-galactopyranoside and resorufin acetate. These are examples of fluorogenic substrates produced by derivitizing resorufin. They are used to assay ß-galactosidase and esterase activity respectively.

escein, resorufin, or other organic compounds containing aromatic rings (Fig. 3.5). For example, adding a substrate such as galactose or acetate to resorufin (Fig. 3.6) creates a relatively non-fluorescent structure that becomes fluorescent when the substrate is cleaved from the ring. These fluorogenic substrates can be used to measure ß-galactosidase and esterase activity. The fluorescence efficiency associated with the π electrons of these and other aromatic rings is frequently very high and this makes them ideal candidates for use as fluorescent molecular probes.

Genetic expression and regulation is imaged by using organisms containing lacZ fusions. For example, to visualize the expression of specific starvation genes (coding for proteins produced in response to starvation conditions) during the formation of communities along a microstat gradient of substrate concentration, a lacZ fusion in the regulatory region of a specific starvation gene can be used (Matin 1992). The organism containing the fusion is added to the irrigation solution of a flow cell or microstat and colonizes along with members of a natural community. When it attempts to express a specific starvation gene it instead produces ß-galactosidase. This enzyme is assayed using a fluorogenic substrate that translates gene expression into an optical signal. The fluorogenic substrates often used to assay ß-galactosidase activity, X-gal (5-bromo-4-

Fluorescein

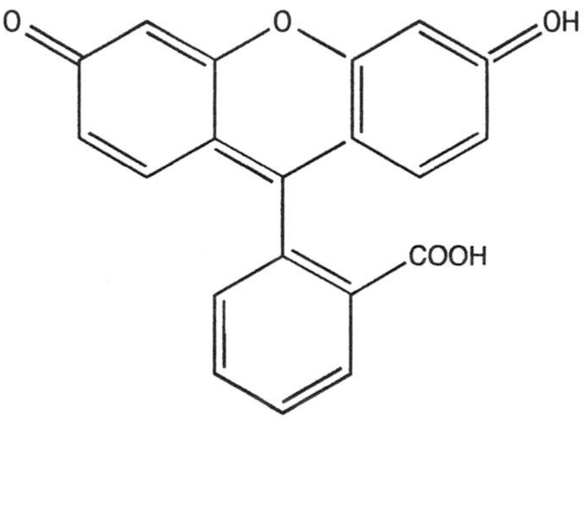

Resorufin

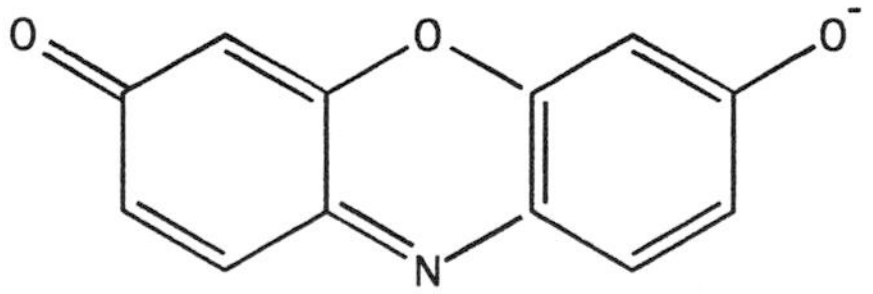

Fig. 3.5. The structure of fluorescein and resorufin, two of the primary reference standards from which other fluorescent molecular probes are synthesized. The π electrons associated with their polycyclic aromatic rings is responsible for their high quantum yield (fluorescence efficiency).

chloro-3-indolyl ß-D-galactopyranoside) and MUG (4-methylumbelliferyl ß-D-galactopyranoside) are not optimal for the 488 and 514 excitation wavelength of most laser microscopes. Consequently, other substrates can be used that liberate either fluorescein (fluorescein mono-ß-D-galatopyranoside) or resorufin (resorufin ß-D-galactopyranoside), which have excitation spectra that compliment the emission bands of the argon laser commonly used in flow cytometery and laser microscopy. The design and selection of these and other probes is discussed by Haugland (Haugland 1992).

Community level auxanography

Auxanography was originally described by Beijerinck (1889) and used to create auxanograms showing the growth response of pure cultures to gradients of nutrient in gelatin.

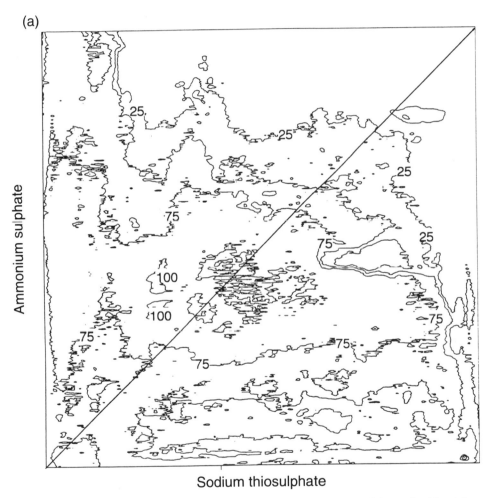

(a)

Ammonium sulphate

Sodium thiosulphate

Fig. 3.7. (a) Auxanograph showing the ecotone (diagonal line) between an ammonia oxidizing and a thiosulphate oxidizing biofilm community obtained from the South Saskatchewan River Basin. The biofilm developed on the surface of a two dimensional gradient of ammonium (supplied from left edge of gel) and thiosulphate (supplied from bottom edge of gel) irrigated with river water. The contours of growth shown are in units of grey level ranging from 0 to 256 (256 representing maximum growth). The biofilm in the upper left corner of the gel is supported entirely by chemoautotrophic ammonia oxidizers. The biofilm in the lower right is supported entirely by chemoautotrophic sulphur oxidizers. The region of the biofilm between these two extremes corresponds to the transition or ecotone between ammonia and sulphur-oxidizing communities. The resolution of this auxanograph is 0.05 mm (50 μm per pixel). The size of the gel shown was 25 × 25 mm.

(b)

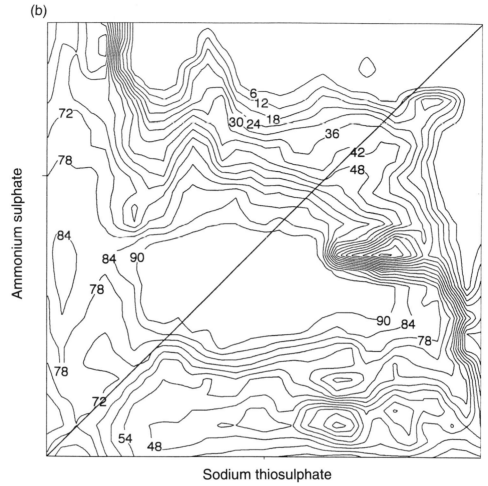

Sodium thiosulphate

Fig. 3.7 *(cont.)*   (b) Auxanograph of the same biofilm shown in (a) at a resolution of 0.8 mm (806 μm per pixel). The resolution was reduced by reducing the image size of an Adobe Photoshop™ (Aldus Corp., Seattle, Washington) image and then enlarging again to its original size for analysis using Spyglass™ (Spyglass Inc., Champaign, Illinois). Note that the features of the biofilm tend to change depending upon the scale used for image analysis. Consequently it is important to perform these analyses as well as to determine the statistically representative area (Korber *et al.* 1992) when deciding upon the spatial scale used for more detailed study or sampling of biofilm communities.

Beijerinck envisioned other applications as he explains with the following remarks:

I would like to point out that the principle outlined above can be applied to other problems. For instance, if an organism carries out a certain function that is visible directly or can be made visible by chemical means, such as excreting an acid or an enzyme, forming a pigment or emitting light, actions which are not necessarily linked to growth, but which are carried out solely under the influence of definite substances, such processes may be studied by the auxanographic procedure.

The auxanographic method has been modified by substituting two dimensional steady state diffusion gradients for the one dimensional non-steady state gradients devised by Beijerinck (Caldwell & Hirsch 1973; Caldwell *et al.* 1973) for use in studying the response of microbial communities as opposed to pure cultures (Caldwell *et al.* 1975, 1992a). Using this approach, biofilm communities form on the surface of two dimensional steady state diffusion gradients. This is visualized as a pseudocolour or contour map representing chemistry, gene expression, growth, metabolism, etc., as determined using laser microscopy. By using a computer controlled stage, the surface of these gradients can be repeatedly scanned along a uniform sampling grid. This reveals the time course of community development as a function of controlled environmental gradients. An auxanograph of biofilm development on the surface of a two dimensional gradient of thiosulphate versus ammonia is shown in Fig. 3.7.

The advantage of studying the distribution of microbial populations and communities along environmental gradients has been exploited in the laboratory (Caldwell & Hirsch 1973; Ho & McCurdy 1979; Shimkets *et al.* 1979; Wimpenny 1981; Wimpenny *et al.* 1988) and in the field (Caldwell *et al.* 1973; Caldwell & Tiedje 1975; Ramaley & Bitinger 1975; Jorgensen 1979). Similar gradient studies have been conducted by plant and animal ecologists (Walker 1970; Terborgh 1971) working primarily in the field. This is undoubtedly one of the most effective approaches to understanding the role of environmental variations in the dynamics of community interactions. It should also be useful for studying relationships between biological adaptation and reproductive success.

Until the development of digital imaging and scientific visualization software, there was no way of efficiently collecting, processing or understanding all of the information that an auxanographic method is capable of generating. The application of the auxanographic approach using these digital technologies should thus be one of the more promising new approaches to biofilm ecology.

Community hypotheses

Laser microscopy, on-line digital image analysis, continuous flow slide culture, and microstat culture make it possible to avoid disrupting microbial communities within biofilms when studying the molecular and behavioural aspects of their ecology. This analytical capability makes it feasible to begin systematic laboratory studies of microbial communities in biofilms as units of ecological activity. The concept is to define microbial communities by defining the environment and allowing natural selection to act. By controlling both the concentration and flux of molecules within each microenvironment,

defined climax communities are formed and can be sustained in a dynamic state (quasi-steady state).

As more community level studies of biofilms begin, several obvious questions and hypotheses will need to be addressed. How does the population structure of cultured biofilm communities compare with *in situ* communities? Are the principles learned from cultivated biofilms applicable with *in situ* biofilms although the two may not be identical? What are the successional processes required for the formation of specific biofilms? Is there only one climax community that can be established in a specific physical environment or are there several potentially stable sets of populations? Are climax communities stable given a stable environment or are there periodic oscillations in otherwise stable communities? Are spatial environmental pathways or periodic temporal variations required for the formation of some biofilms? Is the habitat range of isolated cell lines significantly increased through the formation of biofilms? All of these questions are not only intriguing, but of critical importance in developing a systematic understanding of microbial community interactions in medicine, industry, agriculture, and the environment.

References

Beijerinck, M. W. (1889). L'auxanographie, ou la methode de l'hydrodiffusion dans la gelatine appliquée aux recherches microbiologiques. *Archives Néerlandaises des Sciences Exactes et Naturelles, Haarlem, Bol.*, **23**, 367–72.

Beijerinck, M. W. (1901). Enrichment culture studies with urea bacteria. *Centralblatt für Bakteriologie*, Part II, **7**, 33–61.

Brefeld, O. (1881). *Botanische Untersuchungen über Schimmelpilze: Culturemethoden*. Leipzig.

Caldwell, D. E. (1977). The planktonic microflora of lakes. In *Critical Reviews in Microbiology*, **5**, 306–70. Boca Raton, Fla: CRC Press.

Caldwell, D. E. (1993). The microstat: steady state microenvironments for subculture of steady state consortia, communities, and microecosystems. In *Trends in Microbial Ecology*, ed. R. Guerrero & C. Pedros-Alio, pp. 123–28. Barcelona: Spanish Society for Microbiology.

Caldwell, D. E., Brannan, D. E., Morris, M. E. & Betlach, M. R. (1981). Quantitation of microbial growth on surfaces. *Microbial Ecology*, **7**, 1–11.

Caldwell, D. E., Caldwell, S. J. & Tiedje, J. M.

(1975). An ecological study of sulphur-oxidizing bacteria from the littoral zone of a Michigan lake and a sulphur spring in Florida. *Plant and Soil*, **43**, 101–14.

Caldwell, D. E. & Germida, J. J. (1985). Evaluation of difference imagery for visualizing and quantitating microbial growth. *Canadian Journal of Microbiology*, **31**, 35–44.

Caldwell, D. E. & Hirsch, P. (1973). Growth of microorganisms in two-dimensional steady state diffusion gradients. *Canadian Journal of Microbiology*, **19**, 53–8.

Caldwell, D. E., Korber, D. R. & Lawrence, J. R. (1992a). Confocal laser microscopy and digital image analysis in microbial ecology. In *Advances in Microbial Ecology*, **12**, 1–67.

Caldwell, D. E., Korber, D. R. & Lawrence, J. R. (1992b). Imaging of bacterial cells by fluorescence exclusion using scanning confocal laser microscopy. *Journal of Microbiological Methods*, **15**, 249–61.

Caldwell, D. E., Korber, D. R. & Lawrence, J. R. (1993). Analysis of biofilm formation using 2-D versus 3-D digital imaging. *Journal of Applied Bacteriology Symposium Supplement*, **74**, 52S–66S.

Caldwell, D. E., Lai, S. H. & Tiedje, J. M. (1973). A two-dimensional steady state diffusion gradient for ecological studies. In *Modern Methods in the Study of Microbial Ecology*, 17, ed. T. Rosswell, pp. 151–8. Stockholm: Swedish National Research Council.

Caldwell, D. E. & Lawrence, J. R. (1988). Study of attached cells in continuous-flow slide culture. In *A handbook of model systems for microbial ecosystem research*, ed. J. W. T. Wimpenny, pp. 117–38. Boca Raton, Fla: CRC Press.

Caldwell, D. E. & Lawrence, J. R. (1990). Microbial growth and behavior within surface microenvironments. In *Proceedings of the Fifth International Symposium on Microbial Ecology*, ed. T. Hattori, Y. Ishida, Y. Maruyama, R. Y. Morita & A. Uchida, pp. 140–5. Tokyo: Japan Scientific Societies Press.

Caldwell, D. E., Malone, J. A. & Kieft, T. L. (1983). Derivation of a growth rate equation describing microbial surface colonization. *Microbial Ecology*, **9**, 1–6.

Caldwell, D. E. & Tiedje, J. M. (1975). The structure of anaerobic bacterial communities in the hypolimnia of several lakes in Michigan. *Canadian Journal of Microbiology*, **21**, 377–85.

DeLong, E. F., Schmidt, T. M., & Pace, N. R. (1990). Analysis of single cells and oligotrophic picoplankton populations using 16S rRNA sequences. In *Proceedings of the Fifth International Symposium on Microbial Ecology*, ed. T. Hattori, Y. Ishida, Y. Maruyama, R. Y. Morita & A. Uchida, pp. 697–701. Tokyo: Japan Scientific Societies Press.

Haugland, R. P. (1992). *Molecular probes: handbook of fluorescent probes and research chemicals*. Eugene, Oregon: Molecular Probes Inc.

Herbert, D. (1956). The continuous culture of bacteria: a theoretical and experimental study. *Journal of General Microbiology*, **14**, 601–771.

Hirsch, P. (1980). Some thoughts on and examples of microbial interactions in the natural environment. In *Aquatic microbial ecology*, ed. R. R. Colwell & A. J. Foster, pp. 36–54. Maryland: University of Maryland.

Hirsch, P. (1984). Microcolony formation and consortia. In *Microbial adhesion and aggregation*, ed. K. C. Marshall, pp. 373–93. New York: Springer-Verlag.

Ho, J. & McCurdy, H. D. (1979). Demonstration of positive chemotaxis to cyclic GMP and 5'-AMP in *Myxococcus xanthus* by means of a simple apparatus for generating practically stable concentration gradients. *Canadian Journal of Microbiology*, **25**, 1214–18.

Hobson, P. N. & Mann, S. O. (1957). Some studies on the identification of rumen bacteria with fluorescent antibody. *Journal of General Microbiology*, **16**, 463–71.

Jannasch, H. W. (1967). Enrichment of aquatic bacteria in continuous culture. *Archiv für Mikrobiologie*, **59**, 165–73.

Jannasch, H. W.(1974). Steady state and the chemostat in ecology. *Limnology and Oceanography*, **5**, 716–20.

Jørgensen, B. B. (1979). Diurnal cycle of oxygen and sulfide microgradients and microbial photosynthesis in a cyanobacterial mat sediment. *Applied and Environmental Microbiology*, **38**, 46–58.

Kieft, T. L. & Caldwell, D. E. (1983). A computer simulation of surface microcolony formation during microbial colonization. *Microbial Ecology*, **9**, 7–13.

Kieft, T. L. & Caldwell, D. E. (1984). Chemostat and *in situ* colonization kinetics of *Thermothrix thiopara* on calcite and pyrite surfaces. *Geomicrobiology Journal*, **3**, 217–29.

Koch, R. (1881). Methods for the study of pathogenic organisms. *Mittheilungen aus dem Kaiserlichen Gesundheitsamte*, **1**, 1-48.

Koch, R. (1884). The etiology of tuberculosis. *Mittheilungen aus dem Kaiserlichen Gesundheitsamte*, **2**, 1–88.

Kolenbrander, P. E. & London, J. (1992). Ecological significance of coaggregation among oral bacteria. In *Advances in Microbial Ecology*, **12**, 183–217.

Korber, D. R., Lawrence, J. R., Hendry, M. J. & Caldwell, D. E. (1992). Programs for determining statistically representative areas of microbial biofilms. *Binary*, **4**, 204–10.

Korber, D. R., Lawrence, J. R., Sutton, B. & Caldwell, D. E. (1989). Effects of laminar flow velocity on the kinetics of surface recolonization by Mot$^+$ and Mot$^-$ *Pseudomonas fluorescens*. *Microbial Ecology*, **18**, 1–19.

Korber, D. R., Lawrence, J. R., Zhang, Lu & Caldwell, D. E. (1991) Effect of gravity on bacterial deposition and orientation in laminar flow environments. *Biofouling*, **2**, 335–50.

Lawrence, J. R. & Caldwell, D. E. (1987). Behavior of bacterial stream populations within the hydrodynamic boundary layers of surface microenvironments. *Microbial Ecology*, **14**, 15–27.

Lawrence, J. R., Delaquis, P. J., Korber, D. R. & Caldwell, D. E. (1987). Behavior of *Pseudomonas fluorescens* within the hydrodynamic boundary layers of surface microenvironments. *Microbial Ecology*, **14**, 1–14.

Lawrence, J. R., Korber, D. R., Hoyle, B. D., Costerton, J. W., & Caldwell, D. E. (1991). Optical sectioning of microbial biofilms. *Journal of Bacteriology*, **173**, 6558–67.

Lawrence, J. R., Korber, D. R. & Caldwell, D. E. (1992). Behavioral analysis of *Vibrio parahaemolyticus* variants in high and low viscosity microenvironments using digital image processing. *Journal of Bacteriology*, **174**, 5732–39.

Malone, J. A. & Caldwell, D. E. (1983). Evaluation of surface colonization kinetics in continuous culture. *Microbial Ecology*, **9**, 299–305.

Margulis, L. (1981). *Symbiosis in cell evolution: life and its environment on the early earth*. San Francisco: W. H. Freeman.

Matin, A. (1992). Physiology, molecular biology and applications of the bacterial starvation response. *Journal of Applied Bacteriology* (Symposium Supplement), **74**, 49S–57S.

Monod, J. (1949). The growth of bacterial cultures. *Annual Reviews of Microbiology*, **3**, 371–94.

Pringsheim, E. G. (1946). The biphasic or soil–water culture method for growing algae and flagellata. *Journal of Ecology*, **33**, 193–204.

Ramaley, R. & Bitinger, K. (1975). Types and distribution of obligate thermophilic bacteria in man-made and natural thermal gradients. *Applied Microbiology*, **30**, 152–5.

Schlegel, H. G. & Jannasch, H. W. (1967). Enrichment cultures. *Annual Reviews of Microbiology*, **21**, 49–66.

Schmidt, E. L. (1972). Fluorescent antibody techniques for the study of microbial ecology. In *Modern Methods in the Study of Microbial Ecology*, **17**, ed. T. Rosswall, pp. 67–76. Stockholm: Swedish Natural Science Research Council.

Shimkets, L. J., Dworkin, M. & Keller, K. H. (1979). A method for establishing stable concentration gradients in agar suitable for studying chemotaxis on a solid surface. *Canadian Journal of Microbiology*, **25**, 1460–7.

Sonea, S. & Panisset, M. (1983). *A new bacteriology*. Boston, Mass: Jones and Bartlett Inc.

Terborgh, J. (1971). Distribution on environmental gradients: theory and a preliminary interpretation of distributional patterns in the Avifauna of the Cordillera Vilcabamba. *Ecology*, **52**, 23–40.

Veldkamp, H. (1977). Ecological studies with the chemostat. *Advances in Microbial Ecology*, **1**, 59–94.

Walker, B. H. (1970). Relationships between derived vegetation gradients and measured environmental variables in Saskatchewan wetlands. *Ecology*, **6**, 230–6.

Ward, D. M., Bateson, M. M., Weller, R. & Ruff-Roberts, A. L. (1992). Ribosomal RNA analysis of microorganisms as they occur in nature. *Advances in Microbial Ecology*, **12**, 219–75.

Wimpenny, J. W. T. (1981). Spatial order in microbial ecosystems. *Biological Reviews*, **93**, 1–48.

Wimpenny, J. W. T., Waters, P. & Peters, A. (1988). Gel-plate methods in microbiology. In *A handbook of model systems for microbial ecosystem research*, Vol. 1, pp. 229–51. Boca Raton, Fla: CRC Press.

Wolfaardt, G. M., Lawrence, J. R., Hendry, M. J., Robarts, R. D. & Caldwell, D. E. (1993). Development of steady state diffusion gradients for the cultivation of degradative microbial consortia. *Applied and Environmental Microbiology*, **59**, 2388–96.

Wolfaardt, G. M., Lawrence, J. R., Robarts, R. D. & Caldwell, D. E. (1994a). The role of interactions, sessile growth and nutrient amendments on the degradative efficiency of a microbial consortium. *Canadian Journal of Microbiology* (in press).

Wolfaardt, G. M., Lawrence, J. R., Robarts, R. D. & Caldwell, D. E. (1994b). Bioaccumulation of carbon sources in EPS and subsequent utilization by biofilm bacteria. *Applied and Environmental Microbiology* (in press).

4

Genetic Responses of Bacteria at Surfaces

Amanda E. Goodman and Kevin C. Marshall

Introduction

The majority of studies on biofilms are concerned with the formation and structure of biofilms (see Korber *et al.*, Chapter 1), the direct effects of microbial activities on surfaces, such as in metal corrosion (see Hamilton, Chapter 9), dental caries (see Marsh, Chapter 18) tissue invasion, or indirect effects of biofilms on fluid frictional resistance, heat exchange across metal surfaces, substrate transformations and biocide resistance of the biofilm organisms on biomaterials implanted in human patients (see McLean *et al.*, Chapter 16) and on water distribution pipeline surfaces (Characklis & Marshall 1990; Anwar *et al.* 1992). Extensive reviews have dealt with the activities of microorganisms at surfaces (Marshall 1971; Stotzky 1986; van Loosdrecht *et al.* 1990; Fletcher 1991), but little on the actual genetic responses of microorganisms at surfaces. The aim of this chapter is to consider the environmental conditions existing at gel–air and solid–water interfaces, as well as in biofilms, and to relate these conditions to possible genetic responses in the immobilized bacteria.

Properties of surfaces

Bacteria at surfaces or, more broadly, at interfaces are exposed to environmental conditions not found in an aqueous phase. An interface is defined, in physicochemical terms, as the boundary between two phases in a heterogeneous system. The most common interfaces to which bacteria are exposed in nature and under laboratory conditions are solid–water interfaces (stones, soil particles, ship hulls, pipeline surfaces, glass and plastic culture vessels), although air–water (surface of bodies of water, bubbles), oil–water (oil spills) and solid–air (intertidal zones, agar surfaces) interfaces also are encountered. The concept of bacterial behaviour at various interfaces has been treated in detail by Marshall (1976).

Why do these interfaces provide novel environmental conditions? In general, the molecules located within the bulk of a solid, liquid or gas are subjected to equal forces of attraction in all directions, whereas those located at an interface experience unbalanced attractive forces. These unbalanced forces give rise to the unusual physicochemical characteristics observed at interfaces, such as surface charge and interfacial tension, expressed as the more fundamental quantity – surface free energy (Shaw 1980). We have a tendency to think of interfaces as two dimensional surfaces, neglecting their thickness of several molecular diameters. This third dimension is the most significant feature of interfaces because it is here that rapid changes in molecular composition, density and other properties occur (Miller & Neogi 1985). These gradients across an interface can provide bacteria with an array of environmental conditions different from those found in the bulk liquid phase.

The following discussion on ion distribution near an interface will be restricted to solid–water interfaces, although similar principles apply to other interfaces. The charge existing at a solid surface results in the establishment of a potential between the surface and the bulk aqueous phase. Ions of the opposite charge (counter- or gegen-ions) are attracted towards the interface to counterbalance the surface charge. Thermal agitation of the gegen-ions results in a loose association with the surface, termed the diffuse electrical double layer. Some of the gegen-ions are held at the surface by forces strong enough to overcome thermal agitation, and the ions in this layer (termed the Stern layer) reduce the potential at the surface. The number of ions in this Stern layer increases with both increasing concentration and valency of the electrolyte, such that a charge reversal can occur at a surface in the presence of sufficient trivalent gegen-ions (Shaw 1980).

Most surfaces in nature, whether they are initially negatively or positively charged, acquire a relatively low negative charge by the adsorption of macromolecules, humic acids and low molecular weight, hydrophobic molecules (Neihof & Loeb 1972, 1974). These adsorbed molecules, often termed conditioning films, also alter the surface free energy of the surface in question (Baier 1980). Thus, a bacterium approaching a surface encounters a gradient of gegen-ions, alterations in surface free energy and a concentration of organic molecules not normally found in the bulk aqueous phase (Marshall 1988). Other differences in the bacterial microenvironment at the three dimensional solid–water interface could include lower pH levels (resulting from proton accumulation), altered water structure (and, hence, altered viscosity and/or water availability) and altered rates of gas exchange. All of these environmental conditions may influence genetic responses of bacteria at surfaces.

Conditions within biofilms

Biofilms have been regarded in the past as consisting of a uniform distribution of bacteria embedded in a relatively homogeneous polymer matrix of bacterial origin (Characklis & Marshall 1990). Recent evidence, based mainly on scanning confocal laser microscopy, has revealed an entirely different structure in which the bacteria are surrounded by the polymer matrix in discrete columns surrounded by extensive water filled voids (Lawrence *et al.* 1991; Caldwell *et al.* 1992a, b; see Korber *et al.*, Chapter 1). The bacteria tend to develop as a series of microcolonies within these columnar structures. New

models will have to be developed to account for the unexpected patterns of oxygen distribution and mass transport recently observed within both the columnar bacterial structures and associated voids of the biofilms (de Beer *et al.* 1993).

These evolving concepts of biofilm structure must have important ramifications in terms of genetic responses in bacteria present in various parts of the biofilm. Studies with microelectrodes have revealed that oxygen penetrates to the substratum level in voids (de Beer *et al.* 1993). Bacterial metabolism, along with restricted gas diffusion, results in rapid O_2 utilization and CO_2 buildup in the centre of the columnar structures, and these changes in gas concentrations can alter the expression of certain genes in bacteria. The effects would not be homogeneous throughout the columns, however, as entirely different gas concentrations should exist at the column–void interface. Caldwell *et al.* (1992b) have clearly shown the development of pH gradients around bacterial microcolonies within bacterial columns and the altered pH of the microenvironment may influence gene expression of bacteria in adjacent microcolonies. Limitations to the mass transport of energy substrates and nutrients into the bacterial columns should result in the differential starvation of those bacteria near the centre of the columns, with the induction of starvation specific genes in these bacteria. Many of the above conditions are density dependent functions, occurring in compact, fairly mature biofilms but not in thin or newly developing biofilms. Examples of genetic responses induced by conditions found within biofilms are dealt with later in this chapter.

Observations of physiological differences between cells at surfaces compared with those in liquid

Abundant evidence is available to indicate that cells grown on solid surfaces or on media solidified by a gelling agent, such as agar, have properties different from those of cells grown in liquid media. An historical description of the use of solids and gelling agents for the growth of microorganisms has been presented by Codner (1969).

Growth on semi-solid versus liquid media

One of the most obvious differences between certain bacteria grown on semi-solid media compared with the same organisms grown in liquid media is the form of their motility. Species of *Serratia*, *Vibrio* and *Proteus* produce swarmer cells on agar media, whereas in broth they produce swimmer cells (de Boer *et al.* 1975). This difference in motility involves gross morphological changes, whereby short (about 1 μm) swimmer cells, motile by means of one or a few flagella, differentiate at an agar surface into long (several to hundreds of μm) cells covered by hundreds of lateral flagella (Belas *et al.* 1986). The molecular genetics of this differentiation process and its regulation are discussed in more detail later in this chapter. The heterogeneous group of gliding bacteria also produce a motile phenotype on agar and solid surfaces, but show no motility in liquid culture (Burchard 1981). To date the mechanism for the motility exhibited at a surface is unknown, as there are no obvious morphological differences between surface- and liquid grown gliding bacteria. However, physiological differences between such cells have been reported.

Electron microscopic observations have revealed a number of differences in cell structure between agar grown and liquid grown cells (Lorian 1989). Lorian (1986) showed that cells of *Escherichia coli* grown in broth were thicker than those grown on a membrane overlaying an agar surface, whereas those of *Staphylococcus aureus* showed thicker cell walls and cross walls when grown on the membrane surface (Lorian 1986). In the presence of penicillin, on the other hand, staphylococci grown on an agar surface were approximately 12 times the mass, when compared to cells grown in liquid (Lorian *et al.* 1985a), and were held together by thick cell walls (Lorian *et al.* 1985b).

Another difference noted between agar grown and liquid grown cells is that yields of many biological products are superior from surface grown cells. Staphylococci and pneumococci, for example, produce haemolysins when grown on agar surfaces, but not in liquid (Lorian 1971; Lorian & Popoola 1972). Similarly, an inhibitor of *Neisseria gonorrhoeae* produced by streptococci has been isolated only from agar grown cells (Dubrevil *et al.* 1985).

Growth at solid surfaces versus that in liquid media

In general, the growth rates of bacteria at surfaces are different from those of the same cells grown in the liquid phase. Growth rates have been reported as being stimulated, inhibited or unaffected depending on the organism, the substrate provided and the substratum with which the organism and the substrate must interact (Marshall 1971; Stotzky 1986; Fletcher 1991). Many studies have concentrated on growth in the presence of low molecular weight compounds that may not adsorb at solid surfaces or should be rapidly utilized in the bulk aqueous phase, particularly in oligotrophic environments, and ignore those substrates that rapidly partition at surfaces, such as macromolecules and smaller hydrophobic molecules (Marshall 1988). The nature of the substratum is a key factor in determining the degree of substrate adsorption, the orientation of the substrate molecules and, hence, their availability to the variety of bacteria associated with the surface (Marshall 1971; Stotzky 1986; Fletcher 1991).

The importance of adsorbed nutrient accumulation for the growth of surface associated bacteria was illustrated by the demonstration of growth and cellular reproduction of starved marine bacteria at surfaces exposed to very low levels of energy substrate (2 mg L^{-1} of both yeast extract and tryptone) in the medium flowing past the surface (Kjelleberg *et al.* 1982), when the same level of substrate in the aqueous phase failed to support growth. Similarly, growth of surface associated bacteria has been observed with a low molecular weight, hydrophobic substrate, such as stearic acid (Power & Marshall 1988), or a macromolecular substrate, such as protein (Samuelsson & Kirchman 1990), adsorbed to the substratum. Marine bacteria starving at a surface were found to become smaller and exhibit higher oxygen uptake and heat output than for cells starving in the aqueous phase, effects that were indicative of a triggering of endogenous metabolism in the surface associated bacteria (Humphrey *et al.* 1983; Kjelleberg *et al.* 1983; Humphrey & Marshall 1984). Any triggering, however, may have been induced by surfactants present at the surface (Humphrey & Marshall 1984).

Davies & McFeters (1988) found that

Klebsiella oxytoca cells attached to granular activated carbon (GAC) particles had a growth rate 10 times higher than cells in liquid in the presence of glutamate, a substrate that adsorbed to the surface. No difference was observed in the presence of glucose, a substrate that did not adsorb to the GAC. Uptake of labelled thymidine, an estimate of DNA synthesis, and labelled uridine, an estimate of RNA turnover, by attached cells in a glutamate medium were five and 11 times more, respectively, than by cells grown in liquid medium (Davies & McFeters 1988). A shorter lag time and higher specific activity in the degradation of nitrilotriacetate by bacteria attached to sand particles than by bacteria in the bulk aqueous phase has been reported by McFeters *et al.* (1990). Bacteria associated with particles exhibited higher relative phosphorus uptake rates per cell than free living bacteria, with the differences being greatest under oligotrophic conditions and least under eutrophic conditions (Paerl & Merkel 1982). Vandevivere & Kirchman (1993) have shown that exopolymer production by bacterial cells is stimulated at a solid surface. These authors grew bacteria, originally isolated from deep aquifer samples, in continuous flow columns packed with sand particles, and showed that exopolymer production was increased in these cells over that produced by cells grown in liquid. When surface grown cells were resuspended in liquid medium the amount of exopolymer produced by the cells decreased to levels similar to those found for liquid grown cells.

Pertsovskaya *et al.* (1972) investigated several bacterial species attached to soil particles and glass surfaces and found that ultrastructural appendages, such as fimbriae, were present on strongly attached cells, and were absent on non-attached ones. Fimbriae on cells were generally located peritrichously, with numbers and length dependent upon the organism, although unusual appendages, such as microcapsular outgrowths and braids or bundles of fimbriae, were also noted (Pertsovskaya *et al.* 1972). It is feasible, of course, that these authors studied species with heterogeneous populations and that only fimbriated cells within the populations attached to surfaces. A more definite morphological difference in cells growing at a surface is the observation by McCoy & Costerton (1982) that a *Pseudomonas* sp. formed long filamentous cells when grown as a biofilm on solid test surfaces in a Robbins device, whereas cells grown on agar plates or in liquid were normal rods, about 1 μm long.

The observed resistance of biofilm bacteria to the effects of antibiotics and other biocides (Costerton *et al.* 1987) depends on a complex interplay between the bacterial species involved, the antibiotic under test, the specific growth rate, the phase in the division cycle and whether some daughter cells are released into the aqueous phase (Gilbert *et al.* 1989, 1990; Evans *et al.* 1990a, b, 1991; see Gilbert & Brown, Chapter 6). Ageing biofilms possess very different environmental conditions from those found at a newly colonized surface, and apparent resistance of the biofilm bacteria may result from altered gene expression including altered permeability of the cell membrane, antibiotic adsorption by the polymer matrix of the biofilm and enzymatic degradation of the antibiotics (Anwar *et al.* 1992). The question as to whether antibiotic diffusion is limited by the polymer matrix remains controversial (Nichols *et al.* 1989; Anwar & Costerton 1992). The phenomenon of altered ultrastructure may also be a factor in the difference in susceptibility to antibiotics displayed by pathogenic bacteria. This may depend on whether they occur in biofilms or in liquid culture, as staphylococci isolated from infections *in vivo* possess an ultrastructure resembling that of cells grown on solid, rather than those in liquid, medium (Ernst *et al.* 1985; Lorian *et al.* 1982, 1984). The altered susceptibility of biofilm bacteria to antibiotic treatment eventually may be overcome by selecting antibiotics that are effective against such immobilized cells (Anwar & Costerton 1992). Another method involves the simultaneous application of antibiotics or biocides and an electrical current (electrical enhancement) to biofilms formed in difficult clinical and industrial situations (Blenkinsopp *et al.* 1992).

Such physiological differences between surface- and liquid grown cells are not confined to heterotrophic microorganisms. *Thiobacillus ferrooxidans*, a chemolithotroph utilizing the oxidation of iron(II) and reduced sulphur compounds to obtain energy, simultaneously oxidizes iron(II) and elemental sulphur when grown in liquid (Espejo *et al.* 1988). When attached to either pyrite (Wakao *et al.* 1984) or chalcopyrite (Shrihari *et al.* 1991) ore particles, *T. ferrooxidans* preferentially oxidizes the sulphide moiety and the iron oxidizing ability is inhibited. Interestingly, *T. ferrooxidans*, which does not require organic carbon sources for growth, was found to have a growth rate 70 times lower than that of freely suspended cells when grown as a biofilm on glass beads in an inverse fluidized bed reactor with ferrous iron as substrate (Karamanev 1991). Chemolithotrophic oxidation of positively charged ammonium ions is either stimulated or inhibited by the presence of various soil constituents, depending on the degree of adsorption of the ammonium by the particulate materials (Goldberg & Gainey 1955). McLaren & Skujins (1963) reported that the pH optimum for nitrite oxidation by *Nitrobacter agilis* was about 0.5 units higher in soil and negatively charged ion exchange resins than in liquid culture; a shift that may be related to altered physiology of the attached bacteria or to repulsion of the nitrite ions from the negatively charged surfaces. The rate of nitrification in soil was shown by Macura & Stotzky (1980) to be enhanced by montmorillonite clay and these authors concluded that the stimulation was the result of the clay maintaining the pH at a level favoured by the nitrifying bacteria. However, Underhill & Prosser (1987) concluded, from studies using ion exchange resins, that substrate adsorption and cell attachment were more important in increased activity at surfaces than pH maintenance. Surface enhancement of bacterial manganese oxidation has been observed (Marshall 1980; Nealson & Ford 1980). Again, this may be related to changes in the physiology of the adherent manganese oxidising bacteria and/or to the fact that the soluble Mn(II) species tend to adsorb to surfaces and, more specifically, to previously oxidized Mn(IV) species (Marshall 1979).

Characterization of bacterial responses to different types of surfaces

Recently, attention has focused on defining the differences in physiological changes between cells grown at a surface and those grown in liquid. This has been approached in two ways: first, by examining differences in structural components (proteins, exopolymers) produced during the two modes of growth, and secondly, by investigating the responses of various genes in organisms confronted by surfaces.

Agar surfaces

Agar is a hydrated, hydrophilic gelling agent that produces a semi-solid support for growing bacterial colonies. The percentage of agar may be varied and, hence, the amount of support provided may be modified. The observed phenotype may require a specific agar concentration; for example, *Serratia marcescens* cells swarm on an 0.85% agar surface, but not on 0.35% or 1.5% agar surfaces (Alberti & Harshey 1990).

A change in the pattern of cellular appendages is the most easily discernible difference between cells grown on agar and in liquid. The best understood example of this phenomenon is the differentiation from swimmer to swarmer cells of *Vibrio parahaemolyticus* taken from liquid culture to an agar surface. In aqueous culture, this organism is motile by means of a single, sheathed, polar flagellum. At an agar surface, the cells produce hundreds of unsheathed lateral flagella which give the cells their characteristic swarming ability. The regulation of lateral flagella (*laf*) gene transcription has been elucidated by the use of transposon mutagenesis to insert the promoterless *lacZ* and *lux* reporter genes into the appropriate target genes of *V. parahaemolyticus* (Belas *et al.* 1984, 1986; McCarter *et al.* 1988; McCarter & Silverman 1990). Expression of the *laf* gene is controlled by the combination of two external factors: (i) a deficiency of iron and (ii) a high viscosity of the agar surface. In fact, *laf* genes could be induced in cells suspended in a highly viscous liquid medium, or when flagellar antibodies bound the polar flagellum thus interfering with its rotation in normal liquid medium (McCarter *et al.* 1988). It has been proposed that the polar flagellum acts as a dynamometer, with its very slow rotation or lack of rotation in a viscous fluid providing the signal to initiate lateral flagella production (McCarter *et al.* 1988; McCarter & Silverman 1990). A similar molecular genetics approach has been used by Belas *et al.* (1991a, b) and Alberti & Harshey (1990) to investigate the genetic regulation of the swarming phenotype in *Proteus mirabilis* and *S. marcescens*, respectively. A further example of the effect of surfaces on the production of specific cell surface appendages is that of the production of S fimbrial adhesins (*sfa*) by pathogenic *E. coli* strains. These adhesins enable the bacterial cells to bind specific receptors on eukaryotic tissue cells. Using a promoterless *lacZ* reporter gene fused to the promoter region of the *sfaA* gene, Schmoll *et al.* (1990) found that expression of the gene was about four times higher in cells grown on an agar surface than in cells grown in liquid, and that the gene was further regulated by catabolite repression, osmolarity and temperature.

Humphrey *et al.* (1979) reported that the gliding bacterium *Flexibacter* (now *Cytophaga*) BH3 failed to form extracellular polymer in liquid media but formed copious quantities on agar media. Analyses of the polymer formed during growth directly on the agar medium revealed the presence of large quantities of galactose, a major component of agar. When *Cytophaga* BH3 was grown on sterile dialysis membrane placed on top of the agar medium, the exopolymer was found to contain protein, glucose, fucose, galactose and some uronic acid, but no lipid, 2-keto-3-deoxyoctonate (KDO) or nucleic acids were detected (Humphrey *et al.* 1979).

The gliding bacterium *Myxococcus xanthus* displays intercellular cooperation whereby cells aggregate, forming mounds which develop into fruiting bodies. Conditions necessary for such behaviour are high cell density, nutrient depletion and a surface. Orndorff & Dworkin (1982) investigated the induction of several membrane proteins during aggregation of *M. xanthus*. Synthesis of a 59 kDa protein was induced by growth at an agar surface under conditions of both high and low cell density, as well as by suspension in nutrient depleted liquid medium. The synthesis of a 110 kDa protein was induced only by growth on the agar surface. Similarly, the synthesis of specific proteins of 120–220 kDa, located in the cell wall of several *Staphylococcus aureus* strains, is induced by growth of the organism on agar or nitrocellulose membrane surfaces (Cheung & Fischetti 1988). In another gliding bacterium, *Cytophage johnsonae*, Abbanat *et al.* (1988) showed that cells grown on agar or on a membrane produce two novel sulphonolipids which are more polar than those of liquid grown cells. Synthesis of these two sulphonolipids is induced and maintained by growth on the surface, but ceases when cells are transferred to liquid medium. These specific sulphonolipids are localized in the outer membrane of *C. johnsonae* (Godchaux & Leadbetter 1988). In a non-motile mutant of *C. johnsonae*, the addition of cysteate (a specific sulphonolipid precursor) restored the

lipid content and motility, indicating that the sulphonolipid is essential for gliding (Gorski *et al.* 1991). Recently, Pitta *et al.* (1993) found that a 50 kDa protein is localised in the outer membrane of liquid grown cells. Upon contact with an agar surface this 50 kDa protein becomes associated with the peptidoglycan sacculi of the cells. The authors suggest that *C. johnsonae* is able to sense contact with the surface and modify the structure of its envelope in response.

Electron microscope and time lapse video studies have shown that bacterial colony formation on an agar surface is a highly ordered affair, apparently under genetic control, and involves cell–cell signalling and developmental regulation (Shapiro & Higgins 1989; Shapiro & Trubatch 1991). *E. coli* cells growing and dividing on an agar surface during the initial stages of colony formation do so asymmetrically and nonsynchronously, forming the familiar circular-shaped colony (Shapiro & Hsu 1989). During this stage, cells in nearby microcolonies (that is, those consisting of only a few cells each) elongate towards each other so that young microcolonies can fuse to give rise to a single colony. As the edges of older colonies, consisting of multilayers of cells, approached one another, single cells were seen to 'leap' across colony boundaries before the growing edges joined each other. Colonies arising from a mutant of *E. coli* containing a Mu d(*lac*) fusion element were found to show 'spatially localized replication' zones of the Mu fusion element depending upon the growth history of the bacterial population (Shapiro & Higgins 1989). A mutant of *P. mirabilis*, PRM2007, in which swarming ability was defective, was also unable to form circular colonies on agar (Shapiro & Trubatch 1991). When grown on agar, PRM2007 grew along lines of irregularity on the surface, forming multi-branched colonies. Shapiro & Trubatch (1991) postulate that genetic control of colony formation at surfaces produces the regular circular shape of bacterial colonies, and is able to compensate for irregularities in the surface structure.

Eukaryotic cell surfaces

Pathogenic and symbiotic associations between microorganisms and eukaryotic cells involve attachment and then invasion of the host eukaryotic cells by the particular microorganisms. As with non-specific bacterial adhesion (Marshall *et al.* 1971), this attachment stage has been shown in several cases to consist of a binding, or reversible, stage, followed by an irreversible, attachment step (Jones *et al.* 1981). Such attachment is most often mediated by site specific lectin/receptor interactions (Switalski *et al.* 1989).

In an investigation of the specific adhesion to, and invasion of, epithelial cells by *Salmonella typhimurium* and *S. choleraesius*, Finlay *et al.* (1989) reported that few bacteria attached within the first 2 h, after which binding increased exponentially up to 6 h. Synthesis of several new bacterial proteins was shown to be a requirement for firm attachment. If bacteria were attached to epithelial cells for 3 h and then removed and presented to fresh epithelial cells, stable binding occurred within 1 h at levels similar to those with bacteria continuously exposed to epithelial cells for in excess of 2 h (Finlay *et al.* 1989). The authors found that host cell invasion was not essential for the production of these proteins, and that modification of the surface of epithelial cells by protease or neuraminidase treatment resulted in poor adhesion by bacteria and a lack of synthesis of the specific binding proteins.

Agrobacterium tumefaciens is found worldwide in soils, and infects plants through wounds or natural openings, causing tumours in susceptible hosts (Matthysse 1986). A prerequisite for infection is attachment to the surface of exposed plant cells. The initial binding of the bacterial cells is to a vitronectin-like protein in the plant host, as shown for carrot cells (Wagner & Matthysse 1992). This binding is reversible, and bacteria may be washed off plant tissue by gentle shear at this stage. Bound cells are then induced to produce cellulose fibrils which cause irreversible binding of the bacteria to plant tissue cells and also aggregation or clumping of bacterial cells to each other (Matthysse 1986; Wagner & Matthysse 1992). The moiety on the bacterium responsible for binding the vitronectin-like protein in the plant cell has not yet been identified, although evidence suggests that it consists of lipopolysaccharide (LPS) and protein and is located on the surface of the bacterial cell (Matthysse 1986). This irreversible binding stage is essential for efficient tumour production. It is possible that production of cellulose fibrils by *A. tumefaciens* may be induced by contact with specific components of the plant cell, or by chemicals released from plant wounds, since the formation of fibrils can be induced by soluble plant extracts as well as by cellulose filter paper (Matthysse *et al.* 1981).

A similar phenomenon occurs with *Rhizobium* species which form symbiotic nitrogen fixing root nodules on leguminous plants. In this case, the bacterial adhesin responsible for the initial binding to the plant root hair has been identified as the protein rhicadhesin, although the plant lectin receptor remains to be identified (Smit *et al.* 1992). Following initial attachment of bacterial cells to root hairs, the production of cellulose fibrils by the bacteria is induced (Smit *et al.* 1992). These fibrils cause firm attachment of the bacteria to the plant cells, as well as promoting aggregation of bacteria at the root hair tips. Smit *et al.* (1986) were able to isolate heavily fibrillated strains of *R. leguminosarum* by repeated selection of cells growing in surface pellicles at the air–water interface of unshaken, non-aerated liquid cultures. In contrast to the situation with *A. tumefaciens*, which does not invade the plant cell and thus must stay attached to the outside surface, the initial, reversible attachment of *Rhizobium* cells is essential for root hair infection and subsequent nodule formation, and the production of cellulose fibrils is not essential for infection (Smit *et al.* 1992).

Solid surfaces

The physiochemical conditions existing at a solid–liquid interface are certainly different from those found in an aqueous phase or at the surfaces of cells and agar media. As a result, it is feasible that different suites of genes would be 'switched on' or 'switched off' at such interfaces.

Reporter gene technology has been employed to demonstrate the switching on of genes at a solid surface such as polystyrene in mutants that failed to express the genes in either liquid or semi-solid media (Dagostino *et al.* 1991). The transposon miniMu, containing the promoterless reporter gene *lacZ*, was transferred into the marine *Pseudomonas* sp. strain S9 using the plasmid vector pJO100 (Östling *et al.* 1991) and mutants were selected that failed to express ß-galactosidase (the product of the *lacZ* gene) in liquid or on semi-solid media but produced the enzyme at a solid–liquid interface. It should be possible also to identify genes that are switched off at surfaces, by selecting mutants that express the *lacZ* gene on semi-solid media but fail to express at a solid surface.

Similar technology has been employed to demonstrate increased activity of an alginate gene of *P. aeruginosa* at a solid surface. An 85-fold increase in expression of the *algC* gene in cells at a solid surface (Teflon), compared with cells in the aqueous phase, was found in a mucoid cell line derived from a cystic fibrosis isolate (Davies *et al.* 1993). Similar surface activation of alginate production may be of significance in the accumulation of mucoid material on lung tissue in cystic fibrosis patients. Recently, it has been found that the *algC* gene is induced in some bacteria within about 30 min of the cells attaching to the Teflon surface and that, as the biofilm develops, the majority of alginate production occurs in cells located at the substratum interface (G. G. Geesey, personal communication).

Hudson & Curtiss (1990), also using reporter gene technology, found that expression of the glucosyltransferase (*gtf*) operon was induced several-fold in *Streptococcus mutans* cells that had been allowed to attach to solid hydroxyapatite beads for only 3 h. The *gtf* operon is involved in sucrose metabolism and encodes enzymes thought to be important virulence factors in *S. mutans*, which is implicated as the principle organism producing dental caries (Hudson & Curtiss 1990).

Conditions inducing genetic responses at surfaces and within biofilms

Physicochemical conditions in the aqueous phase are very different from those at surfaces and within biofilms, where conditions vary in time and space as the biofilm develops. Bacteria in these unique environments must alter the regulation of expression of certain genes, and hence their overall physiology, in order to capitalize on the changed conditions. Conditions within biofilms will be heterogeneous and, in some instances, will be somewhat similar to those within bacterial colonies growing on an agar surface. Two important differences between these environments, however, are that (i) biofilms are normally exposed to a flowing aqueous phase and colonies on agar to an air phase, and (ii) biofilms in most natural environments are composed of many different microorganisms, each contributing only a portion to the overall metabolism of

the community, whereas colonies normally are pure cultures. In order to understand the complex question of how bacteria may respond to changing physicochemical conditions within biofilms, however, we must first understand more simple systems, and determine whether, in fact, bacteria can alter the expression of their genes to adapt to changing environmental conditions. Responses to the types of environmental conditions affecting bacterial physiology at surfaces and in biofilms are discussed below.

Gas diffusion

Oxygen availability to cells in biofilms is limited to that diffusing from the surrounding aqueous phase, with conditions below the surface of the polymer matrix being anaerobic or, at best, microaerophilic. Similarly, diffusional limitations would result in higher concentrations of CO_2, produced by cell metabolism, within the biofilm than in the aqueous phase. Kepkay *et al.* (1986) used microelectrodes to demonstrate steep gradients of both oxygen and carbon dioxide within biofilms.

Regulation of bacterial physiology by O_2 tension has been intensively studied in the nitrogenase systems, which are inactivated by O_2 and, hence, require anaerobic conditions for activity (Hill 1988). Oxygen concentration affects gene expression in *Rhodobacter capsulatus*, with low levels of expression of *nifA* and no expression of *nifH* and *nifR4* in the presence of O_2 (Hübner *et al.* 1991). Many genes have been identified, mostly in facultatively anaerobic *E. coli* and *Salmonella* strains, that are induced or repressed by anaerobic conditions. In such organisms, most genes involved in anaerobic metabolism are not expressed in the presence of O_2 and, similarly, the expression of many genes involved in aerobic respiration are suppressed in the absence of O_2 (Smith & Neidhardt 1983; Barrett *et al.* 1984; Clark 1984; Jones & Gunsalus 1987; Iuchi *et al.* 1990; Spiro & Guest 1990). The *fnr* gene, involved in regulation of expression of anaerobically induced genes in *E. coli*, has significant homology with the catabolite activator protein, suggesting that *fnr* may be a DNA binding protein (Shaw *et al.* 1983). The expression, in *E. coli* and *S. typhimurium*, of genes not directly involved in anaerobic respiration has been found to be induced by anaerobic conditions; for example, lysyl-tRNA synthetase (lysU) (Lévêque *et al.* 1991), HPII catalase (Meir & Yagil 1990), genes involved in catabolite repression and nitrogen regulation (Winkelman & Clark 1986) and the vitamin B_{12} biosynthetic genes (Richter-Dahlfors & Andersson 1991). Genes whose expression is induced by anaerobic conditions have been studied in other organisms, for example *P. aeruginosa* (Zimmerman *et al.* 1991) and *Rhodobacter capsulatus* (Bauer *et al.* 1988). It is common practice to study anaerobic conditions by replacing O_2 with mixtures of other gases, in particular high CO_2 concentrations (e.g. Schellhorn & Hassan 1988), and it is possible that expression of some genes thought to be regulated by lack of O_2 may in fact, be regulated, by CO_2 instead.

It has been known for some time that CO_2 concentration is a critical factor for differentiation in fungi (Graafmans 1973). Carbon dioxide levels higher than ambient are required for fruiting and α-1,3-glucan metabolism in *Aspergillus nidulans* (Zonneveld 1988). Cotty (1987) found that many aspects of *in vitro* sporulation of *Alternaria tagetica* are regulated by CO_2 levels, in contrast to the widespread view that sporulation is regulated by the nutritional status of the cell. It was proposed that high respiration rates on nutrient media produce high levels of CO_2 which, in turn, inhibit sporulation (Cotty 1987). Recent work has shown that CO_2 concentration also may be an important regulatory signal in bacteria. High CO_2 concentrations stimulated toxin production by *Staphylococcus aureus* (Kass *et al.* 1987) and cholera enterotoxin by *Vibrio cholerae* (Shimamura *et al.* 1985), with 10% CO_2 being optimal for maximum enterotoxin production. Transcription of the protective antigen gene (*pag*), a major virulence factor of *Bacillus anthraci*s, is induced in cells grown in minimal media containing bicarbonate, or under CO_2 (Bartkus & Leppla 1989). The bicarbonate may inhibit sporulation (Bartkus & Leppla 1989), and capsule production *in vitro* requires 5–20% CO_2 (Makino *et al.* 1988). Cell surface structure and adhesion potential are different in staphylococci grown in the presence or absence of 5% CO_2 (Denyer *et al.* 1990). In the microalga *Chlorella regularis*, low CO_2 levels induce, and high levels inhibit, cell-surface carbonic anhydrase activity (Umino & Shiraiwa 1991). Using *lacZ* as a reporter gene, Scanlan *et al.* (1990) identified CO_2-regulated promoters in the cyanobacterium

Synechococcus R2 PCC7942. We have identified *lacZ* fusion mutants in *Pseudomonas* sp. strain S9 in which gene expression is induced by increased levels of CO_2 (A. E. Goodman, unpublished data). Such an induction could occur within biofilms where CO_2 tends to accumulate as a result of microbial metabolic activity.

pH

Negatively charged surfaces in nature attract cations, including protons. Proton accumulation could lead to significantly lower pH values near the surface. This may directly affect gene expression, or may have indirect effects by altering the apparent pH optimum for enzymatic reactions (McLaren & Skujins 1963) or altering the transmembrane potential of bacteria adhering to these surfaces (Ellwood *et al.* 1982). Within biofilms, bacterial metabolism will change the pH of the external milieu heterogeneously around each microcolony. Caldwell *et al.* (1992b), using confocal laser microscopy, have clearly demonstrated the production of pH gradients around microcolonies and individual cells in a *V. parahaemolyticus* biofilm. Both acid- and alkali-inducible genes have been demonstrated in various enteric bacteria (Slonczewski 1992). The inducible acid tolerance response of *S. typhimurium* requires *de novo* protein synthesis (Foster & Hall 1991). In *E. coli*, expression of the porin gene *ompC* (Thomas & Booth 1992) and the lysine decarboxylase gene (*cadA*) (Watson *et al.* 1992) are induced by acidification of the external medium. A specific cell surface antigen of *Rhizobium leguminosarum* is produced at pH values below 5.3 under aerobic conditions, and at neutral pH under low O_2 tension conditions (Kannenberg & Brewin 1989). In addition, a condition required for optimal induction, *in vitro*, of the *vir* genes of *A. tumefaciens* is an external pH below 5.2 (Vernade *et al.* 1988). Amaro *et al.* (1991) have shown that pH shifts between 1.5 and 3.5 result in changes in the general protein synthesis patterns of *T. ferrooxidans*. An alkaline inducible locus, *alx*, has been identified in *E. coli* (Bingham *et al.* 1990).

Cell density and metabolite accumulation

Compared with the surrounding aqueous phase, the cell density within a biofilm is high. Removal of metabolites produced by cellular activity would occur by consumption by other organisms or by diffusion out of the biofilm and, hence, it is probable that metabolites would accumulate within the biofilms. Production of ß-haemolysins by pneumococci on agar, but not in broth, is thought to be triggered by the accumulation of a metabolite within a small space surrounding the colony (Lorian & Popoola 1972). Transcription of the bacterial luminescent system in *Vibrio fischeri* is regulated by a diffusible, sensory auto-inducer, N-(ß-ketocaproyl) homoserine lactone (KHL), which must accumulate to sufficiently high levels to induce expression of the genes encoding luminescence (Nealson & Hastings 1979; Meighen 1991). Thus, visible lumines-cence only occurs when *V. fischeri* populations reach high density in a confined environment, such as in biofilms on surfaces of organic mater-ial. KHL is produced by a wide range of bacteria, and induces the expression of genes controlling carbapenem antibiotic synthesis in *Erwinia caro-tovora* in a growth phase dependent manner (Bainton *et al.* 1992). Gadkari (1990) found that soil particles, agar strands and alginate beads doubled the rate of ammonium oxidation by nitrifying bacteria, whereas sand particles had no affect. It was concluded that polysaccharides secreted by microorganisms are important in the regulation of nitrification in natural environments (Gadkari 1990). This conclusion is over-simplis-tic when viewed in the light of the more detailed analysis of the effects of soil particles on ammo-nium adsorption by Goldberg & Gainey (1955). Regulation of several bacterial functions has been attributed to diffusible metabolites, the majority of which are as yet unidentified; for example, transcription of *katE* and *katF* in *E. coli* (Mulvey *et al.* 1990), and transfer of the Ti plasmid between *A. tumefaciens* cells (Zhang & Kerr 1991). Intercellular signalling by metabolites plays an important role in the differentiation process of individual cells of myxobacteria into fruiting bod-ies (Kaiser 1989). Metabolites may also be impor-tant signals that regulate the shape of bacterial colonies as they develop on agar surfaces (Shapiro & Hsu 1989; Shapiro & Trubatch 1991).

Nutrients

ZoBell (1943) suggested that surface associated bacteria gained a growth advantage in olig-otrophic waters by utilizing organic molecules adsorbed at the surfaces as sources of energy. Various researchers have shown positive, negative and no responses by adherent bacteria to added nutrients (Marshall 1971; Stotzky 1986; Fletcher 1991). Most of these studies used low molecular weight substrates that almost certainly are rapidly utilized in the aqueous phase in nature and never adsorb to surfaces (Marshall 1988). ZoBell (1943) stressed the importance of macromole-cules as substrates at surfaces because these require the production of extracellular enzymes for breakdown prior to utilization by bacteria and, unlike most low molecular weight organic substrates, they are very prone to adsorb to sur-faces. Starved adhesive and non-adhesive bacte-ria metabolized adsorbed organic substrates (Kefford *et al.* 1982; Hermansson & Marshall 1985; Samuelsson & Kirchman 1990), then exhibited cellular growth and reproduction at surfaces even when little or no energy substrate was available in the aqueous phase (Kjelleberg *et al.* 1982; Power & Marshall 1988). Bacterial gene expression, and hence protein production, is altered by changing nutrient levels. Nutrient depletion and subsequent starvation produce programmed response patterns in non-differen-tiating *E. coli* (Matin 1991) and the marine *Vibrio* sp. strain S14 (Nyström *et al.* 1990). A gene in *Vibrio* S14, induced by carbon starva-tion, was induced to a greater extent by ultra-violet irradiation, but was not induced by starvation for other nutrients nor by anaerobic conditions (Östling *et al.* 1991). Competition between bacteria within biofilms would ensure that many cells are in a starved state for variable time periods.

Surfactants

Many bacteria in natural habitats are capable of producing surfactants (Rosenberg 1986) that are likely to adsorb to surfaces. Surface bound sur-factants trigger increases in heat output, oxygen consumption and size reduction in starving marine bacteria associated with the surface (Humphrey & Marshall 1984). Surfactant pro-duction by surface colonizing and biofilm form-ing bacteria could be a significant factor in regulating gene expression in other immobilized bacteria.

Viscosity

McCarter & Silverman (1990) have demonstrated that an agar surface or increased liquid medium viscosity can induce the switch from polar to lateral flagellation in *V. parahaemolyticus*. The polar flagellum appears to act as a dynamometer, with the slowing of its rotation in a viscous medium triggering expression of the *laf* gene. The likelihood that fluid viscosity at a solid–liquid interface is greater than that in the bulk aqueous phase, along with the highly viscous polymer matrix of the biofilm, suggests that viscosity can be an important factor in altering gene expression in bacteria at surfaces.

Osmolarity and water activity

Using reporter genes fused to the promoter for the epidermolytic toxin A (*eta*) gene of *Staphylococcus aureus*, Sheehan *et al.* (1992) found that expression of *eta* is osmotically regulated, and that the gene product, Agr, of the *agr* gene is involved in transmission of the osmolarity signal to *eta*. The *ompR* gene of Gram negative bacteria regulates genes involved in virulence in pathogenic species, such as *Salmonella* (Dorman *et al.* 1989; Galán & Curtiss 1990), and is itself regulated by osmolarity. The response to alterations in extracellular osmotic strength by *E. coli* cells, characterized by changes in levels of the outer membrane porin proteins, OmpC and OmpF (pores by which small hydrophilic molecules move across the outer membrane), is one of the best studied of the environmentally regulated gene systems in bacteria (Csonka & Hanson 1991). OmpC porin is abundant under conditions of high osmolarity, whereas OmpF porin predominates at low osmolarity. Only two proteins are involved in the signal transduction response. EnvZ is a histidine protein kinase, capable of autophosphorylation, in which the periplasmic domain 'senses' the external environment and controls the state of phosphorylation of the cytoplasmic domain. Phospho-EnvZ is able to phosphorylate OmpR, the response regulator which controls porin gene transcription (Stock *et al.* 1990). The state of phosphorylation of OmpR determines which porin gene, *ompC* or *ompF*, is transcribed. Dephosphorylation of phospho-OmpR is effected by EnvZ.

Pseudomonas aeruginosa produced higher amounts of exopolymer when grown at low water activities in a sand medium than when grown at higher water activities (Roberson & Firestone 1992). It is feasible that, at solid–liquid interfaces and in biofilms, the liquid osmolarity is higher and the water availability lower than in the bulk aqueous phase, and that these altered physical conditions may trigger changes in gene expression in attached bacteria.

Inorganic ions

Since most surfaces in nature possess a net negative charge, cations tend to accumulate near the surfaces as gegen-ions. Similarly, ions tend to be bound by the biofilm polymer matrix, with Ca^{2+} ions being essential for stability of the polymer matrix (Turakhia *et al.* 1983). Expression of a light inducible gene, *lipA*, in the corynebacterium *Arthrobacter photogonimos* is regulated by Ca^{2+} concentration, such that synthesis of the LIP protein is repressed at concentrations greater than 1 mM Ca^{2+} (Phinney & Hoober 1992). The LIP protein is located on the external surface of the cell and is thought to be a pilin protein (Phinney & Hoober 1992). Ca^{2+} has also been implicated as a regulator of gene expression in other bacteria. Transcription of one of the major promoters of the cell wall gene operon, P2, of *Bacillus brevis* was inhibited in stationary phase cultures by millimolar concentrations of either Ca^{2+} or Mg^{2+} (Adachi *et al.* 1991). It was proposed that synthesis of cell wall protein in *B. brevis* is coordinated with the integrity of cell wall structure, since the presence of Ca^{2+} and Mg^{2+} prevents loss of cell wall layers in early stationary phase. In the absence of either Ca^{2+} or Mg^{2+}, layers of cell wall are lost during early stationary phase, and synthesis of new proteins to replace these is highly induced (Adachi *et al.* 1991). Millimolar concentrations of Ca^{2+} induce synthesis of a haemolytic protein in *Actinobacillus pleuropneumoniae* (Frey & Nicolet 1988), whereas synthesis of several proteins (including outer membrane proteins) involved in virulence in *Yersinia* species is repressed by this concentration of Ca^{2+} (Straley & Bowmer 1986).

Multiple signals

It is probable that for most, if not all, environmentally regulated gene functions, gene expression will be found to be modulated by multiple signals, such as the *eta* gene of *S. aureus* (Sheehan *et al.* 1992), and the *sfaA* gene of pathogenic *E. coli* (Schmoll *et al.* 1990) as described above. Further, coordinated regulation of gene expression by two (or more) environmental conditions will undoubtedly be revealed by future investigations, as has been found, for example, for *V. parahaemolyticus* in which simultaneous cessation of flagellum rotation and low iron concentration are required for the induction of *laf* gene expression (McCarter & Silverman 1989).

How do external physicochemical changes alter gene regulation?

In order to respond to changing external physicochemical conditions, bacteria must have mechanisms for sensing the external environment and then relaying this information to effect differential gene expression at the chromosomal level.

Signal transduction

The majority of well studied examples of bacterial responses to environmental stimuli can be accounted for by the mechanism of signal transduction. This process involves two families of proteins, the so-called signal transduction proteins, in which one protein type acts as a sensor of the environment, and the second type responds to this message by relaying the information to the appropriate chromosomal response elements (Stock *et al.* 1990), as described above for the synthesis of OmpF and OmpC. Members of these protein families have been identified in more than 10 different bacterial species, and are known to regulate in excess of 20 different types of responses. It is thought that in *E. coli* alone there may be more than 50 pairs of these regulatory proteins, each responsible for a specific response to a particular environmental stimulus, although 'cross-talk' between different pairs may also occur (Stock *et al.* 1990). Processes known to be so regulated include motility, chemotaxis, phosphorus accumulation, nitrogen fixation and membrane transport. Response to environmental conditions by means of the signal transduction mechanism is of major importance in pathogenic bacteria for coordinately regulated expression of virulence determinants (Miller *et al.* 1992).

DNA conformation

It is possible that the three dimensional structure of the chromosome itself regulates the expression of some genes, and that regional chromosome twist and supercoiling can change in response to external environmental stimuli (Wang & Syvanen 1992). DNA in the bacterial cell is in a supercoiled state (Worcel & Burgi 1972), with negative supercoiling favouring strand separation for reactions such as transcription, replication, recombination and transposition (Dorman 1991). DNA gyrase, whose subunits are encoded by the genes *gyrA* and *gyrB*, introduces negative supercoils into DNA, whereas topoisomerase I, encoded by *topA*, removes these supercoils (Dorman 1991). It is thought that the topology of chromosomal regions is maintained in a state that best suits bacterial function mainly by the coordinated activities of topoisomerase I and DNA gyrase (Pruss *et al.* 1982). Environmental conditions affect the degree of DNA supercoiling (Dorman 1991). Such conditions include oxygen availability and nutrient levels (Dorman *et al.* 1988); and osmolarity, whereby high osmolarity increases supercoiling (Higgins *et al.* 1988). Galán & Curtiss (1990) showed that the apparent osmotic regulation of the expression of genes involved in invasion of epithelial cells by *S. typhimurium* was independent of *ompR*, and that the regulation was in fact caused by the level of DNA supercoiling. Ni Bhriain *et al.* (1989) identified a set of stress-regulated genes in *E. coli* and *S. typhimurium*, whose primary control was effected by the level of DNA supercoiling, which changed in response to environmental conditions. Regulation of the *eta* gene in *S. aureus* is controlled by *agr* (in response to osmolarity and growth phase), as described above (p. 90), as well as by environmentally induced changes in DNA supercoiling (Sheehan *et al.* 1992).

DNA sequence changes

A further mechanism of regulating bacterial gene expression involves changes to the DNA sequence itself. A specific sequence may be inverted, giving rise to a different phenotype as found for the switching on or off of flagellar genes in *Salmonella* (Silverman & Simon 1980). Such phase variation of cell surface appendages is thought to be an important mechanism whereby pathogenic bacteria escape host controls. Mobile genetic elements, such as transposons and insertion sequences, may insert into, or excise from, genes (including regulatory sequences), switching them between inactive or active. Bartlett *et al.* (1988) showed that the reversible switching off/on of extracellular polysaccharide (encoded by the *eps* gene) production by *Pseudomonas atlantica* is caused by insertion/excision of a specific insertion sequence (IS*492*, Bartlett & Silverman 1989) in the *eps* gene. Bartlett *et al.* (1988) argued that the ability of bacteria to respond directly to environmental change 'must surely be limited', and that this type of dynamic genomic heterogeneity is important in creating diversity within a bacterial population in a random way. Thus, a certain percentage of cells would be 'pre-adapted' to environmental changes and would be better able to survive new conditions. Regulation of such DNA rearrangements may be caused by environmental changes, however, as for the phase variation shown by pyelonephritis associated (pap) pili production in *E. coli* (Maluszynska *et al.* 1992). In *S. typhimurium* and *E. coli*, *osmZ* regulates the expression of a number of unrelated genes, as well as the frequency of DNA sequence inversions that cause phase variation of fimbriae (Higgins *et al.* 1988). The *osmZ* gene is known to be involved in environmentally induced (osmolarity sensing) control of the degree of *in vivo* supercoiling of the chromosome, and the pleiotropic effects seen in *osmZ* mutants are thought to arise from this role of *osmZ* in control of DNA topology (Higgins *et al.* 1988). It is known that transposition functions of mobile genetic elements can be induced by conditions of stress (Shapiro & Leach 1990), and that *E. coli* cells growing as colonies on agar display developmental control of the activity of transposable elements (Shapiro & Higgins 1989).

Relevance to biofilms

The fact that physicochemical conditions at solid–liquid interfaces and within biofilms are different from those existing in the bulk aqueous phase means that bacteria exposed to these conditions will experience changes in the regulation of expression of certain genes. Thus, some physiological characteristics of immobilized bacteria may be different from those of the same organisms in liquid media. It is imperative, consequently, that more consideration be given to studies on the physiology of bacteria both attached to surfaces and existing in biofilms. Caution needs to be observed in the methods employed in such studies. Removal of the bacteria from the surface or from the biofilm will result in a change in the unique physicochemical conditions existing *in situ*, and the likelihood of changes in gene expression leading to artefacts.

Molecular biological methods should prove useful in: (i) observing alterations in gene expression at surfaces and in biofilms, (ii) examining the nature of the genes triggered at surfaces and in biofilms, (iii) providing convenient methods to study, *in situ*, the physicochemical factors responsible for triggering changes in gene expression, and (iv) determining how these physicochemical factors are recognized and translated into altered gene regulation by the bacteria at surfaces or in biofilms.

Acknowledgements

This work was supported by the Australian Research Council and the US Office of Naval Research (grant no. N00014–93–1–0230).

References

Abbanat, D. R., Godchaux, W. III & Leadbetter, E. R. (1988). Surface-induced synthesis of new sulfonolipids in the gliding bacterium *Cytophaga johnsonae*. *Archives of Microbiology*, **149**, 358–64.

Adachi, T., Yamagata, H., Tsukagoshi, N. & Udaka, S. (1991). Repression of the cell wall protein gene operon in *Bacillus brevis* 47 by magnesium and calcium ions. *Journal of Bacteriology*, **173**, 4243–5.

Alberti, L. & Harshey, R. M. (1990). Differentiation of *Serratia marcescens* 274 into swimmer and swarmer cells. *Journal of Bacteriology*, **172**, 4322–8.

Amaro, A. M., Chamorro, D., Seeger, M., Arredondo, R., Peirano, I. & Jerez, C. A. (1991). Effect of external pH perturbations on *in vivo* protein synthesis by the acidophilic bacterium *Thiobacillus ferroxidans*. *Journal of Bacteriology*, **173**, 910–15.

Anwar, H. & Costerton, J. W. (1992). Effective use of antibiotics in the treatment of biofilm-associated infections. *American Society for Microbiology News*, **58**, 665–8.

Anwar, H., Strap, J. L. & Costerton, J. W. (1992). Establishment of aging biofilms: possible mechanism of bacterial resistance to antibiotic therapy. *Antimicrobial Agents and Chemotherapy*, **36**, 1347–51.

Baier, R. E. (1980). Substrata influences on adhesion of microorganisms and their resultant new surface properties. In *Adsorption of microorganisms to surfaces*, ed. G. Bitton & K. C. Marshall, pp. 59–104. New York: Wiley.

Bainton, N. J., Bycroft, B. W., Chhabra, S. R. *et al.* (1992). A general role for the *lux* autoinducer in bacterial cell signalling: control of antibiotic biosynthesis in *Erwinia. Gene*, **116**, 87–91.

Barrett, E. L., Kwan, H. S. & Macy, J. (1984). Anaerobiosis, formate, nitrate, and *pyrA* are involved in the regulation of formate hydrogenlyase in *Salmonella typhimurium. Journal of Bacteriology*, **158**, 972–7.

Bartkus, J. M. & Leppla, S. H. (1989). Transcriptional regulation of the protective antigen gene of *Bacillus anthracis. Infection and Immunity*, **57**, 2295–300.

Bartlett, D. H. & Silverman, M. (1989). Nucleotide sequence of IS*492*, a novel insertion sequence causing variation in extracellular polysaccharide production in the marine bacterium *Pseudomonas atlantica. Journal of Bacteriology*, **171**, 1763–6.

Bartlett, D. H., Wright, M. E. & Silverman, M. (1988). Variable expression of extracellular polysaccharide in the marine bacterium *Pseudomonas atlantica* is controlled by genome rearrangement. *Proceedings of the National Academy of Sciences USA*, **85**, 3923–7.

Bauer, C. E., Young, D. A. & Marrs, B. L. (1988). Analysis of the *Rhodobacter capsulatus puf* operon. Location of the oxygen-regulated promoter region and the identification of an additional *puf*-encoded gene. *Journal of Biological Chemistry*, **263**, 4820–7.

Belas, R., Erskine, D. & Flaherty, D. (1991a). *Proteus mirabilis* mutants defective in swarmer cell differentiation and multicellular behavior. *Journal of Bacteriology*, **173**, 6279–88.

Belas, R., Erskine, D. & Flaherty, D. (1991b). Transposon mutagenesis in *Proteus mirabilis. Journal of Bacteriology*, **173**, 6289–93.

Belas, R., Mileham, A., Simon, M. & Silverman, M. (1984). Transposon mutagenesis of marine *Vibrio* spp. *Journal of Bacteriology*, **158**, 890–6.

Belas, R., Simon, M & Silverman, M. (1986). Regulation of lateral flagella gene transcription in *Vibrio parahaemolyticus. Journal of Bacteriology*, **167**, 210–18.

Bingham, R. J., Hall, K. S. & Slonczewski, J. L. (1990). Alkaline induction of a novel gene locus, *alx*, in *Escherichia coli. Journal of Bacteriology*, **172**, 2184–6.

Blenkinsopp, S. A., Khoury, A. E. & Costerton, J. W. (1992). Electrical enhancement of biocide efficacy against *Pseudomonas aeruginosa* biofilms. *Applied and Environmental Microbiology*, **58**, 3770–3.

Burchard, R. P. (1981). Gliding motility of prokaryotes: ultrastructure, physiology, and genetics. *Annual Review of Microbiology*, **35**, 497–529.

Caldwell, D. E., Korber, D. R. & Lawrence, J. R. (1992a). Imaging of bacterial cells by fluorescence exclusion using scanning confocal laser microscopy. *Journal of Microbiological Methods*, **15**, 249–61.

Caldwell, D. E., Korber, D. R. & Lawrence, J. R. (1992b). Confocal laser microscopy and digital image analysis in microbial ecology. *Advances in Microbial Ecology*, **12**, 1–67.

Characklis, W. G. & Marshall, K. C. (ed.) (1990). *Biofilms*. New York: Wiley.

Cheung, A. L. & Fischetti, V. A. (1988). Variation in the expression of cell wall proteins of *Staphylococcus aureus* grown on solid and liquid media. *Infection and Immunity*, **56**, 1061–5.

Clark, D. P. (1984). The number of anaerobically regulated genes in *Escherichia coli. FEMS Microbiology Letters*, **24**, 251–4.

Codner, R. C. (1969). Solid and solidified growth media in microbiology. In *Methods in microbiology*, Vol. 1, ed. J. R. Norris & D. W. Ribbons, pp. 427–54. London: Academic Press.

Costerton, J. W., Cheng, K.-J., Geesey, G. G. *et al.* (1987). Bacterial biofilms in nature and disease. *Annual Review of Microbiology*, **41**, 435–64.

Cotty, P. J. (1987). Modulation of sporulation of *Alternaria tagetica* by carbon dioxide. *Mycologia*, **79**, 508–13.

Csonka, L. N. & Hanson, A. D. (1991). Prokaryotic osmoregulation: genetics and physiology. *Annual Reviews of Microbiology*, **45**, 569–606.

Dagostino, L., Goodman, A. E. & Marshall, K. C. (1991). Physiological responses induced in bacteria adhering to surfaces. *Biofouling*, **4**, 113–19.

Davies, D. G., Chakrabarty, A. M. & Geesey, G. G. (1993). Exopolysaccharide production in biofilms: substratum activation of alginate gene expression by *Pseudomonas aeruginosa. Applied and Environmental Microbiology*, **59**, 1181–6.

Davies, D. G. & McFeters, G. A. (1988). Growth and comparative physiology of *Klebsiella oxytoca* attached to granular activated carbon particles and in liquid media. *Microbial Ecology*, **15**, 165–75.

de Beer, D., Stoodley, P., Roe, F. & Lewandowski, Z. (1993). Effects of biofilm structure on oxygen distribution and mass transport. *Biotechnology and Bioengineering*, **43**, 1131–8.

de Boer, W. E., Golten, C. & Scheffers, W. A. (1975). Effects of some physical factors on flagellation and swarming of *Vibrio alginolyticus. Netherlands Journal of Sea Research*, **9**, 197–213.

Denyer, S. P., Davies, M. C., Evans, J. A. *et al.* (1990). Influence of carbon dioxide on the surface characteristics and adherence potential of coagulase-negative staphylococci. *Journal of Clinical Microbiology*, **28**, 1813–17.

Dorman, C. J. (1991). DNA supercoiling and environmental regulation of gene expression in pathogenic bacteria. *Infection and Immunity*, **59**, 745–9.

Dorman, C. J., Barr, G. C., Ni Bhriain, N. & Higgins, C. F. (1988). DNA supercoiling and the anaerobic and growth phase regulation of *tonB* gene expression. *Journal of Bacteriology*, **170**, 2816–26.

Dorman, C. J., Chatfield, S., Higgins, C. F., Hayward, C. & Dougan, G. (1989). Characterization of porin and *ompR* mutants of a virulent strain of *Salmonella typhimurium*: *ompR* mutants are attenuated *in vivo*. *Infection and Immunity*, **57**, 2136–40.

Dubrevil, D., Bisaillon, J. G., Beaudet, R. & Portelance, V. (1985). *In vitro* inhibition of *Neisseria gonorrhoeae* growth by a urogenital strain of *Streptococcus faecalis*. *Experimental Biology*, **43**, 243–50.

Ellwood, D. C., Keevil, C. W., Marsh, P. D., Brown, C. M. & Wardell, J. N. (1982). Surface-associated growth. *Philosophical Transactions of the Royal Society Series, B*, **297**, 517–32.

Ernst, J., Sy, E., Lorian, V. & Kim, Y. (1985). Ultrastructure of staphylococci in respiratory infections treated with nafcillin. *Drugs in Experimental Clinical Research*, **11**, 357–60.

Espejo, R. T., Escobar, B., Jedlicki, E., Uribe, P. & Badilla-Ohlbaum, R. (1988). Oxidation of ferrous iron and elemental sulfur by *Thiobacillus ferroxidans*. *Applied and Environmental Microbiology*, **54**, 1694–9.

Evans D. J., Allison, D. G., Brown, M. R. W. & Gilbert, P. (1990a). Effect of growth-rate of Gram-negative biofilms to cetrimide. *Journal of Antimicrobial Chemotherapy*, **26**, 473–8.

Evans, D. J., Allison, D. G., Brown, M. R. W. & Gilbert, P. (1991). Susceptibility of *Pseudomonas aeruginosa* and *Escherichia coli* biofilms towards ciprofloxacin: effect of specific growth rate. *Journal of Antimicrobial Chemotherapy*, **27**, 177–84.

Evans, D. J., Brown, M. R. W., Allison, D. G. & Gilbert, P. (1990b). Susceptibility of bacterial biofilms to tobramycin: role of specific growth rate and phase in the division cycle. *Journal of Antimicrobial Chemotherapy*, **25**, 585–91.

Finlay, B. B., Heffron, F. & Falkow, S. (1989). Epithelial cell surfaces induce *Salmonella* proteins required for bacterial adherence and invasion. *Science*, **243**, 940–3.

Fletcher, M. (1991). The physiological activity of bacteria attached to solid surfaces. *Advances in Microbial Physiology*, **32**, 53–85.

Foster J. W. & Hall, H. K. (1991). Inducible pH homeostasis and the acid tolerance response of *Salmonella typhimurium*. *Journal of Bacteriology*, **173**, 5129–35.

Frey, J. & Nicolet, J. (1988). Regulation of hemolysin expression in *Actinobacillus pleuropneumoniae* serotype 1 by Ca^{2+}. *Infection and Immunity*, **56**, 2570–5.

Gadkari, D. (1990). Nitrification in the presence of soil particles, sand, alginate beads and agar strands. *Soil Biology and Biochemistry*, **22**, 17–21.

Galán, J. E. & Curtiss, R. (1990). Expression of *Salmonella typhimurium* genes required for invasion is regulated by changes in DNA supercoiling. *Infection and Immunity*, **58**, 1879–85.

Gilbert, P., Allison, D. G., Evans, D. J., Handley, P. S. & Brown, M. R. W. (1989). Growth rate control of adherent bacterial populations. *Applied and Environmental Microbiology*, **55**, 1308–11.

Gilbert, P., Collier, P. J. & Brown, M. R. W. (1990). Influence of growth rate on susceptibility to antimicrobial agents: biofilms, cell cycle, dormancy and stringent response. *Antimicrobial Agents and Chemotherapy*, **34**, 1865–8.

Godchaux, W. III & Leadbetter, E. R. (1988). Sulfonolipids are localized in the outer membrane of the gliding bacterium *Cytophaga johnsonae*. *Archives of Microbiology*, **150**, 42–7.

Goldberg, S. S. & Gainey, P. L. (1955). Role of surface phenomena in nitrification. *Soil Science*, **80**, 43–53.

Gorski, L., Leadbetter, E. R. & Godchaux, W. III (1991). Temporal sequence of the recovery of traits during phenotypic curing of a *Cytophaga johnsonae* motility mutant. *Journal of Bacteriology*, **173**, 7534–9.

Graafmans, W. D. J. (1973). The influence of carbon dioxide on morphogenesis in *Penicillium isariiforme*. *Archiv für Mikrobiologie*, **91**, 67–76.

Hermansson, M. & Marshall, K. C. (1985). Utilization of surface localized substrate by non-adhesive marine bacteria. *Microbial Ecology*, **11**, 91–105.

Higgins, C. F., Dorman, C. J., Stirling, D. A. *et al.* (1988). A physiological role for DNA supercoiling in the osmotic regulation of gene expression in *S. typhimurium* and *E. coli*. *Cell*, **52**, 569–84.

Hill, S. (1988). How is nitrogenase regulated by oxygen? *FEMS Microbiological Reviews*, **54**, 111–30.

Hübner, P., Willison, J. C., Vignais, P. M. & Bickle, T. A. (1991). Expression of regulatory *nif* genes in *Rhodobacter capsulatus*. *Journal of Bacteriology*, **173**, 2993–9.

Hudson, M. C. & Curtiss, R. (1990). Regulation of expression of *Streptococcus mutans* genes important to virulence. *Infection and Immunity*, **58**, 464–70.

Humphrey, B. A., Dickson, M. R. & Marshall, K. C. (1979). Physicochemical and *in situ* observations on the adhesion of gliding bacteria to surfaces. *Archives of Microbiology*, **120**, 231–8.

Humphrey, B. A., Kjelleberg, S. & Marshall, K. C. (1983). Responses of marine bacteria under starvation conditions at a solid–water interface. *Applied and Environmental Microbiology*, **45**, 43–7.

Humphrey, B. A. & Marshall, K. C. (1984). The triggering effect of surfaces and surfactants on heat output, oxygen consumption and size reduction of a starving marine *Vibrio*. *Archives of Microbiology*, **140**, 166–70.

Iuchi, S., Matsuda, Z., Fujiwara, T. & Lin, E. C. C. (1990). The *arcB* gene of *Escherichia coli* encodes a sensor-regulator protein for anaerobic repression of the *arc* modulon. *Molecular Microbiology*, **4**, 715–27.

Jones, H. M. & Gunsalus, R. P. (1987). Regulation of *Escherichia coli* fumarate reductase (*fdrABCD*) operon expression by respiratory electron acceptors and the *fnr* gene product. *Journal of Bacteriology*, **169**, 3340–9.

Jones, G. W., Richardson, L. A. & Uhlman, D. (1981). The invasion of HeLa cells by *Salmonella typhimurium*: reversible and irreversible bacterial attachment and the role of bacterial motility. *Journal of General Microbiology*, **127**, 351–60.

Kaiser, D. (1989). Multicellular development in myxobacteria. In *Genetics of bacterial diversity*, ed. D. A. Hopwood & K. F. Chater, pp. 243–63. London: Academic Press.

Kannenberg, E. L. & Brewin, N. J. (1989). Expression of a cell surface antigen from *Rhizobium leguminosarum* 3841 is regulated by oxygen and pH. *Journal of Bacteriology*, **171**, 4543–8.

Karamanev, D. G. (1991). Model of the biofilm structure of *Thiobacillus ferrooxidans*. *Journal of Biotechnology*, **20**, 51–64.

Kass, E. H., Kendrick, M. I., Tsai, Y.-C. & Parsonnet, J. (1987). Interaction of magnesium ion, oxygen tension, and temperature in the production of toxic-shock-syndrome toxin-1 by *Staphylococcus aureus*. *The Journal of Infectious Diseases*, **155**, 812–15.

Kefford, B., Kjelleberg, S. & Marshall, K. C. (1982). Bacterial scavenging: utilization of fatty acids localized at a solid–liquid interface. *Archives of Microbiology*, **133**, 257–260.

Kepkay, P. E., Schwinghamer, P., Willar, T. & Bowen, A. J. (1986). Metabolism and metal binding by surface-colonizing bacteria: results of microgradient measurements. *Applied and Environmental Microbiology*, **51**, 163–70.

Kjelleberg, S., Humphrey, B. A. & Marshall, K. C. (1982). Effect of interfaces on small, starved marine bacteria. *Applied and Environmental Microbiology*, **43**, 1166–72.

Kjelleberg, S., Humphrey, B. A. & Marshall, K. C. (1983). Initial phases in the starvation and activity of bacteria at surfaces. *Applied and Environmental Microbiology*, **46**, 978–84.

Lawrence, J. R., Korber, D. R., Hoyle, B. D., Costerton, J. W. & Caldwell, D. E. (1991). Optical sectioning of microbial biofilms. *Journal of Bacteriology*, **173**, 6558–67.

Lévêque, F., Gazeau, M., Fromant, M., Blanquet, S. & Plateau, P. (1991). Control of *Escherichia coli* lysyl-tRNA synthetase expression by anaerobiosis. *Journal of Bacteriology*, **173**, 7903–10.

Lorian, V. (1971). Effects of antibiotics on staphylococcal hemolysin production. *Applied Microbiology*, **22**, 106–9.

Lorian, V. (1986). Effect of low antibiotic concentration. In *Antibiotics in laboratory medicine*, 2nd edn, ed. V. Lorian, pp. 596–661. Baltimore, Md: Williams & Wilkins.

Lorian, V. (1989). *In vitro* simulation of *in vivo* conditions: physical state of the culture medium. *Journal of Clinical Microbiology*, **27**, 2403–6.

Lorian, V., Atkinson, B., Waluschka, A. & Kim, Y. (1982). Ultrastructure, *in vitro* and *in vivo*, of staphylococci exposed to antibiotics. *Current Microbiology*, **7**, 301–4.

Lorian, V. & Popoola, B. (1972). Pneumococci producing beta hemolysis on agar. *Applied Microbiology*, **24**, 44–7.

Lorian, V., Tosch, W. & Joyce, D. (1985a). Weight and morphology of bacteria exposed to antibiotics. In *The influence of antibiotics on the host-parasite relationship*, Vol. II, ed. D. Adam, H. Hahn & W. Opferkuch, pp. 65–72. Berlin: Springer-Verlag.

Lorian, V., Zak, O., Kunz, S. & Vaxelaire, J. (1984). Staphylococcal endocarditis in rabbits treated with a low dose of cloxacillin. *Antimicrobial Agents and Chemotherapy*, **25**, 311–15.

Lorian, V., Zak, O., Suter, J. & Bruecher, C. (1985b). Staphylococci, *in vitro* and *in vivo*. *Diagnostic Microbiology and Infectious Diseases*, **3**, 433–44.

McCarter, L., Hilmen, M. & Silverman, M. (1988). Flagellar dynamometer controls swarmer cell differentiation of *V. parahaemolyticus*. *Cell*, **54**, 345–51.

McCarter, L. & Silverman, M. (1989). Iron regulation of swarmer cell differentiation of *Vibrio parahaemolyticus*. *Journal of Bacteriology*, **171**, 731–6.

McCarter, L. & Silverman, M. (1990). Surface-induced swarmer cell differentiation of *Vibrio parahaemolyticus*. *Molecular Microbiology*, **4**, 1057–62.

McCoy, W. F. & Costerton, J. W. (1982). Fouling biofilm development in tubular flow systems. *Developments in Industrial Microbiology*, **23**, 551–8.

McFeters, G. A., Egli, T., Wilberg, E. *et al.* (1990). Activity and adaptation of nitrilotriacetate (NTA)-degrading bacteria: Field and laboratory studies. *Water Research*, **24**, 875–81.

McLaren, A. D. & Skujins J. J. (1963). Nitrification by *Nitrobacter agilis* on surfaces and in soil with respect to hydrogen ion concentration. *Canadian Journal of Microbiology*, **9**, 729–31.

Macura, J. & Stotzky, G. (1980). Effect of montmorillonite and kaolinite on nitrification in soil. *Folia Microbiologia*, **25**, 90–105.

Makino, S., Sasakawa, C., Uchida, I., Terakado, N. & Yoshikawa, M. (1988). Cloning and CO_2-dependent expression of the genetic region for encapsulation from *Bacillus anthracis*. *Molecular Microbiology*, **2**, 371–6.

Maluszynska, G. M., Magnusson, K.-E. & Rosenquist, Å. (1992). Reduced environmental redox potential affects both transcription and expression of the pap pili gene. *Microbial Ecology in Health and Disease*, **5**, 257–67.

Marshall, K. C. (1971). Sorptive interactions between soil particles and microorganisms. In *Soil biochemistry*, Vol. 2, ed. A. D. McLaren & J. J. Skujins, pp. 409–45. New York: Marcel Dekker.

Marshall, K. C. (1976). *Interfaces in microbial ecology*. Cambridge, Mass: Harvard University Press.

Marshall, K. C. (1979). Bigeochemistry of manganese minerals. In *Biogeochemical cycling of mineral-forming elements*, ed. P. A. Trudinger & D. J. Swaine, pp. 253–92. Amsterdam: Elsevier.

Marshall, K. C. (1980). The role of surface attachment in manganese oxidation by freshwater hyphomicrobia. In *Biogeochemistry of ancient and marine environments*, ed. P. A. Trudinger, M. R. Walter & B. J. Ralph, pp. 333–7. Canberra: Australian Academy of Science.

Marshall, K. C. (1988). Adhesion and growth of bacteria at surfaces in oligotrophic habitats. *Canadian Journal of Microbiology*, 34, 503-6.

Marshall, K. C., Stout, R. & Mitchell, R. (1971). Mechanism of the initial events in the sorption of marine bacteria to surfaces. *Journal of General Microbiology*, **68**, 337–48.

Matin, A. (1991). The molecular basis of carbon-starvation-induced general resistance in *Escherichia coli*. *Molecular Microbiology*, **5**, 3–10.

Matthysse, A. G. (1986). Initial interactions of *Agrobacterium tumefaciens* with plant host cells. *CRC Critical Reviews in Microbiology*, **13**, 281–307.

Matthysse, A. G., Holmes, K. V. & Gurlitz, R. H. G. (1981). Elaboration of cellulose fibrils by *Agrobacterium tumefaciens* during attachment to carrot cells. *Journal of Bacteriology*, **145**, 583–95.

Meighen, E. A. (1991). Molecular biology of bacterial bioluminescence. *Microbiological Reviews*, **55**, 123–42.

Meir, E. & Yagil, E. (1990). Regulation of *Escherichia coli* catalases by anaerobiosis and catabolic repression. *Current Microbiology*, **20**, 139–43.

Mekalanos, J. J. (1992). Environmental signals controlling expression of virulence determinants in bacteria. *Journal of Bacteriology*, **174**, 1–7.

Miller, J. F., Mekalanos, J. J. & Falkow, S. (1989). Coordinate regulation and sensory transduction in the control of bacterial virulence. *Science*, **243**, 916–22.

Miller, C. A. & Neogi, P. (1985). *Interfacial phenomena. Equilibrium and dynamic effects*. New York: Marcel Dekker.

Mulvey, M. R., Switala, R., Borys, J. & Loewen, P. C. (1990). Regulation of transcription of *katE* and *katF* in *Escherichia coli*. *Journal of Bacteriology*, **172**, 6713–20.

Nealson, K. H. & Ford, J. (1980). Surface enhancement of bacterial manganese oxidation: implications for aquatic environments. *Geomicrobiology Journal*, **2**, 21–37.

Nealson, K. H. & Hastings, J. W. (1979). Bacterial bioluminescence: its control and ecological significance. *Microbiological Reviews*, **43**, 496–518.

Neihof, R. A. & Loeb, G. I. (1972). The surface charge of particulate matter in seawater. *Limnology and Oceanography*, **17**, 7–16.

Neihof, R. & Loeb, G. (1974). Dissolved organic matter in seawater and the electric charge of immersed surfaces. *Journal of Marine Research*, **32**, 5–12.

Ni Bhriain, N., Dorman, C. J. & Higgins, C. F. (1989). An overlap between osmotic and anaerobic stress responses: a potential role for DNA supercoiling in the coordinate regulation of gene expression. *Molecular Microbiology*, **3**, 933–42.

Nichols, W. W., Evans, M. J., Slack, M. P. E. & Walmsley, H. L. (1989). The penetration of antibiotics into aggregates of mucoid and non-mucoid *Pseudomonas aeruginosa*. *Journal of General Microbiology*, **135**, 1291–1303.

Nyström, T., Albertson, N. H., Flärdh, K. & Kjelleberg, S. (1990). Physiological and molecular adaptation to starvation and recovery from starvation by the marine *Vibrio* sp. S14. *FEMS Microbiology Ecology*, **74**, 129–40.

Orndorff, P. E. & Dworkin, M. (1982). Synthesis of several membrane proteins during developmental aggregation in *Myxococcus xanthus*. *Journal of Bacteriology*, **149**, 29–39.

Östling, J., Goodman, A. & Kjelleberg, S. (1991). Behaviour of IncP-1 plasmids and a miniMu transposon in a marine *Vibrio* sp: isolation of starvation inducible *lac* operon fusions. *FEMS Microbiology Ecology*, **86**, 83–94.

Paerl, H. W. & Merkel, S. M. (1982). Differential phosphorus assimilation in attached vs. unattached microorganisms. *Archiv für Hydrobiologie*, **93**, 125–34.

Pertsovskaya, A. F., Duda, V. I. & Zvyagintsev, D. G. (1972). Surface ultrastructures of adsorbed microorganisms. *Soviet Soil Science*, **4**, 684–9.

Phinney, D. G. & Hoober, J. K. (1992). Regulation of expression by divalent cations of a light-inducible gene in *Arthrobacter photogonimos*. *Archives of Microbiology*, **158**, 85–92.

Pitta, T., Godchaux, W. III & Leadbetter, E. R. (1993). Protein content of peptidoglycan of liquid-grown cells differs from that of surface-grown, gliding *Cytophaga johnsonae*. *Archives of Microbiology*, **160**, 214–22.

Power, K. & Marshall, K. C. (1988). Cellular growth and reproduction of marine bacteria on surface-bound substrate. *Biofouling*, **1**, 163–74.

Pruss, G. J., Manes, S. H. & Drlica, K. (1982). *Escherichia coli* DNA topoisomerase mutants: increased supercoiling is corrected by mutations near gyrase genes. *Cell*, **31**, 35–42.

Richter-Dahlfors, A. A. & Andersson, D. I. (1991). Analysis of an anaerobically induced promoter for the cobalamin biosynthetic genes in *Salmonella typhimurium*. *Molecular Microbiology*, **5**, 1337–45.

Roberson, E. B. & Firestone, M. K. (1992). Relationship between desiccation and exopolysaccharide production in a soil *Pseudomonas* sp. *Applied and Environmental Microbiology*, **58**, 1284–91.

Rosenberg, E. (1986). Microbial surfactants. *CRC Critical Reviews in Biotechnology*, **3**, 109–32.

Samuelsson, M.-O. & Kirchman, D. L. (1990). Degradation of adsorbed protein by attached bacteria in relationship to surface hydrophobicity. *Applied and Environmental Microbiology*, **56**, 3643–8.

Scanlan, D. J., Bloye, S. A., Mann, N. H., Hodgson, D. A. & Carr, N. G. (1990). Construction of *lacZ* promoter probe vectors for use in *Synechococcus*: application to the identification of CO_2-regulated promoters. *Gene*, **90**, 43–9.

Schellhorn, H. E. & Hassan, H. M. (1988). Transcriptional regulation of *katE* in *Escherichia coli* K-12. *Journal of Bacteriology*, **170**, 4286–92.

Schmoll, T., Ott, M., Oudega, B. & Hacker, J. (1990). Use of a wild-type gene fusion to determine the influence of environmental conditions on expression of the S fimbrial adhesin in an *Escherichia coli* pathogen. *Journal of Bacteriology*, **172**, 5103–11.

Shapiro, J. A. & Higgins, N. P. (1989). Differential activity of a transposable element in *Escherichia coli* colonies. *Journal of Bacteriology*, **171**, 5975–86.

Shapiro, J. A. & Hsu, C. (1989). *Escherichia coli* K-12 cell–cell interactions seen by time-lapse video. *Journal of Bacteriology*, **171**, 5967–74.

Shapiro, J. A. & Leach, D. (1990). Action of a transposable element in coding sequence fusions. *Genetics*, **126**, 293–9.

Shapiro, J. A. & Trubatch, D. (1991). Sequential events in bacterial colony morphogenesis. *Physica D*, **49**, 214–23.

Shaw, D. J. (1980). *Introduction to colloid and surface chemistry*, 3rd edn. London: Butterworths.

Shaw, D. J., Rice, D. W. & Guest, J. R. (1983). Homology between CAP and Fnr, a regulator of anaerobic respiration in *Escherichia coli*. *Journal of Molecular Biology*, **166**, 241–7.

Sheehan, B. J., Foster, T. J., Dorman, C. J., Park, S. &

Stewart, G. S. A. B. (1992). Osmotic and growth-phase dependent regulation of the *eta* gene of *Staphylococcus aureus*: a role for DNA supercoiling. *Molecular and General Genetics*, **232**, 49–57.

Shimamura, T., Watanabe, S. & Sasaki, S. (1985) Enhancement of enterotoxin production by carbon dioxide in *Vibrio cholerae*. *Infection and Immunity*, **49**, 455–6.

Shrihari, Kumar, R., Gandhi, K. S. & Natarajan, K. A. (1991). Role of cell attachment in leaching of chalcopyrite mineral by *Thiobacillus ferrooxidans*. *Applied Microbiology and Biotechnology*, **36**, 278–82.

Silverman, M. & Simon, M. (1980). Phase variation: genetic analysis of switching mutants. *Cell*, **19**, 845–54.

Slonczewski, J. L. (1992). pH-regulated genes in enteric bacteria. *American Society for Microbiology News*, **58**, 140–4.

Smit, G., Kijne, J. W. & Lugtenberg, B. J. J. (1986). Correlation between extracellular fibrils and attachment of *Rhizobium leguminosarum* to pea root hair tips. *Journal of Bacteriology*, **168**, 821–7.

Smit, G., Swart, S., Lugtenberg, B. J. J. & Kijne, J. W. (1992). Molecular mechanisms of attachment of *Rhizobium* bacteria to plant roots. *Molecular Microbiology*, **6**, 2897–903.

Smith, M. W. & Neidhardt, F. C. (1983). Proteins induced by anaerobiosis in *Escherichia coli*. *Journal of Bacteriology*, **154**, 336–43.

Spiro, S. & Guest, J. R. (1990). FNR and its role in oxygen-regulated gene expression in *Escherichia coli*. *FEMS Microbiology Reviews*, **75**, 399–428.

Stock, J. B., Stock, A. M. & Mottonen, J. M. (1990). Signal transduction in bacteria. *Nature*, **344**, 395–400.

Stotzky, G. (1986). Influence of soil mineral colloids on metabolic processes, growth, adhesion, and ecology of microbes and viruses. In *Interactions of soil minerals with natural organics and microbes*, Special Publication No. 17, pp. 305–428. Madison, Wis: Soil Science Society of America.

Straley, S. C. & Bowmer, W. S. (1986). Virulence genes regulated at the transcriptional level by Ca^{2+} in *Yersinia pestis* include structural genes for outer membrane proteins. *Infection and Immunity*, **51**, 445–54.

Switalski, L., Höök, M. & Beachey, E. (ed.) (1989). *Molecular mechanisms of microbial adhesion*. New York: Springer-Verlag.

Thomas, A. D. & Booth, I. R. (1992). The regulation of expression of the porin gene *ompC* by acid pH. *Journal of General Microbiology*, **138**, 1829–35.

Turakhia, M. H., Cooksey, K. E. & Characklis, W. G. (1983). Influence of a calcium-specific chelant on biofilm removal. *Applied and Environmental Microbiology*, **46**, 1236–8.

Umino, Y. & Shiraiwa, Y. (1991). Effect of

metabolites on carbonic anhydrase induction in *Chlorella regularis*. *Journal of Plant Physiology*, **139**, 41–4.

Underhill, S. E. & Prosser, J. I. (1987). Surface attachment of nitrifying bacteria and their inhibition by potassium ethyl xanthate. *Microbial Ecology*, **14**, 129–39.

Vandevivere, P. & Kirchman, D. L. (1993). Attachment stimulates exopolysaccharide synthesis by a bacterium. *Applied and Environmental Microbiology*, **59**, 3280–6.

van Loosdrecht, M. C. M., Lyklema, J., Norde, W. & Zehnder, A. J. B. (1990). Influence of interfaces on microbial activity. *Microbiological Reviews*, **54**, 75–87.

Vernade, D., Herrera-Estrella, A., Wang, K. & van Montagu, M. (1988). Glycine betaine allows enhanced induction of the *Agrobacterium tumefaciens vir* genes by acetosyringone at low pH. *Journal of Bacteriology*, **170**, 5822–9.

Wagner, V. T. & Matthysse, A. G. (1992). Involvement of a vitronectin-like protein in attachment of *Agrobacterium tumefaciens* to carrot suspension cells. *Journal of Bacteriology*, **174**, 5999–6003.

Wakao, N., Mishina, M., Sakurai, Y. & Shiota, H. (1984). Bacterial pyrite oxidation III. Adsorption of *Thiobacillus ferroxidans* cells on solid surfaces and its effect on iron release from pyrite. *Journal of General and Applied Microbiology*, **30**, 63–77.

Wang, J.-Y. & Syvanen, M. (1992). DNA twist as a transcriptional sensor for environmental changes. *Molecular Microbiology*, **6**, 1861–6.

Watson, N., Dunyak, D. S., Rosey, E. L., Slonczewski, J. L. & Olson, E. R. (1992). Identification of elements involved in transcriptional regulation of the *Escherichia coli cad* operon by external pH. *Journal of Bacteriology*, **174**, 530–40.

Winkelman, J. W. & Clark, D. P. (1986). Anaerobically induced genes of *Escherichia coli*. *Journal of Bacteriology*, **167**, 362–7.

Worcel, A. & Burgi, E. (1972). On the structure of the folded chromosome of *Escherichia coli*. *Journal of Molecular Biology*, **71**, 127–47.

Zhang, L. & Kerr, A. (1991). A diffusible compound can enhance conjugal transfer of the Ti plasmid in *Agrobacterium tumefaciens*. *Journal of Bacteriology*, **173**, 1867–72.

Zimmermann, A., Reimmann, C., Galimand, M. & Haas, D. (1991). Anaerobic growth and cyanide synthesis of *Pseudomonas aeruginosa* depend on *anr*, a regulatory gene homologous with *fnr* of *Escherichia coli*. *Molecular Microbiology*, **5**, 1483–90.

ZoBell, C. E. (1943). The effect of solid surfaces upon bacterial activity. *Journal of Bacteriology*, **46**, 39–56.

Zonneveld, B. J. M. (1988). Effect of carbon dioxide on fruiting in *Aspergillus nidulans*. *Transactions of the British Mycological Society*, **91**, 625–9.

5

Biochemical Reactions and the Establishment of Gradients within Biofilms

Julian W. T. Wimpenny and Sarah L. Kinniment

Biofilms and the physiology of microorganisms within them

Microbiologists have been accustomed until recently to think of microbes as homogeneous cultures grown in well mixed containers ranging in size from shake flasks to large industrial continuously stirred tank reactors. The investigative tool of choice has been the chemostat which is homogeneous not only in spatial terms but, when operating at a steady state, in time as well.

The only other traditional badge of the microbiologist is the colony. This is much more representative of the natural ecosystem since it is a microbiological aggregate dominated by diffusion gradients. For example oxygen only penetrates about 25–35 µm into a rapidly growing young colony (Wimpenny & Coombs 1983).

Microbial ecosystems are generally spatially heterogeneous implying that solutes move down concentration gradients between sources and sinks. Such gradients are found over a huge range of dimensions, from nanometres for pH gradients around clay lattices to hundreds of metres in the case of oxygen gradients in the Black Sea. These scale factors are illustrated, generally for oxygen, in Table 5.1.

Biofilms have been defined in many different ways; however, perhaps the simplest view is that it is a microbial aggregate that forms at phase interfaces. The most common biofilms appear at solid–water interfaces epitomized by the epilithon that forms on submerged rocks in streams and other water bodies. Generally such biofilm development follows a fairly standard life history. Clean surfaces become coated with a conditioning film consisting of organic molecules, for example proteins or polysaccharides. A little later individual cells attach to the surface, first loosely and reversibly and then firmly and irreversibly. These cells start to proliferate forming microcolonies which spread and eventually coalesce. At the same time extracellular polysaccharides are produced which help form a matrix within which the primary colonizing bacteria and later secondary colonizers exist and grow. Additional materials from the environment (clay and other mineral particles, organic detritus, etc.) may be incorporated into the biofilm adding to its mechanical structure. As the film deepens so diffusion gradients form, the most important of which is due to oxygen tension. Below the point where oxygen disappears through respiration, anaerobic species can develop. These may cause corrosion if the substratum is steel and the anaerobes sulphate reducers or they may be fermentative species generating acids from sugars leading to demineralization of dental enamel. Often the anaerobes lyse and generate gas, both processes leading to destabilization of the film which then sloughs off leaving a relatively clean surface which can be colonized again.

The intact biofilm is a community whose development is strongly influenced by solute diffusion gradients. It is important to understand the complex physiological changes that can take place in biofilm.

Table 5.1. Gradient and scale in some structured microbial communities

Scale	Scale factor in terms of μm	System
nm	$c.\ 10^{-3}$	Diffuse double layer around clay lattices leading to short range pH gradients
20–100 μm	$2 \times 10^1 – 1 \times 10^2$	Oxygen gradients across a bacterial colony profile
>100 μm	$>10^2$	Oxygen gradients in bacterial biofilms
0.5–10 mm	$5 \times 10^2 – 1 \times 10^4$	Oxygen gradients into an aquatic sediment
2–5 mm	$2–5 \times 10^3$	Oxygen gradients across a saturated soil crumb
10–100 cm	$1 \times 10^5 – 1 \times 10^6$	Oxygen gradient across a soil profile
50–500 cm	$5 \times 10^5 – 5 \times 10^6$	Oxygen gradients across a thermally stratified lake water profile
5–500 m	$5 \times 10^6 – 5 \times 10^8$	Oxygen gradients across a stably stratified water body (Lake Gek' Gel, the Black Sea

Some physiological considerations

Bacteria have a wide variety of sophisticated regulatory procedures which ensure that they can adapt to large changes in environmental conditions. Some of these systems are only becoming obvious through current research. These include the responses of bacteria to stress and the possibility of signal molecules leading to changes amongst neighbouring groups of bacteria. Many other regulatory mechanisms have been known and well understood for many years. These remarks are presented here to illustrate the complex problems microbial physiologists have in understanding regulatory processes in systems showing significant spatial heterogeneity over a very short distance.

To take just one example, a biofilm consisting of a facultative anaerobe such as *Escherichia coli* exposed at its surface to oxygen and thick enough to become anaerobic at some point will be subject to control by oxygen tension. As we showed in our earlier chemostat work (Wimpenny 1969; Wimpenny & Necklen 1971), at least four adaptive phases can be recognized though the changes represent a continuum rather than discrete steps. In the absence of oxygen characteristically anaerobic enzymes are induced. These include formate hydrogen lyase and hydrogenase; also fumarate reductase and elevated amounts of other fermentative enzymes. Meanwhile aerobic TCA cycle enzymes are at a low level. In the presence of limiting amounts of oxygen anaerobic enzymes are repressed and the terminal electron transport chain consisting of cytochromes and cytochrome oxidase enzymes proliferates, presumably to capture oxygen molecules as effectively as possible.

TCA cycle enzymes remain low. At saturating oxygen concentrations the terminal electron transport chain falls to a lower level whilst the TCA cycle enzymes increase since now reducing equivalents rather than electron acceptor becomes limiting. The final phase is seen at higher oxygen tensions where the cells show signs of oxygen toxicity. Growth yield and the levels of most enzymes all fall. If nitrate or nitrite is present as well as oxygen other adaptive changes to derepress nitrate reduction systems will become important once the oxygen tension falls to low enough values.

If such changes take place in a biofilm they represent only *one* aspect of regulation over the film profile. In addition there may be other nutrient gradients, for example of carbon and energy source or of nitrogen. Excess glucose at the film surface may lead to catabolite repression here. This will be relieved at some point within the film and in the end nutrient depletion may lead to starvation at lower points. Starvation may lead to the derepression of a family of proteins concerned with maintaining the cell under stress conditions.

Some of the possible physiological control mechanisms that may affect cells growing in a biofilm are listed in Table 5.2.

The point is that the existence of a structured community such as a biofilm means that a number of different factors vary simultaneously as a function of depth. These include electron donating and accepting energy sources, especially the ratio between the two; other growth substrates; products manufactured from these by cells at different positions; excreted materials such as polysaccharides and extracellular enzymes; and

Table 5.2. Some of the many physiological phenomena which may be expressed by microorganisms in a biofilm

Phenomenon	Determinants
Growth rate	Nutrient type and concentration. Abiotic variables (pH, T, a_w, pressure, etc.) Toxic agents, signal molecules
Electron transport/energy metabolism	Presence of electron donors and acceptors (O_2, NO_3^-, NO_2^-, SO_4^{2-}, CO_2, etc.)
Catabolic systems	Catabolites (induction of specific pathways); catabolite repression in the presence of 'good' substrates such as glucose
Anabolic systems	Biosynthetic precursors (repression)
Expression of stress genomes	Abiotic variables
EPS production	Growth rate (?), adhesion (?), nutrients
Resistance to antimicrobials	Growth rate (?), contact (?)
Motility	Catabolite derepression, chemotactic agent
Swarming	Surface contact, chemotactic agent
Sporulation	Nutrient deprivation, stress
Morphological differentiation	Signal molecules, abiotic variables
Death	Nutrient deprivation, unbalanced growth, toxic compounds, predation, parasitism, etc.
Coaggregation	Specific binding sites, propinquity
Corrosion	Specific chemical activity
Depolymerisation	Specific enzyme activity
Sloughing	Starvation, death, lysis, anaerobiosis and gas generation
Competition	Propinquity, presence of another organism
Cooperation	Propinquity, presence of another organism

products of cell lysis, particularly from cells at the base of the film that may be starved. The response of the cell to these gradients is multivalent. Control systems identified in homogeneous systems like the chemostat (i) may not operate in the same way in a biofilm (ii) may not operate in the same way as *one component* in a number of different control mechanisms all changing simultaneously.

The task of dissecting the biochemistry and physiology of such systems may seem almost insurmountable put like this. It seems to follow that whatever investigative procedure is used it should generate reproducible results. This is true whether the experimenter is working *in situ* with natural biofilms or whether he is using laboratory model systems. Our contention is that an important advantage is to employ a biofilm generator capable of operating under steady state conditions.

Biofilm fermentors

There are numerous different biofilm model systems available. As expressed by a recent Dahlem Conference on the *Structure and function of biofilms* (Characklis & Wilderer 1989, p. 179):

The variety of experimental model systems in use is enormous. The development of models is a dynamic area. Each scientist or engineer seems to have his or her own favourite system. The reason for this is that no single model or group of models satisfies the requirements for every experimental question that is asked.

A few of the more important model systems are briefly reviewed in Table 5.3. For further details of biofilm models see Characklis & Wilderer 1989; Peters & Wimpenny 1988.

The importance of steady state systems

The chemostat, described independently in 1950 by Monod and by Novick and Szillard, has proved an immensely powerful tool in microbiology. Its main advantage has been that a cell population can be maintained for long periods (indefinitely?) in the same state of exponential growth controlled either by nutrient supply (the chemostat) or by some other parameter such as cell density (the turbidostat). Multistage chemostats have also been developed to model unidirectionally flowing systems where conditions are allowed to change from vessel to vessel. A more general example of multistage systems is the gradostat described by Lovitt & Wimpenny

Table 5.3. Some representative biofilm fermentors

Type of biofilm model	Description	Advantages	Disadvantages	Reference
Robbins device	Tubular flow system with removable film plugs	Good analogue of real flow systems. A 'natural' biofilm model	Not true steady state, growth may be position dependent, depending on flow rate	McCoy et al. 1981; Nickel et al. 1985
The Rototorque	Concentric system with rotating inner cylinder and outer equipped with removable sampling slides	Establish biofilm in a constant shear field	Not suitable for many biofilms. Steady states may be possible	Kornegay & Andrews, 1967; Characklis, 1989
The 'baby factory'	Growth on a nutrient perfused membrane	Models attachment, steady state operation at completely controlled growth rates, separates growth from spatial heterogeneity	Does not incorporate diffusion gradient; nutrient flow by percolation not typical of natural biofilms	Gilbert et al. 1989, 1990
Irrigated discs	Growth on discs, e.g. hydroxyapatite, etc. upon which nutrients drop	Simple to operate	Not steady state; not very reproducible???	Jones 1994
Constant depth film fermentor	Film formed in pans maintained at constant depths by scraper blade	Steady state operation, reproducible	Not characteristic of most natural biofilms	Coombe et al. 1984; Peters & Wimpenny 1988

(1981). This is a model of diffusion gradient systems since flow is bidirectional from sources and sinks at each end of a linear array of vessels.

There is a semantic problem concerning the term 'steady state'. It tends to imply that nothing changes: however, this would be a mistake because even in a pure monoculture in a single stage chemostat where all generally measured parameters such as cell concentration, protein, viable numbers, absorbance, product formation, etc. seem to be constant there are continual slight genetic changes due to periodic selection and elimination of fitter or less fit cell lines.

This problem is more serious in a biofilm system since the latter is developing in the framework of significant gradients in environmental physicochemistry. Matters are much more seri-ous if a community of different bacterial species are used from which to construct the biofilm.

There is no easy solution to the problem except to qualify the words by terms such as 'quasi' steady state. A much simpler alternative is that the term steady state should be defined for each application. For example we may define a steady state biofilm as one in which the recovered viable cell numbers, the total protein content, the dry weight and perhaps the total carbohydrates within the film are constant over a period of time. This implies nothing about the metabolic state of cells or the distribution of bacteria within the film if it is a mixed community.

Finally, with these strictures in mind it seems essential to work with a system that provides as constant a background for experimentation as possible. If it falls short of a *perfect* steady state system this is to be deplored but in a pragmatic sense should still be accepted!

The constant depth film fermentor (CDFF)

Design criteria

Each of the many types of biofilm fermentor available has been designed for a particular purpose. It is important to set out design criteria for a system which is capable of use in investigating the structural and functional organization of a biofilm. The following seem important:

1. Steady state operation. For the reasons set out above the system should generate a biofilm which develops under conditions that are close to steady state since this constant background allows the investigator to perform unequivocal experiments by altering key factors one at a time and recording responses of the system to the change.
2. The biofilm formed must be reproducible, minimizing the possibility of variations that could lead to misinterpreting results.
3. Sufficient samples must be available to enable the researcher to assess the statistical significance of the data generated.
4. The system must be capable of operating under closely controlled environmental conditions.
5. It must be possible to use different substrata on which the film grows.
6. Biofilm must be formed in a geometrically regular form suitable for further processing. In particular the depth must be constant to allow the accurate deployment of microelectrodes. In addition constant dimensions allow the film to be cryosectioned to determine the distribution of film components.
7. It should be possible to produce different depth biofilms.
8. The system must be able to operate under aseptic conditions for prolonged periods. This also entails the removal and replacement of biofilm samples without contamination.

Design

The earliest type of constant depth fermentors were described by Atkinson & Fowler (1974). These were based on inclined plates with scraper blades that moved over the surfaces. They were not intended for aseptic operation or for the easy removal of samples but did make a significant step forward in the design of steady state systems. The constant depth film fermentor (CDFF) was designed with the above criteria in mind (Coombe *et al.* 1981, 1984; Peters & Wimpenny

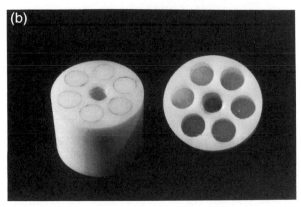

Fig. 5.1. The constant depth film fermentor (CDFF). (a) The CDFF plus associated tools for placing and removing film pans, for setting the depth of the film plugs and for undoing the assembly bolts. (b) Film pans containing six film plugs.

1988). It is illustrated in Fig. 5.1a and b. A borosilicate glass tubing section is fitted with top and bottom plates fabricated from stainless steel. A small 12 V electric motor and reduction gear drives a shaft which enters the fermentor via seals and a bearing assembly and connects to a rotating steel turntable in which 15 removable film pans are located. Each pan contains six removable film plugs. These are recessed a measured amount (usually 300 μm) using a steel template. Biofilm forms in the space between the surface of the film plug and a stationary spring-loaded PTFE scraper blade located above the turntable. The scraper blade removes excess biofilm growth as it forms above the surface of the film pan. The fermentor can be sterilized by autoclaving and is operated under aseptic conditions at constant temperature, rotation speed and gas composition.

Nutrient medium is pumped into the fermentor via grow-back traps and is then fed to the top of the turntable by an 'S' shaped stainless steel tube where it is distributed over the film plugs by the scraper blade. In general the fermentor is inoculated with an overnight culture of the organism or organisms required. This allows organisms to attach to the plug surfaces. The system is then connected to an aspirator containing fresh medium, which is pumped into the fermentor via a peristaltic pump at a selected constant rate.

To sample the CDFF the drive is stopped so that the pan required is located below a sampling port. The latter is sprayed with alcohol, flamed and opened to allow the removal of a complete film pan which can then be replaced with a sterile film pan. Removal and replacement are carried out with a stainless steel tool which screws into a threaded hole at the centre of the pan.

Film geometry
The biofilm measures approximately 4.75 mm in diameter and is set to 0.3 mm deep. This gives a volume of 5.32 mm^3 (say, about 5 mm^3) or 5×10^9 μm^3. If a microbe has a volume of about 1 μm^3, then given perfect packing, each film plug could contain up to 5×10^9 bacteria. Assuming full pans and a water content in living cells of 80% w/v, the maximum amount of dry cell mass present will be about 1 mg. If dry bacterial biomass is about 50% protein each film disc would contain no more than 500 μg protein. In practice values around half this are found, which is not unreasonable considering that no allowance is made in these calculations for loose packing or interstitial polysaccharide.

The CDFF was employed in a number of experiments which will be described in more detail in later sections of this chapter.

Biofilms and solute transport
Transport processes in biofilm

Microbial film, consisting as it does of a community of bacteria held within a matrix of extracellular polysaccharide, is a system which at one extreme is dominated by molecular diffusion. At the other end of the biofilm spectrum transport processes are much more irregular. In such systems pores in the gel structure, columns of cells and irregular colonization of the substratum mean that solutes are carried by bulk transport processes combined with molecular diffusion into and out of the film.

Where diffusion is the most important process, solutes are transferred down a concentration gradient from sources to sinks. The rate of transfer is determined by source and sink concentrations, the diffusion coefficient, and the rate at which the solute is removed from the system, for example by metabolism. In addition the existence of stagnant films at liquid biofilm interfaces interposes a further diffusion resistance to solute transfer.

Other factors affect solute diffusion: these include the presence of insoluble particulate matter which increases the tortuosity of the diffusion path. In addition the number and type of fixed charged groups in the biofilm matrix can be vitally important to the transfer of charged molecules. If predominantly anionic charged groups are present these can absorb protons formed by the ionization of acidic fermentation products for example lactic or acetic acid. Where, as in dental plaque systems, acidity is generated from fermentable sugars, the resulting protonated fixed groups maintain a low pH in the film. It is gradually neutralized as diffusing carbonate and phosphate anions enter the structure mopping up protons. The undissociated acids form a gradient from inside to outside the film and diffuse away from it. In the saliva where the pH is more alkaline, these acids dissociate and the protons are carried away leaving the anions to repeat the process of proton removal from the acid groups in the plaque. The charged groups behave as weak ion exchangers and can establish Donnan equilibria between the inside and the outside of the film. These in turn will affect the flux and distribution of ionized species (Dibdin 1990, 1992, 1994).

The diffusive boundary layer

Mass transfer across a water biofilm surface is due to eddy diffusion under turbulent flow conditions. Close to the film surface, surface friction retards the flow to the extent that it is dominated by viscous forces. The viscous sublayer produced shows little vertical mixing and at the interface itself water flow rates approach zero. Molecular rather than eddy diffusion determines solute transport through this layer. It is this region that has been called the diffuse boundary layer (DBL) or the diffusive or viscous sublayer. The importance of the DBL has been stressed by Jørgensen & Des Marais (1990), who examined this layer in microbial mat from saline ponds. The mean DBL thickness varied from 590 to 160 μm as the water flow velocity increased from 0.3 to 7.7 cm s^{-1} measured at a point 1 cm above the mat. Over this range of flow rates the oxygen consumption rate of the film rose from 3.9 nmol to 9.4 nmol cm^{-2} min^{-1}. Three dimensional maps of the sediment surface revealed a serious level of surface roughness on the μm scale. This was reflected by a damped profile for the surface of the DBL. The surface roughness increased the oxygen uptake rate across the sediment water interface by some 49% compared with calculations from one dimensional profiles of oxygen tension.

Diffuse boundary layers are also present in biofilms where they also provide a barrier to oxygen transfer into or out of the biofilm.

Determining diffusion coefficients and porosities

As was stressed above a knowledge of the diffusion coefficient for oxygen is essential in determinations of respiration rate profiles. Unfortunately diffusion coefficients can vary and what is needed is a method for determining diffusion coefficients at different points across the film profile. Revsbech (1989) has decribed such a technique applied to sediments. The sediments were killed by exposure to 5 mM HgCl$_2$. Apparent diffusion coefficients were determined at constant temperatures by measuring oxygen profiles as a function of time across samples exposed to air saturated water alternating with oxygen or nitrogen saturated water. Data were analysed using a computer simulation to determine the diffusion coefficients.

The calculation of diffusive flux requires a knowledge of both the porosity and the diffusion

Table 5.4. Characteristic penetration times expressed as distance squared divided by the diffusion coefficient for glucose at 30 °C

Distance	Time
0.1 μm	5 μs
1 μm	1.5 ms
10 μm	0.15 s
100 μm	15 s
1 mm	25 min
1 cm	4 h
10 cm	24 weeks
1 m	47 years

From Nichols (1994).

coefficient. A method was developed to determine the two together. It relies on establishing a steady state diffusion gradient across a thin sample of material. The sample is placed on a supported silicone membrane which is freely oxygen permeable and is placed over oxygen saturated water. The upper surface is exposed to air so that there is a stable gradient from top to bottom. Generally an agar layer is also incorporated into the system to provide calibrating data since the diffusion coefficient in agar is well known. It is then very simple to calculate the diffusivity–porosity term simply as a ratio of the slopes of the gradient in agar and in the sample respectively. If the diffusion coefficient is known for the sample separately, for example by computer simulation, the porosity can easily be determined as a separate function. Fuller details for these methods can be found in Revsbech (1989).

Rates of transfer by molecular diffusion

Molecular diffusion is a process where molecules move through Brownian motion from sources to sinks down a concentration gradient. Movement is by a random walk where molecules within a 'cage' of neighbouring molecules change places. Diffusion, as Table 5.4. shows, is very fast on a small scale and extremely slow on larger scales.

Reaction and diffusion

Depths of solute penetration

A simple relationship beween rate of substrate utilization, concentration gradient and diffusion coefficient is as follows:

$$x = \sqrt{2Dc_x/Q_x}$$

where x is the distance to which solute x will penetrate, D its diffusion coefficient, c_x the driving concentration of x and Q_x the rate at which x is removed from the system.

Where the cell concentration is high as in a bacterial colony Pirt (1967) has calculated that oxygen diffuses about 40 μm into the structure. Pirt assumed the following values: D for oxygen, 1.9×10^{-5} cm^2 s^{-1}; c_x at the colony surface, 8×10^{-6} g cm^{-3}; Q_x 2×10^{-5} g O$_2$ cm^{-3} colony s^{-1}. These correspond well with experimental measurements of oxygen penetration into colonies of *E. coli*, *Bacillus cereus* and *Staphylococcus albus* (Wimpenny & Coombs 1983; Peters *et al.* 1987). Pirt also calculated that if cells were respiring endogenously this figure would rise to 127 μm. We have observed that old colonies of *B. cereus* are fully aerobic since by this time they have oxidized all available carbon and energy source in the medium.

The cell concentration in biofilms may be generally lower than that in colonies; however, the cell concentration was high in a *Pseudomonas aeruginosa* biofilm grown on a triethanolamine–fatty acid medium whilst oxygen diffused about 150 μm into the structure, suggesting that the film was energy limited.

Transfer by flow through channels

The trouble with solute transfer in many biofilms is that it cannot be said to be by molecular diffusion alone. For example, polysaccharides can generate a matrix which may contain pores through which liquids can flow. Many biofilms are highly irregular in structure. Some may contain columns of cells surrounded by an aqueous region (Keevil 1994). Where there is significant flow as in a stream or in a water conduit, solute transfer will be partly by eddy diffusion, partly by molecular diffusion. At all events it will be significantly speeded up.

The balance of oxidant and reductant

The balance between oxidant and reductant entering the system is of utmost importance to the physiology of a biofilm. Reductant (often but not always a carbon and energy source), may be very restricted in some natural ecosystems, in particular in epilithic films growing on surfaces associated with oligotrophic water bodies. In the mouth, resting saliva might contain 5 mg l^{-1} of glucose and oxygen might penetrate to the base of a plaque biofilm. High sucrose concentrations will stimulate respiration and the plaque may become anaerobic at some point across its profile. In these anaerobic regions fermentation will take place and the pH will drop. Three of many possible scenarios are illustrated in Fig. 5.2a and b. The first set indicates possibilities with a monoculture of aerobic bacteria. Here the depletion of oxygen will lead to death of bacteria as oxygen becomes exhausted. If nutrients are limited and excess oxygen is available, cells below the point at which nutrients disappear will be aerobic but starved. There may be regrowth of bacteria at the expense of other lysed cells in this region so the situation is not necessarily symmetrical.

Where mixed cultures containing aerobes and anaerobes or facultatively anaerobic species are present growth may be possible throughout the film even when oxygen disappears above the base of the film. Obligate anaerobic species commonly find a niche in many biofilms under such conditions. Naturally the fate of such a biofilm will also be determined by the availability of carbon

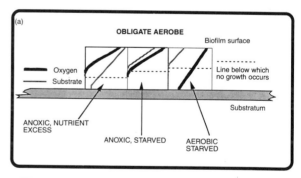

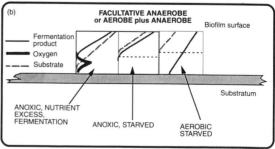

Fig. 5.2. Diagram of oxygen and substrate penetration into a biofilm of (a) an obligate aerobe, (b) a facultative anaerobe or mixture of aerobe and anaerobe.

and energy sources. A shortage of the latter may allow oxygen to diffuse to the base of the film with potentially disastrous consequences to anaerobes present.

Microelectrode studies

One of the most significant advances in the last decade has been the application of microelectrodes to structured microbial ecosystems. Whilst microprobes have been employed by physiologists since at least the 1950s, their application in microbiology arrived comparatively late on the scene.

Oxygen

One of the earliest applications of micro-oxygen electrodes in microbiology was to microbial film. Bungay *et al.* (1969) and Whalen *et al.* (1969) grew a natural biofilm on glass slides to a depth of about 200 μm and determined oxygen profiles within the film. They used the data to determine diffusivities and respiration rates for these biofilms. Jørgensen and his colleagues started work with microelectrodes in the late 1970s (see, for example, Jørgensen *et al.* 1979) and have applied this technology with increasing sophistication to algal mat communities (e.g. Revsbech *et al.* 1983), marine sediments (Revsbech *et al.* 1980) and biofilms (Revsbech 1989a). Revsbech (1989b) has reviewed the application of microelectrodes in microbial ecology. Our own group has used oxygen microelectrodes since 1983 to investigate oxygen gradients in microbial colonies (Wimpenny & Coombs 1983; Peters *et al.* 1987).

Determination of oxygen consumption rates
The simplest method for determining the respiration rates of a non-photosynthetic sediment or biofilm is to enclose it in a watertight container containing oxygen saturated water and to measure the rate of oxygen disappearance using a Clark type oxygen electrode. In flow systems measuring the difference between the oxygen partial pressure in influent versus effluent water can give the same data and is commonly used in performing mass balances for biofilms important in process engineering. Using microelectrodes the net oxygen uptake rate can be calculated from the slope of the linear oxygen concentration pro-

file across the diffuse boundary layer above the film surface using the one dimensional version of Fick's first law (Revsbech & Jørgensen 1986). These methods say nothing about the vertical distribution of respiration within the structure (sediment, algal mat, biofilm, etc.) itself.

In the simplest case, where a steady state oxygen profile is determined in the absence of photosynthesis and where the diffusion coefficient for oxygen and the porosity of the matrix is known, it is quite simple to calculate respiration at different depths using equations derived from Fick's first and second laws (see, for example, equations 7 and 8 in Revsbech & Jørgensen 1986). Determining respiration rates where conditions are dynamically changing, including of course during light–dark shifts to determine photosynthetic activity, it is necessary to have recourse to computer simulations to calculate respiration rates.

In an interesting set of experiments which did *not* use the oxygen microelectrode, Patel & Bott (1991) determined the effects of medium flow rate on oxygen penetration into a *P. aeruginosa* biofilm. These workers constructed a system resembling the Robbins device in which a bacterial culture flowed through a tapered tube section containing removable test pieces in which an oxygen electrode could be fitted flush with the section surface. Biofilm grew over the test surface and electrode and measurements of oxygen tension could be recorded at regular intervals. The taper ensured that flow rates were a function of the tube cross-sectional area. Biofilm grew well in the system increasing to more than 500 μm after 14 days. They confirmed that oxygen penetration increased as a function of flow rate. However, the amount of oxygen available to cells at the base of the film is very small and above about 300 μm thickness, flow velocity has very little effect on oxygen tension in this region.

Photosynthetic oxygen generation in biofilms.
Many naturally occurring biofilms contain a wide range of physiological types of microorganism including eukaryotic and/or prokaryotic oxygen generating photosynthetic species. Switching between light and dark allows one to determine the rate of photosynthetic oxygen generation. This is because under light conditions there is a steady state where oxygen production just balances oxygen removal. Thus the rate of oxygen removal immediately after changing to dark con-

ditions will be the same as the light generated oxygen supply. The experiment can be done at different depths so that a profile of photosynthetic oxygen formation is easily produced. This value gives the gross photosynthetic rate. Net oxygen production can be determined by measuring the total oxygen flux away from the photosynthetic region and subtracting this from the gross photosynthetic rate. The difference between net and gross oxygen fluxes is assumed to be total respiration. The experimental technique just described underestimates photosynthetic oxygen generation in the most active regions and overestimates it further away from this position. This is because of the rapid diffusion of oxygen away from the active zone during the transition from light to dark conditions. Computer modelling was therefore used to determine both gross photosynthesis and respiration including photorespiration. For a full discussion of this approach, see Revsbech *et al.* (1986) and Glud *et al.* (1992).

Kuenen *et al.* (1986) determined oxygen gradients across trickling filter biofilms containing photosynthetic and other organisms and in non-photosynthetic bacterial films. In the former, 60–70% of the oxygen generated by photosynthesis was consumed by respiratory process in the biofilm. About one third of this oxygen removal

was thought to be due to photorespiration.

Wimpenny *et al.* (1993) measured oxygen profiles in *P. aeruginosa* biofilm that was grown on an amine:carboxylate medium in the CDFF. One example of such a profile is shown below (Fig. 5.3). The figure suggests that oxygen penetrates to about 150 μm into the film.

Nitrous oxide:oxygen microelectrodes

Understanding reactions of the nitrogen cycle is important in the ecology and physiology of biofilms, especially those involved in effluent treatment. The development of combined oxygen–nitrous oxide microelectrodes by Revsbech *et al.* (1988) has provided a route for investigating the spatial organization of denitrification. The secret is to block nitrous oxide reduction to nitrogen gas with acetylene and then to measure the presence of N_2O. Dalsgaard & Revsbech (1992) used such a probe to determine the N_2O and O_2 profiles in a trickling filter biofilm. The data were used to determine the flux of each reactant and the mass balances involved. Modelling of the microelectrode profiles used programs developed by Revsbech *et al.* (1986). No N_2O was detected in the absence of added nitrate or nitrite. Denitrification, generally an anaerobic process, only occurred at oxygen partial pressures less than 20 μM. The addition of NH_4^+ increased the rate of denitrification, presumably by sparing demand by the cells for a source of biosynthetic nitrogen. The biofilm contained photosynthetic species. In the dark oxygen penetrated 200–300 μm into the film whilst in the light oxygen was measured down to 1200 μm from the surface. In this region denitrification ceased, but recovered immediately when the dark oxygen levels were restored.

Sulphide

Sulphide microelectrodes are comparatively simple to construct. They can be made from platinum glass microelectrodes where the platinum is etched away into a recess in the glass. The platinum is silver plated and the silver coated with sulphide by dipping the tip on to ammonium sulphide for a short period (for fuller details see Revsbech & Jørgensen, 1986).

These probes have been used to investigate sulphide gradients in natural ecosystems

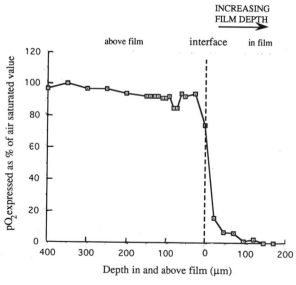

Fig. 5.3. Oxygen penetration into a 'steady state' *Pseudomonas aeruginosa* biofilm growing on an amine:carboxylate medium.

(Revsbech *et al.* 1983; Nelson *et al.* 1986); however, they have not so far been used extensively in biofilm research. This is a great pity since a variety of economically important biofilms are anaerobic at their bases and, if the substratum is steel, can lead to the formation of corrosion cells. A key activity in this process is the anaerobic reduction of sulphate to sulphide by sulphate reducing bacteria (Hamilton, 1985).

pH

pH has been determined at the base of a biofilm by growing the film on the surface of a full size electrode fitted into the wall of a cylinder in which the film formed (Szeverinski *et al.* 1986). In this system nitrification in the film led to a drop in pH of about a unit at its base.

However, pH measurements using microelectrode methods have been known for many years and can give useful information. There are two main types. (i) A family of different designs based on the use of pH sensitive glass. These can be accurate, long lasting and often very hard to make. (ii) Liquid ion exchange (LIX) microelectrodes. These are simple glass capillaries whose tips are rendered hydrophobic before filling the tip with ion exchanger. These electrodes are easy to make, tedious to calibrate and last only a short time!

Revsbech *et al.* (1983) have used pH microelectrodes in conjunction with oxygen and sulphide microprobes to determine changes over the diurnal cycle in an algal mat. During maximum solar irradiance the pH near the surface of the mat rose to >9.5!

LIX pH microelectrodes have been used by us (Robinson *et al.* 1991) to determine pH profiles in colonies of *B. cereus* grown on an amino acid containing medium in the presence and absence of glucose. Without the fermentation substrate the pH rose as the organism oxidized amino acids. When glucose was present the pH fell but only in thicker parts of the colony where oxygen was likely to be scarce.

A preliminary investigation of pH gradients across a monoculture biofilm of *P. aeruginosa* revealed only small pH changes when the film was grown on a triethanolamine:fatty acid mixture (S. L. Kinniment, unpublished data).

Glucose

Recently enzyme electrodes have been modified for use as microprobes. These electrodes were described by Cronenberg *et al.* (1991, 1993). A glucose microelectrode having a 9 μm diameter tip was used to measure glucose profiles in upflow anaerobic sludge blanket reactor pellets by Lens *et al.* (1993). This electrode had a 95% response time of 1 min and a linear calibration from 0 to 5 mM. If stored at room temperature in water saturated air the electrode could be used for up to 5 days.

Others

There are other possible microelectrodes that should be used in biofilm research. The E_h (redox potential) electrode is easy to construct since it consists only of a bare platinum wire etched to the appropriate dimensions and requiring only a sensitive millivoltmeter to operate. E_h is a useful measure of the availability of oxidant and has been used to determine the point at which oxygen becomes exhausted in water bodies and in gel models of oil storage tank water bases (Wimpenny *et al.* 1981). It is more sensitive than an oxygen electrode since it monitors the *response* of the system to oxygen reduction rather than to oxygen itself.

A growing family of ion exchangers is now available for use in liquid ion exchange microelectrodes such as the pH electrode described above. These include Na^+, K^+ and Ca^{2+} probes. Other probes for CO_2 and for specific organic chemicals using immobilized enzymes, have been mooted (see Revsbech & Jørgensen 1986).

Sectioning studies

Spatial heterogeneity in biofilm systems can be determined for some species using microelectrode methods; however, a key enabling technology is the ability physically to separate 'layers' within the film. This of course can be done in either the vertical or the horizontal dimension. If the former it is possible to visualize organisms across the whole profile. The advantages here may lie in the application of fluorescent or gold labelled antibodies to determine the position of different members of the biofilm community. In addition there is also the possibility of assaying specific enzyme activities using enzyme histo-chemistry techniques to determine the position of specific markers of the prevailing physiology at that point.

Sectioning in the horizontal direction gives slices of a uniform composition reflecting a position in the film array. Such slices can be assayed further using almost any method provided it is sensitive enough. It is not normally possible to section fresh material, though this was actually achieved in experiments using a tissue slicer by Watanabe (Characklis & Wilderer 1989). Normally the material is fixed and embedded for electron or visible microscopy and using some methods for identifying organisms using imm-unological agents. A valuable alternative is cryosectioning which means cutting sections from the frozen material. This can, of course, be done in the vertical or the horizontal directions. Horizontal cryosections retain much of their biological activities and many assays can be performed on this material as the next section indicates.

Cryosectioning

Ritz (1969) used freeze sectioning coupled with fluorescent antibody labelling to investigate the location of bacteria in dental plaque. He showed that the aerobic *Neisseria* was located in the upper part of the film whilst the anaerobic *Veilonella* appeared at deeper points within the structure.

Some time was spent in our laboratory, developing methods for freeze-sectioning biofilm, in particular to examine the distribution of viability and of the adenylates ATP, ADP and AMP (Kinniment & Wimpenny 1992; Wimpenny *et al.* 1993).

A major problem emerged when attempting to determine viable counts on sectioned material. Freezing, sectioning, thawing and dispersal of the cells led to a high and variable degree of cell death, viable counts often falling by four orders of magnitude. Therefore to reduce the cell death dextran (60–90 kDa) at a concentration of 25% w/v, was finally used as a cryoprotectant (Ashwood-Smith & Warby 1971). Freezing and thawing in liquid nitrogen indicated reasonable protection, with a maximum drop in viability of up to five-fold. The distribution of viable cells across the 300 µm (Fig. 5.4), shows that viability is highest a little below the surface of the film and

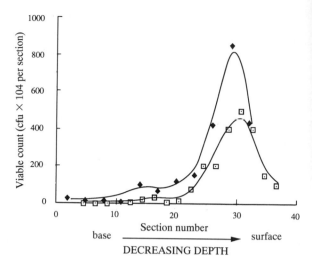

Fig. 5.4. Viable counts across a 'steady state' *Pseudomonas aeruginosa* biofilm growing on an amine:carboxylate medium. The film was exposed to a dextran cryoprotectant before freeze-sectioning.

falls off to about two orders of magnitude less near the film base.

It is unlikely that the gradients in viability were due to a lack of dextran penetration through the film, since separate experiments, where dextran was determined in biofilm sections, indicated that it had penetrated to the base of the biofilm. Indeed higher levels of dextran were found near the base of the film than near its surface. This was probably because the dextran was located in intercellular spaces which would be lower where viable cells were packed closely a little below the surface. On the other hand, the presence of many dead and lysed cells near the film base would allow more dextran to accumulate there.

Changes in viability can be related to other evidence concerning film heterogeneity. Viability results correspond to transmission electron micrographs (Fig. 5.5) which suggest that there were many dead and lysed cells near the base of the film, whilst cell density is highest at a point a little below the film surface.

Adenylates and the energy status of biofilm

Another important experimental approach is to examine the energy status of biofilm. Scourfield (1990) and Wimpenny *et al.* (1989) have examined biofilms of oral bacteria grown in the CDFF. Adenylates were extracted in perchloric acid and determined with standard firefly lantern assays in a luminometer. Adenylate energy charge values were then calculated from the data at different points in the development of a biofilm. In addition, glucose was added to the basic amino acid containing medium at concentrations from 0 to 100 mg L^{-1}. In most cases the average energy charge value for the whole biofilm was between about 0.25 and 0.45. However, at the earliest times and with the highest amount of glucose, these values rose to between *c.* 0.6 and just under 1.0 (Fig. 5.6).

These results suggest that the biofilm as a whole may be energy limited, especially where glucose concentrations were low but also as the film thickened and diffusion barriers became important. These data fit well with transmission electron micrographs of the films grown on the same medium containing 10 mg L^{-1} glucose, which show a sharp transition between what appear to be healthy cells and the largely empty walls of dead and lysed cells. Scourfield (1990) also examined changes in adenylates in a biofilm as a function of time. When the biofilm was exposed to a sudden increase in sucrose to emulate what might happen to a natural plaque on being 'fed'!, the results showed a transitory but rapid rise in energy charge of the film (Fig. 5.7).

More recently the distribution of adenylates across a biofilm profile has been determined

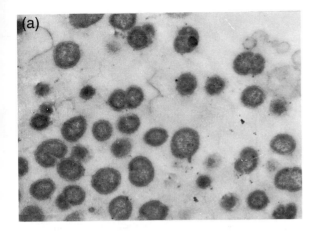

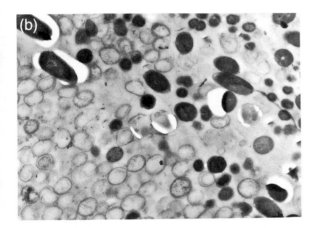

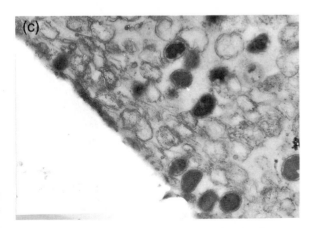

Fig. 5.5. Transmission electron micrographs of cells present in an oral biofilm grown in the CDFF. (a) Surface of biofilms; (b) region in lowe two thirds of biofilm; (c) base of biofilm. Magnification × 12 000 (Scourfield 1990).

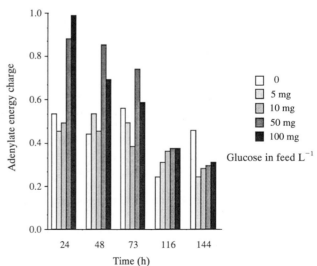

Fig. 5.6. Adenylate energy charge values present in oral biofilms as a function of age and of increasing concentrations of glucose (Scourfield 1990).

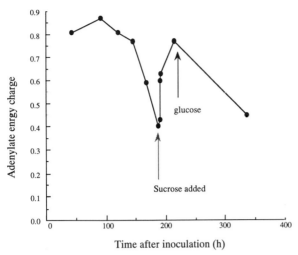

Fig. 5.7. Changes in energy charge value on pulsing an oral biofilm with sucrose (Scourfield 1990).

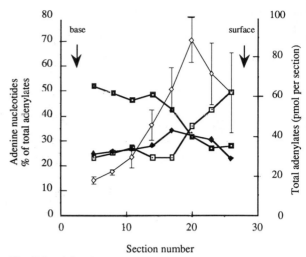

Fig. 5.8. Adenylates present across the profile of a 'steady state' *Pseudomonas aeruginosa* biofilm growing on an amine:carboxylate medium. □, AMP; ◆, ADP; ▫, ATP; ◇, total adenylates.

(Kinniment & Wimpenny 1992). The results of these experiments on steady state biofilm showed that total adenylates peaked in a region a little below the film surface, which again coincided with the point at which cell density was highest when measured by viability distribution and transmission electron microcroscopy. The distribution of individual adenylates expressed as a percentage of the total, showed a trend where AMP fell from the base to the surface of the film, ATP rose and ADP, though variable, tended to be approximately constant throughout the profile. Fig. 5.8 shows one example of these measurements. The data shown were subject to a three point moving average, since the errors involved in freezing, sectioning and assaying the nucleotides generated a noisy graph.

Energy charge values of steady state biofilm (Fig. 5.9, shows one example) showed a trend that was quite compatible with the earlier dental plaque work. Values were low at the base of the film, ranging between 0.25 to 0.35, rising to 0.4–0.5, occasionally 0.6, at the surface of the film.

These results must be seen in conjunction with viability data which had a difference of up to two orders of magnitude across the profile. The exact status of adenine nucleotides near the base of the film is uncertain. They may be associated only with live cells or alternatively may be dispersed in the film matrix outside lysed cells or bound to other cellular components.

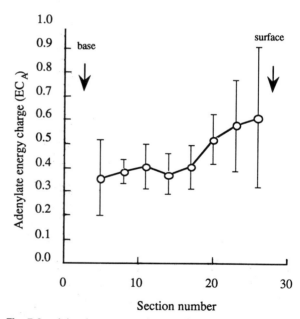

Fig. 5.9. Adenylate energy charge values across the profile of a 'steady state' *Pseudomonas aeruginosa* biofilm growing on an amine:carboxylate medium.

Microscopy

Microscopy is to some extent peripheral to the problems discussed in this chapter, since it is not easy to adapt such a technique to determinations of biochemistry of physiology. Having said that, an advantage here could be use of enzyme histo-chemical techniques, which, when combined with microscopy, will reveal the presence of specific activities and their location within the biofilm. So far almost nothing has been done in this area.

Biofilm generated in the CDFF has been examined by both scanning and transmission electron microscopy. When the surface of a steady state *P. aeruginosa* biofilm grown on amine:carboxylate medium was examined using SEM it appeared to be densely packed with rod shaped bacteria (Fig. 5.10). Examination by TEM of the same film, and a film, generated from a five membered dental plaque community, showed that there appeared to be healthy cells present in the upper two thirds of the film, below this a large number of cells appeared to be lysed (Fig. 5.11). This suggests that in steady state biofilm nutrients diffuse downwards to a reproducible position below which cells are starved and/or anaerobic.

More recently a dental plaque biofilm community of nine organisms has been examined using electron microscopy. The nine organisms were representative of bacteria commonly found in the oral cavity: *Neisseria subflava, Lactobacillus casei, Actinomyces viscosus, Streptococcus mutans, S. oralis, S. gordonii, Fusobacterium nucleatum, Veillonella dispar, Porphyromonas gingivalis*. A similar community has been studied in the chemo-

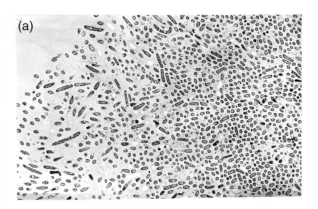

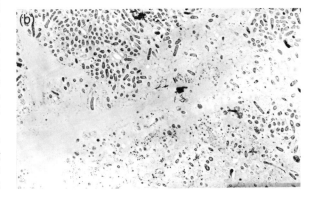

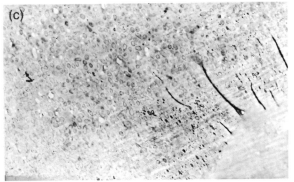

Fig. 5.11. Transmission electron micrographs across a steady state *Pseudomonas aeruginosa* biofilm growing on an amine:carboxylate medium. (a)–(c) Upper, middle and basal layers: (a) Surface of biofilm; (b) near visible interface between viable and apparently lysing cells; (c) base of film.

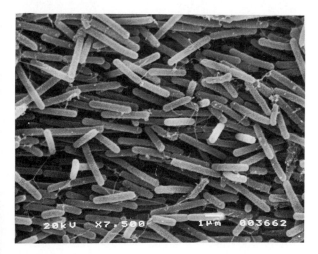

Fig. 5.10. The surface of a steady state *Pseudomonas aeruginosa* biofilm growing on an amine:carboxylate medium examined using SEM. Bar = 1 μm.

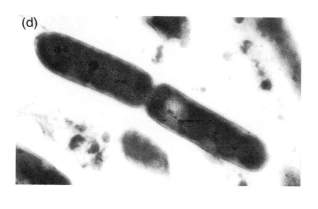

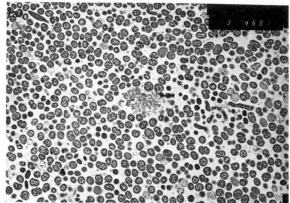

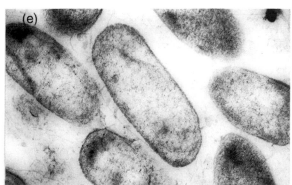

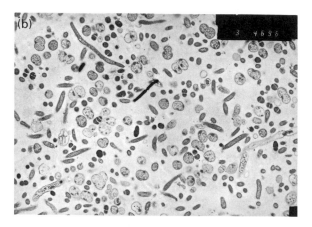

Fig. 5.11 *(cont.).*   (d), (e) Details of cells taken from the upper and basal layers: (d) 'viable' cell from near the surface of film; (e) empty 'dead' cell from near the film base. Magnification × 2500.

stat by Marsh and colleagues (Bradshaw *et al.* 1989, 1990), yielding valuable information on population dynamics. Electron micrographs indicated, judging purely from bacterial morphology, that *F. nucleatum* (a filamentous anaerobic organism) may be mainly confined to the lower regions of the biofilm whilst the upper layers may be mostly *N. subflava* (an aerobic 'bean shaped' organism) (Fig. 5.12).

We are conscious that because of the paucity of experimental information, little is known of the biochemistry and the establishment of gradients within biofilms. We have tried to identify key 'enabling' technologies that can contribute to the investigation of heterogeneity within biofilms and believe in the importance of experimental film models capable of generating highly reproducible, preferably steady state biofilm samples. Using this material, there are three main lines of investigation that seem important.

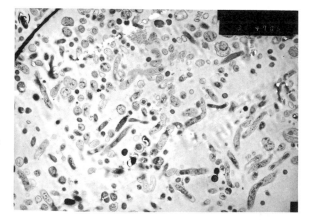

Fig. 5.12.   Transmission electron micrographs of (a) upper, (b) middle and (c) basal layers of an biofilm produced by a community of nine oral bacteria. Magnification × 3800.

1. The deployment of microelectrodes coupled with mathematical simulations capable of giving rate data for the production and utilization of solutes for which electrodes are available.
2. Sectioning, in particular cryosectioning. The opportunities here are almost unlimited. The distribution of viable bacteria (with appropriate care taken to protect cells from freezing and thawing) will help to reveal the spatial organization of mixed population biofilms. In addition the location of substrates, products, inhibitors, signal molecules etc. should be relatively simple to determine, given that sensitive enough assays are available. The use of radioactive tracers here will also be helpful.
3. Microscopy is a generally important tool. Sectioning, either cryosectioning or fixation and embedding can be used to locate cells, cell fractions and presumably enzymes using fluorescent or gold labelled antibodies. Finally, enzyme histochemistry could be used to determine the position of particular enzyme activities within the film.

References

Ashwood-Smith, M. J. & Warby, C. (1971). Studies on the molecular weight and cryoprotective properties of polyvinylpyrolidone and dextran with bacteria and erythrocytes. *Cryobiology*, **8**, 453–464.

Atkinson, B. & Fowler, H. W. (1974). The significance of microbial film in fermenters. *Advances in Biochemical Engineering*, **3**, 224–77.

Bradshaw, D. J., McKee, A. S. & Marsh, P. (1989). The use of defined inocula stored in liquid nitrogen for mixed culture chemostat studies. *Journal of Microbiological Methods*, **9**, 123–8.

Bradshaw, D. J., McKee, A. S. & Marsh, P. (1990). Prevention of population shifts in oral microbial communities *in vitro* by low fluoride concentrations. *Journal of Dental Research*, **69**, 436–441.

Bungay, H. R., Whalen, W. J. & Sanders, W. M. (1969). Microprobe technique for determining diffusivities and respiration rates in microbial slimes. *Biotechnology and Bioengineering*, **11**, 765–772.

Characklis, W. G. (1989). Laboratory biofilm reactors. In *Biofilms*, ed. W. G. Characklis & K. C. Marshall, New York: Wiley.

Characklis, W. G. & Wilderer, P. A. (ed.) (1989). *Structure and Function of Biofilms*. Chichester: Wiley.

Coombe, R. A., Tatevossian, A. & Wimpenny, J. W. T. (1981). Bacterial thin films as *in vitro* models for dental plaque. In *Surface and colloid phenomena in the oral cavity: methodological aspects*, ed. R. M. Frank & S. A. Leach, pp. 239–49. London: Information Retrieval.

Coombe, R. A., Tatevossian, A. & Wimpenny, J. W. T. (1984). Factors affecting the growth of thin bacterial films *in vitro*. In *Bacterial adhesion and preventative dentistry*, ed. J. M. ten Cate, S. A. Leach & J. Arends, p. 193. Oxford: IRL Press.

Cronenberg, C. C. H., van den Heuvel, J. C. & Ottengraf, S. P. P. (1993). Direct measurement of glucose profiles in immobilised yeast gels with a pH-insensitive microelectrode under anaerobic conditions. *Biotechnology Technology*, **7**, 237–42.

Cronenberg, C., van Groen, B., de Beer, D. & van den Heuvel, H. (1991). Oxygen independent glucose microsensor based on glucose oxidase. *Analytica Chimica Acta*, **242**, 275–8.

Dalsgaard, T. & Revsbech, N. P. (1992). Regulating factors of denitrification in trickling filter biofilms as measured with the oxygen/nitrous oxide microsensor. *FEMS Microbiology Ecology*, **101**, 151–64.

Dibdin, G. H. (1990). Plaque fluid and diffusion: study of the cariogenic challenge by computer modelling. *Journal of Dental Research*, **69**, 1324–31.

Dibdin, G. H. (1992). A finite difference computer model of solute diffusion in bacterial films with simultaneous metabolism and chemical reaction. *Computer Applications in the Biological Sciences*, **8**, 489–500

Dibdin, G. (1994). Computer modelling of diffusion/reaction in biofilms. In *Bacterial biofilms and their control in medicine and industry*, ed. W. Nichols, J. W. T. Wimpenny, D. J. Stickler & H. M. Lappin-Scott, 77–81. Cardiff: Bioline.

Gilbert, P. J., Allison, D. G., Evans, D. J., Handley, P. S. & Brown, M. R. (1989). Growth rate control of adherent bacterial populations. *Applied and Environmental Microbiology*, **55**, 1308–11.

Gilbert, P., Collier, P. J. & Brown, M. R. (1990). Influence of growth rate on susceptibility to antimicrobial agents: biofilms, cell cycle, dormancy, and stringent response. *Antimicrobial Agents and Chemotherapy*, **34**, 1865–8.

Glud, R. N., Ramsing, N. B. & Revsbech, N. P. (1992). Photosynthesis and photosynthesis-coupled respiration in natural biofilms quantified with oxygen microsensors. *Journal of Phycology*, **28**, 51–60.

Hamilton, W. A. (1985). Sulphate reducing bacteria and anaerobic corrosion. *Annual Reviews of Microbiology*, **39**, 195–217

Jones, M. V. (1994). Biofilms in the food industry. In *Bacterial Biofilms and their control in medicine and industry*, ed. W. Nichols, J. W. T. Wimpenny, D. J. Stickler & H. M. Lappin-Scott, pp. 113–16. Cardiff: Bioline.

Jørgensen, B. B. & Des Marais, D. J. (1990). The diffusive boundary layer of sediments: oxygen microgradients over a microbial mat. *Limnology and Oceanography*, **35**, 1343–55.

Jørgensen, B. B., Kuenen, J. G. & Cohen, Y. (1979).

Microbial transformations of sulfur compounds in a stratified lake (Solar Lake, Sinai). *Limnology and Oceanography*, **24**, 799–822.

Keevil, C. W. (1994). Methods for assessing the activity of biofilm microorganisms *in situ*. In *Bacterial biofilms and their control in medicine and industry*, ed. W. Nichols, J. W. T. Wimpenny, D. J. Stickler & H. M. Lappin-Scott, pp. 45–7. Cardiff: Bioline.

Kinniment, S. L. & Wimpenny, J. W. T. (1992). Measurements of the distribution of adenylate concentrations and adenylate energy charge across *Pseudomonas aeruginosa* biofilms. *Applied and Environmental Microbiology*, **58**, 1629–35.

Kornegay, B. H. & Andrews, J. F. (1967). *Characteristics and kinetics of fixed-film biological reactors*. Final report, grant WP-01181, Washington, DC: Federal Water Pollution Control Administration.

Kuenen, J. G., Jørgensen, B. B. & Revsbech, N. P. (1986). Oxygen microprofiles of trickling filter biofilms. *Water Research*, **20**, 1589–98.

Lens, P., de Beer, D., Cronenberg, C., Houwen, F., Ottengraf, S. & Verstraete, W. (1993). Inhomogenic distribution of microbial activity in UASB aggregates: pH and glucose microprofiles. *Applied and Environmental Microbiology*, **59**, 3803–15.

Lovitt, R. W. & Wimpenny, J. W. T. (1981). The gradostat, a bidirectional compound chemostat, and its applications in microbiological research. *Journal of General Microbiology*, **127**, 261–8.

McCoy, W. F., Bryers, J. D., Robbins, J. & Costerton, J. W. (1981). Observations in fouling biofilm formation. *Canadian Journal of Microbiology*, **27**, 910–17.

Monod, J. (1950). La technique de culture continue. Théorie et applications. *Annales de l'Institut Pasteur*, **79**, 390–409.

Nelson, D. C., Jørgensen, B. B. & Revsbech, N. P. (1986). Growth pattern and yield of a chemoautotrophic *Beggiatoa* sp. in oxygen-sulphide microgradients. *Applied and Environmental Microbiology*, **52**, 225–33.

Nichols, W. W. (1994). Biofilm permeability to antibacterial agents. In *Bacterial biofilms and their control in medicine and industry*, ed. J. W. T. Wimpenny, W. Nichols, D. J. Stickler & H. M. Lappin-Scott, pp. 141–9. Cardiff: Bioline.

Nickel, J. C., Ruseska, I., Wright, J. B. & Costerton, J. W. (1985). Tobramycin resistance of cells of *Pseudomonas aeruginosa* growing as a biofilm on urinary catheter material. *Antimicrobial Agents and Chemotherapy*, **27**, 619–24.

Novick, A. & Szilard, L. (1950). Description of the chemostat. *Science*, **112**, 715–16.

Patel, T. D. & Bott, T. R. (1991). Oxygen diffusion through a developing biofilm of *Pseudomonas*

aeruginosa. *Journal of Chemical Technology and Biotechnology*, **52**, 187–99.

Peters, A. C. & Wimpenny, J. W. T. (1988). A constant depth laboratory model film fermentor. *Biotechnology and Bioengineering*, **32**, 263–70.

Peters, A. C., Wimpenny, J. W. T. & Coombs, J. P. (1987). Oxygen profiles in, and in the agar beneath, colonies of *Bacillus cereus*, *Staphylococcus albus* and *Escherichia coli. Journal of General Microbiology*, **133**, 1257–63.

Pirt, S. J. (1967). A kinetic study of the mode of growth of surface colonies of bacteria and fungi. *Journal of General Microbiology*, **47**, 181–97.

Revsbech, N. P. (1989a). Diffusion characteristics of microbial communities determined by use of oxygen microsensors. *Journal of Microbiological Methods*, **9**, 111–22.

Revsbech, N. P. (1989b). Microsensors: spatial gradients in biofilms. In *Structure and function of biofilms*, ed. W. G. Characklis & P. A. Wilderer, pp. 129–44. Chichester: Wiley.

Revsbech, N. P., Jørgensen, B. B. & Blackburn, T. H. (1983). Microelectrode studies of the photosynthesis and O_2, H_2S, and pH profiles of a microbial mat. *Limnology and Oceanography*, **28**, 1062–74.

Revsbech, N. P. & Jørgensen, B. B. (1986). Microelectrodes: their use in microbial ecology. *Advances in Microbial Ecology*, **9**, 293–352.

Revsbech, N. P., Madsen, B. & Jørgensen, B. B. (1986). Oxygen production and consumption in sediments determined at high spatial resolution by computer simulation of oxygen microelectrode data. *Limnology and Oceanography*, **31**, 293–304.

Revsbech, N. P., Nielsen, L. P., Christensen, P. B. & Sørensen, J. (1988). Combined oxygen and nitrous oxide microsensor for denitrification studies. *Applied and Environmental Microbiology*, **54**, 2245–9.

Revsbech, N. P., Sørensen, J., Blackburn, T. H. & Lomholt, J. P. (1980). Distribution of oxygen in marine sediments measured with microelectrodes. *Limnology and Oceanography*, **25**, 403–11.

Ritz, H. L. (1969). Fluorescent antibody labelling of *Neisseria*, *Streptococcus* and *Veillonella* in frozen sections of human dental plaque. *Archives of Oral Biology*, **14**, 1073–83.

Robinson, T., Wimpenny, J. W. T. & Earnshaw, R. G. (1991). pH gradients through colonies of *Bacillus cereus* and the surrounding agar. *Journal of General Microbiology*, **137**, 2285–9.

Scourfield, M. A. (1990). An investigation into the structure and function of model dental plaque communities using a laboratory film fermenter. PhD thesis, University College Cardiff, Wales.

Szwerinski, H., Arvin, E. & Harremoës (1986). pH-decrease in nitrifying biofilms. *Water Research*, **20**, 971–6.

Whalen, W. A., Bungay, H. R. & Sanders, W. M.

(1969). Microelectrode determination of oxygen profiles in microbial slimes. *Environmental Science and Technology*, **3**, 1297–8.

Wimpenny, J. W. T. (1969). Oxygen and carbon dioxide as regulators of microbial growth and metabolism. *Symposia of the Society for General Microbiology*, **19**, 161–97.

Wimpenny, J. W. T. & Coombs, J. P. (1983). The penetration of oxygen into bacterial colonies. *Journal of General Microbiology*, **129**, 1239–42.

Wimpenny, J. W. T., Coombs, J. P., Lovitt, R. W. & Whittaker, S. G. (1981). A gel-stabilized model ecosystem for investigating microbial growth in spatially ordered solute gradients. *Journal of General Microbiology*, **127**, 277–87.

Wimpenny, J. W. T., Kinniment, S. L. & Scourfield, M. A. (1993). The physiology and biochemistry of biofilm. *Society for Applied Bacteriology Technical Series*, **30**, 51–94.

Wimpenny, J. W. T. & Necklen, D. K. (1971). The redox environment and microbial physiology. I. The transition from anaerobiosis to aerobiosis in continuous cultures of facultative bacteria. *Biochimica et Biophysica Acta*, **253**, 352–9.

Wimpenny, J. W. T., Peters, A. C. & Scourfield, M. A. (1989). Modelling spatial gradients. In *Structure and function of biofilms*, ed. W. G. Characklis & P. A. Wilderer, pp. 111–27. Chichester: Wiley.

6

Mechanisms of the Protection of Bacterial Biofilms from Antimicrobial Agents

Peter Gilbert and Michael R. W. Brown

Introduction

Growth of microorganisms in close association with soft animal and plant tissue surfaces or upon non-living surfaces such as pipework, submerged materials or particulates in soil (Costerton *et al.* 1987), leads to the generation of microbial biofilms. *In vivo* biofilms often represent mono-cultures (for example, *Staphylococcus epidermidis* infections of indwelling medical devices), whilst in other situations they are more likely to be comprised of diverse species and represent not only bacteria but also algae, protozoa and fungi. Organization of microbial populations into biofilms is thought to confer many advantages upon the component organisms (Costerton *et al.* 1985). These include *in vivo* protection from host immune defences, and conferment of some degree of homeostasis with respect to the physicochemical environment for growth. The extent to which the glycocalyx may modify the growth environment of the cells depends greatly upon the thickness of the biofilm and the density of the cells within it. In this latter respect, whilst the glycocalyx can function as an ion exchange column and exclude large, highly charged molecules, most solutes will equilibrate across it to be accessed by the resident cell population. Diffusion limitation by the glycocalyx, together with localized high densities of cells, will create gradients across the biofilm; thus, for biofilms established upon impervious surfaces, biological demand for oxygen creates within them microaerophilic/anaerobic conditions within the depth of the film yet aerobic conditions at the surface. Similar secondary metabolite, pH and nutrient gradients will also be established across the thickness of the biofilm.

Cells at different parts of the biofilm will experience different nutrient and physicochemical environments. It is well established that microbial cells are adaptive to their environments and will express different physiologies in response to varied growth conditions (Brown *et al.* 1990). Such plasticity of the microbial cell is particularly evident with respect to their envelopes. Biofilms will therefore consist not only of mixed species but also of different phenotypes of each given species according to its spatial location in the consortium. Growth rates are likely to be slow in the depths of biofilms owing to shortages of particular nutrients. Cooperative relationships may also exist between the component species within a biofilm, whereby the growth of one organism is dependent upon nutrient provision or modification by another. Biofilms therefore have a spatial organization and structure which obliges microbiologists to consider them as dynamic microcosms of life rather than alternatives to planktonic growth. Such biofilms can be extensive and achieve biomasses which cause biofouling and physical blockage of industrial pipework, heat exchangers and fuel lines (Costerton & Lashen 1984).

In other circumstances, less extensive biofilms may act as sources of infection of humans, as is the case of growth of *Legionella* spp. (Lee & West 1991) in cooling and hot water systems. Biofilms formed on soft tissue surfaces (Lippincott & Lippincott 1969) may liberate toxins to the body (e.g. diphtheria), or may cause physical obstructions within vital organs with associated immune mediated tissue damage (e.g. *Pseudomonas* lung infections in cystic fibrosis patients). In recent years, infections associated with indwelling medical devices such as cardiac pacemakers, catheters and intra-arterial shunts (Bisno & Waldvogel 1989) have become noteworthy, not only because they often involve organisms not previously recognized as pathogens (e.g. S. *epidermidis*) but also because they resist antibiotic treatment and cause persistent, chronic infections. Resistance of biofilms and attached organisms to antimicrobial treatments is not a problem which is unique to the biomedical field (Nickel *et al.* 1985; Gristina *et al.* 1987) but is also widely experienced in the environmental industries where all but the most vigorous antifouling strategies fail to eliminate the problem (Costerton & Lashen 1984). This chapter addresses our current understanding of the mechanisms associated with such resistance.

Mechanisms of resistance

Various suggestions have been made as to the mechanisms responsible for the recalcitrance of bacterial biofilms towards chemical and antibiotic treatments. Evidence exists to support the validity of each hypothesis. In view of the generality of the observation to a wide range of chemical entities, it is unlikely that any single mechanism underlies resistance. Rather, the processes will apply to differing extents according to the nature of the biofilm, its age and location.

Whilst some hypotheses concentrate upon a role for the glycocalyx in moderating access of biocides to the sessile cells, others suggest that the limited availability of key nutrients within the biofilm imposes a slowing of the specific growth rate and expression of phenotypes which are not typical of planktonic cells in the same medium. Nutrient gradients within the biofilm might, in this instance, lead to a dominance of metabolically dormant cells witin its depths. An alternative explanation is that attachment to surfaces causes the cells to derepress/induce genes associated with a sessile existence which coincidentally affect antimicrobial susceptibility (see Goodman & Marshall, Chapter 4).

Central to all of these hypotheses is the principle that attached cells within a biofilm differ significantly from their planktonic laboratory grown counterparts. Initial screening processes leading to the development of antimicrobial systems have generally been developed and optimized in laboratories, and employ planktonic broth cultures as inocula. As a consequence, the possibility must not be overlooked that resistance in biofilms may partially reflect an inappropriate selection of compounds for development during the screening process, rather than simply innate recalcitrance of biofilms to antimicrobial agents. Whilst explanations for the observed lack of sensitivity towards today's antimicrobial agents may well be explicable in physiological terms (Brown *et al.* 1990; Gilbert *et al.* 1990), such propositions cannot be disregarded until primary screens for antimicrobial substances routinely involve realistic, relevant challenge systems (Gilbert *et al.* 1987).

Direct role of the glycocalyx in resistance

Electron microscope studies of antibody stabilized preparations of biofilm, and X-ray crystallographic analysis show the glycocalyx to be an ordered array of fine fibres providing a thick, continuous, hydrated, polyanionic matrix around the cells (Costerton *et al.* 1981). Possession of such an envelope will undoubtedly influence the access of molecules and ions to the cell wall and membrane (Costerton *et al.* 1981). It is not surprising, therefore, that the glycocalyx has been widely suggested to hinder the access of antimicrobial compounds to the cell surface. In this respect it has been proposed that the recalcitrance of biofilms might be solely exclusion (Costerton 1977; Slack & Nichols 1981, 1982; Costerton *et al.* 1987).

The most simple *in vitro* method of generating biofilms to study antimicrobial sensitivity is to inoculate the surface of an agar plate to produce a confluent growth. Such cultures, whilst not fully duplicating the *in vivo* situation, have been suggested to model the close proximity of individual cells to one another and the various gradients found in biofilms. In this respect, colonies grown on agar may be representative of biofilms at solid–air interfaces. The polyanionic nature of agar, however, will affect the availability of antimicrobials included within it. Comparisons of susceptibility between agar grown organisms must therefore involve harvested cells, free from agar residuals. Whilst this negates any protective effects of the glycocalyx, significant increases in minimum inhibitory concentration and minimum bacteriocidal concentration have been reported for agar grown *Haemophilus influenzae* (Bergeron *et al.* 1987), *Pseudomonas aeruginosa* (De Matteo *et al.* 1981), *Escherichia coli* (Al-Hiti & Gilbert 1983; Holhl & Felber 1988) and *S. epidermis* (Kurian & Lorian 1980) relative to broth cultures. Ideally, cells should be removed from the agar surface prior to susceptibility testing yet be maintained as intact biofilm. This has been achieved by inoculating the cultures onto cellulose membranes placed on the agar (Millward & Wilson 1989; Nichols *et al.* 1989). Such studies, whilst lacking physiological control, have led to challenges of the exclusion hypotheses. Such challenges have been made on the basis that differences in the diffusion coefficients, for antibiotics such as tobramycin and cefsoludin, between hydrated exopolymers and water are insufficient to account for the observed changes in minimum inhibitory/bactericidal concentrations for entrapped cells (Nichols *et al.* 1988, 1989).

Gristina *et al.* (1989) examined antibiotic susceptibilities of *S. epidermis* biofilms composed of slime producing and non-slime producing strains. No differences were observed, suggesting that the presence of slime *per se* was not of significance in antibiotic penetrations. Similarly, Evans *et al.* (1990a) showed no difference in control of intact and resuspended biofilms of *E. coli* by tobramycin. Other studies, however, have demonstrated clear reductions in susceptibility associated with the expression of a mucoid phenotype. Thus, Evans *et al.* (1991) assessed the susceptibility towards the quinolone antimicrobial ciprofloxacin, of mucoid and non-mucoid strains of *P. aeruginosa* grown as biofilms (Fig. 6.1). In this instance possession of a mucoid phenotype was associated with decreases in susceptibility which were no longer apparent when the biofilm cells were removed from the glycocalyx. Such data suggest that ciprofloxacin interacts in some manner with the glycocalyx. Similarly, the activities of chemically, highly reactive biocides such as iodine and iodine–polyvinylpyrollidone complexes (Favero *et al.* 1983) are substantially reduced by the presence of protec-

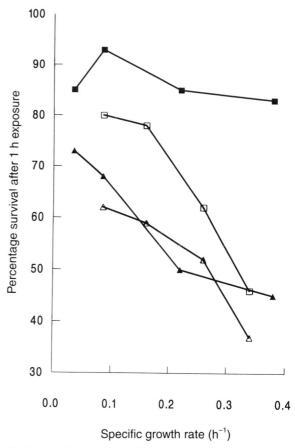

Fig. 6.1. Influence of specific growth rate upon the susceptibility of mucoid (solid symbols) and non-mucoid (open symbols) *Pseudomonas aeruginosa* biofilms towards exposure to ciprofloxacin (0.5 mg L^{-1}). Intact biofilms ($\square$, $\blacksquare$) and cells resuspended from them ($\blacktriangle$, $\triangle$) were exposed for 1 h at 37 °C. (Data from Evans *et al.* 1991a.)

tive exopolymers. In such instances the polymers are not only acting as adsorption sites but will react chemically with, and neutralize, such biocides.

Clearly, whether or not the glycocalyx constitutes a physical barrier to antibiotic penetration depends greatly upon the nature of the antibiotic, the binding capacity of the glycocalyx towards it, the levels of agent used therapeutically and the rate of growth of the microcolony relative to antibiotic diffusion rate. For antibiotics such as tobramycin and cefsoludin, these effects are suggested to be minimal (Nichols *et al.* 1988, 1989), but for agents such as the quinolones and reactive biocides such as iodine they are significant. If the sessile cells are able to produce drug inactivating enzymes such as ß-lactamases (Giwercman *et al.* 1991) within the glycocalyx this will create marked concentration gradients of the antibiotic and protect, to some extent, the underlying cells.

In summary, reductions in diffusion coefficents across the glycocalyx, relative to liquid media, are alone not sufficient to account for antibiotic recalcitrance of biofilms since the glycocalyx is fully hydrated and at equilibrium, concentrations of antimicrobial at the cell surface will be the same as that in the bathing medium. With losses occurring within the biofilm, either through chemical or enzymatic inactivation of the agent or irreversible binding, reduced diffusion might facilitate resistance to the agent. Such protection might also apply where the adsoptive/reactive capacity of the glycocalyx is very high with respect to biocide. In some instances, therefore, possession of an exopolymer glycocalyx will modify susceptibility, whilst in others the effects will be minimal.

Indirect role of the glycocalyx in resistance

During the early stages of biofilm formation, concentration of nutrients at the solid–liquid interface will lead to accelerated growth of attached cells. Such effects will only be apparent in nutritionally poor media containing concentrations of nutrients which are rate limiting. Such observations have become known as the 'bottle effect' (ZoBell 1943). As the biofilms thicken, however, nutrient availability will be reduced relative to the planktonic phase for the cells in immediate contact with the substratum. This is provided that the substratum is not itself a substrate. Bacterial biofilms are functional consortia which influence their microenvironment through the localized concentration of enzymes and metabolic products and the relative exclusion of gases (Costerton *et al.* 1985, 1987). Establishment of gradients within the biofilm reflects the consumption of nutrients and oxygen by the cells and generation of secondary metabolites and waste products (Wimpenny *et al.* 1989). The glycocalyx performs a homeostatic function with respect to these gradients and minimizes fluctuations in microenvironment at any given location. Consequentially, cells deep within the biofilm are subject to growth limitation by substrates and exposed to hydrogen ion concentrations and oxidation potentials which are substantially different

to those experienced by planktonic cultures grown in the same medium (Brown *et al.* 1988). Growth rates will therefore be reduced in biofilms through the imposition of a specific nutrient limitation (Brown & Melling 1969) not necessarily reflecting the composition of the aqueous phase. Growth rate and nutrient deprivation (Brown & Williams 1985a, b; Brown *et al.* 1990; Gilbert *et al.* 1990) are well established as factors which determine the physiological response of cells to stress and also the expression of phenotype. Within biofilms such effects will contribute greatly to the observed biocide resistance (Gilbert *et al.* 1987).

Nutrient limitation

Bacteria are capable of rapid growth and division in optimized conditions. This rapid division will not persist for any significant period in most natural ecosystems since nutrients will be consumed and the extent of growth will become limited by the availability of one essential nutrient. Nutrient deprivation such as this causes cells to reduce their use of the deficient nutrient. Such reductions can be achieved by metabolic changes, through the use of alternative substrates, and/or reductions in cell components which contain these elements. Thus phosphate limitation causes replacement in Gram positive cell walls of teichoic acids by teichuronic acids (Ellwood & Tempest 1972). In addition the cell must become more adept at acquisition of the rate limiting nutrient. This might occur through alterations to the cell surface and induction of more competitive transport mechanisms for the restricted nutrient. In this respect *in vivo* deficiencies of iron will often be associated with production of siderophores and the expression of iron regulated, siderophore binding proteins in the envelope (Griffiths 1983; Brown & Williams 1985a, b). A net effect of the above changes is a reduction in the physiological growth rate. Imposition of growth limitation by different nutrients will therefore give rise not only to populations which divide slowly but also to cells with physiologies and cell envelopes which are peculiar to that limitation (Brown & Melling 1969; Ellwood & Tempest 1972; Holme 1972; Brown 1975, 1977; Lambert 1988). Such cells will also be substantially different from those grown under nutrient rich conditions.

A wide range of Gram positive and Gram negative microorganisms have been reported to be greatly affected in their susceptibility to antimicrobial agents (Gilbert &Wright 1986) and antibiotics (Brown & Williams 1985a) through such effects. These have often been linked to changes in the composition and organization of the cell envelope (Ellwood & Tempest 1972; Gilbert *et al.* 1990; Brown *et al.* 1990). Changes in envelope lipid composition have been related to susceptibility towards bisbiguanide antiseptics (Klemperer *et al.* 1980; Broxton *et al.* 1984; Ikeda *et al.* 1984), gentamicin (Pechey *et al.* 1974) and polymyxin (Finch & Brown 1978; Wright & Gilbert 1987a), the initial binding and action of which are mediated through membrane phospholipids. Susceptibility changes, associated with the nature of the growth limiting nutrient, have also been related to porin protein composition (Harder *et al.* 1977; Williams *et al.* 1984), lipopolysaccharide (Tamaki *et al.* 1971) and cation content (Brown & Melling 1969; Boggis *et al.* 1979; Shand *et al.* 1985) of the envelopes. Such effects have been attributed both to an altered abundance of target and to changes in biocide binding and partitioning through the cell (Brown *et al.* 1990; Gilbert *et al.* 1990).

It is impossible to predict precisely which nutrients will be of limited availability within naturally occurring biofilms or indeed within any part of such a biofilm. It is likely, however, that cation availability in particular, and also carbon and nitrogen utilization, will impose nutrient gradients and nutrient limitations within the population which are different to those confronted by the same organisms growing planktonically within the bathing medium. It is probable that the availability of different nutrients will limit growth at various points within a single biofilm. Biofilm populations will therefore be composed of many different phenotypes, each with a different susceptibility to any given treatment by antimicrobial compounds. If eradication of the biofilm is taken as an end point for its overall susceptibility then the resistance of the biofilm will be that of the most resistant represented phenotype. Simultaneous resistance of the biofilm towards a number of chemically or biochemically distinct agents probably relates, therefore, to different represented phenotypes in each instance at various points within the biofilm. The most resistant phenotype would survive the antimicrobial

treatment in each instance. The general observation of biofilm recalcitrance might therefore relate to their fostered heterogeneity of nutrient deprived phenotypes. In this respect, Al-Hiti & Gilbert (1983) assessed the susceptibility of suspensions prepared from variously seeded nutrient agar plates and noted not only differences in susceptibility towards a number of pharmaceutical preservatives but also an increased heterogeneity of response for suspensions prepared from the larger colonies. Cultures grown on solid media show size variation according to colony density, presumably related, as in a biofilm, to nutrient availability and gradients of oxygen tension (Shapiro 1987). Such effects will also be experienced when dealing with the performance properties of isolated dense biofilms.

Growth rate

Whilst many species of bacteria have the potential for rapid division (2–3 divisions per hour), outside laboratory conditions division times are likely to be much longer than this (Brown *et al.* 1990). Continuous culture techniques (Herbert *et al.* 1956) have been widely employed as means of assessing the effects of growth rate upon a variety of physiological markers. These include susceptibility towards antibiotics, disinfectants and preservatives. Such techniques allow the control of growth rate for cells exposed to a constant physicochemical environment. A general conclusion which can be drawn from such investigations is that slowly growing cells are particularly recalcitrant to inimical agents (Brown 1975; Finch & Brown 1978; Gilbert & Brown 1978, 1980; Tuomanen 1986). As with changes in the growth limiting nutrient already discussed, growth rate has a profound effect upon the composition of the bacterial cell envelope (Holme 1972; Ellwood & Tempest 1972). Specifically, there are quantitative and qualitative changes in fatty acid, phospholipid (Gilbert & Brown 1978), metal cation (Boggis *et al.* 1979; Kenward *et al.* 1979) and protein composition of the envelopes (Brown & Williams 1985a, b) together with changes in the extent of extracellular enzyme and exopolysaccharide production (Sutherland 1982; Ombaka *et al.* 1983). Envelope composition, in turn, influences susceptibility of the cells towards antimicrobial agents. Thus, the action of simple antimicrobial agents, such as substituted phenols

(Gilbert & Brown 1978), biguanides (Gilbert & Brown 1980; Wright & Gilbert 1987b) and quaternary based compounds (Wright & Gilbert 1987c) can be altered by up to 1000-fold through changes in growth rate. Expression of penicillin binding proteins (PBP) is growth rate dependent, influenced profoundly by cation availability and has dramatic consequences for the activities of ß-lactam antibiotics (Turnowsky *et al.* 1983; Brown & Williams 1985a, b). In this respect, since the appropriate PBP is poorly expressed in slow growing *E. coli*, the antibiotics ceftoxidine and ceftriaxone have little activity against them, irrespective of the growth limiting nutrient (Cozens *et al.* 1986; Tuomanen *et al.* 1986a, b). Conversely, the ß-lactam CGP 17520 is particularly effective against cultures with a low specific growth rate, with activity directed against PBP 7 (Cozens *et al.* 1986; Tuomanen & Schwartz 1987). Such results illustrate that low growth rates need not necessarily be associated with antibiotic resistance. Similarly, whilst the activity of polymyxin increases (10-fold), towards *E. coli*, with increasing growth rate imposed under magnesium and phosphate limitation it decreases when growth rate control is exerted through availability of carbon or nitrogen source (Finch & Brown 1975; Wright & Gilbert 1987a). Growth rate, together with the nature of the growth limiting substrate, are therefore fundamental modulators of drug activity. If growth rates in biofilms differ widely from those in planktonic culture then this is likely to be associated with decreased susceptibility towards chemical agents.

Figure 6.2 illustrates the effects of growth rate upon the susceptibility of *E. coli* biofilms and planktonic cultures towards the antiseptic/preservative compound cetrimide (Evans *et al.* 1990b). As previously reported (Wright & Gilbert 1987c), susceptibility was observed to change as a complex function of specific growth rate. Notably, however, the susceptibilities, at any given growth rate, were similar for planktonically grown cells, intact biofilms and cells spontaneously shed from the biofilms during culture. There have been several reports in the literature, of resistance of biofilms comprised of Gram negative bacteria towards quaternary compounds (Ruseska *et al.* 1982; Costerton & Lashen 1984). In such experiments the susceptibility of fast growing planktonic cells was compared with those of biofilms established by submerging surfaces into nutrient

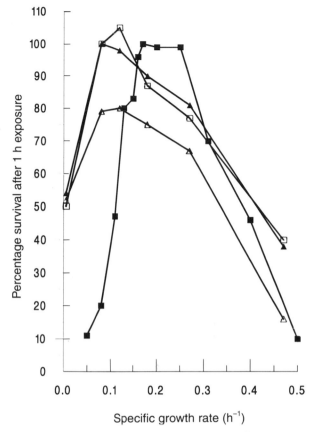

Fig. 6.2. Influence of specific growth rate upon the susceptibility of *E. coli* ATCC 8739 exposed to cetrimide USP (0.7 g ml^{-1}) for 1 h at 37 °C. Suspensions of cells were grown in suspended, chemostat culture (■), as biofilms (△), resuspended from growth rate controlled biofilms (□) or dispersed spontaneously from growing biofilms, (▲). (Data from Evans *et al.* 1990a.)

media. It is easy to see how the relative resistance in the biofilm cells might manifest itself if cells of substantially different rates of growth were compared. It is also possible to question the interpretation of other reports of susceptibility decreases in biofilm populations. Thus, the susceptibility towards amidinocillin, cephaloridine, cefamandole and chlorhexidine, of a variety of Gram negative bacteria have been shown to be substantially reduced (Prosser *et al.* 1987; Dix *et al.* 1988; Stickland *et al.* 1989) and the tobramycin resistance of *P. aeruginosa* and *S. epidermidis* and vancomycin resistance of *S. epidermidis* to be increased 20 to 1000-fold (Gristina *et al.* 1987) for biofilms relative to planktonic populations. As noted by Brown *et al.* (1988), however, these experimental systems were not controlled with respect to growth rate and often make comparisons between slow growing biofilm cells and fast, exponentially growing planktonic cells. As such, the effects of growth rate cannot be distinguished from those associated with adhesion. Nevertheless, biofilms constructed in this manner can be used as a primary screen (Anwar *et al.* 1989a, b) when there are many antibiotics to be tested.

Batch cultures such as these do not, however, adequately model natural ecosystems. These are often continuously perfused with fresh supplies of nutrient and grow slowly over extended periods of time. A possible extension to appoaches such as these is therefore to include submergcd tcst surfaces in a chemostat and to monitor the attached population (Keevil *et al.* 1987; Anwar *et al.* 1989a, b). Control of growth rate in continuous culture depends upon the establishment of steady state within a perfectly mixed culture (Herbert *et al.* 1956). When biofilms are generated within continuous fermentors we do not consider that they achieve steady state, since the adherent population and the total biomass in the fermentor increase with time, and there will be incomplete mixing of added nutrients. Such models are useful in the construction of long-term models of the environment, although their use does not confer growth rate control. We postulate that as the biofilms develop, initially through sequestration of planktonic cells to the substratum and then through growth of the attached cells, then growth rates will decrease. Thus, the properties of the formed biofilm are likely to change substantially with incubation time in the chemostat. Indeed, Anwar *et al.* (1990) reported substantial differences in antibiotic susceptibility of 'old' and 'young' biofilms generated in this manner.

Rather than suspend test surfaces within a continuous flow fermentor, many investigators flow medium through culture devices which provide large surface areas for bacterial attachment. The moving liquid phase imposes shear stresses upon the developing biofilm, causing cells to be shed and possibly influencing exopolymer deposition (Deretic *et al.* 1989). Cells growing planktonically and dispersed from the biofilm are eliminated. One such system is the Modified Robbins Device (McCoy *et al.* 1981). In this, retractable pistons form the inside of a pipe of rectangular section. Inoculation is made by passing a mid-log phase culture of test organisms through the reactor for several hours followed by sterile medium. Alternatively, the device is coupled to the outflow of a continuous fermentation vessel. Biofilms develop and may be sampled by removal of the pistons. The Robbins device has been used to model infections of soft tissues and indwelling medical devices as well as industrial situations (Costerton *et al.* 1987). Typical studies (Nickel *et al.* 1985; Evans & Holmes 1987; Gristina *et al.* 1989) compare the properties of adherent cells with those of the equivalent planktonic population of cells passing through the device. Whilst this technique has provided many valuable demonstrations of resistance to antimicrobial compounds in bacterial biofilms, its use is not intended to distinguish between the effects of growth rate and adherence (Brown *et al.* 1988).

In attempts to duplicate the conditions prevailing in the natural biofilms many *in vitro* studies lack experimental control of the physiologies of the cells. The results which such studies generate, therefore, add little to the observations of resistance which have been made in the field. The extent to which reduced growth rate may explain the observed differences in susceptibility between adherent biofilms and suspension cultures has only recently received direct study.

Gilbert *et al.* (1989) described a technique which establishes a biofilm on the underside of a bacteria proof cellulose membrane, perfused from the sterile side with fresh medium. A steady state is developed where the size of the biofilm population remains constant and dispersed cells collect in the spent medium. At steady state, the rate of perfusion with fresh medium controls the overall growth rate of the culture. We consider that the method has proved to be a useful model of *in vivo* infections involving attached bacteria and biofilms. The technique has been sucessfully employed for several bacterial species, including clinical, mucoid isolates of *P. aeruginosa* (Evans *et al.* 1991) and Gram positive organisms such as *S. epidermidis* (DuGuid *et al.* 1992a).

Biocide (Evans *et al.* 1990b) and antibiotic (Evans *et al.* 1990a, 1991; DuGuid *et al.* 1992a, b) susceptibility of bacterial biofilms have been investigated with such models by our research groups. A number of important observations were made in these studies. First, for the species studied, with cetrimide (Fig. 6.2), ciprofloxacin

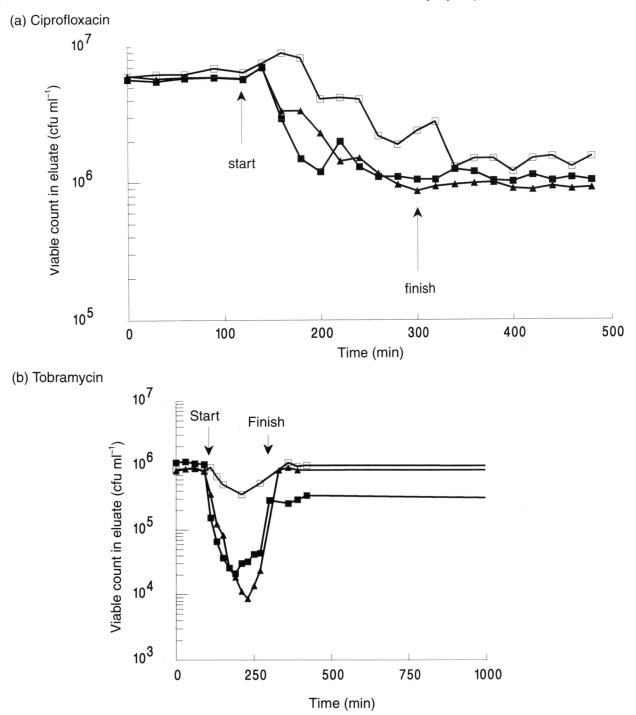

Fig. 6.3. Perfusion of steady-state *Staphylococcus epidermidis* biofilms with (a) Ciprofloxacin (5 mg L^{-1}) or (b) Tobramycin (25 µg L^{-1}). Changes in the viability of the cells and the numbers of cells spontaneously shed form the biofilm with the perfusate are indicated. Antibiotics were included in the perfusing medium at 120 min and removed at 300 min. The three curves correspond to slow ($\square$, 0.1 h^{-1}), medium ($\triangle$, 0.32 h^{-1}) and fast ($\triangle$, 0.41 h^{-1}) growth rates. (Data taken from Duguid *et al.* 1992a, b.)

and tobramycin, susceptibility increased with growth rate not only for the planktonic chemostat cultures but also for the biofilms. With the exception of ciprofloxacin, at any given growth rate, the susceptibilities of the resuspended biofilms and planktonic cells were similar. We suggest that a major contributor to the resistance of biofilms towards antimicrobial agents must therefore be their reduced growth rate. Secondly, when intact biofilms were exposed and their susceptibilities related to those of planktonic cells and cells resuspended from biofilms then, at any given growth rate, organization of cells as a biofilm increased resistance, at least to some extent. For ciprofloxacin, the susceptibility of mucoid pseudomonads was significant (Fig. 6.1), but for *E. coli* and tobramycin was less so. Organization of the cells within a glycocalyx does therefore contribute towards their resistance, but the extent of such protection depends upon the amount of glycocalyx and the nature of the agent. Finally, perfusion of the intact biofilms *in situ* with antibiotics (Fig. 6.3) has enabled the effectiveness of specific chemotherapeutic regimens to be compared and evaluated (Duguid *et al.* 1992a, b).

Growth rate has also been reported to influence profoundly not only the production of extracellular virulence factors (Ombaka *et al.* 1983) but also the immunogenicity of microbes (Anwar *et al.* 1985). Bacterial sensitivity to phagocytosis (Finch & Brown 1978; Griffiths 1983), lysozyme (Gilbert & Brown 1980) and serum killing (Anwar *et al.* 1983) are also affected. Such growth rate mediated effects will also be associated with slowly growing biofilms.

Adherence phenotypes

Surface induced stimulation of bacterial activity has often been observed (ZoBell 1943) and is epitomized by the 'bottle effect', that is the rapid growth of bacteria in stored waters (ZoBell & Anderson 1936). There is still much debate, however, as to whether such stimulation results from derepression/induction, by contact with the surface, of specific operons/genes or are indirect and reflect physicochemical influences of the surface on the surroundings of the cell (van Loosdrecht *et al.* 1990). Degradation of nitrilotriacetate is enhanced when the degradative organisms are attached to inert surfaces to which the substrate does not adsorb (McFeters *et al.* 1990), suggesting an increased production of degradative enzymes by attached cells. Similarly, gliding bacteria lack extracellular polymer biosynthesis when grown in suspension culture (Humphrey *et al.* 1979; Abbanat *et al.* 1988). Work on the regulation of lateral flagella gene transcription in *Vibrio parahaemolyticus* has shown it to produce a single polar flagellum in liquid, and numerous lateral, unsheathed flagella on solid culture media (Belas *et al.* 1986; McCarter *et al.* 1988). This is thought to reflect increased viscosity restricting the movement of the polar flagellum. The net effect, however, was a switching on of the *laf* genes through contact with a surface. Similarly, Lee & Falkow (1990) recognized that reduced oxygen tensions such as might be experienced by cells enveloped within a biofilm or in association with a surface causes the expression of *Salmonella* invasins. A number of studies have been reported which attempt to demonstrate directly the existence of touch sensors in microorganisms. Dagostino *et al.* (1991) utilized transposon mutagenesis to insert randomly into the chromosome of *E. coli* a marker gene which lacked its own promoter element. A number of mutant cell lines were isolated which did not express the gene either in association with an agar surface or in liquid culture, but which did so when cells were attached to polystyrene (Dagostino *et al.* 1991).

Whether or not bacteria possess touch receptors remains to be proven. If they do, then they will clearly be of immense importance not only in the initiation and formation of biofilms but also in their expressed physiology.

Acknowledgements

We wish to acknowledge grants from the Medical Research Council (UK) and the Cystic Fibrosis Research Trust (UK) which has supported much of the work from our laboratories. We also gratefully acknowledge help from Bayer and Ciba-Giegy.

References

Abbanat, D. R., Godchaux, W. & Leadbetter, E. R. (1988). Surface induced synthesis of new sulfonolipids in the gliding bacterium *Cytophaga johnsonae*. *Archives of Microbiology*, **149**, 358–64.

Al-Hiti, M. M. A. & Gilbert, P. (1983). A note on inoculum reproducibility: solid culture versus liquid culture. *Journal of Applied Bacteriology*, **55**, 173–6.

Anwar, H., Brown, M. R. W., Day, A. & Weller, P. H. (1985). Outer membrane antigens of mucoid *Pseudomonas aeruginosa* isolated directly from the sputum of a cystic fibrosis patient. *FEMS Microbiology Letters*, **24**, 235–9.

Anwar, H., Brown, M. R. W. & Lambert, P. A. (1983). Effect of nutrient depletion on the sensitivity of *Pseudomonas cepacia* to phagocytosis and serum bactericidal activity at different temperatures. *Journal of General Microbiology*, **129**, 2021–7.

Anwar, H., Dasgupta, M. K. & Costerton J. W. (1989b). Tobramycin resistance of mucoid *Pseudomonas aeruginosa* biofilm grown under iron limitation. *Journal of Antimicrobial Chemotherapy*, **24**, 647–55.

Anwar, H., Dasgupta, M. K. & Costerton, J. W. (1990). Testing the susceptibility of bacteria in biofilms to antibacterial agents. *Antimicrobial Agents and Chemotherapy*, **34**, 2043–6.

Anwar, H., van Biesen, T., Dasgupta, M. K., Lam, K. & Costerton, J. W. (1989a). Interaction of biofilm bacteria with antibiotics in a novel *in vitro* chemostat system. *Antimicrobial Agents and Chemotherapy*, **33**, 1824–16.

Belas, R., Simon, M. & Silverman, M. (1986). Regulation of lateral flagella gene transcription in *Vibrio parahaemolyticus*. *Journal of Bacteriology*, **167**, 210–18.

Bergeron, M. G., Simard, P. & Provencher, P. (1987). Influence of growth medium and supplement on growth of *Haemophilus influenzae* and on antibacterial activity of several antibiotics. *Journal of Clinical Microbiology*, **25**, 650–5.

Bisno, A. L. & Waldvogel, F. A. (ed.) (1989). *Infections associated with indwelling medical devices*. Washington, DC: American Society of Microbiology.

Boggis, W., Kenward, M. A. & Brown, M. R. W. (1979). Effects of divalent metal cations in the growth medium upon the sensitivity of batch grown *Pseudomonas aeruginosa* to EDTA or polymyxin B. *Journal of Applied Bacteriology*, **47**, 477–88.

Brown, M. R. W. (1975). The role of the envelope in resistance. In *Resistance of Pseudomonas aeruginosa*, ed. M. R. W. Brown, pp. 71–107. London: Wiley.

Brown, M. R. W. (1977). Nutrient depletion and antibiotic susceptibility. *Journal of Antimicrobial Chemotherapy*, **3**, 198–201.

Brown, M. R. W., Allison. D. G. & Gilbert, P. (1988). Resistance of bacterial biofilms to antibiotics: a growth rate related effect. *Journal of Antimicrobial Chemotherapy*, **22**, 777–89.

Brown, M. R. W., Collier, P. J. & Gilbert, P. (1990). Influence of growth rate on susceptibility to antimicrobial agents: modifications of the cell envelope and batch and continuous culture studies. *Antimicrobial Agents and Chemotherapy*, **34**, 1623–1628.

Brown, M. R. W. & Melling, J. (1969). Role of divalent cations in the action of polymyxin B and EDTA on *Pseudomonas aeruginosa*. *Journal of General Microbiology*, **59**, 263–74.

Brown, M. R. W. & Williams, P. (1985a). Influence of substrate limitation and growth phase on sensitivity to antimicrobial agents. *Journal of Antimicrobial Chemotherapy*, **15**, (Suppl. A), 7–14.

Brown, M. R. W. & Williams, P. (1985b). The influence of the environment on envelope properties affecting survival of bacteria in infections. *Annual Reviews of Microbiology*, **39**, 527–56.

Broxton, P., Woodcock, P. M. & Gilbert, P. (1984). Interaction of some polyhexamethylene biguanides and membrane phospholipids in *Escherichia coli*. *Journal of Applied Bacteriology*, **57**, 115–24.

Costerton, J. W. (1977). Cell-envelope as a barrier to antibiotics. In *Microbiology*, ed. F. Kavanagh, pp. 151–7. London: Academic Press.

Costerton, J. W., Cheng, K. J., Geesey, K. G. *et al.* (1987). Bacterial biofilms in nature and disease. *Annual Reviews of Microbiology*, **41**, 435–64.

Costerton, J. W., Irwin, R. T. & Cheng, K.-T. (1981). The bacterial glycocalyx in nature and disease. *Annual Reviews of Microbiology*, **35**, 399–424.

Costerton, J. W. & Lashen, E. S. (1984). Influence of biofilm efficacy of biocides on corrosion-causing bacteria. *Materials Performance*, **23**, 34–7.

Costerton, J. W., Marrie, T. J. & Cheng, K.-J. (1985). Phenomena of bacterial adhesion. In *Bacterial adhesion: mechanisms and physiological significance*, ed. D. C. Savage & M. Fletcher, pp. 3–43. New York: Plenum Press.

Cozens, R. M., Tuomanen, E., Tosch, W., Zak, O., Suter, J. & Tomasz, A. (1986). Evaluation of the bactericidal activity of ß-lactam antibiotics upon slowly growing bacteria cultured in the chemostat. *Antimicrobial Agents and Chemotherapy*, **29**, 797–802.

Dagostino, L., Goodman, A. E. & Marshall, K. C. (1991). Physiological responses induced in bacteria adhering to surfaces. *Biofouling*, **4**, 113–19.

DeMatteo, C. S., Hammer, M. C., Baltch, A. L., Smith, R. P., Sutphen, N. T. & Michelsen, P. B. (1981). Susceptibility of *Pseudomonas aeruginosa* to serum bactericidal activity. *Journal of Laboratory and Clinical Medicine*, **98**, 511–18.

Deretic, V., Dikshit, R., Konyecsni, W. M., Chakrabarty, A. M. & Misra, T. K. (1989). The *algR* gene, which regulates mucoidy in *Pseudomonas aeruginosa*, belongs to a class of environmentally responsive genes. *Journal of Bacteriology*, **171**, 1278–83.

Dix, B. A., Cohen, P. S., Laux, D. C. & Cleeland, R.

(1988). Radiochemical method for evaluating the effects of antibiotics on *Escherichia coli* biofilms. *Antimicrobial Agents and Chemotherapy*, **32**, 770–2.

Duguid, I. G., Evans, E., Brown, M. R. W. & Gilbert, P. (1992a). Growth-rate independent killig by ciprofloxacin of biofilm-derived *Staphylococcus epidermidis*: evidence for cell-cycle dependency. *Journal of Antimicrobial Chemotherapy*, **30**, 791–802.

Duguid, I. G., Evans, E., Brown, M. R. W. & Gilbert, P. (1992b). Effect of biofilm culture upon the susceptibility of *Staphylococcus epidermidis* towards tobramycin. *Journal of Antimicrobial Chemotherapy*, **30**, 803–10.

Ellwood, D. C. & Tempest, D. W. (1972). Effects of environment on bacterial cell wall content and composition. *Advances in Microbial Physiology*, **7**, 83–117.

Evans, D. J., Allison, D. G., Brown, M. R. W. & Gilbert, P. (1990b). Effect of growth-rate on resistance of Gram negative biofilms to cetrimide. *Journal of Antimicrobial Chemotherapy*, **26**, 473–8.

Evans, D. J., Allison, D. G., Brown, M. R. W. & Gilbert, P. (1991). Susceptibility of *Pseudomonas aeruginosa* and *Escherichia coli* biofilms towards ciprofloxacin: effect of specific growth rate. *Journal of Antimicrobial Chemotherapy*, **27**, 177–84.

Evans, D. J., Brown, M. R. W., Allison, D. G. & Gilbert, P. (1990a). Susceptibility of bacterial biofilms to tobramycin: role of specific growth rate and phase in division cycle. *Journal of Antimicrobial Chemotherapy*, **25**, 585–91.

Evans, R. C. & Holmes, C. J. (1987). Effect of vancomycin hydrochloride on *Staphylococcus epidermidis* biofilm associated with silicone elastomer. *Antimicrobial Agents and Chemotherapy*, **31**, 889–94.

Favero, M. S., Bond, W. W., Peterson, N. J. & Cook, E. H. Jr. (1983). Scanning electron microscopic study of bacteria resistant to iodophor solutions. In *Proceedings of the International Symposium on Povidone Iodine*, ed. G. A. Digenis & J. Ansell, pp. 158–66. Kentucky: University of Kentucky Press.

Finch, J. E. & Brown, M. R. W. (1975). The influence of growth rate and nutrient limitation in a chemostat on the sensitivity of *Pseudomonas aeruginosa* to polymyxin B and EDTA. *Journal of Antimicrobial Chemotherapy*, **1**, 379–86.

Finch, J. E. & Brown, M. R. W. (1978). Effect of growth environment upon *Pseudomonas aeruginosa* killing by rabbit polymorphonuclear leucocytes and cationic proteins. *Infection and Immunity*, **20**, 340–6.

Gilbert, P., Allison, D. G., Evans, D. J., Handley, P. S. & Brown, M. R. W. (1989). Growth rate control of adherent bacterial populations. *Applied and Environmental Microbiology*, **55**, 1308–11.

Gilbert, P. & Brown, M. R. W. (1978). Influence of growth rate and nutrient limitation on the gross

cellular composition of *Pseudomonas aeruginosa* and its resistance to some uncoupling phenols. *Journal of Bacteriology*, **133**, 1062.

Gilbert, P. & Brown, M. R. W. (1980). Cell-wall mediated changes in the sensitivity of *Bacillus megaterium* to chlorhexidine and 2-phenoxyethanol associated with growth rate and nutrient-limitation. *Journal of Applied Bacteriology*, **48**, 223–30.

Gilbert, P., Brown, M. R. W. & Costerton, J. W. (1987). Inocula for antimicrobial sensitivity testing: a critical review. *Journal of Antimicrobial Chemotherapy*, **20**, 147–54.

Gilbert, P., Collier, P. J. & Brown, M. R. W. 1990. Influence of growth rate on susceptibility to antimicrobial agents: biofilms, cell cycle, dormancy, and stringent response. *Antimicrobial Agents and Chemotherapy*, **34**, 1865–8.

Gilbert, P. & Wright, N. E. (1986). Non-plasmidic resistance towards preservatives in pharmaceutical and cosmetic products. *Society for Applied Bacteriology Technical Series*, **22**, 255–79.

Giwercman, B., Jensen, E. T., Høiby, N., Kharazmi, A. & Costerton, J. W. (1991). Induction of b-lactamase production in *Pseudomonas aeruginosa* biofilms. *Antimicrobial Agents and Chemotherapy*, **35**, 1008–10.

Griffiths, E. (1983). Availability of iron and survival of bacteria in infection. In *Medical microbiology*, Vol. 3, ed. C. S. F. Earsmon, J. Jeljasewicz, M. R. W. Brown & P. A. Lambert. London: Academic Press.

Gristina, A. G., Hobgood, C. D., Webb, L. X. & Myrvik, Q. N. (1987). Adhesive colonisation of biomaterials and antibiotic resistance. *Biomaterials*, **8**, 423–6.

Gristina, A. G., Jennings, R. A., Naylor, P. T., Myrvik, Q. N., Barth, E. & Webb, L. X. (1989). Comparative *in vitro* antibiotic resistance of surface colonising coagulase negative staphylococci. *Antimicrobial Agents and Chemotherapy*, **33**, 813–16.

Harder, W., Kuenen, J. G. & Matin, S. (1977). Microbial selection in continuous culture. *Journal of Applied Bacteriology*, **43**, 1–24.

Herbert, D., Elsworth, R. & Telling, R. C. (1956). The continuous culture of bacteria; a theoretical and experimental study. *Journal of General Microbiology*, **14**, 601–22.

Humphrey, B. A., Dickson, M. R. & Marshall, K. C. (1979). Physiological and *in situ* observations on the adhesion of gliding bacteria to surfaces. *Archives of Microbiology*, **120**, 231–8.

Hohl, P. & Felber, A. M. (1988). Effect of method, medium, pH and inoculum on the *in vitro* antibacterial activities of fleroxacin and norfloxacin. *Journal of Antimicrobial Chemotherapy*, **22** (Suppl. D), 71–80.

Holme, T. (1972). Influence of the environment on the content and composition of bacterial envelopes.

Journal of Applied Chemical Biotechnology, **22**, 391–9.

Ikeda, T., Ledwith, C. H., Bamford, C. H. & Hann, R. A. (1984). Interaction of a polymeric biguanide with phospholipid membranes. *Biochimica et Biophysica Acta*, **769**, 57–66.

Keevil, C. W., Bradshaw, D. J., Dowsett, A. B. & Feary, T. W. (1987). Microbial film formation: dental plaque deposition on acrylic tiles using continuous culture techniques. *Journal of Applied Bacteriology*, **62**, 129–38.

Kenward, M. A., Brown, M. R. W. & Fryer, J. J. (1979). The influence of calcium or magnesium on the resistance to EDTA, polymyxin or cold shock, and the composition of *Pseudomonas aeruginosa* in glucose or magnesium depleted culture. *Journal of Applied Bacteriology*, **47**, 489–503.

Klemperer, R. M. M., Ismail, N. T. A., & Brown, M. R. W. (1980). Effect of R-plasmid RP1 and nutrient depletions on the resistance of *Escherichia coli* to cetrimide, chlorhexidine and phenol. *Journal of Applied Bacteriology*, **48**, 349–57.

Kurian, S. & Lorian, V. (1980). Discrepencies between results obtained by agar and broth techniques in testing of drug combinations. *Journal of Clinical Microbiology*, **11**, 527–9.

Lambert, P. A. (1988). Enterobacteriacae: composition, structure and function of the cell envelope. *Journal of Applied Bacteriology*, **65** (Suppl.), 21S–34S.

Lee, C. A. & Falkow, S. (1990). The ability of *Salmonella* to enter mammalian cells is affected by bacterial growth state. *Proceedings of the National Academy of Sciences USA*, **87**, 4304–8.

Lee, J. V. & West, A. A. (1991). Survival and growth of *Legionella* species in the environment. *Journal of Applied Bacteriology*, **70** (Suppl.), 121S–130S.

Lippincott, B. B. & Lippincott, J. A. (1969). Bacterial attachment to a specific wound site as an essential stage in tumour initiation by *Agrobacterium tumefaciens*. *Journal of Bacteriology*, **97**, 620–8.

McCarter, L., Hilmen, M., Silverman, M. (1988). Flagellar dynamometer controls swarmer cell differentiation of *Vibrio parahaemolyticus*. *Cell*, **54**, 345–51.

McCoy, W. F., Bryers, J. D., Robbins, J., & Costerton, J. W. (1981). Observations in fouling biofilm formation. *Canadian Journal of Microbiology*, **27**, 910–17.

McFeters, G. A., Egli, T., Wilberg, E., Adler A., Schneider, R., Snozzi, M. & Giger, W. (1990). Activity and adaptation of nitrilotriacetate (NTA) degrading bacteria: field and laboratory studies. *Water Research*, **24**, 875–81.

Millward, T. A. & Wilson, M. (1989). The effect of chlorhexidine on *Streptococcus sangius* biofilms. *Microbios*, **58**, 155–64.

Nichols, W. W., Dorrington, S. M., Slack, M. P. E. & Walmsley, H. L. (1988). Inhibition of tobramycin diffusion by binding to alginate. *Antimicrobial Agents and Chemotherapy*, **32**, 518–23.

Nichols, W. W., Evans, M. J., Slack, M. P. E. & Walmsley, H. L. (1989). The penetration of antibiotics into aggregates of mucoid and non-mucoid *Pseudomonas aeruginosa*. *Journal of General Microbiology*, **135**, 1291–1303.

Nickel, J. C., Ruseska, I., Wright, J. B., & Costerton, J. W. (1985). Tobramycin resistance of cells of *Pseudomonas aeruginosa* growing as a biofilm on urinary catheter material. *Antimicrobial Agents and Chemotherapy*, **27**, 619–24.

Ombaka, A., Cozens, R. M. & Brown, M. R. W. (1983). Influence of nutrient limitation of growth on stability and production of virulence factors of mucoid and non-mucoid strains of *Pseudomonas aeruginosa*. *Reviews in Infectious Diseases*, **5**, 5880–8.

Pechey, D. T., Yau, A. O. P. & James, A. M. (1974). Total and surface lipids of *Pseudomonas aeruginosa* and their relationship to gentamycin resistance. *Microbios*, **11**, 77–86.

Prosser, B. T., Taylor, D., Dix, B. A. & Cleeland, R. (1987). Method of evaluating effects of antibiotics upon bacterial biofilms. *Antimicrobial Agents and Chemotherapy*, **31**, 1502–6.

Ruseska, I., Robbins, J., Costerton, J. W. & Lashen, E. S. (1982). Biocide testing against corrosion causing oil-field bacteria helps control plugging. *Oil and Gas Journal*, **80**, 253–4.

Shand, G. H., Anwar, H., Kadurugamuwa, J., Brown, M. R. W., Silverman, S. H. & Melling, J. (1985). *In vivo* evidence that bacteria in urinary tract infections grow under iron restricted conditions. *Infection and Immunity*, **48**, 35–9.

Shapiro, J. A. (1987). Organisation of developing *Escherichia coli* colonies viewed by scanning electron microscopy. *Journal of Bacteriology*, **169**, 142–56.

Slack, M. P. E. & Nichols, W. W. (1981). The penetration of antibiotics through sodium alginate and through the exopolysaccharide of a mucoid strain of *Pseudomonas aeruginosa*. *Lancet*, **ii**, 502–3.

Slack, M. P. E., & Nichols, W. W. (1982). Antibiotic penetration through bacterial capsules and exopolysaccharides. *Journal of Antimicrobial Chemotherapy*, **10**, 368–72.

Stickland, D., Dolman, J., Rolfe, S. & Chawla, J. (1989). Activity of antiseptics against *Escherichia coli* growing as biofilms on silicone surfaces. *European Journal of Clinical Microbiology and Infectious Diseases*, **8**, 974–8.

Sutherland, I. W. (1982). Biosynthesis of microbial exopolysaccharide. *Advances in Microbial Physiology*, **23**, 79–150.

Tamaki, S., Sato, T. & Matsuhashi, M. (1971). Role of lipopolysaccharide in antibiotic resistance and bacteriophage adsorption of *Escherichia coli* K-12. *Journal of Bacteriology*, **105**, 968–75.

Tuomanen, E. (1986). Phenotypic tolerance: The search for ß-lactam antibiotics that kill non-growing bacteria. *Reviews in Infectious Disease*, **8**, (Suppl.), S279-S291.

Tuomanen, E., Cozens, R. M., Tosch, W., Zak, O. & Tomasz, A. (1986b). The rate of killing of *Escherichia coli* by ß-lactam antibiotics is strictly proportional to the rate of bacterial growth. *Journal of General Microbiology*, **132**, 1297–1304.

Tuomanen, E., Durack, D. & Tomasz, A. (1986a). Antibiotic tolerance among clinical isolates of bacteria. *Antimicrobial Agents and Chemotherapy*, **30**, 521–7.

Tuomanen, E. & Schwartz, J. (1987). Penicillin binding protein 7 and its relationship to lysis of non-growing *Escherichia coli*. *Journal of Bacteriology*, **169**, 4912–15.

Turnowsky, F., Brown, M. R. W., Anwar, H. & Lambert, P. A. (1983). Effect of iron limitation of growth rate on the binding of penicillin G to penicillin binding proteins of mucoid and non-mucoid strains of *Pseudomonas aeruginosa*. *FEMS Microbiology Letters*, **17**, 243–5.

van Loosdrecht, M. C. M., Lyklema, J., Norde, W. & Zinder, A. J. B. (1990). Influence of interfaces on microbial activity. *Microbiological Reviews*, **54**, 75–87.

Williams, P., Brown, M. R. W. & Lambert, P. A.

(1984). Effect of iron-deprivation on the production of siderophores and outer membrane proteins in *Klebsiella aerogenes*. *Journal of General Microbiology*, **130**, 2357–65.

Wimpenny, J. W. T., Peters, A. & Scourfield, M. (1989). Modeling spatial gradients. In *Structure and function of biofilms*, ed. W. G. Charackalis & P. A. Wilderer, pp. 111–27. Chichester: Wiley.

Wright, N. E. & Gilbert, P. (1987a). Influence of specific growth rate and nutrient limitation upon the sensitivity of *Escherichia coli* towards polymyxin B. *Journal of Antimicrobial Chemotherapy*, **20**, 303–12.

Wright, N. E. & Gilbert, P. (1987b). Influence of specific growth rate and nutrient limitation upon the sensitivity of *Escherichia coli* towards chlorhexidine diacetate. *Journal of Applied Bacteriology*, **62**, 309–14.

Wright, N. E. & Gilbert, P. (1987c). Antimicrobial activity of *n*-alkyl trimethylammonium bromides: influence of specific growth rate and nutrient limitation. *Journal of Pharmacy and Pharmacology*, **39**, 685–90.

ZoBell, C. E. (1943). The effect of solid surfaces upon bacterial activity. *Journal of Bacteriology*, **46**, 39–56.

ZoBell, C. E. & Anderson, D. Q. (1936). Observations on the multiplication of bacteria in different volumes of stored seawater and the influence of oxygen tension and solid surfaces. *Biological Bulletin*, **71**, 324–42.

Part II BIOFILMS AND INERT SURFACES

7

Biofilm Development in Purified Water Systems

Marc W. Mittelman

Introduction

The use of purified water in various industrial and medical applications has increased dramatically over the past 20 years. These different applications often require varying levels of water quality. Each industry sets specifications for the acceptance of purified water quality based upon their product or process demands. The aggressive nature of ionically ultrapure water (18.2 Mohm cm) can contraindicate its use in, for example, stainless steel distribution systems which are often employed in the pharmaceutical industry. While the presence of trace levels of silica in condensate polishing loop waters creates great concern in the power industry, low level silica contamination of purified waters used in the production of medical devices or photomasks, for example, may not constitute cause for alarm. If any one group of contaminants can be viewed as 'universal' in their distribution, significance and recalcitrance, it is the bacteria and their by-products. Their role as purified water contaminants appears to cross all boundaries of purified water application and usage.

In the semiconductor industry, the demand for contaminant free water has, to a great extent, driven ultrapure water technologies. Indeed, water purification technologies have advanced to the point whereby levels of ionic, organic, and particulate contaminants can be reduced to concentrations below current analytical detection limits. Unfortunately, this industry's ability to detect and remove biological particulates and organics has not kept pace with the needs of today's sophisticated devices. Yang & Tolliver (1989)

have noted that, at the 1 megabit level where minimum circuit feature size is typically 1 μm, device yields are limited by 0.1 μm and larger particles.

Bacteria may represent the greatest fraction of total particles in this size range. However, current analytical capabilities for ultrapure water analysis have generally limited the detection of bacteria in this size range to those capable of growth on microbiological media. Deficiencies associated with viable count techniques have been reviewed previously (Morita 1985; Novitsky 1984; Reasoner & Geldreich 1985; Mittelman *et al*. 1987).

In addition to the difficulties associated with bacterial detection, treatment of biological particulate contamination has proven difficult. Bacterial penetration of 0.2 μm microporous membrane filters, which are the current industry standard, has been reported by a number of workers in the pharmaceutical industry (Simonetti & Schroeder 1984; Anderson *et al*. 1985; Christian & Meltzer 1986; Howard & Duberstein 1986). In addition, the effects of biocide treatments in these systems have been rather short lived, owing in part to the presence of recalcitrant bacteria associated with surfaces (Mittelman 1986a).

Bacteria associated with surfaces are the source of both product and process biological contaminants. It is biofilm formation which gives rise to the myriad of problems in various process waters. The absence of any antagonistic agents, coupled with the extremely oligotrophic environment associated with purified water systems, promotes bacterial adhesion and fouling biofilm formation.

This chapter describes some of the design strategies which have been employed for limiting biological growth and biofilm formation. The microbial growth cycle in purified water environments will be described as it relates to the formation of biological particles. In addition, emerging technologies for *in situ* detection of particulate and biofilm bacteria and their by-products are explored. Finally, some novel approaches to controlling biofilm populations in purified waters will be introduced.

Bacterial growth and biofilm formation

Growth promoting factors

Virtually all bacteria isolated from purified water systems are Gram negative rods. The Gram negative cell wall consists of multiple layers of phospholipids surrounded by a lipopolysaccharide structure (LPS). This multi-laminate structure may afford the cell protection from the extremely hypotonic environment which is intrinsic to purified water systems. These bacteria are heterotrophic in nature, requiring the presence of reduced organic compounds as energy sources. Therefore, these compounds serve as limiting growth factors in purified water systems. The term oligotroph has been assigned to organisms which are capable of growth in media containing <1 mg L^{-1} organic carbon (Ishida & Kadota 1981). In general, a positive correlation exists between assimilable organic carbon levels and planktonic bacterial numbers in purified waters.

In addition to organic energy and carbon sources, bacteria require a number of other nutrients such as nitrogen, phosphorus, sulphur, and trace metals and salts in order to carry out normal metabolic and replicative functions (Mittelman *et al.* 1987). Table 7.1 lists the major factors and some common reservoirs.

Starvation survival

When one or more of these essential growth factors is limited, as is often the case in benthic environments and purified water systems, bacteria utilize a number of strategies designed to assure their survival. In the short term, starved bacteria tend to utilize endogenous energy reserves for replication; therefore, one strategy for survival involves replication to increase the probability of species survival once additional energy

Table 7.1. Growth-promoting factors in purified water systems

Substrate	Source(s)
Carbon	Pipe extractables Microbial by-products Airborne dust Treatment chemicals (RO pretreat) Personnel (skin flakes, etc.) Lubricants Membrane filter surfactants Storage tank entry hatch gaskets Membrane filter media
Water	Vent filters Degasifier HEPA filters Storage tank sight tubes Dead-legs
Nitrogen	Feedwaters (humic and fulvic acids, nitrites, etc.) Airborne dust Nitrogen blankets Microbial by-products
Phosphorous/sulphur	Feedwaters (phosphates, sulphates) Sulphuric acid (RO pretreatment) Airborne dust Membrane surfactants Microbial by-products
Trace metals and salts	Airborne dust (degasifiers) Piping extractables Membrane filter extractables RO pretreatment chemicals Metallic system components
Light (algae/diatoms)	Storage tank sight tubes Entry hatch gaskets Vent filters

sources become available (Novitsky & Morita 1978). In marine systems limited in assimilable nutrients, some bacteria reduce their cell volume, forming ultramicrobacteria. Tabor *et al.* (1981) demonstrated that these organisms, many of which were <0.3 μm in diameter, were capable of passing through 0.45 μm membrane filters. Their findings were supported by the work of Christian & Meltzer (1986) and others reporting bacterial penetration of 'sterilizing grade' microporous membrane filters.

Biofilm formation

Perhaps the most significant adaptive mechanism utilized by nutrient limited purified water bacteria involves adhesion to surfaces. Indeed, the majority of bacteria in nutrient limited environments, such as an ultrapure water system, are attached to surfaces. Surfaces afford these organisms three major advantages over the bulk phase: (i) concentration of trace organics, which can serve as nutrient sources, occurs on clean surfaces shortly after their immersion; (ii) surface associated bacteria tend to produce extracellular polymeric substances which can further concentrate trace growth factors; (iii) bacteria in biofilms are afforded some protection from antagonistic agents such as biocides, heat treatments, and other inhibitory factors (see Gilbert & Brown, Chapter 6). Several workers have shown that nutrient limiting environments promote the attachment of bacteria to surfaces (Marshall 1988; Mittelman *et al.* 1987).

The relatively high surface area:volume ratio associated with industrial water systems provides ample space for bacterial attachment. Any one of the many systems utilized in the purification of feedwaters is therefore a potential reservoir for contamination. In terms of biological contaminant generation, however, granular activated carbon columns (Johnston & Burt 1976; Collentro 1986), reverse osmosis (RO) membranes (Ridgway *et al.* 1986), ion exchange systems (Flemming 1987), degasifiers (Mittelman *et al.* 1987), RO/DI water storage tanks (Youngberg 1985), and microporous membrane filters (Meltzer 1987; Mittelman & White 1989) are the most significant areas of concern. Sources of contamination in industrial water systems have been reviewed by Mittelman & Geesey (1987).

Bacterial biofilm formation has significant implications for both the treatment and detection of contaminating purified water populations. LeChevallier *et al.* (1988) showed, for example, that attachment of *Klebsiella pneumoniae* to glass slides increased resistance to free chlorine disinfection by a factor of 150. It has been suggested that the presence of polyanionic extracellular polymeric substances affords cells some measure of protection by inhibiting biocide diffusion. Since biofilm associated bacteria are not detected by current test methodologies, serious underestimates of contaminating populations result. Conversely, this inability to detect the major source of bulk phase biological contaminants results in an overestimate of treatment efficacies.

Bacterial contamination case histories

Pharmaceuticals and medical devices

A number of product recalls have been initiated for contaminated solutions marketed as sterile. Anderson *et al.* (1985) described contamination of sterile distilled water respiratory care solutions as a result of *Pseudomonas pickettii* passage through sterilizing grade 0.2 μm membrane filters. Eye care solutions have been contaminated by *Pseudomonas* sp. originating in purified water feed systems (Wilson *et al.* 1981). The origin of contaminated povidone iodine solutions by *P. cepacia* was found to be a make-up purified water system (Michels 1981). Novitsky *et al.* (1986) described problems with removal of endotoxins emanating from the purified water system adsorbed to parenteral fluid containers.

Roberts (1988) has described contamination of medical device sterilant solutions with *Mycobacterium chelonae*. Klein *et al.* (1990) showed that 19% of the dialysate samples taken from acute dialysis centres in the USA were out of compliance with the Association for the Advancement of Medical Instrumentation bacteria standards. Murphy *et al.* (1987) have described problems with endotoxacmia during high flux dialysis procedures.

A listing of recalls and other USA Food and Drug Administration (FDA) actions related to water is presented at the end of this chapter.

Microelectronic devices

Bacterial particulates have been implicated in a number of device failures in the USA and elsewhere. In one study performed at a Hewlett Packard facility in California, Dial & Chu (1987) reported a direct correlation between purified water bacterial levels and device defect frequencies. Craven *et al.* (1986) have attributed failures in back side metallization processes to bacterial adhesion processes. Eisenmann & Ebel (1988) described horizontal and vertical defects built in during the photoresist steps resulting from contamination by bacteria and other particulates. Conductance, electromigration, and corrosion in oxide layers were also noted. Sodium and potassium ions leaching from bacteria during metabolism or cell death can diffuse into silicon wafer surfaces to change the device field effects. DePaiva *et al.* (1992) utilized Auger electron spectroscopy to demonstrate that carbon contamination of GaSb and Si semiconductor surfaces could be traced to bacterial contamination of deionized tap water.

Patterson *et al.* (1991) described the role of biofilms associated with polyvinylidene fluoride (PVDF) pipeline surfaces as a source of bacterial particulates at an IBM semiconductor manufacturing facility. While viable and direct bacteria counts of <1 per 100 ml were often reported, these workers found significant levels of a number of species associated with pipeline surfaces. Morphological examination of these surfaces demonstrated that many of the microcolonies were dispersed throughout a thick extracellular polymeric matrix. Their data suggested that the overall water systems bioburden had previously been grossly underestimated owing to sampling bias.

Organic contamination resulting from bacterial activities may be one source of total oxidizable carbon (TOC). Carbon containing particulates, such as bacteria, have the potential to act as dopants. Among other effects, high TOC levels create problems with photolithographic deposition processes and hazing of quartz furnace tubes.

Harned (1986) presented scanning electron microscopic (SEM) evidence of bacterial contamination of MOS devices in a Motorola facility. Similar evidence of semiconductor device contamination was presented earlier by Crooke & Lutsch (1976). They surmised that bacterial contamination was responsible for alkali attack of 'high packing density' semiconductor logic circuits. They found that high bacteria levels ($>8 \text{ ml}^{-1}$) in deionized processing waters were the source of contamination on the IC devices. Conductance phenomena resulting from bridging of adjoining circuit paths by bacteria present a potentially devastating problem.

Other purified water systems

In addition to the serious product contamination problems reviewed above, biological contamination of purified water system production and distribution systems can result in significant material and labour costs. Permanent losses in reverse osmosis flux and salt rejection, plugging of submicron membrane and ultrafilters, contamination of primary and polishing deionization resins, and

fouling of piping and storage systems have all been attributed to bacterial growth and biofilm formation (Mittelman *et al.* 1987). Stoecker & Pope (1986) have described corrosion processes in a high temperature water system resulting from microbially influenced corrosion activities (see Hamilton, Chapter 9). Biofilms were found to be the source of contaminated dental unit waters (Whitehouse *et al.* 1991). Scanning electron microscopy of dental unit water system pipelines revealed bacteria embedded within an amorphous matrix. These workers suggested that adherent microorganisms within the pipelines accounted for the continuous dissemination of oral microbes throughout the distribution systems.

Design and preventative maintenance considerations

Microporous membrane filters

In addition to increases in transmembrane pressure, membrane degradation, and pyrogen production, a major consequence of membrane biofouling is bacterial penetration or 'grow-through'. Bacterial passage through microporous membrane filters has been described by several workers (Christian & Meltzer 1986; Howard & Duberstein 1986). Bacterial penetration occurs under both dynamic and quiescent flow conditions. Bacteria have been shown to penetrate membranes whose pore size is smaller than that of the smallest cellular aspect (Meltzer 1987). Membranes with relatively narrow pore size distributions are also susceptible to grow-through (Howard & Duberstein 1986). Wolf & Schoppmann (1989) have described the penetration of 0.2 μm membranes (polycarbonate and cellulose nitrate) by streptomycetes. They suggest that passage is mediated by constriction of vegetative filaments rather than deformation of pores.

The current FDA (FDA 1987), Health Industry Manufacturers Association (HIMA 1982), and American Society for Testing and Materials (ASTM 1983) requirements for viable particle retention are designed to evaluate filtration efficacy. The assay involves a challenge of 10^7 cells per cm^2 of feed water surface. The challenge organism, *Pseudomonas diminuta*, is grown in a defined saline lactose broth such that a fairly uniform population of 0.2 μm cells results. The cells are passed through the membrane on a one-time basis and a determination of viable particle passage is made using standard cultural techniques. Passage of >1 cell per 10^7 cm^2 constitutes membrane failure.

Membrane fouling or grow-through potential are not, however, addressed by these standards. Perhaps of more concern to manufacturers is the apparent lack of insight into what factors might contribute to the development of conditions promoting grow-through. Despite deficiencies inherent in the current assays, it is important that microporous membranes be evaluated for retention efficacy prior to their installation. The large surface area and surface charge associated with pleated and disc-type membranes can provide a suitable substratum for bacterial biofilm development.

Materials of construction and hydraulic effects

The pharmaceutical industry has long recognized the importance of purified water systems design and materials of construction in limiting bacterial growth. Dead-leg areas, threaded fittings, and low flow distribution sections are minimized or eliminated. While flow rates and surface inhomogeneities do appear to play some role in adhesion and biofouling processes, biofilms can and do form on smooth surfaces exposed to high shear environments. The advantages of higher flow rates and reduced dead-leg areas are in mitigating nutrient concentration, polarization and increasing the frequency of ultraviolet and filtration systems contact.

Experiments in oligotrophic marine environments suggest that shear forces induced by increased flow rates may not be sufficient to prevent bacterial colonization and/or remove adhered cells. Mittelman *et al.* (1990) showed that colonization of polished stainless steel surfaces by *Pseudomonas atlantica* was enhanced at higher fluid shear forces up to a critical shear of approximately 12 N m^{-2}. Characklis & Marshall (1990) have suggested that fluid hydraulics may affect the initial rate of colonization; however, following this initial lag phase, biofilm development appears to proceed somewhat independently of fluid flow. Vanhaecke & Haesevelde (1991) and Vanhaecke *et al.* (1990) have described the mechanisms of bacterial adhesion

to stainless steel, primarily in pharmaceutical systems. Their work has also shown that surface smoothness may not play an important role in colonization processes. In particular, they showed that for hydrophobic strains of *Pseudomonas* spp., electropolished stainless steel was no more resistant to colonization than was 120 grit-polished stainless steel. Their work suggested that the colonizing strain rather than the substratum physicochemical characteristics was the major determinant of colonization efficacy.

The US Navy has been evaluating various surfaces for their resistance to biofilm formation for over 100 years. As yet, they have failed to develop materials or surface characteristics which prevent microbial adhesion. In addition, most of the biocidal coatings which have been employed are effective for relatively short periods of time (Goldberg 1986). In any case, these coatings all have intrinsic release rates, which militate against their use in purified water systems. With regard to materials of construction, there is as yet no reproducible experimental evidence to suggest the influence of material type on colonization rate. This is likely to be a function of experimental procedure, organism selection and surface preparation. A preliminary field study of various polymeric and stainless surfaces exposed to 18 Mohm cm water failed to demonstrate any significant influence of material type on biofilm formation (Martyak *et al.* 1993).

One approach to limiting attachment processes would be to interfere with those physicochemical forces which mediate the initial adhesion event. Reviews of efforts in this direction have been made by Baier *et al.* (1968), Absolon *et al.* (1983), and Rosenberg & Kjelleberg (1987). In general, hydrophobic surfaces (e.g. Teflon) promote bacterial adhesion in purified water. Conversely, the rate of attachment is lower for higher energy, hydrophilic surfaces (e.g. glass). However, once cells have colonized these two types of surfaces, they tend to desorb from the hydrophobic surfaces more rapidly; attachment to high energy surfaces appears to be less reversible.

In this regard, some preliminary work suggests that incorporating alternating hydrophobic/hydrophilic regions in polymers used in distribution system piping may retard bacterial adhesion. Clearly, more research is needed to characterize further those factors influencing both the initial stages of bacterial adhesion and the stability of the adhered cells and biomass.

Point-of-use distribution

Personnel contact with product water dispensing systems is a major source of contaminants. Attention should therefore be given to the most common types of systems: sink waste traps, hand spray rinsers, hoses, cascading sinks, and dump rinsers. Contaminants such as skin oils, hairs and respiratory droplets can serve as a direct source of bacteria as well as a source of growth promoting substrates. The low flow and turnover rates associated with some of these dispensing systems can also create dead-leg areas. Weekly treatments of dump rinsing and cascading sinks with a 10% (v/v) hydrogen peroxide solution for 30 min is an effective preventative treatment for these systems.

Several types of end use treatment devices have been employed to control growth at the nozzle orifice. A 254 nm UV light has been used on the orifices of laboratory sink waste traps to limit adventitious contamination from the outside environment. The system, which is no longer in production (EPCO Corporation, Minnesota), did appear to be effective in this regard. However, there is no evidence to suggest that external contamination of taps by cleanroom air is a major source of bacteria. A microbial 'barrier tap' has been developed for the US space transportation systems (shuttle) employing a laminar flow air disinfectant flush system (Sauer 1981).

Hand spray rinsers, by virtue of their discontinuous operation and large internal surface area, can be a significant dead-leg reservoir for bacteria. Fluoroware Corporation (Chaska, Minnesota) has developed a recirculating hand sprayer designed to limit the effects of nutrient and bacterial polarization.

Ultraviolet systems

Ultraviolet sterilization systems are typically employed for the control of planktonic bacterial populations within purified water systems. When these systems are properly installed and maintained, they provide greater than a 99.9999% reduction in viable bacterial numbers (Powitz & Hunter 1985). The poor penetrating power of UV light renders this type of treatment relatively ineffective against bacteria in biofilms. Where

bacteria (and other microorganisms) are inactivated by UV treatment, they are often removed by filtration. UV inactivated cells may then serve as nutrients for viable cells and/or a source of pyrogenic contaminants.

Newly installed UV lamps are usually designed to provide approximately 60 000 μW s cm^{-2}; after one year of service, the lamps should provide a minimum of approximately 20 000 μW s cm^{-2}. Lamps based on mercury/argon mixtures exhibit the highest efficiency. Following a 1 year service time, the lamps should be replaced and quartz sleeves cleaned. Many manufacturers recommend a quarterly inspection with a UV dosimeter to insure proper lamp operation. Since UV light efficacy is primarily a function of energy and contact time, it is important that the design flow rates not be exceeded.

Novel detection methods

Deficiencies associated with cultural techniques

The classical evaluation methods that rely on the culturing of organisms suffer from several serious deficiencies. Organisms from extreme environments (ultrapure water is an extreme environment) are notoriously difficult to culture and the standard plate counts will not give an accurate estimate. The problem of viable but non-culturable bacteria is a well accepted concept in public health microbiology. When starved, *Vibrio cholerae* forms minicells that are often not culturable but readily initiate the disease when the water is consumed.

Not only do the microorganisms not grow out readily on attempted isolation, but the growth of many oligotrophic organisms is extremely slow. This delay is clearly not acceptable for process waters in which continual control is essential.

Several of the classical microscopic techniques lack sensitivity. They often require at least 10^3 organisms per ml in small samples for reliable detection. Concentration by centrifugation or filtration with vital staining and microscopic examination requires that the water system be sampled (thereby creating a potential contamination), is time consuming, and requires trained operators for monitoring. Usually, the type(s) of bacteria involved in the contamination are not readily determined with the microscopic examination. It is possible to identify specific microbes using fluorescent labelled probes. However, these techniques require that the organisms commonly contaminating the system be well known so that antibody or gene probes can be developed. Despite these deficiencies, the recoveries associated with direct counting techniques are often significantly greater than those obtained with cultural techniques.

Direct count techniques with applications for bacterial biofilms in purified waters have been described by a number of investigators (Mittelman *et al.* 1987; Newby 1991; Patterson *et al.* 1991). Rodriguez *et al.* (1992) have described a direct count technique employing two different fluorochromes for assessing *in situ* bacterial activity. Yu & McFeters (1993) employed this technique to assess bacterial viability on stainless steel surfaces in nutrient limited environments. Both the total and the respiring fraction of attached bacteria may be measured using this technique. Adams *et al.* (1991) described the application of direct count techniques in ozonated waters. These techniques may prove particularly useful in estimating populations of damaged and viable, but non-culturable bacteria in biofilms.

In situ biofilm monitoring

In any process, it is important to monitor possible contamination so that corrective action can be instigated as rapidly as possible. The ideal monitors should be non-destructive so that they will not inhibit or damage the biofilm. Thus the sensors will provide a true picture of the problem. If the biosensors can be placed in supply lines or in the system ahead of purification/disinfection systems then treatments can be modified to maintain the desired level of purity.

FTIR

Fourier transforming infrared spectrometry (FTIR) operated in the attenuated total reflectance mode allows the detection of bacterial biofilms as they form on a crystal of zinc selenide or germanium (Nichols *et al.* 1985). The amide stretching of the proteins and ether stretch of the carbohydrates are clearly detectable when bacteria attach to surfaces. Often the nutrients attracted to the surface from the bulk phase are also clearly indicated by their infrared fingerprint.

If the water system contained IR 'windows', biofilm formation could be monitored and an indication of the chemical nature of the contamination would be apparent. Periodically, some mechanism for cleaning the 'window' would need to be provided. A detection limit of c. 5×10^5 cells per cm^2 *Caulobacter crescentus* cells was obtained using a germanium substratum monitored in an ATR (attenuated total reflectance) flow cell (Nivens *et al.* 1993a). This non-destructive technique holds promise as an online monitoring tool for bacterial colonization in purified water systems.

The technique also provides information on the nutritional status of the biofilm microbes as the formation of the endogenous storage polymer poly ß-hydroxyalkanoate (PHA) can be detected in the spectra. PHA forms during conditions of unbalanced growth when some critical shortage prevents bacterial cell division. This ability to monitor biomass signature compounds may be of particular utility in assessing chemical treatment efficacies against purified water system foulants.

Resonance Raman spectroscopy

Lasers which provide high power at frequencies between 190 and 260 nm can differentially activate components in the bacterial cell walls and cytoplasm and produce a shift in the wavelength of the scattered photons which can be detected as a resonance Raman spectrum (Manoharan *et al.* 1990). With a highly focused pulsed laser it is possible to detect a single cell. Selective excitement of bacterial cellular components can be achieved using a tunable laser. By progressively shortening the incident wavelength, nucleic acids, calcium dipicolinate in endospores, tyrosine and tryptophan in the proteins, and the other aromatic amino acids can be distinguished by their vibronic fingerprint spectra. The patterns are specific for groups of bacteria and can be utilized for identification. Information on the nutritional status of the bacteria can be obtained (Dalterio *et al.* 1987) as they pass through UV transmissive capillaries.

In the presence of precious metal microstructures, the electron plasmon can markedly potentiate the Raman scattering of molecules intimately associated with the surface in surface enhanced Raman spectroscopy (SERS). In theory, it should be possible to use this technique to detect non-destructively the attachment of bacteria or bacterial derived products to the metal microbases. The Raman fingerprint should be diagnostic of the chemical composition of the attaching species. Since this can be a cumulative effect, the sensitivity should be very great (D.C. White, personal communication).

Fluorometry

The use of fluorometry for on-line monitoring of biofilm development has been described for monitoring of antifouling (Mittelman *et al.* 1993) and uncoated, stainless steel surfaces (Khoury *et al.* 1992). Reduced NAD compounds along with the aromatic amino acids (for example, tyrosine and tryptophan) can be monitored as indicators of cell energy charge and biomass, respectively. As with FTIR and quartz crystal microgravimetry (described below), this technique holds promise for application to monitoring of purified water system fouling. Detection limits remain relatively high, however, in the range of c. 10^5 cells per cm^2.

Electrochemical monitoring

If electrically isolated electrodes can be incorporated into a stainless steel system it is possible to monitor biofilm formation and persistence electrochemically. Electrochemical methods have, for the most part, been applied towards monitoring of microbially influenced corrosion of metallic substrata. However, potential exists for application of electrochemical monitoring to purified water system biofilm detection.

There are two types of electrochemical monitoring systems. In the first system, the potential between a standard electrode and a working electrode is monitored. The open circuit potential becomes more negative as the biofilm forms on the electrode surface. The potential shifts as metabolic activity and/or biomass associated with the biofilm changes (Mittelman *et al.* 1992).

The effects of the biofilm may also be monitored on the same coupon using a frequency spectrum of the noise (electrochemical noise). Perturbation response methods may also be utilized to show changes in the surface chemistry as a result of biofilm activity. In electrochemical impedance spectroscopy, a small sinusoidal potential is induced in the system over a 5 decade frequency range and the induced changes in current and phase shift angle determined (Dowling *et al.* 1988).

The presence of material of intrinsic viscosity significantly different from that of the water can be detected as a change in frequency of a vibrating quartz crystal in quartz crystal microbalance (QCM) technology. The surface can also be used as an electrode to produce both the microbalance and electrochemical signals (Deakin & Buttry 1989). This technique was applied to monitoring biofilm formation in a simulated purified water system for the space station project. Detection limits of c. 10^5 cells per cm^2 were obtained using an AT-cut quartz crystal (Nivens *et al.* 1993b).

Significant insight into the nature of contamination of ultrapure water can be gained from destructive analysis of biofilms or filter retentates recovered from the system. The most important contaminants of pure water systems are Gram negative bacteria or the lipopolysaccharide (LPS) products from these organisms. Novitsky (1984, 1987) has reviewed the significance and detection of endotoxins in semiconductor grade and other ultrapure waters. Lack of sensitivity appears to be the greatest problem for the semiconductor industry. It is unlikely that bacterial LPS contributes significantly to the total oxidizable carbon load in purified water systems.

Bacterial pyrogens (endotoxins)

The lipid A of the LPS of Gram negative bacteria evokes a most powerful physiological response when injected into human beings. The mildest response is fever (pyrogen) which can progress to anaphylactic shock and death. The massive inflammatory response provoked by endotoxin represents the human body's natural response to the danger of Gram negative septicaemia. Consequently it is of paramount importance that water for intravenous infusion and medical devices for implantation be free of Gram negative bacteria or their endotoxic products.

Currently, endotoxins in purified waters are most often tested by inducing a biological response with the *Limulus* amoebocyte lysate (LAL) test. This test allows detection of as few as approximately 10^3 cell equivalents of endotoxin. The test is a bioassay that depends on the activation of proteins by the endotoxin. It is not effective in solutions containing alcohol, chelating agents, urea, high salts or some proteins (Novitsky 1987). Various methods for measuring endotoxin levels in pharmaceutical waters have been reviewed by Pearson (1991). This test is most effective against endotoxin or bacteria recovered from the bulk phase as endotoxin attaches to surfaces quite readily (Novitsky *et al.* 1986).

Novitsky (1987) found that a significant positive correlation exists between free and free + bound endotoxin and bulk phase bacterial concentrations in ultrapure water. The methods described are rapid (requiring approximately 1 h incubation) and amenable to automation. The extrapolated detection limit for bacteria in ultrapure water is in the order of 100–1000 cells per ml.

A new endotoxin assay that is bacterial family/species specific, more sensitive, and less subject to interferences has been developed (White & Mittelman 1991). The test will also detect endotoxin attached to surfaces. The technique depends on the extraction of lipids from the biofilms or filter retentates, hydrolysis of the lipid-extracted residue, and re-extraction with recovery of the previously ester or amide linked hydroxy fatty acids in the lipid A of the LPS. Derivatization of the carbonyl and hydroxyl moieties allows the separation of the different hydroxy fatty acids by capillary gas chromatography, and detection by chemical ionization mass spectrometry of the negative ions provides sensitivity to the equivalency of a few cells. The composition patterns of the different LPS hydroxy fatty acids provide identification of bacterial families. Recent advances have led to the possibility of automating this analysis.

Treatment considerations

Influence of biofilms on treatment efficacy

The so-called 'bacterial regrowth' phenomenon following chlorine treatments of purified and potable water systems is likely a function of biofilm formation rather than resistance in the classical sense of antibiotic resistance (Ridgway *et al.* 1984; Wolfe *et al.* 1988). A combination of the polyanionic nature of many bacterial polysaccharides coupled with the inherent biocide demand associated with extracellular slime matrices may act as an inhibitor of biocide actions. The fact that most power plant condenser waters are treated on a continuous or frequent slug dosage basis is an indication of the recalcitrance of fouling biofilms. A treatment strategy which focused on modifications of the primary conditioning film and/or initial colonizing bacterial population might obviate the need for toxic biocides, thus removing a significant threat to aquatic ecosystems and labile product formulations.

Inactivation and removal of bulk phase bacteria following the most commonly employed chemical and physical treatments is usually complete. Within days or weeks of treatment, however, bulk phase bacteria levels are often as high as or higher than pretreatment levels. Bacteria existing in sessile biofilms are protected from the effects of biocides and, to some extent, physical treatments (see Gilbert & Brown, Chapter 6). A number of studies have shown that repopulation of purified and other industrial water systems is mediated by the presence of adherent bacterial populations (Mittelman & Geesey 1987). The key to effective treatments, therefore, is inactivation and removal of this population of contaminating biological particulates.

Biocide (chemical) treatments

The development of process and product compatible biofouling treatments has long evaded manufacturers of critical products which require purified water as a raw material (Mittelman 1986b). Of all the physical and biocide treatments applied to semiconductor grade purified water systems, ozone appears to hold the greatest promise as an effective, product compatible biocide. The technology is currently available for inactivation of residual ozone levels *in situ* via UV light (Meltzer 1993). Most of the problems with ozone application to purified water systems have been in the realm of component compatibility. Since ozone is an extremely reactive oxidant, its use is confined to systems constructed of such inert materials as Teflon and other fluoropolymers. Table 7.2 lists some of the more commonly applied chemical biocides along with typical dosage regimes.

Hydrogen peroxide, a commonly employed purified water system biocide, can also be readily degraded by UV light. However, it has relatively

Table 7.2. An overview of purified water biocide treatments

Treatment	Dosage regime	Treatment duration
Quaternary ammonium compounds	300–1000 ppm	2–3 h
Chlorine	50–100 ppm	2–3 h
Peracetic acid/peroxide	0.02–0.05% (v/v)	2 h
Iodine	50–100 ppm	1–2 h
Hydrogen peroxide	10% (v/v)	2–3 h
Formaldehyde	1% (v/v)	2–3 h

little activity against attached bacteria in biofilms. There is also evidence which suggests that hydrogen peroxide requires catalytic concentrations of Fe^{2+}, Ni^{2+} or Cu^{2+} for optimal biocide efficacy (Block 1991).

Surfactants such as the quaternary ammonium compounds have excellent biocide activity in addition to their intrinsic detergency. Synergistic combinations of chlorinated or brominated compounds with surfactants may provide additional activity against bacterial biofilm populations. Their ability to interact with surfaces does, however, create problems related to removal of these compounds following application. Larger volumes of rinse water are required for the surfactant compounds than, for example, hydrogen peroxide.

The phenomenon of RO membrane biofouling has been reviewed by Ridgway et al. (1984). Several workers have described chemical treatments for various RO membrane systems (Maltais & Stern 1990; Ridgway et al. 1986). Membrane compatibility is a contentious issue which often is the driving force in treatment selection.

Effective regeneration with sodium hydroxide and sulphuric acid sterilizes contaminated ion exchange resins. However, some users have chosen to employ chemical sterilants between regenerations. Formaldehyde, although rarely employed in North America, has previously been used for sanitization of mixed bed resins. Trichloromelamine has been used for both ion exchange and granular activated carbon (GAC) disinfection. This compound is chemically less reactive than hypochlorite and thus has proven more compatible with resin materials. Peracetic acid, at 0.02% (v/v), was found to be an effective sanitizer for both resins and associated water system components (Flemming 1984).

Resin bonded quaternary ammonium triiodide has been used to preserve ion exchange systems. Several workers have described the application of iodide–resin combinations for on-line use (Flemming 1987). The current joint US–Canadian–European–Japanese space station design utilizes an iodine resin treatment system for its potable supply (Pyle & McFeters 1989). Itabashi et al. (1992) have described the application of a low cross-linked poly-bromide resin for removing bacteria from process water. In studies with E. coli, the resin removed greater than 99.9999% of the organisms within a 2 h contact time. Silver ions bonded to solid supports have been used to disinfect process waters since the 1930s (Shapiro & Hale 1937). Concerns over problems ranging from development of silver resistant bacterial populations to reactions with products have limited its application in the pharmaceutical and semiconductor industries. However, silver and silver–copper combinations released through electrolysis have found applications as hospital water systems (Muraca et al. 1990) and swimming pool disinfectants (Landeen et al. 1989).

Physical treatments

The use of high temperature water or steam flushes for biofilm removal has proven effective in pharmaceutical grade purified water systems. Hot water maintained at or above 80 °C will prevent the growth of all bacteria encountered in a purified water system. Purified water-for-injection used in the production of injectable drugs is typically stored and recirculated through stainless steel storage tanks which are maintained at a constant temperature of 80 °C. Flowing steam is often employed in the biotechnology industry to sterilize stainless steel fermentation vessels and attendant piping. So called sterilize-in-place systems employing hot water or steam have been used for many years in the dairy and food products industries. The utility and economy of such systems in the semiconductor industry is, however, questionable. Husted & Rutkowski (1991) have described the successful application of high temperature deionized water for treating biofilms associated with mixed bed deionizer resins.

An electrochemical technique for *in situ* sterilization of GAC columns has been described (Matsunaga *et al.* 1992). Several types of bacteria and one yeast species associated with GAC surfaces were successfully inactivated with application of voltages ranging from 0.65 to 1.0 V (SCE). Significant decreases in both cell viability and numbers of cells sorbed to the GAC surfaces were reported.

Blenkinsopp and co-workers have described a bioelectric enhancement of biocide efficacy against biofilm associated bacteria (Blenkinsopp *et al.* 1992). Killing efficacy was enhanced by several orders of magnitude when glutaraldehyde or isothiazalones were applied in the presence of low level electrical fields ($<100 \ \mu A \ cm^{-2}$). The bioelectric effect appears to be useful only in the presence of antimicrobial agents; low level electrical current application alone is not an effective antimicrobial treatment. While this technique has not yet been utilized in purified water systems, it may hold promise as an adjuvant in treating fouling of pipelines, DI resin beds and GAC columns.

Future treatment strategies will undoubtedly focus on the prevention of primary bacterial colonization. This will entail alterations in the surface chemistry of both the bulk fluid and the substratum.

The ability of bacteria to self-replicate distinguishes these contaminants from other, abiological, particulates. Their growth, replication, and production of various ionic and organic by-products in otherwise contaminant free purified waters presents a tremendous challenge to production and quality assurance personnel. The same metabolic and structural properties which enable survival in such an otherwise hostile environment create special problems for bacterial detection and treatment. In purified water systems and other industrial water circuits, the majority of bacteria are associated with surfaces. Biofilm development in purified water systems is

a major cause of product contamination problems in the pharmaceutical, medical device, food processing, and microelectronic industries. Evolving on-line monitoring tools, including FTIR spectroscopic fluorometric and electrochemical methods, will enable better control of fouling in purified water systems. An understanding of the mechanisms which lead to the transition from a planktonic to an attached state in these extremely oligotrophic environments is essential to the development of targeted treatments.

Although great progress has been made in developing advanced fluid handling materials of construction, mankind still has not developed a surface to which bacteria will not adhere.

FDA recalls and court actions related to water

(1980) *Pseudomonas cepacia* contamination concerns FDA. *Quality Control Reports*, 14(12), 1–4.

(1981) FDA stresses deionizer validation. *Quality Control Reports*, 15(7), 3–4.

(1984a) 18 recalls in 2 years are water related. *Quality Control Reports*, 18(1), 4–8.

(1984b) Analytical Products initiated recall by letter of its calibration buffer pH 7.383, in 3 ml ampules, aqueous sodium and potassium phosphate buffer for *in vitro* diagnostic use, because bacterial contamination of water may result in miscalibration. *Devices And Diagnostics Letter*, 11(46).

(1984c) *Class III recall: Bel-Mar Laboratories brand bacteriostatic water for injection.* US FDA, D-155-4.

(1984d) FDA inspection program emphasizing purified water system validation. *FDC Reports*, 46(43), 1–2.

(1984e) FDA views on water treatment. *Quality Control Reports*, 18(1), 4–6.

(1985) Water in halothane vaporizers. *Health Devices*, 14(10), 326–7.

(1986a) Cosmetic manufacturer with portable deionizer system switched to purchasing bottled purified water. *FDC Reports*, 7(8), 5.

(1986b) FDA inspections of deionized water systems. *Quality Control Reports*, 20, 4–8.

(1986c) FDA wants recirculating water systems. *Quality Control Reports*, 20, 1–4.

(1987) Recall letters cite contamination controls; validate from microbiological perspective cooling and holding of bulk product questioned; training procedures also emphasized by FDA complaint; investigation may require microbial testing; comparison of assays with supplier is not validation; labels must identify name/quality of preservative contaminant. *Quality Control Reports*, 21(10), 1–6.

(1988a) Clinical safety and product hazards; contaminated dialysis water: FDA alert warns of chloramine hazard. *Biomedical Safety and Standards*, 18(7), 50.

(1988b) Packets of Povadyne povidone-iodine solution, USP 10% in kits were recalled by Acme/Chaston because of bacterial contamination. *Devices And Diagnostics Letter*, 15(17).

(1989) Clinical safety & product hazards; sodium azide contamination of dialysis water: FDA issues warning. *Biomedical Safety Standards*, 19(8), 58–9.

(1990a) 'Wet Cote' contact lens solution contaminated with harmful bacteria. *PR Newswire*, 17 January, 1990.

(1990b) Class I recall: Fisons Opticrom 4% ophthalmic solution in 10 ml bottles due to bacterial contamination. *Washington Drug Letter*, 22(37).

(1990c) *Class III recall: Quad Pharmaceuticals bacteriostatic water for injection.* US FDA, D-307/311-0.

(1990d) Clinical safety and product hazards; should more sensitive culturing methods be used to check dialysis water contamination? *Biomedical Safety Standards*, 20(9), 65–7.

(1990e) Opticrom recalled. *FDA Consumer*, 24(8), 4–5.

(1991) Contaminated bottled water dumped. *FDA Consumer*, 25(1), 39–40.

References

Absolon, D. R., Lamberti, F. V., Policova, Z., Zingg, W., Oss, C. V. & Neumann, W. W. (1983). Surface thermodynamics of bacterial adhesion. *Applied and Environmental Microbiology*, 46, 90–7.

Adams, J. C., Lytle, M. S., Dickman, D. G. & Bressler, W. R. (1991). Use of direct viable count methodology with ozonation in drinking water. *Ozone Science Eng.*, 13, 1–10.

Anderson, R. L., Bland, L. A., Favero, M. S., McNeil, M. M., Davis, B. J., Mackel, D. C. & Gravelle, C. R. (1985). Factors associated with *Pseudomonas pickettii* intrinsic contamination of commercial respiratory solutions marketed as sterile. *Applied and Environmental Microbiology*, 50, 1343–8.

American Society for Testing and Materials (1983). *Bacterial retention of membranes utilized for liquid filtration.* ASTM F838-83.

Baier, R. E., Shafrin, E. G. & Zisman, W. A. (1968). Adhesion: mechanisms that assist or impede it. *Science*, 162, 1360–3.

Blenkinsopp, S. A., Khoury, A. E. & Costerton, J. W. (1992). Electrical enhancement of biocide efficacy against *Pseudomonas aeruginosa* biofilms. *Applied and Environmental Microbiology*, 58, 3770–3.

Block, S. S. (1991). Peroxygen compounds. In *Disinfection, sterilization, and preservation*, 4th edn, ed. S. S. Block, pp. 167–81. Malvern, Pa: Lea & Febiger.

Characklis, W. G. & Marshall, K. C. (ed.) (1990). *Biofilms*. New York: Wiley.

Christian, D. A. & Meltzer, T. H. (1986). The penetration of membranes by organism growthrough and its related problems. *Ultrapure Water*, 3, 39–44.

Collentro, W. V. (1986). Pretreatment part II: activated carbon filtration. *Ultrapure Water*, 3, 39–44.

Craven, R. A., Ackerman, A. J. & Tremont, P. L. (1986). High purity water technology for silicon wafer cleaning in VLSI production. *Microcontamination*, 4, 1421–2.

Crooke, M. & Lutsch, A. G. K. (1976). *Process evaluation of high packing density semiconductor logic circuits by a scanning electron microscope.* Johannesburg: Electron Microscopy Society of Southern Africa.

Dalterio, R. A., Nelson, W. H., Britt, D. & Sperry, J. F. (1987). An ultraviolet (242 nm excitation) resonance Raman study of live bacterial components. *Applied Spectroscopy*, 41, 417–2.

Deakin, M. R. & Buttry, D. A. (1989). Electrochemical applications of the quartz crystal microbalance. *Analytical Chemistry*, 61, 1147–54.

DePaiva, E. T., DaSilva, F. W. O., Raisin, C. & Lassabatere, L. (1992). Contribution of the contamination of deionized water by bacteria to the adsorption of carbon on Si and GaSb. *Journal of Material Science*, 27, 1585–8.

Dial, F. & Chu, T. (1987). The effect of high bacteria levels with low TOC levels on bipolar transistors: a case study. *Proceedings of the Semiconductor Pure Water Conference*, pp. 178–93. San José, Calif.

Dowling, N. J. E., Stansbury, E. E., White, D. C., Borenstein, S. W. & Danko, J. C. (1988). On-line electrochemical monitoring of microbially induced corrosion. In *Microbial Corrosion*, pp. 5–17. Palo Alto, Calif.

Eisenmann, D. E. & Ebel, C. J. (1988). Sulfuric acid and DI point of use particle counts and resultant silicon wafer FM levels. *Annual Meeting of the Institute for Environmental Sciences*, pp. 547–59. Los Angeles, USA.

Food and Drug Administration (1987). *Guidelines on sterile drug products produced by aseptic processing.*

Code of Federal Regulations, 21 CFR 10.90, US FDA.

Flemming, H. C. (1984). Bakterienwachstum auf Ionenaustauscher-Harz. Untersuchungen an einem stark sauren Kationen-Austauscher. Teil III. Desinfektion mit Peressigsäure. *Zeitung Wasserbau Abwasser Forscheit*, **17**, 229–34.

Flemming, H. C. (1987). Microbial growth on ion exchangers. *Water Research*, **21**, 745–56.

Goldberg, E. D. (1986). TBT, an environmental dilemma. *Environment*, **28**, 17–20.

Harned, W. (1986). Bacteria as a particle source in wafer processing equipment. *Journal of Environmental Science*, **24**, 33–4.

Health Industry Manufacturers Association (1982). *Microbiological evaluation of filters for sterilizing liquids*, HIMA 3(4).

Howard, G. & Duberstein, R. (1986). A case of penetration of 0.2 µm-rated membrane filters by bacteria. *Journal of the Parenteral Drug Association*, **34**, 95–102.

Husted, G. & Rutkowski, A. (1991). Control of microorganisms in mixed bed resin polishers by thermal sanitization. *Proceedings of Watertech 1991*, pp. 43–51. San José, Calif.

Ishida, Y. & Kadota, H. (1981). Growth patterns and substrate requirements of naturally occurring obligate oligotrophs. *Microbial Ecology*, **7**, 123–30.

Itabashi, O., Goto, T., Yoshida, T., Kamata, S., Kudo, M. & Arai, K. (1992). A new adsorbent for removing bacteria from water: low cross-linked poly (4-vinylpyridinium bromide)-porous substance composite. *Chemistry and Industry* (15 June), 450.

Johnston, P. R. & Burt, S. C. (1976). Bacterial growth in charcoal filters. *Filtration and Separation*, **13**, 240–4.

Khoury, A. E., Nicholov, R., Soltes, S., Bruce, A. W., Reid, G. & DiCosmo, F. (1992). A preliminary assessment of *Pseudomonas aeruginosa* biofilm development using fluorescence spectroscopy. *International Biodeterioration and Biodegradation*, **30**, 187–99.

Klein, E. T., Pas, G. B. H., Wright, R. & Million, C. (1990). Microbial and endotoxin contamination of water and dialysate in the central United States. *Journal of Artificial Organs*, **14**, 85–94.

Landeen, L. K., Yahya, M. T. & Gerba, C. P. (1989). Efficacy of copper and silver ions and reduced levels of free chlorine in inactivation of *Legionella pneumophila*. *Applied and Environmental Microbiology*, **55**, 3045–50.

LeChevallier, M. W., Cawthon, C. D. & Lee, R. G. (1988). Factors promoting survival of bacteria in chlorinated water supplies. *Applied and Environmental Microbiology*, **54**, 649–54.

Maltais, J. B. & Stern, T. (1990). An evaluation of various biocides for disinfection of reverse osmosis membranes and water distribution systems. *Ultrapure Water*, **7**, 37–40.

Manoharan, R., Ghiamati, E., Dalterio, R. A., Britton, K. A. & Nelson, W. H. (1990). Ultraviolet resonance Raman spectra of bacteria, bacterial spores, protoplasts, and calcium dipicolinate. *Journal of Microbiological Methods*, **11**, 1–15.

Marshall, K. C. (1988). Adhesion and growth of bacteria at surfaces in oligotrophic environments. *Canadian Journal of Microbiology*, **34**, 503–6.

Martyak, J. E., Carmody, J. C. & Husted, G. R. (1993). Characterizing biofilm growth in deionized ultrapure water piping systems. *Microcontamination*, **11**, 39–44.

Matsunaga, T., Nakasono, S. & Masuda, S. (1992). Electrochemical sterilization of bacteria absorbed on granular activated carbon. *FEMS Microbiology Letters*, **72**, 255–9.

Meltzer, T. H. (1987). *Filtration in the pharmaceutical industry*. New York: Marcel Dekker.

Meltzer, T. H. (1993). *High-purity water preparation for the semiconductor, pharmaceutical, and power industries*. Littleton, Colo: Tall Oaks Publishing.

Michels, D. L. (1981). Validation and control of deionized water systems. *FDA Abstracts* (August).

Mittelman, M. W. (1986a). Biological fouling of purified water systems: part III, treatment. *Microcontamination*, **4**, 30–40.

Mittelman, M. W. (1986b). Trends in the detection and control of biological fouling in purified water systems. *Ultrapure Water*, **3**(6), 22–3.

Mittelman, M. W. & Geesey, G. G. (1987). *Biological fouling of industrial water systems: a problem solving approach*. San Diego, Calif: Water Micro Associates.

Mittelman, M. W., Islander, R. & Platt, R. M. (1987). Biofilm formation in a closed-loop purified water system. *Medical Device and Diagnostics Industry*, **10**, 50–5;75.

Mittelman, M. W., Kohring, L. L. & White, D. C. (1992). Multipurpose laminar-flow adhesion cells for the study of bacterial colonization and biofilm formation. *Biofouling*, **6**, 39–51.

Mittelman, M. W., Nivens, D. E., Low, C. & White, D. C. (1990). Differential adhesion, activity, and carbohydrate:protein ratios of *Pseudomonas atlantica* monocultures attaching to stainless steel in a linear shear gradient. *Microbial Ecology*, **19**, 269–78.

Mittelman, M. W., Packard, J., Arrage, A. A., Bean, S. L., Angell, P. & White, D. C. (1993). Test systems for evaluating antifouling coating efficacy using on-line detection of bioluminescence and fluorescence in a laminar-flow environment. *Journal of Microbiological Methods*, **18**, 1–10.

Mittelman, M. W. & White, D. C. (1989). The role of biofilms in bacterial penetration of microporous membranes. *Proceedings of Pharmaceutical Technology 1989*, pp. 211–21. Philadelphia, USA.

Morita, R. Y. (1985). Starvation and miniaturization of heterotrophs, with special emphasis on maintenance of the starved viable state. In *Bacteria in the natural environments: the effect of nutrient conditions*, ed. M. Fletcher & G. Floodgate, pp. 111–30. London: Society for General Microbiology.

Muraca, P. W., Yu, V. L. & Goetz, A. (1990). Disinfection of water distribution systems for *Legionella*: a review of application procedures and methodologies. *Infection Control and Hospital Epidemiology*, 11, 79–88.

Murphy, J. J., Bland, L. A., Davis, B. J., Maxey, R. W., Light, A., Favero, M. S. & Solomon, S. L. (1987). *Pyrogenic reactions associated with high-flux hemodialysis*. Proceedings of ICAAC '87. Washington, DC: American Society of Microbiology.

Newby, P. J. (1991). Analysis of high-quality pharmaceutical grade water by a direct epifluorescent filter technique microcolony method. *Letters in Applied Microbiology*, 13, 291–3.

Nichols, P. D., Henson, J. M., Guckert, J. B., Nivens, D. E. & White, D. C. (1985). Fourier transform–infrared spectroscopic methods for microbial ecology: analysis of bacteria, bacteria–polymer mixtures, and biofilms. *Journal of Microbiological Methods*, 4, 79–94.

Nivens, D. E., Chambers, J. Q., Anderson, T. R., Tunlid, A., Smit, J. & White, D. C. (1993a). Monitoring microbial adhesion and biofilm formation by attenuated total reflection/Fourier transform infrared spectroscopy. *Journal of Microbiological Methods*, 17, 199–213.

Nivens, D. E., Chambers, J.Q., Anderson, T.R. & White, D.C. (1993b). Long-term, on-line monitoring of microbial biofilms using a quartz crystal microbalance. *Analytical Chemistry*, 65, 65–9.

Novitsky, T. J. (1984). Monitoring and validation of high purity water systems with the *Limulus* amebocyte lysate test for pyrogens. *Pharmaceutical Engineering*, 4, 21–33.

Novitsky, T. J. (1987). Bacterial endotoxins (pyrogens) in purified waters. In *Biological fouling of industrial water systems: a problem solving approach*, ed. M. W. Mittelman & G. G. Geesey, pp. 77–96. San Diego, Calif: Water Micro Associates.

Novitsky, J. A. & Morita, R. Y. (1978). Possible strategy for the survival of marine bacteria under starvation conditions. *Marine Biology*, 48, 289–95.

Novitsky, T. J., Schmidt-Gengenbact, J. & Remillard, J. F. (1986). Factors affecting recovery of endotoxin adsorbed to container surfaces. *Journal of Parenteral Science and Technology*, 40, 284–6.

Patterson, M. K., Husted, G. R., Rutkowski, A. & Mayett, D. C. (1991). Bacteria: isolation, identification, and microscopic properties of biofilms

in high-purity water distribution systems. *Ultrapure Water*, 8(4), 18–24.

Pearson, F. C. (1991). *Limulus* amebocyte lysate testing: comparative methods and reagents. In *Sterile pharmaceutical manufacturing*, Vol. 2, ed. M. J. Groves, W. P. Olsen & M. H. Anisfeld, pp. 185–97. Buffalo Grove, Ill: Interpharm Press.

Powitz, R. W. & Hunter, J. (1985). Design and performance of single-lamp, high-flow ultraviolet disinfectors. *Ultrapure Water*, 2, 32–4.

Pyle, B. H. & McFeters, G. A. (1989). Iodine sensitivity of bacteria isolated from iodinated water systems. *Canadian Journal of Microbiology*, 35, 520–3.

Reasoner, D. J. & Geldreich, E. E. (1985). New medium for the enumeration and subculture of bacteria from potable water. *Applied and Environmental Microbiology*, 49, 1–7.

Ridgway, H. F., Justice, C. A., Whittaker, C., Argo, D. G. & Olsen, B. H. (1984). Biofilm fouling of RO membranes—its nature and effect on treatment of water for reuse. *Journal of the American Water Works Association*, 76, 94–102.

Ridgway, H. F., Rogers, D. M. & Argo, D. G. (1986). Effect of surfactants on the adhesion of mycobacteria to reverse osmosis membranes. *Proceedings of the Semiconductor Pure Water Conference*, pp. 133–64. San Francisco.

Roberts, C. (1988). Direct detection and enumeration of mycobacteria in disinfectants by epifluorescence microscopy. *Annual Meeting American Society for Microbiology*, Miami Beach, Fla.

Rodriguez, G. G., Phipps, D., Ishiguro, K. & Ridgway, H. F. (1992). Use of a fluorescent redox probe for direct visualization of actively respiring bacteria. *Applied and Environmental Microbiology*, 58, 1801–8.

Rosenberg, M. & Kjelleberg, S. (1987). Hydrophobic interactions: role in bacterial adhesion. *Advances in Microbial Ecology*, 9, 353–93.

Sauer, R. L. (1981). *The potable water*. NASA, N82-15711 06-51.

Shapiro, R. & Hale, F. E. (1937). An investigation of the Katadyn treatment of water with particular reference to swimming pools. *Journal of the New England Water Works Association*, 51, 113–24.

Simonetti, J. A. & Schroeder, H. G. (1984). Evaluation of bacterial grow-through. *Journal of Environmental Science*, 27, 27–32.

Stoecker, J. G. & Pope, D. H. (1986). Study of biological corrosion in high temperature demineralized water. *Proceedings of the National Association of Corrosion Engineers Annual Meeting*, Houston, Texas.

Tabor, P. S., Ohwada, K. & Colwell, R. R. (1981). Filterable marine bacteria found in the deep sea: distribution, taxonomy, and response to starvation. *Microbial Ecology*, 7, 67–83.

Vanhaecke, E. & Haesevelde, K. V. D. (1991). Bacterial contamination of stainless steel equipment. In *Sterile pharmaceutical manufacturing*, Vol. 2, ed. M. J. Groves, W. P. Olson & M. H. Anisfeld, pp. 141–62. Buffalo Grove, Ill: Interpharm Press.

Vanhaecke, E., Remon, J. P., Raes, F., Moors, J., DeRudder, D. & Peteghen, A. V. (1990). Kinetics of *Pseudomonas aeruginosa* adhesion to 304 and 316-L stainless steel: role of cell surface hydrophobicity. *Applied and Environmental Microbiology*, **56**, 788–95.

White, D. C. & Mittelman, M. W. (1991). *Detection of endotoxins of Gram negative bacteria*. United States Patent, 5,059,527.

Whitehouse, R. L., Peters, E., Lizotte, J. & Lilge, C. (1991). Influence of biofilms on microbial contamination in dental unit water. *Journal of Dentistry*, **19**, 290–5.

Wilson, L. A., Schlitzer, A. B. & Ahearn, D. G. (1981). *Pseudomonas* corneal ulcers associated with soft contact lens wear. *American Journal of Ophthalmology*, **92**, 546–54.

Wolf, H. & Schoppmann, H. (1989). Streptomycetes can grow through small filter capillaries. *FEMS Microbiology Letters*, **57**, 259–64.

Wolfe, R. L., Ward, N. R. & Olsen, B. H. (1988). Inorganic chloramines as drinking water disinfectants: a review. *Journal of the American Water Works Association*, **76**, 74–88.

Yang, M. & Tolliver, D. L. (1989). Ultrapure water particle monitoring for advanced semiconductor manufacturing. *Journal of Environmental Science* (July/August), 35–42.

Youngberg, D. A. (1985). Sterilizing storage tanks in a pure water system. *Ultrapure Water*, **2**, 45.

Yu, F. P. & McFeters, G. A. (1993). Rapid *in situ* physiological assessment of bacteria in biofilms using fluorescent probes. *American Society for Microbiology Annual Meeting*, pp. 359. Atlanta, USA.

8

Mineralized Bacterial Biofilms in Sulphide Tailings and in Acid Mine Drainage Systems

Gordon Southam, F. Grant Ferris and Terrance J. Beveridge

Introduction

The disposal of sulphide bearing mine wastes (tailings) as open deposits results in the formation of unique acid generating ecosystems (Mills 1985). Acid formation occurs via biooxidation of mineral sulphides which generates dilute solutions of sulphuric acid containing soluble metal ions. The microorganisms (predominantly acidophilic thiobacilli) responsible for this mineral oxidation exist as discrete biofilms obtaining energy for their metabolism from the sulphide mineral surfaces which they colonize (Norman & Snyman 1987; Southam & Beveridge 1992). Most references to biofilms conjure up images of thick layers of bacteria (>100 μm) encased in capsular material. In this chapter we will employ the term 'discrete biofilm' to refer to the colonization of mineral surfaces by *Thiobacillus ferrooxidans* which typically exist as a single cell layer that is devoid of capsule. In this context a biofilm is simply a metabolically active surface at a solid solution interface.

Bacterial leaching produces acid mine drainage (AMD) which seeps into the surrounding aquatic and terrestrial environments. AMD is of environmental concern not only because of its acidity but also because of the soluble heavy metals which may enter the eukaryotic food chain (Schofield 1976; Tamm 1976; Koryak *et al.* 1979; Cloutier *et al.* 1986). Sometimes bacterial leaching can be exploited for industrial purposes, for example in the recovery of copper or uranium (Brierly 1982). Bacterial oxidation is, then, both

an economic liability due to the legislated costs associated with AMD neutralization and an economic benefit as evidenced in industrial leaching systems. Clearly, an understanding of the fundamental aspects of bacteria mineral interaction is essential for the prevention of this deleterious environmental impact and for the promotion of these industrial bio oxidative processes.

In both aerobic and anaerobic aquatic environments receiving AMD, bacteria which are different from the leaching microorganisms play an important role in the immobilization of metals and could possibly provide a bioremediation solution to the problem of the toxic leachate. These bacteria exist in planktonic and sessile forms. However, it is the sessile forms existing in biofilms that play a predominate role in the neutralization of these acidic effluents (Gyure *et al.* 1990) and in the immobilization of toxic heavy metals (Ferris *et al.* 1989b). In an industrial setting, the surfaces of pipes or tanks which develop biofilms can be easily identified. However, in the natural environment, we need to describe more precisely the surfaces which develop biofilms. Simply stated, any surface which possesses biological activity can constitute a biofilm. Bacteria coated with minerals, sorb nutrients so that other bacteria can adhere and grow to form new multi-μm surfaces for sorption of dissolved species in solution. We refer to these as micron sized biofilms.

Within the framework of this chapter, we will offer a biogeochemical perspective of the specific biofilms formed by the colonization of the sulphide mineral surface by *T. ferrooxidans*. In addition, we will examine the interaction between bacteria other than *T. ferrooxidans* and AMD effluents in the natural environment and assess the potential outcome of these mineralization processes in terms of geological time scales.

Colonization of sulphide tailings by *Thiobacillus* spp. and acid mine drainage generation

T. ferrooxidans oxidizes reduced sulphur compounds and ferrous iron to produce sulphuric acid and iron(III) as by-products of their metabolism (Ingledew 1982; Harrison 1984). The primary oxidation reactions mediated by *T. ferrooxidans* are (1) (2):

$$FeS_2 + 3\tfrac{1}{2}O_2 \rightarrow FeSO_4 + H_2SO_4 \quad (1)$$

$$2FeSO_4 + \tfrac{1}{2}O_2 + H_2SO_4 \rightarrow Fe_2(SO_4)_3 + H_2O \quad (2)$$

Carbon nutrients are provided through carbon dioxide fixation (Leathen *et al.* 1956). Most thiobacilli are capable of growth on a variety of pyritic minerals (sources of reducing power) (Harrison 1984) as well as synthetic laboratory media (Silverman & Lundgren 1959).

In addition to the leaching effect on the minerals by sulphuric acid, the oxidation of iron(II) (Fe^{2+}) to iron(III) (Fe^{3+}) yields a chemically reactive cation which can scavenge electrons from less electronegative metal species (Keller & Murr 1982). Leaching of toxic heavy metals from tailings, then, is a chemical process enhanced by biocatalysis (Singer & Stumm 1970; Hutchins *et al.* 1986; Baldi *et al.* 1992).

Physical and chemical characterization of tailings material

This biocatalysis, which generates acidity and soluble heavy metals, is controlled by the physical and chemical characteristics of the tailings impoundment. For this reason, it is important to have a clear view of how tailings dumps are formed and how they promote the formation of AMD.

The extraction of base metals by grinding, aeration and flotation of base metal bearing rock produces a fine grained waste material (typically in the order of 200 mesh or approximately

50 μm), which is transported as a slurry from the mill to a tailings impoundment. The coarser tailings material settles near the point of discharge while the finer tailings settle more distally. With the eventual building up of the coarse tailings, the discharge point is often moved to another location resulting in irregular vertical and horizontal stratigraphy of the tailings material (Blowes & Jambor 1990). The multiple discharge point as well as the intrinsic differences in mineral composition of the ore processed through the mill cause the differences in mineralogy throughout the tailings impoundment (Cherry *et al.* 1986a, b).

This stratigraphic diversity controls the hydrologic flow pattern of each tailings impoundment by creating conduits which promote water drainage from the tailings deposit. In general terms, the coarse tailings surrounding the perimeter of the impoundment will be hydrogeologically conductive in comparison to the fine tailings material at the distal regions of the discharge points (Dave *et al.* 1986). This provides downward movement of pore water at the tailings perimeter as well as promoting its horizontal movement from the central or distal tailings regions towards the perimeter of the tailings impoundment (Dave *et al.* 1986).

Historically, once mining was complete, the tailings were allowed to drain in an unrestricted manner. In this granular material, water is retained at the grain boundaries and at the pores between grains because of the capillary forces counteracting gravity (Nicholson *et al.* 1990). This area of partial water saturation is referred to as the vadose zone. Rainfall promotes the release of AMD effluents from sulphide tailings undergoing biooxidation. This reduces the concentration of toxic heavy metals encountered by thiobacilli and probably stimulates further metabolic activity and mineral oxidation.

The vadose zone provides all of the essential physical and chemical requirements for growth of *Thiobacillus* spp. The capillary border on the mineral surface supplies water for bacterial growth (that is, protection against drying) and the pore spaces allow for the influx of gaseous oxygen (for mineral oxidation) and carbon dioxide (for carbon fixation). The final requirement for growth of *T. ferrooxidans* is a source(s) of reducing power which is provided by the sulphide minerals present in the tailings (Harrison 1984). These sulphides may exhibit subtle chemical variations within a single tailings impoundment and will vary extensively between tailings deposits depending on the mineralogy of the base metal being mined. Accordingly, the progression of tailings colonization by *Thiobacillus* spp. also reflects the differences in general mineralogy and hydrokinetics. Sulphide composition in tailings can vary from a few per cent by weight (for example, Lemoine, Quebec, Canada: Southam & Beveridge 1992) up to enormous concentrations (for example, 85% S by weight at Waite-Amulate, Quebec, Canada: Blowes & Jambor 1990). The non-sulphurous remainder is referred to as gangue (or waste) material and may contain carbonates (important for buffering AMD) but is typically dominated by silicates such as quartz, muscovite, feldspars and pyroxenes.

The sulphide type present in a tailings impoundment can affect its colonization by *Thiobacillus* spp. Certain sulphides (e.g. pyrrotite) are more easily oxidized than other more recalcitrant mineral sulphides (e.g. pyrite). This can be explained in geochemical terms by galvanic conversion. Since pyrite possesses the highest standard potential of all metal sulphides (Mehta & Murr 1982), it is protected from oxidation (solubilization) in all multi-sulphide leaching systems. Once it is released in a soluble form, the toxicity of the base metal contained in the metal sulphide (e.g. Ni in millerite) may also decrease the growth of some thiobacilli.

Tailings which are actively leaching consist of three physical zones: an upper sulphide depleted zone in which the sulphides have been extensively removed and oxidation is largely complete; an intermediate zone where sulphide oxidation and acid neutralization reactions are occurring and from where AMD is originating; and an unoxidized zone below the depth of oxygen penetration (often coinciding with saturated pore spaces immediately below the vadose zone) where sulphides remain. Leaching of dissolved solids into the unoxidized zone can result in secondary mineral formation, typically as carbonates, if intrinsic neutralization capacity exists (Blowes & Jambor 1990).

In regions where evapotranspiration rates are high relative to the average annual rainfall or in regions with extensive drying periods and a low water table, a dry crust may form on the surface of the tailings impoundment. Drying at the surface of a sulphide tailings does not allow for the growth or survival of *Thiobacillus* spp. (Southam & Beveridge 1992). Yet this phenomenon does not reduce acidity, probably due to capillary action drawing the sulphuric acid upward through the fine grained tailings during sporadic episodes of water infiltration and drying. This acidity, coupled with the presence of the toxic heavy metals, reduces the likelihood of decommissioning tailings impoundments by revegetation (Heale 1991). Although freezing over winter prevents the release of AMD (Southam & Beveridge 1992), near freezing conditions do not (Ahonen & Tuovinen 1992). The capillary capabilities of fine grained materials are being employed by Nicholson *et al.* (1990), who are promoting the addition of a coarse grained layer followed by a fine grained layer to the top of tailings impoundments to form a vertical hydrostatic cap. This cap will create an upper artificial groundwater table within the added fines which should serve as a barrier to oxygen and to the downward flow of water. Therefore, in unoxidized tailings, it should prevent the generation of AMD by producing low oxygen tension. In oxidized tailings, it should prevent the continued generation and subsequent release of acid and soluble heavy metals.

Initiation of tailing colonization by *T. ferrooxidans*

Even though tailings do not possess large populations of thiobacilli immediately after discharge from the mill, they quickly become inoculated because they are open to the natural environment once in the impoundment (Southam & Beveridge 1993). *Thiobacillus* spp. are then able actively to colonize unoxidized tailings in the vadose zone and persist in the water saturated tailings below this zone (Southam & Beveridge 1992). The role of *Thiobacillus* spp. in the generation of AMD is unequivocal (Norman & Snyman 1987), yet the mechanism of its colonization of tailings is not well understood (Dispirito *et al.* 1983).

Growth on minerals is facilitated by close bacteria–mineral interaction (Duncan & Drummond 1973; Bennet & Tributsch 1978; Norman & Snyman 1987; Southam & Beveridge 1992; see Fig. 8.1) and it is usually assumed that these bacteria are tightly associated with the mineral surface (Suzuki *et al.* 1990; Southam & Beveridge

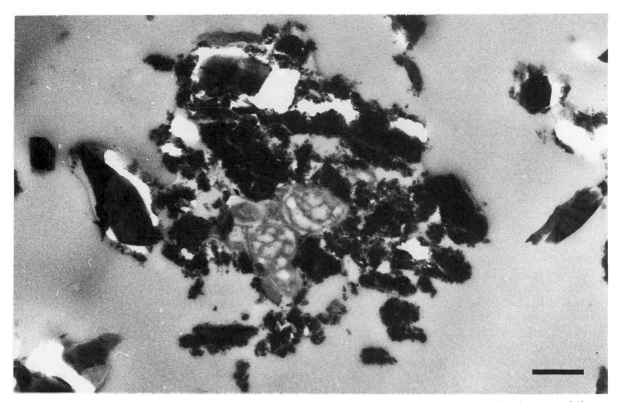

Fig. 8.1. An unstained thin section of a tailings sample from the Lemoine mine which demonstrates the close association between *T. ferrooxidans* and chalcopyrite mineral surfaces. Bar = 0.5 µm. (From Southam & Beveridge 1992, with permission.)

1992). The details surrounding this bacteria–mineral interaction are poorly defined, although a combination of ionic forces and hydrophobicity may play a role in mineral colonization (Southam & Beveridge 1992). It is clear, however, that the overriding factor in the establishment of active, growing *Thiobacillus* populations in tailings, and the subsequent acidification and leaching of resident toxic metals lies in the association between the bacterium and the mineral (Tributsch 1976; Norman & Snyman 1987). Since the enzymes responsible for electron flow are not released extracellularly by the bacterium but are associated with the cell envelope (Harrison 1984), the juncture must be close enough for iron or sulphur utilization and transfer of constituent electrons (Wiersma & Rimstidt 1984).

The bacterial surface component most likely responsible for a close mineral interaction in this Gram negative *Thiobacillus* is lipopolysaccharide (LPS), since this molecule is situated on the outer membrane surface and its side chains extend beyond the usual bilayer structure of a membrane. *Thiobacillus* spp. possess no capsular material (Shively *et al.* 1970; Wang *et al.* 1970; Vestal *et al.* 1973; Hirt & Vestal 1975; Rodriguez *et al.* 1986; Yokota *et al.* 1988), which is conventionally viewed as being important for cellular adherence; however, Dispirito *et al.* (1983) found that extraction of LPS resulted in a decline in the rate and total amount of cell adsorption to pyrite. Extracellular surface activating substances (of unknown composition) have also been implicated in the colonization of reduced sulphur compounds (Jones & Starkey 1961).

For *Pseudomonas aeruginosa*, a genetically related Gram negative eubacterium (Lane *et al.* 1985), LPS heterogeneity reflects the molecule's different chemistry and chain length as it resides on the cell's surface (Chester & Meadow 1975). If this analogy holds true for *T. ferrooxidans*, the different LPS profiles observed for various *T. ferrooxidans* isolates (Southam & Beveridge 1993) would correspond to different LPS chemistry. Differences in LPS chemistry, especially in *O*-side-chains, may confer a different cell surface charge character or hydrophobicity which could affect the ability of the strain to colonize mineral surfaces. Since charge character is also pH dependent, the acidity of the environmental pH would also affect the ability of a *T. ferrooxidans* strain to colonize tailings as seen with acidophilic heterotrophic bacteria (Chakrabarti & Banergee 1991). Lizama & Suzuki (1991) characterized a *T. ferrooxidans* strain which could oxidize either pyrite or chalcopyrite but not sphalerite. A second strain was studied which could oxidize either pyrite or sphalerite but not chalcopyrite. The ability to differentiate between copper (chalcopyrite) or zinc (sphalerite) minerals suggests that a form of recognition exists in these strains for these specific minerals.

Cells grown on ore are difficult to dissociate from the ore particles (Gormley & Duncan 1974; Suzuki *et al.* 1990; Southam & Beveridge 1992), demonstrating that a tight association must occur between *Thiobacillus* spp. and the mineral surfaces. The development of a tight association between the bacteria and the mineral surface may proceed via a hypothetical two phase mechanism. First, either ionic, salt bridging or hydrophobic interactions would anneal the bacterium to the mineral surface so that an interface could be established to sustain chemolithotrophy. Phase two encompasses the subsequent cementing of the bacterium to the mineral surface with iron oxyhydroxides (Bhatti *et al.* 1993), forming an even more acidic microenvironment (that is, precipitation of $Fe(OH)_3$ yields H^+) to support bacterial growth and multiplication. LPS may be responsible for the initial bacteria–mineral association and it is likely that ferric oxyhydroxides are responsible for the second phase of association, since the solubilization of these precipitates results in the release of bacteria from the mineral surfaces (Southam & Beveridge 1993). Strong adherence of *T. ferrooxidans* to minerals via iron precipitates (Southam & Beveridge 1992, 1993) may have an important ecological role in reducing the diffusion of metabolic products (for example, Fe^{3+} and sulphuric acid) away from the cell–mineral interface. This would help maintain an acidic microenvironment at the mineral sur-

face, and provide a potential source of soluble ferrous iron through repeated chemical oxidation of the sulphide, thereby promoting the growth of *T. ferrooxidans* (Singer & Stumm 1970; Wiersma & Rimstidt 1984).

Bennet & Tributsch (1978) and Norman & Snyman (1987) have demonstrated that *T. ferrooxidans* chooses to colonize fracture lines and dislocations on pyritic mineral surfaces. These may represent preferred bacterial attachment sites; possibly they provide convenient physical recesses for bacterial colonization. For aquatic bacteria, Ferris *et al.* (1989a) demonstrated that surface topology is important for colonization of mineral surfaces. For *T. ferrooxidans*, fracture lines containing alternating metal and sulphide atomic boundaries may also provide unique surface charge characteristics for attachment and this is one way in which certain strains may exhibit mineral selectivity.

T. ferrooxidans strains are known to adapt to various types of sulphide ores prior to the initiation of active microbial leaching (Suzuki *et al.* 1990). This adaptation mechanism is not well understood although it might have something to do with activation by chemical oxidation (Moses & Herman 1991). Although phenotypic switching has recently been demonstrated in *T. ferrooxidans*, it has not been related to LPS chemistry (Schrader & Holmes 1988). Early chemical analysis of LPS from *T. ferrooxidans* strains grown on iron (Fe-LPS) and sulphur (S-LPS) has produced controversial ideas on the nutritional modulation of LPS chemistry. Fe-LPS and S-LPS were found to be chemically different (Vestal *et al.* 1973; Hirt & Vestal 1975) or identical (Yokota *et al.* 1988) for different *T. ferrooxidans* strains. This discrepancy in LPS chemistry may be related to the presence of contaminating acidophilic heterotrophs which have been found in many *T. ferrooxidans* cultures (Harrison *et al.* 1980). Our data from the Copper Rand tailings (Chibougamau, Quebec, Canada) suggests that LPS modulation does not occur in a *T. ferrooxidans* isolate grown on minerals (primarily pyrite, chalcopyrite) and in iron(II) based media (Southam & Beveridge 1993). Therefore, bacterial adaptation to different mineral surfaces owing to modulation of LPS composition is unlikely. Further study of the molecular interactions between *T. ferrooxidans* and mineral surfaces should provide a better understanding of the mechanism of sulphide mineral colonization.

The development and release of acid mine drainage

The rapid development of AMD requires extensive colonization of sulphide tailings by *Thiobacillus* spp. Norman & Snyman (1987) demonstrated that thiobacilli direct their oxidative attack at the point of mineral contact. This is usually at fine grooves (0.2–0.5 µm) in the mineral surface which eventually develop into corrosion pits, widening and enlarging until the mineral grain is destroyed. Since this colonization is controlled by the physical and chemical environment of the tailings, each tailings deposit represents a unique system.

Growth of *T. ferrooxidans* on S (as the source of electrons) and Fe^{3+} (as the terminal electron acceptor) under anaerobic conditions (Pronk *et al.* 1992) is a mechanism which explains the survival of *T. ferrooxidans* in saturated tailings material (Southam & Beveridge 1992). However, for active growth of thiobacilli and the resultant acid generating reactions, the development of a vadose zone is required in the tailings. It is unclear how the initial acidification occurs in such tailings dumps but some researchers have implicated *Gallionella* spp. (Ivarson & Sojak 1978) and *Leptothrix* spp. (Crerar *et al.* 1979). Although one of the optimal conditions for growth of *Thiobacillus* spp. is pH <3 (Trafford *et al.* 1973; Amaro *et al.* 1991), viable thiobacilli can be cultured from tailings possessing neutral environmental pH values (Southam & Beveridge 1992). The suggestion that *T. ferrooxidans* can colonize a neutral tailings and develop an acid environment (pH <3) by itself lends a new insight to tailings ecology. Colonization of sulphide mineral and resulting chemolithotrophy then is possible under 'neutral' pH conditions and probably occurs through the development of an acidic interface between the bacteria and the mineral surfaces. In a related system, Blais *et al.* (1992) demonstrated that cooperation between two *Thiobacillus* strains can reduce the pH from 7.0–8.5 to pH 4.0–4.5 and, then, to below pH 2.0. Thus, it seems likely that the development of certain acidified tailings may occur through a succession of different *Thiobacillus* species.

In a study of 2 month old oxidized zones in the Copper Rand mine tailings, a newly oxidized zone (surface area 1 m², depth *c.* 0.01 m and pH <3) was established by a single (or dominant) strain of *T. ferrooxidans* (Southam & Beveridge 1993). In practical terms, older tailings are probably colonized by several strains of *T. ferrooxidans* since it is almost impossible to maintain pure cultures in the natural environment. The Lemoine tailing is an example of an older tailings dump (8 years) which has been colonized and acidified by at least six strains of *T. ferrooxidans* (Southam & Beveridge 1992).

Acidification of tailings material initiates at the surface of the impoundment. Horizontal and vertical (downward) migration from acid 'hot' spots will depend on the mineralogy, mineral grain size and hydrogeological flow at the site. In tailings which contain low levels of sulphides (5–10% by weight) and low buffering capacity, seepage typically occurs downward into the saturated zone from where it then migrates horizontally through the groundwater system (Morin *et al.* 1988). However, Morin *et al.* (1988) determined that the horizontal acidic flow from tailings was $\frac{1}{440}$ the rate of groundwater flow since the AMD was retained within a more slowly migrating neutralization zone. This neutralization zone is produced by the sum of the geochemical (for example, alkali minerals such as carbonates) and biogeochemical processes present in the system (discussed later in the chapter).

In massive sulphides, hardpan formation has been observed throughout the oxidation zone and consists of tailings material cemented with goethite, lepidocrite, ferrihydrite and jarosite. This remineralization in the vadose zone has been observed to initiate at bacterial cell surfaces (most likely thiobacilli since it is the predominant bacterium: Southam & Beveridge 1992). In tailings containing lower levels of sulphides (*c.* 10% by weight), hardpan formation may occur below the oxidation zone if high levels (*c.* 10–12% by weight) of carbonates are present (Blowes *et al.* 1987). In these systems, hardpan inhibits the downward movement of water and oxygen and promotes the horizontal movement of water above the hardpan. Microaerophilic to anaerobic environments form below, thereby reducing the growth of thiobacilli (Pronk *et al.* 1992), limiting subsequent bioleaching to within the upper oxidized zone and inhibiting biooxidation below the hardpan.

The accurate enumeration of *T. ferrooxidans* and the identification of different subspecies are essential for the study of population growth in natural tailings environments. However, the growth of thiobacilli as a mineralized biofilm makes their study in tailings environments difficult. Several methods exist for studying this growth but not all can be used for natural samples. Protein assays are unsuitable for the study of population growth because they measure dead cells and do not allow for the characterization of bacterial subspecies. The measurement of iron(II) oxidation (Ramsay *et al.* 1988; Suzuki *et al.* 1990) is a good technique to study metabolic activity, but it fails to account for metabolic variability among subspecies and to provide bacteria for the identification of the subspecies. Most Probable Number (MPN) analysis, in addition to being a quantitative measure of viable thiobacilli, provides a sample of cells which can be later used to identify subspecies (Southam & Beveridge 1993). Enumeration of *T. ferrooxidans* by MPN analysis also eliminates the problem of low plating efficiency, making it the best available technique to estimate populations of *T. ferrooxidans* in tailings environments even though it does underestimate the actual population. Since growth of *Thiobacillus* spp. on sulphide mineral surfaces occurs as iron cemented biofilms, it is not surprising that their enumeration is difficult and that 100% accuracy is likely impossible.

Soluble metal interactions with bacteria in natural systems

One of the most dramatic responses of bacteria to soluble metals is the precipitation (i.e. biomineralization) of these metals on the cell surface resulting in fine grained mineral formation. This precipitation may be due to the physicochemical nature of the bacterial cell envelope and can be affected by metabolic activity (Urrutia *et al.* 1992; Schultze-Lam *et al.* 1992). In metal contaminated environments, certain bacteria are more extensively mineralized than other adjacent bacteria. It is possible that these mineralized cells have sacrificed themselves by cleansing toxic heavy metals from the surroundings so that the remaining bacterial population can survive. At the cytoplasmic level, resistance to toxic heavy metals can also occur through plasmid mediated

membrane porters which transport metals from the cell (Silver & Walderhaus 1992). This chapter will concentrate only on the biomineralization phenomena involving bacteria. For an explanation of biomineralization processes involving eukaryotic organisms as well as prokaryotes, we refer the reader to the Proceedings of the Society for General Microbiology Symposium, *Metal microbe interactions* (ed. Poole & Gadd 1989).

Fundamental aspects of bacteria metal interactions

Bacteria, as a prokaryotic group, possess a wide array of surfaces which can interact with soluble metals in the environment. Among the eubacteria, which dominate tailings environments, bacteria are divided into two major groupings: Gram positive and Gram negative types (although Gram variable organisms also exist).

Gram positive eubacteria are characterized by a thick (typically 15–25 nm), amorphous cell wall containing peptidoglycan which provides the framework to which the secondary polymers (teichoic acids or teichuronic acids) attach (Beveridge 1981). Peptidoglycan, the major shape determining structure for the organism, consists of repeating ß(1,4)-linked *N*-acetyl-glucosamine-*N*-acetylmuramic acid dimers. The *N*-acetylmuramic acid residues, which possess short peptide stems (4–5 amino acids), may be covalently bonded via these peptide stems to other muramic acid residues on neighbouring strands. This results in a three dimensional macromolecule in the shape of the bacterium (Beveridge 1989). These peptide stems also possess carboxylate groups which dominate the charge density of this structure (Beveridge & Murray 1980).

Teichoic acids are composed of polyalcohol based chains joined by phosphodiester linkages (for example, polyglycerol phosphate in *Bacillus subtilis* 168). The phosphate moieties confer a net negative electrical charge on the polymer. Teichuronic acids are generally found under phosphate limiting conditions and are composed of uronic acid polymers which also possess anionic reactive sites (Beveridge *et al.* 1982).

Gram negative cell envelopes are structurally and chemically more complicated than those of Gram positive bacteria. External to the plasma membrane is a thin (2–3 nm) layer of peptidoglycan contained within a periplasm possibly having a gel-like consistency (Hobot *et al.* 1984). The periplasm is bounded by an outer membrane which is an LPS–phospholipid–protein mosaic in which the LPS and phospholipid occur on opposing membrane faces of the bilayer (Beveridge 1981). The LPS is anchored to the outer membrane via its lipid moiety and extends its polysaccharide chains outward from the bacterial surface (Lam *et al.* 1992). The outer membrane is often cemented to the peptidoglycan via salt bridging or covalent bonding of the outer membrane proteins. One class of these proteins forms hydrophilic pores or channels (see Beveridge 1981 for more details). The LPS and the peptidoglycan of Gram negative bacteria possess a net electronegative charge which allows their interaction with soluble cations (Hoyle & Beveridge 1983, 1984).

Additional wall layers, for example capsules, S layers and sheaths, can exist singly or in combination and occur external to the peptidoglycan based cell wall of Gram positive eubacteria or external to the LPS of Gram negative eubacteria.

Capsules are highly hydrated, amorphous assemblages of polysaccharides or polypeptides that are chemically linked to the cell surface and may extend up to 1 μm from the cell (Beveridge 1988). They usually possess a net negative charge due to carboxylate groups and they may also possess additional anionic reactive groups due to the presence of phosphate moieties in polysaccharide chains. The highly hydrated nature of bacterial capsules and their cell surface location allows for extensive interaction between capsular material and soluble metal cations (Mittelman & Geesey 1985; Geesey *et al.* 1992). Our own observations of encapsulated bacteria in sediments, soils and biofilms indicates that capsules also stimulate mineral development from the precipitation of absorbed metallic ions.

S layers are paracrystalline assemblages consisting of protein or glycoprotein with p2, p4 or p6 symmetry (Sleytr & Messner 1983, 1988). They self-assemble and associate with the underlying wall though non-covalent interaction (Koval & Murray 1986; Koval 1988). S layers generally possess an acidic pI and thus exhibit a net negative charge. These anionic sites are often employed for intersubunit salt bridging and for subunit cell surface interaction which neutralize the acidic groups (Koval 1988). S layers have

recently been implicated in fine grain mineral development; a *Synechococcus* species uses its S layer to produce either gypsum or calcite on its S layer in a freshwater environment (Thompson & Ferris 1990; Schultze-Lam *et al.* 1992). In this system, calcite formation is mediated by an increase in alkalinity that arises because of the cyanobacterium's photosynthetic activity.

Sheaths which surround chains of cells are less frequently encountered. They are usually recalcitrant structures composed of homo- or heteropolymers and usually remain intact even after cell degradation. Sheaths of *Sphaerotilis* (Rodgers & Anderson 1976a, b) and *Leptothrix* (Crerar *et al.* 1979; Adams & Ghiorse 1987; Boogert & deVrind 1987; Corstjens *et al.* 1992) are important since they are responsible for an oxidative enzyme mediated precipitation of metals. Enzyme mediated metal precipitation has been identified as an important biogeochemical process. Two additional systems include the formation of metal phosphates through phosphatase activity (Macaskie *et al.* 1987) and metal precipitation by dissimilatory metal reduction such as chromium(VI) to chromium(III) (Ishibashi *et al.* 1990), selenate to selenium (Oremland *et al.* 1989) and uranium(VI) to uranium(IV) (Lovley *et al.* 1991; Gorby & Lovley 1992).

The existence of anionic bacterial cell surface polymers is important for the immobilization of dissolved metal species. However, a second feature of the bacterial cell that is also important is its high surface-to-volume ratio. From the perspective of a bacterium, this facilitates its growth because access to nutrients is based on diffusion and the greater this ratio, the greater the diffusion of nutrients into cells (Purcell 1977; Beveridge 1988). In terms of mineralization, high surface-to-volume ratios also provide a tremendous surficial biomineralization potential when compared with eukaryotic microorganisms (e.g. eukaryotic algae) which typically possess a surface-to-volume ratio one tenth that of bacteria. Ion exchange phenomena play an important role in the initiation of mineral formation through competition between hydrogen ions, alkaline earth ions and heavy metals for anionic reactive sites on the bacterial surface. Divalent cations (usually Ca^{2+} and Mg^{2+}) are important for the stability of teichoic and teichuronic acid polymers (Beveridge *et al.* 1982), LPS (Ferris & Beveridge 1986) and S layers (Beveridge &

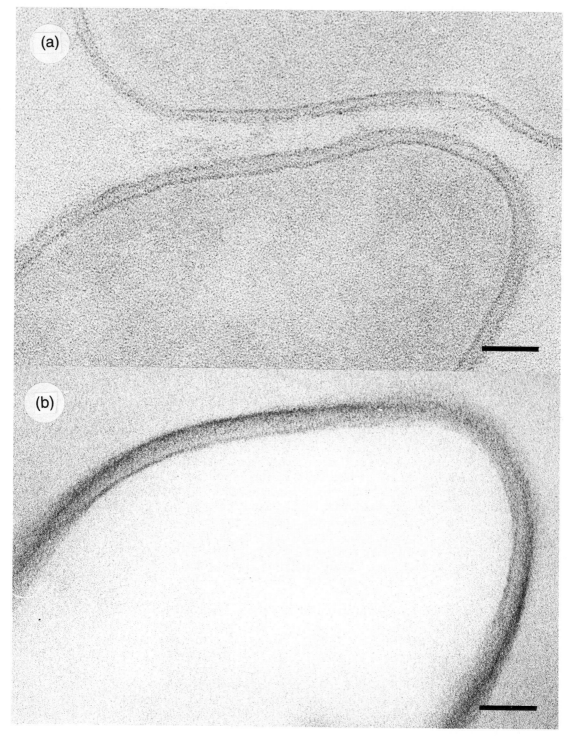

Fig. 8.2.   Transmission electron micrographs of uranyl acetate treated control (a) and gamma irradiated (b) *Bacillus subtilis* cells. Bars = 50 nm. (From Urrutia & Beveridge 1993, with permission.)

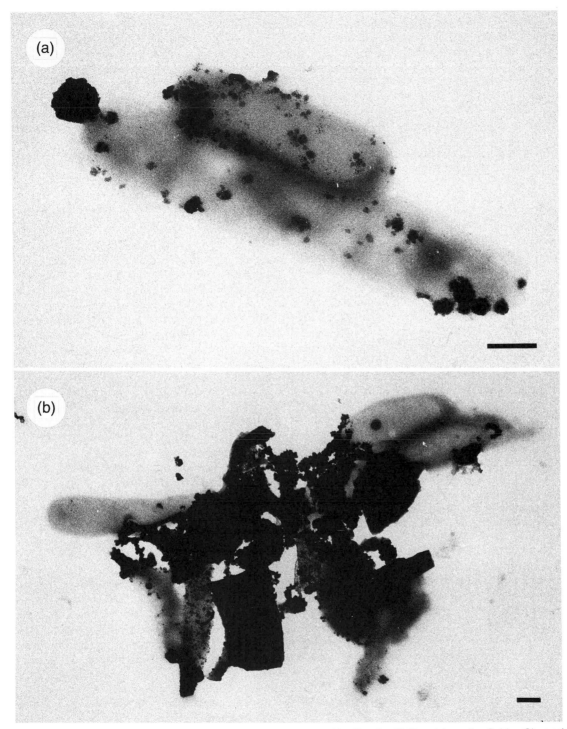

Fig. 8.3.   An unstained whole mount of a particulate fraction obtained by filtration (0.45 μm) from the Golden Giant mine tailings pond (Hemlo gold region, Marathon, Ontario, Canada) demonstrating the development of fine grained minerals at the bacterial cell surface (a) leading towards cell mineral aggregates (b). Bars = 0.5 μm.

Murray 1976). It is these cation stabilized anionic sites that are replaced by heavy metals which serve as nucleation sites for the formation of minerals at the surface of bacteria.

In natural biofilms exposed to heavy metal contaminants, individual bacterial cells may or may not be extensively mineralized (Ferris *et al.* 1989a). The precise reason for this is not clear; however, in laboratory metal binding experiments, living *B. subtilis* cells bound less metal than non-living cells (Urrutia *et al.* 1992). These experiments determined that the membrane induced proton motive force, which pumps protons into the wall fabric, reduces the metal binding ability of the cell walls, probably through competition of protons with metal ions for anionic wall sites (Fig. 8.2). The metabolic activity and energized membranes of individual bacterial cells may therefore help explain why one bacterium in a natural population is not mineralized whereas a second adjacent bacterium is extensively mineralized (Fig. 8.3; Fig. 8.4a). The non-mineralized cell is likely to have possessed high metabolic activity which creates an acidic nanoenvironment at the bacterial cell surface and prevents heavy metal binding, while the other two cells possessed little or no metabolic activity, no acidic nanoenvironment and are therefore mineralized.

Bacterial metal precipitation typically exceeds the stoichiometry expected per chemical reaction site within the cell envelope (Mullen *et al.* 1989). While it is difficult to explain the high sorption capacity that results in metallic precipitates, most metal binding occurs after neutralization of the chemically active site and must proceed via nucleation of these sorbed metals (Beveridge & Murray 1979). These critical nuclei, stabilized by the wall, are less prone to remobilization by dissolving because the wall reduces the interfacial tension between the mineral nucleus and the bulk water phase. Mineral growth then is most active at the outer surface of the bacterium where these nuclei are formed and where space constraints by the envelope polymers do not inhibit metal precipitation. When iron is present in the reaction system, walls interact with one another to form visible flocs which immobilize a variety of metals and then sediment by gravity (Mayers & Beveridge 1989). These iron oxyhydroxide wall matrices enhance the immobilization of metals such as Cr^{3+} and Al^{3+}. Cooperative binding

between iron and other metals has also been observed for *B. licheniformis* which contains an extensive anionic capsule (McLean *et al.* 1990).

Metal precipitates on bacterial surfaces are generally hydrous, amorphous aggregates; however, poorly crystallized phases may also reorder and become more crystalline with time (that is, by solution redeposition or solid state transformation). In geological time scales, the precipitates found on bacterial surfaces represent the early stages in authigenic mineral formation (Beveridge & Fyfe 1985).

Microbe metal interactions in tailings deposits

Most environmental impact studies on pyritic mine tailings concentrate on the acidification of toxic heavy metals and the release of leachates to the watershed surrounding the tailings. In these environments, biomineralization or the biopromoted concentration of heavy metals by bacteria (Ferris *et al.* 1986; Bigham *et al.* 1990), fungi (Glenn & Gold 1985; Milodowski *et al.* 1990; Mullen *et al.* 1992), algae (Nakajima *et al.* 1981; Sakaguchi *et al.* 1981; Mann & Fyfe 1985; Mann *et al.* 1987) and higher plants (Hammer 1990) has been observed and could be used as a natural bioremediation control to heavy metal pollution. In addition to tailings environments, biomineralization is also known to occur in a diverse variety of natural environments such as downstream from pyritic soils (Trafford *et al.* 1973; Ivarson & Sojak 1978), in bog iron deposits (Crerar *et al.* 1979), around hydrothermal vents (Ferris *et al.* 1986; Holm 1987) and at the ocean floor (Ehrlich 1975).

Biomineralization by thiobacilli within tailings has received little attention since the acid environment both extracts and mobilizes metals from the existing mineral phase. Yet, since acidic holding ponds downstream from tailings are recognized for their microbial induced precipitation of leachates (Ferris *et al.* 1989b, c), biomineralization can occur at low pH. Indeed, in laboratory culture, growth of *T. ferrooxidans* in an iron(II) based medium can be measured by the presence of jarosite precipitates (ferric hydroxysulphates) associated with the thiobacilli and the surfaces of the culture vessel (Ramsay *et al.* 1988; McCready & Gould 1990).

Although corrosion products (iron hydroxides)

are commonly observed during the growth of *T. ferrooxidans* in synthetic laboratory media (Ramsay *et al.* 1988; G. Southam, unpublished data), their identification within tailings has received little attention. A laboratory scale leaching study carried out by McCready & Gould (1990) demonstrated that *T. ferrooxidans* was capable of precipitating both ferric oxyhydroxide and ferric hydroxysulphate. Biomineralization of thiobacilli in the natural environment was observed in the upper 3 cm of the Lemoine tailings. In this region, secondary mineral formation resulted from the transition of mineral sulphides to iron chlorides and iron phosphates (Southam & Beveridge 1992). The mineralized bacteria (*T. ferrooxidans* and, to a lesser degree, indigenous heterotrophs) encased within the secondary mineral aggregates, make up an organic fraction of *c.* 0.03% (w/w) dry weight.

The development of secondary minerals from the primary pyritic phase, with the concomitant production of organic matter through microbial growth and decay, may represent the earliest phase in the formation of a soil in the tailings. We do not know the time scale required for the natural development of a productive soil from pyritic tailings, but it will depend on tailings depth, the input of natural soil nutrients, and the amount and availability of the sulphur within the pyrite. Presumably, leaching of all available sulphide minerals is required prior to the development of a stable neutral topsoil and, for this reason, long time frames would be necessary. Certain pyritic soils have caused major problems in the agricultural industry for decades because of acidification (Trafford *et al.* 1973; Ivarson & Sojak 1978).

The response of indigenous bacteria to acid mine drainage

Aquatic systems which encounter AMD are dominated by iron based precipitates in both aerobic (iron oxyhydroxides: Ferris *et al.* 1989b, c) and anaerobic (Ferris *et al.* 1987) environments due to the high concentration of iron in AMD effluents (typically at least three orders of magnitude above that of any other base metal present in the system). The immobilization of metals by reactive groups on microorganisms typically proceeds through a succession of metals ($Fe^{2+} \gg Co^{2+} > Cu^{2+} > Ni^{2+} > Zn^{2+} > UO_2^{2+}$: Förstner 1982).

Planktonic and sessile bacteria nucleate metals from solution resulting, initially, in fine grained mineral development (Fig. 8.3). Iron mineralization of microbial surfaces in aquatic systems produces minerals which are at least of micrometre dimensions (Fig. 8.4). Therefore, photoreductive dissolution of iron oxides is not likely to play a role in the remobilization of these iron encrusted bacteria in natural aquatic systems since it requires a high colloidal surface area (<0.1 µm colloids: Walte & Morel 1984). As metal precipitation proceeds in the planktonic phase, larger aggregates form containing insoluble particulate material and mineralized bacteria (Walker *et al.* 1989; Figs. 8.3, 8.4). In the water column, these aggregates tend to settle owing to their increased density, thus promoting the transfer of once soluble metals into sediments (Mayers & Beveridge 1989). In sediments receiving oxidized metals, dissimilatory metal reduction can potentially result in the resolubilization of metals such as Mn and Fe (Lovley & Phillips 1988; Myers & Nealson 1988; Nealson *et al.* 1991; Nealson & Myers 1992); however, these reduced metals are commonly reprecipitated as biogenic sulphides. The immobilization of soluble heavy metals by microbial mechanisms, then, produces sediments rich in these metals (Ferris *et al.* 1987). This can be used in exploration geochemistry to trace the source of the heavy metal found in solution (Timperly & Allan 1974).

Ferris *et al.* (1989b) demonstrated that natural biofilms, consisting of sessile bacteria, form rapidly (within a month) in both acidic (pH 3.1) and neutral (pH 7.0) waters located downstream from oxidized mine tailings. These biofilms, containing bacterial microcolonies (Ferris *et al.* 1989a), possess bacterial populations at least one order of magnitude higher (in terms of CFU cm^{-2}) than that of the planktonic organisms (CFU ml^{-1}). This increased biomass occurring in the biofilms resulted in increased metal adsorption (Mn, Fe, Cu, Ni).

Metal immobilization in aerobic surface waters occurs by two modes of biofilm growth. The first involves the formation of mineralized bacterial flocs or micrometre scale biofilms which would concentrate nutrients, thereby promoting bacterial growth and metal immobilization. The second involves biofilm formation and mineralization on macroscopic surfaces such as rocks or plant detritus. Ferris *et al.* (1989a) determined that

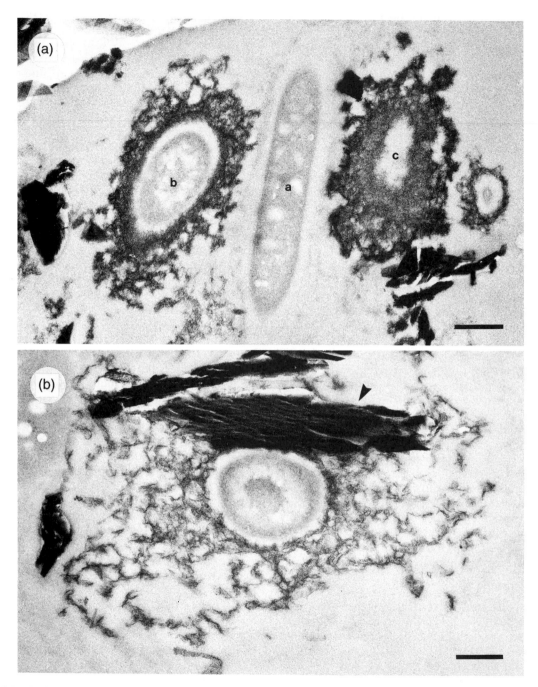

Fig. 8.4. (a) An unstained thin section of a mineralized biofilm from the Copper Rand tailings pond which demonstrates a range of mineralized bacterial forms. 'a' is probably a viable cell possessing exopolysaccharide material which is not mineralized; 'b' is a cell which is extensively mineralized yet still possesses good cytoplasmic integrity; 'c' is a microfossil containing no obvious organic material. Energy dispersive electron microscopy for elemental analysis demonstrated that these mineralized forms consisted of iron oxides. (b) This is an unstained thin section of a bacterium from the same tailings pond as in (a) which has adsorbed onto an iron aluminum silicate particle (arrow head) and has been mineralized with an iron manganese oxide. Bars = 0.5 μm.

population densities of the epilithic microorganisms present on different rocks were inversely related to mineral hardness. Limestone (Mohs hardness, 3.8) exhibited 10 to 100-fold higher bacterial populations over either the alkali feldspar or plagioclase based rocks (Mohs hardness, 6.3), or the quartz (Mohs hardness, 7). The difference in population size probably reflects differences in the surface topology of the rocks (Ferris *et al.* 1989a).

The role of sulphate reducing bacteria (SRB) in AMD bioremediation

For other discussions of the activities of SRB in biofilms, see Hamilton (Chapter 9). In aquatic systems, the major processes which generate alkalinity include biological reduction of sulphate and nitrate, and the exchange of H^+ for Ca^{2+} (Cook *et al.* 1986; Schindler *et al.* 1986; Henrot & Wieder 1990). Of these three processes only sulphate reduction is important for natural (Henrot & Wieder 1990) or engineered (Dvorak *et al.* 1992) systems receiving the high levels of sulphate consistent with AMD.

One of the original systems studied was that of a sawdust based ecosystem receiving AMD at pH 3 (Tuttle *et al.* 1968). In this system, degradation of the cellulosic waste by an uncharacterized mixed population of microorganisms yielded organic by-products which supported SRB activity (sulphate to sulphide) subsequent to the precipitation of FeS (Tuttle *et al.* 1969; see Fig. 8.5). The basic biochemical (3) and geochemical (4) reactions mediated by SRBs are as follows:

$$2CH_2O + SO_4^{2-} \rightarrow H_2S + 2HCO_3^- \qquad (3)$$

$$M^{2+} + S^{2-} \rightarrow MS \downarrow \qquad (4)$$

(where M^{2+} represents a divalent metal ion such as Fe^{2+} or Cu^{2+})

In anoxic aquatic systems receiving increased levels of sulphate, the SRB outcompete methanogens for the available H_2 and acetate and become the dominant, terminal redox organisms (Winfrey & Zeikus 1977; Martens & Goldhaber 1978; Lovley & Klug 1983). Although SRB promote the neutralization of sediments receiving AMD, to overcome the acidity generated by the action of *Thiobacillus* in tailings, tremendous dilution (i.e. distance in the hydrosphere) is often

required resulting in literally thousands of miles of affected waterways (Koryak *et al.* 1979).

In pilot scale operations, researchers have recognized the importance of providing a matrix with high surface area (for example, acid washed mushroom compost: Maree & Strydom 1985; Hammack & Edenborn 1992) to maximize biofilm formation and an organic nutritional supplement to promote SRB activity, for example simple nutrients such as lactate (Tuttle *et al.* 1969) and acetate (Skyring 1988) or more complex nutrients such as molasses (Maree & Strydom 1987). This nutrient source is probably required to maintain a viable population of SRB and compensates for the loss of cells through biomineralization (Fig. 8.5).

In natural sediments, particulate biofilms, similar to the mineralized bacterial flocs described in the water column, play an important role in supporting SRB activity. In an acidic strip mine lake (pH 3), Gyure *et al.* (1987) found that heterotrophic bacterial activity was greatest at the sediment–water interface. In this system, the sediment–water interface is likely to provide a competitive advantage for SRB activity since it concentrates the available dissolved nutrients. In laboratory studies, H_2S at concentrations of 500–550 mg ml^{-1} (Reis *et al.* 1991; Okabe *et al.* 1992) is a by-product which has a reversible toxicity to SRB (Okabe *et al.* 1992; Reis *et al.* 1992). In SRB systems receiving AMD, soluble heavy metals serve as cofactors for continued SRB metabolism since metal sulphide precipitation reduces the toxic effect of free H_2S. The sediment–water interface, via enhanced SRB activity, therefore serves to immobilize and concentrate toxic heavy metals from solution (Okabe & Characklis 1992).

In sediments receiving AMD, sulphate reduction leading to the formation of FeS (acid volatile sulphur; AVS) is at least one order of magnitude higher than in control sediments (Herlihy & Mills 1985). In one sediment receiving high sulphate water associated with AMD, sulphate was directed into organic sulphur (53%), pyrite (41%) and FeS (6%) (H_2S was not detected: Bak & Pfennig 1991a, b). Mineral diagenesis in these systems, then, favours the formation of FeS_2 (non-acid volatile sulphur, NAVS). Therefore, the time required for the formation of FeS_2 may be short in terms of geological time scales (in the order of years).

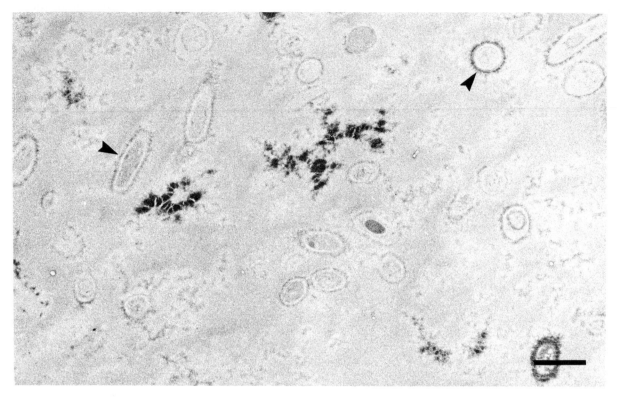

Fig. 8.5. An unstained thin section of a *Desulfotomaculum* sp. which has been cultured in the presence of 100ppm Fe^{2+}, resulting in the precipitation of amorphous FeS (data not shown). Note the extensive mineralization of the bacterial cell surfaces (arrow heads). FeS precipitation at this site is caused by the presence of a HS^-, released as a by-product of SRB metabolism and likely forming a HS^- rich microenvironment at the bacterial surfaces. Bar = 5 µm.

While the metabolic activity of free-living cells was much higher than that of aggregated cells (sediment particles >10 µm: Fukui & Takii 1990a, b), the particle associated SRB (particulate biofilms) were found to be more protected against the deleterious effects of free oxygen than the free living cells in oxygenated sediments. This emphasizes the importance of a microenvironment for the survival of SRB exposed to toxic compounds in nature (Schindler & Turner 1982). SRBs possess a second resistance system to circumvent oxygen toxicity. Dilling & Cypionka (1990) found that *Desulfovibrio* spp. and related genera were capable of aerobic respiration under microaerophilic conditions by using hydrogen as an electron donor. This form of metabolism is only considered to be a momentary fail-safe system and not an alternative major metabolic pathway since it is coupled to ATP production and does not result in cell growth.

In the laboratory, pure cultures of SRB typically do not grow below pH 5 (Tuttle *et al.* 1968) but studies of natural sediments receiving AMD suggest that SRB do possess some mechanism for persisting in bulk acidic environments. For example, a study on microbial sulphate reduction in two acidic strip mine lakes found that sulphate reduction occurred in pH 3.8 sediments (Gyure *et al.* 1990). Also, in a natural bog, a pH reduction to 4.9 resulted in reduced biocatalytic activity when compared with more neutral bogs but did not alter the general carbon and electron flow pathways (Goodwin & Zeikus 1987). Only more neutral microenvironments surrounding the microbiota (associated with particulate biofilms) could account for this observed sulphate reduction. Our experience suggests that biofilms are an efficient way in which microenvironments can be produced and maintained around such cells.

The solubilization and release of base metals

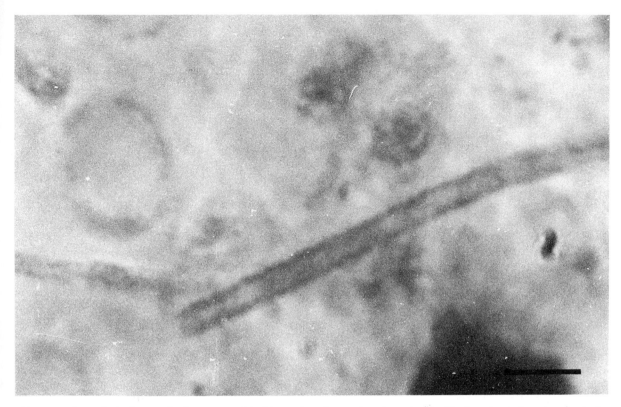

Fig. 8.6. An optical micrograph of a petrographic thin section of grey chert from the Gunflint Iron Formation showing both filamentous and coccoid cyanobacteria. Bar = 5 μm.

from the bulk gangue minerals (silicates) and the subsequent concentration of these metals through sulphate reduction (e.g. FeS precipitation) comprise the first step in transforming base metal sulphide mine tailings into resources. The second step is the formation of crystalline mineral phases (e.g. FeS_2) which can be subjected to conventional mining processes (that is, gravimetric separation by flotation and sedimentation). Clearly, natural wetlands or engineered SRB bioremediation systems must further investigated before unequivocal statements can be made, but we suspect that ultimately, it will be shown that particulate biofilms containing SRB will prove best for the bioremediation of AMD.

Eventual mineral development

The initial observation of microfossils in 2 billion year old Gunflint chert from the Animikie series in the Thunder Bay district of Ontario (Barghoorn & Tyler 1965) has stimulated tremendous interest in the role of microorganisms in the formation of sedimentary mineral deposits (Fig. 8.6). The formation of iron silica deposits (Barghoorn & Tyler 1965; Cloud & Licari 1968; Cloud 1973; Pflug & Jaeschke-Boyer 1979), sulphides (Bubela & McDonald 1969; Trudinger *et al.* 1985), phosphorites (Youssef 1965; Chauhan 1979; O'Brien *et al.* 1981) and carbonates (Barghoorn & Tyler 1965) has been attributed to microorganisms.

In present day sediments, the decomposition of organic matter is mediated principally by bacteria and fungi. Sedimentary diagenesis of microbe immobilized metal can result either in the resolubilization of metals into the overlying water column or groundwater, or in authigenic mineral formation (Lowenstam 1981). Solubilization of metals from sediments (microbe metal aggregates) requires high concentrations of competing cations (e.g. Ca^{2+}) or acid (Flemming *et al.* 1990), or the formation of metal soluble organic complexes, probably involving recalcitrant intracellular (cytoplasmic) contents (Nissenbaum & Swaine 1976).

This decomposition does not include organic matter which has bound heavy metals (most likely bacterial cell envelopes: Degens et al. 1970; Beveridge et al. 1983; Ferris et al. 1987). The importance of bacterial flocs in diagenetic mineral formation was determined by Flemming et al. (1990), who found that large (10 to >100 μm) bacterial metal clay aggregates are less likely than individual mineralized bacteria to release metals in metal leaching experiments. This is probably due to the low surface-to-volume ratio of the metal ladden floc once it is formed when compared to the individual cells.

Sedimentary particulate biofilm complexes are instrumental in crystal formation and growth of authigenic minerals such as sulphides, phosphates, oxides and carbonates (Degens & Ittekkot 1982; Beveridge et al. 1983; Henrot & Wieder 1990). In a low temperature (100 °C) diagenesis study, Beveridge et al. (1983) found that bacteria contributed to the formation of crystalline metal phosphates and polymeric metal coated organic residues by accelerating the formation of authigenic mineral phases.

Strata bound and stratiform base metal sulphide deposits appear to have been formed at moderate (in terms of geological processes) to low temperatures (<200 °C). Since the abiotic formation of sulphides from iron(II), sulphate and organic matter requires temperatures of c. 250 °C, SRB have been attributed with the formation of these deposits (Trudinger et al. 1985).

The Gunflint chert examined by Barghoorn & Tyler (1965) possessed fine grained pyrite (0.1 μm to 3.0–4.0 mm) which was closely associated with the organic matter and spaced along the organic filaments like beads. It is possible that these sulphides resulted from SRB activity and originated as FeS prior to their diagenesis to FeS_2 and subsequent continued crystal growth. The siliceous layers are also considered to be biogenic in origin (Barghoorn & Tyler 1965; Cloud & Licari 1968; Pflug & Jaeschke-Boyer 1979; Tazaki et al. 1992), perhaps resulting from the immobilization of $Si(OH)_4^-$ as amorphous silica (Urrutia & Beveridge 1993) with subsequent transformation to chert. In recent sediments, Ferris et al. (1987) observed metal sulphide and silica minerals initiating from bacterial surfaces. The ability to immobilize both iron sulphide and silica minerals (Birnbaum & Wireman 1985) suggests that SRB may have contributed to the formation of the Archaean banded iron formations.

The unnatural geological setting caused by the deposition of sulphide tailings on the Earth's surface provides us with a unique opportunity to study large populations of bacteria that are important to the biogeochemical cycling of sulphur. Acidophilic thiobacilli which inhabit the tailings generate toxic acid leachates as by-products of their metabolism. Although AMD has been identified as an environmental problem for several decades, we are just now beginning to recognize the importance of microbial biofilms in mineral leaching. The adhesion of thiobacilli to the surfaces of sulphide minerals by iron oxy-hydroxide precipitates promotes the formation of acidic microenvironments favourable for their growth. Clearly, these mineralized biofilms are responsible for the leaching of the mineral face and the generation of acidic conditions.

By reducing sulphate to hydrogen sulphide, the SRB are the most important organisms which can be used for the bioremediation of, and the recovery of metals from, AMD effluents. In order to withstand the extreme acidity and the presence of toxic heavy metals, the SRB grow as particulate biofilms (hundreds of μm in diameter). These particles are at least partly responsible for the resistance of the SRB to oxygen, acidic pH, and probably heavy metals.

Biofilms, then, represent an important mechanism which bacteria employ to optimize their growth and survival in base metal sulphide mine tailing environments.

References

Adams, L. F. & Ghiorse, W. C. (1987). Characterization of extracellular Mn^{2+}-oxidizing activity and isolation of an Mn^{2+}-oxidizing protein from *Leptothrix discophora* SS-1. *Journal of Bacteriology*, **169**, 1279–85.

Ahonen, L. & Tuovinen, O. H. (1992). Bacterial oxidation of sulphide: *Applied and Environmental Microbiology*, **58**, 600–6.

Amaro, A. M., Chamorro, D., Seeger, M., Arrendondo, R., Perrano, I. & Jerez, C. A. (1991). Effect of external pH perturbations on *in vivo* protein synthesis by the acidophilic bacterium *Thiobacillus ferrooxidans*. *Journal of Bacteriology*, **173**, 910–15.

Bak, F. & Pfennig, N. (1991a). Microbial sulphate-reduction in littoral sediment of Lake Constance. *FEMS Microbiology Ecology*, 85, 31–42.

Bak, F. & Pfennig, N. (1991b). Sulphate-reducing bacteria in littoral sediment of Lake Constance. *FEMS Microbiology Ecology*, 85, 43–52.

Baldi, F., Clark, T., Pollack, S. S. & Olson, G. J. (1992). Leaching of pyrites of various reactivities by *Thiobacillus ferrooxidans*. *Applied and Environmental Microbiology*, 58, 1853–6.

Barghoorn, E. S. & Tyler, S. A. (1965). Microorganisms from the Gunflint Chert. *Science*, 147, 563–77.

Bennet, J. C. & Tributsch, H. (1978). Bacterial leaching patterns on pyrite crystal surfaces. *Journal of Bacteriology*, 134, 310–17.

Beveridge, T. J. (1981). Ultrastructure, chemistry, and function of the bacterial wall. *International Reviews of Cytology*, 72, 229–317.

Beveridge, T. J. (1988). The bacterial surface: general considerations towards design and function. *Canadian Journal of Microbiology*, 34, 363–72.

Beveridge, T. J. (1989). Role of cellular design in bacterial metal accumulation and mineralization. *Annual Reviews of Microbiology*, 43, 147–71.

Beveridge, T. J., Forsberg, C. W. & Doyle, R. J. (1982). Major sites of metal binding in *Bacillus licheniformis* walls. *Journal of Bacteriology*, 150, 1438–48.

Beveridge, T. J. & Fyfe, W. S. (1985). Metal fixation by bacterial cell walls. *Canadian Journal of Earth Science*, 22, 1893–8.

Beveridge, T. J., Meloche, J. D. Fyfe, W. S. & Murray, R. G. E. (1983). Diagenesis of metals chemically complexed to bacteria: laboratory formation of metal phosphates, sulphides, and organic condensates in artificial sediments. *Applied and Environmental Microbiology*, 45, 1094–1108.

Beveridge, T. J. & Murray, R. G. E. (1976). Reassembly *in vitro* of the superficial wall components of *Spirillum putridiconchylium*. *Journal of Ultrastructural Research*, 55, 105–18.

Beveridge, T. J. & Murray, R. G. E. (1979). Uptake and retention of metals by cell walls of *Bacillus subtilis*. *Journal of Bacteriology*, 127, 1502–18.

Beveridge, T. J. & Murray, R. G. E. (1980). Sites of metal deposition in the cell wall of *Bacillus subtilis*. *Journal of Bacteriology*, 141, 876–87.

Bhatti, T. M., Bigham, J. M., Carlson, L. & Tuovinen, O. H. (1993). Mineral products of pyrrhotite oxidation by *Thiobacillus ferrooxidans*. *Applied and Environmental Microbiology*, 59, 1984–90.

Bigham, J. M., Schuertmann, U., Carlson, L. & Murad, E. (1990). A poorly crystallized oxyhydroxy sulphate of iron formed by bacterial oxidation of Fe (II) in acid mine waters. *Geochimica et Cosmochimica Acta*, 54, 2743–58.

Birnbaum, S. J. & Wireman, J. W. (1985). Sulphate reducing bacteria and silica solubility: a possible mechanism for evaporite diagenesis and silica precipitation in banded iron formations. *Canadian Journal of Earth Science*, 22, 1904–9.

Blais, J.-F., Auclair, J. C. & Tygai, R. D. (1992). Cooperation between two *Thiobacillus* strains for heavy metal removal from municipal sludge. *Canadian Journal of Microbiology*, 38, 181–7.

Blowes, D. W., Cherry, J. A. & Reardon, E. J. (1987). The hydrogeochemistry of four inactive tailings impoundments: perspectives on tailings pore-water evolution. In *Proceedings of the National Symposium on Mining, Hydrology, Sedimentology, and Reclamation*, pp. 253–62. Lexington: University of Kentucky.

Blowes, D. W. & Jambor, J. L. (1990). The pore-water geochemistry and the mineralogy of the vadose zone of sulphide tailings, Waite Amulet, Quebec, Canada. *Applied Geochemistry*, 5, 327–46.

Boogerd, F. C. & deVrind, J. P. M. (1987). Manganese oxidation by *Leptothrix discophera*. *Journal of Bacteriology*, 169, 489–94.

Brierly, C. L. (1982). Microbiological mining. *Scientific American*, 247, 42–51.

Bubela, B. & McDonald, J. A. (1969). Formation of banded sulphides: metal ion separation and precipitation by inorganic and microbial sulphide sources. *Nature*, 221, 465–6.

Chakrabarti, B. K. & Banerjee, P. C. (1991). Surface hydrophobicity of acidophilic heterotrophic bacterial cells in relation to their adhesion on minerals. *Canadian Journal of Microbiology*, 37, 692–6.

Chauhan, D. S. (1979). Phosphate bearing stromatolites of the Precambrian Aravalli phosphate deposits of the Udaipur region, their environmental significance and genesis of phosphorite. *Precambrian Research*, 8, 95–126.

Cherry, J. A., Morel, F. M. N., Rouse, J. V., Schnoor, J. L. & Wolman, M. G. (1986a). Hydrogeochemistry of sulphide and arsenic rich tailings and alluvium along Whitewood Creek, South Dakota (Part 1). *Mineral and Energy Resources*, 29(4), 1–12.

Cherry, J. A., Morel, F. M. N., Rouse, J. V., Schnoor, J. L. & Wolman, M. G. (1986b). Hydrogeochemistry of sulphide and arsenic rich tailings and alluvium along Whitewood Creek, South Dakota (Part 3). *Mineral and Energy Resources*, 29(6), 1–16.

Chester, I. R. & Meadow, P. M. (1975). Heterogeneity of the lipopolysaccharide from *Pseudomonas aeruginosa*. *European Journal of Biochemistry*, 58, 273–82.

Cloud, P. E. (1973). Paleoecological significance of the banded iron-formation. *Economic Geology*, 68, 1135–43.

Cloud, P. E. Jr. & Licari, G. R. (1968). Microbiotas of the banded iron formations. *Proceedings of the National Academy of Sciences USA*, **61**, 779–86.

Cloutier, N. R., Clulow, F. V., Lim, T. P. & Dane, N. F. (1986). Metal (Cu, Ni, Fe, Co, Zn, Pb) and Ru-226 levels in tissues of meadow voles *Microtus pennsylvanicus* living on nickel and uranium tailings in Ontario, Canada. *Environmental Pollution*, **41**, 295–314.

Cook, R. B., Kelly, C. A., Schindler, D. W. & Turner, M. A. (1986). Mechanisms of hydrogen ion neutralization in an experimentally acidified lake. *Limnology and Oceanography*, **3**, 134–48.

Corstjens, P. L. A. M., de Vrind, J. P. M., Westbroek, P. & de Vrind-de Jong, E. W. (1992). Enzymatic iron oxidation by *Leptothrix discophora*: identification of an iron-oxidizing protein. *Applied and Environmental Microbiology*, **58**, 450–4.

Crerar, D. A., Knox, G. W. & Means, J. L. (1979). Biogeochemistry of bog iron in the New Jersey pine barrens. *Chemical Geology*, **24**, 111–35.

Dave, N. K., Lim, T. P., Siwik, R. S. & Blackport, R. (1986). Geophysical and biohydrogeochemical investigations of an active sulphide tailings basin, Noranda, Quebec, Canada. In *Proceedings of the National Symposium on Mining, Hydrology, Sedimentology and Reclamation*, pp. 13-19. Lexington: University of Kentucky.

Degens, E. T. & Ittekkot, V. (1982). *In situ* metal-staining of biological membranes in sediments. *Nature*, **298**, 262–4.

Degens, E. T., Watson, S. W. & Remsen, C. C. (1970). Fossil membranes and cell wall fragments from a 7000-year-old Black Sea Sediment. *Science*, **168**, 1207–8.

Dilling, W. & Cypionka, H. (1990). Aerobic respiration in sulphate-reducing bacteria. *FEMS Microbiology Ecology*, **71**, 123–8.

Dispirito, A. A., Dugan, P. R. & Tuovinen, O. H. (1983). Sorption of *Thiobacillus ferrooxidans* to particulate material. *Biotechnology and Bioengineering*, **25**, 1163–8.

Duncan, D. W. & Drummond, A. D. (1973). Microbiological leaching of porphyry copper type mineralization: post-leaching observations. *Canadian Journal of Earth Science*, **10**, 476–84.

Dvorak, D. H., Hedin, R. S, Edenborn, H. M. & McIntire, P. E. (1992). Treatment of metal-contaminated water using bacterial sulphate reduction: results from pilot-scale reactors. *Biotechnology and Bioengineering*, **40**, 609–16.

Ehrlich, H. L. (1975). The formation of ores in the sedimentary environment of the deep sea with microbial participation: the case for ferromanganese concretions. *Science*, **119**, 36–41.

Ferris, F. G. & Beveridge, T. J. (1986). Physicochemical roles of soluble metal cations in the outer membrane of *Escherichia coli* K-12. *Canadian Journal of Microbiology*, **32**, 594–601.

Ferris, F. G., Beveridge, T. J. & Fyfe, W. S. (1986). Iron-silica crystallite nucleation by bacteria in a geothermal sediment. *Nature*, **320**, 609–11.

Ferris, F. G., Fyfe, W. S. & Beveridge, T. J. (1987). Bacteria as nucleation sites for authigenic minerals in a metal-contaminated lake sediment. *Chemical Geology*, **63**, 225–32.

Ferris, F. G., Fyfe, W. S., Whitten, T., Schultze, S. & Beveridge, T. J. (1989a). Effect of mineral substrate hardness on the population density of epilithic microorganisms in two Ontario rivers. *Canadian Journal of Microbiology*, **35**, 744–7.

Ferris, F. G., Schultze, S., Witten, T. C., Fyfe, W. S. & Beveridge, T. J. (1989b). Metal interaction with microbial biofilms in acidic and neutral pH environments. *Applied and Environmental Microbiology*, **55**, 1249–57.

Ferris, F. G., Tazaki, K. & Fyfe, W. S. (1989c). Iron oxides in acid mine drainage environments and their association with bacteria. *Chemical Geology*, **74**, 321–30.

Flemming, C. A., Ferris, F. G., Beveridge, T. J. & Bailey, G. W. (1990). Remobilization of toxic heavy metals adsorbed to bacterial wall-clay composites. *Applied and Environmental Microbiology*, **56**, 3191–203.

Förstner, U. (1982). Accumulative phases for heavy metals in limnic sediments. *Hydrobiology*, **91**, 269–84.

Fukui, M. & Takii, S. (1990a). Survival of sulphate-reducing bacteria in oxic surface sediment of a seawater lake. *FEMS Microbiology Ecology*, **73**, 317–22.

Fukui, M. & Takii, S. (1990b). Colony formation of free-living and particle-associated sulphate reducing bacteria. *FEMS Microbiology Ecology*, **73**, 85–90.

Geesey, G. G., Bremer, P. J., Smith, J. J., Muegge, M. & Jang, L. K. (1992). Two-phase model for describing the interactions between copper ions and exopolymers from *Alteromonas atlantica*. *Canadian Journal of Microbiology*, **38**, 785–93.

Glenn, J. K. & Gold, M. H. (1985). Purification and characterization of an extracellular Mn(II)-dependent peroxidase from the lignin degrading basidiomycete *Phanerochaete chrysosporium*. *Archives of Biochemistry and Biophysics*, **242**, 329–41.

Goodwin, S. & Zeikus, J. G. (1987). Ecophysiological adaptations of anaerobic bacteria to low pH: analysis of anaerobic digestion in acidic bog sediments. *Applied and Environmental Microbiology*, **53**, 57–64.

Gorby, Y. A. & Lovley, D. A. (1992) Enzymatic uranium precipitation. *Environmental Science Technology*, **26**, 205–7.

Gormley, L. S. & Duncan, D. W. (1974). Estimation of *Thiobacillus ferrooxidans* concentrations. *Canadian Journal of Microbiology*, **20**, 1454–5.

Gyure, R. A., Konopka, A., Brooks, A. & Doemel, W. (1987). Algal and bacterial activities in acidic (pH 3) strip-mine lakes. *Applied and Environmental Microbiology*, **53**, 2069–76.

Gyure, R. A., Konopka, A., Brooks, A. & Doemel, W. (1990). Microbial sulphate reduction in acidic (pH 3) strip-mine lakes. *FEMS Microbiology Ecology*, **73**, 193–202.

Hammack, R. W. & Edenborn, H. W. (1992). The removal of nickel from mine waters using bacterial sulphate reduction. *Applied Microbiology and Biotechnology*, **37**, 674–8.

Hammer, D. A. (1990). Constructed wetlands for acid water treatment – an overview of emerging technology. In *Acid mine drainage designing for closure*, ed. J. W. Gadsby, J. A. Malick & S. J. Dau, pp. 381–94. Vancouver: BiTech Pub. Ltd.

Harrison, A. P., Jr (1984). The acidophilic thiobacilli and other acidophilic bacteria that share their habitat. *Annual Reviews of Microbiology*, **38**, 265–92.

Harrison, A. P., Jr, Jarvis, B. W. & Johnson, J. L. (1980). Heterotrophic bacteria from cultures of autotrophic *Thiobacillus ferrooxidans*: relationships as studied by means of deoxyribonucleic acid homology. *Journal of Bacteriology*, **143**, 448–54.

Heale, E. L. (1991). Reclamation of tailings and stressed lands at the Sudbury, Ontario operations of Inco Ltd. In *The Proceedings of the Second International Conference on the Abatement of Acidic Drainage*. Montreal, Canada.

Henrot, J. & Wieder, R. K. (1990). Processes of iron and manganese retention in laboratory peat microcosms subjected to acid mine drainage. *Journal of Environmental Quality*, **19**, 312–20.

Herlihy, A. T. & Mills, A. L. (1985). Sulphate reduction in freshwater sediments receiving acid mine drainage. *Applied and Environmental Microbiology*, **49**, 179–86.

Hirt, W. E. & Vestal, J. R. (1975). Physical and chemical studies of *Thiobacillus ferrooxidans* lipopolysaccharides. *Journal of Bacteriology*, **123**, 642–50.

Hobot, J. A., Carlemam, E., Villiger, W. & Kellenberger, E. (1984). The periplasmic gel: a new concept resulting from the re-investigation of bacterial cell envelope ultrastructure by new methods. *Journal of Bacteriology*, **160**, 143–52.

Holm, N. G. (1987). Biogenic influences on the geochemistry of certain ferruginous sediments of hydrothermal origin. *Chemical Geology*, **63**, 45–57.

Hoyle, B. & Beveridge, T. J. (1983). Binding of metallic ions to the outer membrane of *Escherichia coli Applied and Environmental Microbiology*, **46**, 749–52.

Hoyle, B. & Beveridge, T. J. (1984). Metal binding by the peptidoglycan sacculus of *Escherichia coli* K-12. *Canadian Journal of Microbiology*, **30**, 204–11.

Hutchins, S. R., Davidson, M. S., Brierly, J. A. & Brierly, C. L. (1986). Microorganisms in reclamation of metals. *Annual Review of Microbiology*, **40**, 311–36.

Ingledew, W. J. (1982). *Thiobacillus ferrooxidans*: The bioenergetics of an acidophilic chemolithotroph. *Biochimica Biophysica Acta*, **683**, 89–117.

Ishibashi, Y., Cervantes, C. & Silver, S. (1990). Chromium reduction in *Pseudomonas putida*. *Applied and Environmental Microbiology*, **56**, 2268–70.

Ivarson, K. C. & Sojak, M. (1978). Microorganisms and ochre deposits in field drains of Ontario. *Canadian Journal of Soil Science*, **58**, 1–17.

Jones, J. E. & Starkey, R. L. (1961). Surface active substances produced by *Thiobacillus thiooxidans*. *Journal of Bacteriology*, **82**, 788–9.

Keller, L. & Murr, L. E. (1982). Acid-bacterial and ferric sulphate leaching of pyrite single crystals. *Biotechnology and Bioengineering*, **24**, 83–96.

Koryak, M. Stafford, L. J. & Montgomery, W. H. (1979). The limnological response of a West Virginia multipurpose impoundment to acid inflows. *Water Resources Research*, **15**, 929–34.

Koval, S. F. (1988). Paracrystalline protein surface arrays on bacteria. *Canadian Journal of Microbiology*, **34**, 407–14.

Koval, S. F. & Murray, R. G. E. (1986). The superficial protein arrays on bacteria. *Microbiology Science*, **3**, 357–61.

Lam, J. S., Graham, L. L., Lightfoot, J., Dasgupta, T. & Beveridge, T. J. (1992). Ultrastructural examination of lipopolysaccharides of *Pseudomonas aeruginosa* strains and their isogenic mutants by freeze-substitution. *Journal of Bacteriology*, **174**, 7159–67.

Lane, D. J., Stahl, D. A., Olsen, G. J., Heller, D. J. & Pace, N. R. (1985). Phylogenetic analysis of the genera *Thiobacillus* and *Thiomicrospira* by 5S rRNA sequences. *Journal of Bacteriology*, **163**, 75–81.

Leathen, W. W., Kinsel, N. A & Braley, S. A., Sr (1956). Ferrobacillus ferrooxidans: a chemosynthetic autotrophic bacterium. *Journal of Bacteriology*, **72**, 700–4.

Lizama, H. M. & Suzuki, I. (1991). Interaction of chalcopyrite and sphalerite with pyrite during leaching by *Thiobacillus thiooxidans*. *Canadian Journal of Microbiology*, **37**, 304–11.

Lovley, D. R. & Klug, M. J. (1983). Sulphate reducers can outcompete methanogens at freshwater sulphate concentrations. *Applied and Environmental Microbiology*, **45**, 187–92.

Lovley, D. R. & Phillips, E. J. P. (1988). Novel mode of microbial energy metabolism: organic carbon oxidation coupled to dissimilatory reduction of iron or manganese. *Applied and Environmental Microbiology*, **54**, 1472–80.

Lovley, D. R., Phillips, E. J. P., Gorby, Y. A. & Landa,

E. R. (1991). Microbial reduction of uranium. *Nature*, **350**, 413–16.

Lowenstam, H. A. (1981). Minerals formed by organisms. *Nature*, **211**, 1126–31.

Macaskie, L. E., Dean, A. C. R., Cheetham, A. K., Jakeman, R. J. B. & Skarnulis, A. J. (1987). Cadmium accumulation by a *Citrobacter* sp.: the chemical nature of the accumulated metal precipitate and its location on the bacterial cells. *Journal of General Microbiology*, **133**, 539–44.

Mann, H. & Fyfe, W. S. (1985). Uranium uptake by algae: experimental and natural environments. *Canadian Journal of Earth Science*, **22**, 1899–1903.

Mann, H., Tazaki, K., Fyfe, W. S., Beveridge, T. J. & Humphrey, R. (1987). Cellular lepidocrocite precipitation and heavy metal sorption in *Euglena* sp. (unicellular alga): implications for biomineralization. *Chemical Geology*, **63**, 39–43.

Maree, J. P. & Strydom, W. F. (1985). Biological sulphide removal in an upflow packed bed reactor. *Water Resources Research*, **19**, 1101–6.

Maree, J. P. & Strydom, W. F. (1987). Biological sulphide removal from industrial effluent in an upflow packed bed reactor. *Water Resources Research*, **21**, 141–6.

Martens, C. S. & Goldhaber, M. B. (1978). Early diagenesis in transitional sedimentary environments of the White Oak River Eastuary, North Carolina. *Limnology and Oceanography*, **23**, 428–41.

Mayers, I. T. & Beveridge, T. J. (1989). The sorption of metals to *Bacillus subtilis* walls from dilute solutions and simulated Hamilton Harbour (Lake Ontario) water. *Canadian Journal of Microbiology*, **35**, 764–70.

McCready, R. G. L. & Gould, W. D. (1990). Bioleaching of uranium. In *Microbial mineral recovery*, ed. H. L. Ehrlich & C. L. Brierly, pp. 107–25. New York: McGraw-Hill.

McLean, R. J. C., Beauchemin, D., Clapham, L. & Beveridge, T. J. (1990). Metal-binding characteristics of the gamma-glutamyl capsular polymer of *Bacillus licheniformis* ATCC 9945. *Applied and Environmental Microbiology*, **56**, 3671–7.

Mehta, A. P. & Murr, L. E. (1982). Kinetic study of sulphide leaching by galvanic interaction between chalcopyrite, pyrite and sphalerite in the presence of *T. ferrooxidans* (30 °C) and a thermophilic microorganism (55 °C). *Biotechnology and Bioengineering*, **24**, 919–40.

Mills, A. L. (1985). Acid mine waste drainage: microbial impact on the recovery of soil and water ecosystems. In *Soil reclamation processes microbiological analyses and applications*, ed. R. L. Tate & D. A. Klein, pp. 35–81. New York: Marcel Dekker.

Milodowski, A. E., West, J. M., Pearce, J. M., Hyslop, E. K., Basham, I. R. & Hooker, P. J. (1990). Uranium-mineralized micro-organisms associated with uraniferous hydrocarbons in southwest Scotland. *Nature*, **347**, 465–7.

Mittelman, M. W. & Geesey, G. G. (1985). Copper-binding characteristics of exopolymers from a freshwater sediment bacterium. *Applied and Environmental Microbiology*, **49**, 846–51.

Morin. K. A., Cherry, J. A., Nand, K. D., Tjoe, P. L. & Vivyurka, A. J. (1988). Migration of acidic groundwater seepage from uranium-tailings impoundments, 1. Field study and conceptual hydrogeochemical model. *Journal of Contaminant Hydrology*, **2**, 271–303.

Moses, C. O. & Herman, J. S. (1991). Pyrite oxidation at circumneutral pH. *Geochimica et Cosmochimica Acta*, **55**, 471–82.

Mullen, M. D., Wolf, D. C., Beveridge, T. J. & Bailey, G. W. (1992). Sorption of heavy metals by the soil fungi *Aspergillus niger* and *Mucor roux* II. *Soil Biology and Biochemistry*, **24**, 129–35.

Mullen, M. D., Wolf, D. C., Ferris, F. G., Beveridge, T. J., Flemming, C. A. & Bailey, G. W. (1989). Bacterial sorption of heavy metals. *Applied and Environmental Microbiology*, **55**, 3143–9.

Myers, C. P. & Nealson, K. H. (1988). Bacterial manganese reduction and growth with manganese oxide as the sole electron acceptor. *Science*, **240**, 1319–21.

Nakajima, A., Horikoshi, T. & Sakaguchi, T. (1981). Distribution and chemical state of heavy metal ions absorbed by *Chlorella* cells. *Agricultural Biological Chemistry*, **45**, 903–8.

Nealson, K. H. & Myers, C. R. (1992). Microbial reduction of manganese and iron: new approaches to carbon cycling. *Applied and Environmental Microbiology*, **58**, 439–43.

Nealson, K. H., Myers, C. R. & Wimpee, B. B. (1991). Isolation and identification of manganese-reducing bacteria and estimates of microbial Mn(IV)-reducing potential in the Black Sea. *Deep-Sea Research*, **38**, S907–S920.

Nicholson, R. V., Gillham, R. W., Cherry, J. A. & Reardon, E. J. (1990). Reduction of acid generation in mine tailings through the use of moisture-retaining cover layers as oxygen barriers: Reply. *Canadian Geotechnology*, **27**, 402–3.

Nissenbaum, A. & Swaine, D. J. (1976). Organic matter metal interactions in Recent sediments: the role of humic substances. *Geochimica et Cosmochimica Acta*, **40**, 809–16.

Norman, P. F. & Snyman, C. P. (1987). The biological and chemical leaching of an auriferous pyrite/arsenopyrite flotation concentrate: a microscopic examination. *Geomicrobiology Journal*, **6**, 1–10.

O'Brien, G. W., Harris, J. R., Milnes, A. R., & Veeh, H. H. (1981). Bacterial origin of East Australian continental margin phosphorites. *Nature*, **294**, 442–4.

Okabe, S. & Characklis, W. G. (1992). Effects of temperature and phosphorous concentration on microbial sulphate reduction by *Desulfovibrio desulfuricans*. *Biotechnology and Bioengineering*, **39**, 1031–42.

Okabe, S., Nielsen, P. H. & Characklis, W. G. (1992). Factors affecting microbial sulphate reduction by *Desulfovibrio desulfuricans* in continuous culture: limiting nutrients and sulphide concentration. *Biotechnology and Bioengineering*, **40**, 725–34.

Oremland, R. S., Hollibaugh, J. T., Maest, A. S., Presser, T. S., Miller, L. G. & Culbertson, C. W. (1989). Selenate reduction to elemental selenium by anaerobic bacteria in sediments and culture: biogeochemical significance of a novel, sulphate-independent respiration. *Applied and Environmental Microbiology*, **55**, 2333–43.

Pflug, H. D. & Jaeschke-Boyer, H. (1979). Combined structural and chemical analysis of 3,800–Myr-old microfossils. *Nature*, **280**, 483–6.

Poole, R. K. & Gadd, G. M. (ed.) (1989). Metal–microbe interactions. *Symposium of the Society for General Microbiology*, **43**. Oxford: IRL.

Pronk, J. T., deBruyn, J. C., Bos, P. & Kuenen, J. G. (1992). Anaerobic growth of *Thiobacillus ferrooxidans*. *Applied and Environmental Microbiology*, **58**, 2227–30.

Purcell, E. (1977). Life at low Reynolds' number. *American Journal of Physics*, **45**, 3–11.

Ramsay, B., Ramsay, J., de Tremblay, M. & Chavarie, C. (1988). A method for the quantification of bacterial protein in the presence of jarosite. *Geomicrobiology*, **6**, 171–7.

Reis, M. A. M., Lemos, P. C., Almeida, J. S. & Carrondo, M. J. T. (1991). Evidence for the intrinsic toxicity of H_2S to sulphide-reducing bacteria. *Applied Microbiology and Biotechnology*, **36**, 145–7.

Reis, M. A. M., Lemos, P. C., Almeida, J. S. & Carrondo, M. J. T. (1992). Effect of hydrogen sulphide on growth of sulphide-reducing bacteria. *Biotechnology and Bioengineering*, **40**, 593–600.

Rodriguez, M., Campos, S. & Gomez-Silva, B. (1986). Studies on native strains of *Thiobacillus ferrooxidans*. III. Studies on the outer membrane of *Thiobacillus ferrooxidans*. Characterization of the lipopolysaccharide and some proteins. *Biotechnology and Applied Biochemistry*, **8**, 292–9.

Rodgers, S. R. & Anderson, J. J. (1976a). Measurement of growth and iron deposition in *Sphaerotilus discophorus*. *Journal of Bacteriology*, **126**, 257–63.

Rodgers, S. R. & Anderson, J. J. (1976b). Role of iron deposition in *Sphaerotilus discophorus*. *Journal of Bacteriology*, **126**, 264–71.

Sakaguchi, T., Nakajima, A. & Horikoshi, T. (1981). Studies on the accumulation of heavy metal elements in biological systems. *European Journal of Applied Microbiology and Biotechnology*, **12**, 84–9.

Schindler, D. W. & Turner, M. A. (1982). Biological, chemical and physical responses of lakes to experimental acidification. *Water, Air and Soil Pollution*, **18**, 259–71.

Schindler, D. W., Turner, M. A., Stainton, M. P. & Linsey, G. A. (1986). Natural sources of acid neutralizing capacity in low alkalinity lakes of the Precambrian Shield. *Science*, **232**, 844–7.

Schofield, C. L. (1976). Acid precipitation: effects on fish. *Ambio*, 228–30.

Schrader, J. A. & Holmes, D. S. (1988). Phenotypic switching of *Thiobacillus ferrooxidans*. *Journal of Bacteriology*, **170**, 3915–23.

Schultze-Lam, S., Harauz, G. & Beveridge, T. J. (1992). Participation of a cyanobacterial S layer in fine-grain mineral formation. *Journal of Bacteriology*, **174**, 7971–81.

Shively, J. M., Decker, G. L. & Greenwalt, J. W. (1970). Comparative ultrastructure of the thiobacilli. *Journal of Bacteriology*, **101**, 618–27.

Silver, S. & Walderhaus, M. (1992). Gene regulation of plasmid and chromosome determined inorganic ion transport in bacteria. *Microbiological Reviews*, **56**, 195–228.

Silverman, M. P. & Lundgren, D. G. (1959). Studies on the chemoautotrophic iron bacterium *Ferrobacillus ferrooxidans*. I. An improved medium and a harvesting procedure for securing high cell yields. *Journal of Bacteriology*, **77**, 642–7.

Singer, P. C. & Stumm, W. (1970). Acidic mine drainage: the rate determining step. *Science*, **167**, 1121–3.

Skyring, G. W. (1988). Acetate as the main substrate for the sulphate-reducing bacteria in Lake Eliza (South Australia) hypersaline sediments. *FEMS Microbiology Ecology*, **53**, 87–94.

Sleytr, U. B. & Messner, P. (1983). Crystalline surface layers on bacteria. *Annual Reviews of Microbiology*, **37**, 311–39.

Sleytr, U. B. & Messner, P. (1988). Crystalline surface layers in procaryotes. *Journal of Bacteriology*, **170**, 2891–7.

Southam, G. & Beveridge, T. J. (1992). Enumeration of thiobacilli with pH-neutral and acidic mine tailings and their role in the development of secondary mineral soil. *Applied and Environmental Microbiology*, **58**, 1904–12.

Southam, G. & Beveridge, T. J. (1993). Examination of lipopolysaccharide (O-antigen) populations of *Thiobacillus ferrooxidans* from two mine tailings. *Applied and Environmental Microbiology*, **59**, 1283–8.

Suzuki, I., Takeuchi, T. L., Yuthasastrakosol, T. D. & Oh, J. K. (1990). Ferrous iron and sulfur oxidation

Biofilms

Clearly in this particular publication, biofilms *per se* are dealt with in considerable detail, and from a number of different standpoints. None the less it is advantagous to present here a general view of biofilms, in this case stressing those features of special note in the context of biocorrosion.

Biofilms are ubiquitous in nature and are found associated with such diverse substrata as rumen and intestinal epithelia; replacement joints, heart valves and pacemakers; urinary catheters; teeth; sediment particles, stones, and man-made structures in aquatic environments; fluid transportation lines, heat exchangers, and cooling towers (Costerton *et al.* 1987). In these environments, they are major sites of biological activity and are responsible for cellulose digestion, bacterial infection and mechanical failure, dental caries, general biodegradation, corrosion, increased fluid frictional resistance, and impaired heat transfer.

Although much of the laboratory analysis to date has been carried out with biofilms involving single species under closely defined conditions, in most circumstances natural biofilms contain a mixed community of microorganisms. Two consequences follow from this which are of major significance with respect to the mechanisms of biocorrosion: (i) the organisms show close interdependencies so that the biofilm as a whole behaves as a unified consortium; (ii) the organisms, and the microenvironments created by their activities, generate structural and temporal heterogeneities within the biofilm.

Microbial consortia

There are now many examples in the literature of microbial consortia in which groups of otherwise unrelated organisms interact by a mechanism of nutrient succession that ensures both cooperation between species and maximum utilization of the primary nutrients. Such consortia inevitably have a structural component and they exist as aggregates in the form of flocculents or biofilms. Diffusional resistance to oxygen penetration, coupled with its uptake by aerobic species, can result in the creation of anoxic and reduced microenvironments within the central regions of the consortium. A general model for a microbial consortium would therefore be a nutritionally linked group of aerobic, facultative and anaerobic bacterial species displaying heterotrophic hydrolytic and fermentative activity, acetogenesis, and a terminal oxidative stage involving either methanogenesis or sulphate reduction (Fig. 9.1; Parkes 1987). The initial heterotrophic species will be determined by the nature and availability of the primary substrates and by the prevailing conditions of aerobiosis in the bulk liquid phase. The nutritionally dependent facultative and anaerobic organisms will be more constant in character, with the key difference between terminal oxidation by methanogens or SRB being largely determined by the availability of sulphate. Whereas the methanogens are restricted to the oxidation of only methylated amines and methanol in addition to hydrogen and acetate which are their normal substates, the SRB are now known to be capable of growth on an extensive range of short and long chain fatty acids and alcohols, aromatic and heterocyclic compounds,

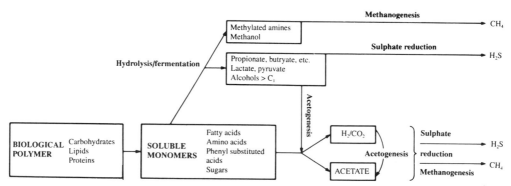

Fig. 9.1. A scheme of carbon flow through an anaerobic ecosystem showing some of the functional metabolic groups and their relationships. (From Parkes 1987, with permission.)

and even hydrocarbons (Widdel 1988; Aeckersberg *et al.* 1991). Even so, the balance of evidence suggests that in most natural environments the SRB, like the methanogens, oxidize hydrogen and acetate as their key energy and carbon substrates (Jørgensen 1982). This implies, therefore, that in biofilms and other microbial consortia the processes of acetogenesis and interspecies hydrogen transfer are of central importance (Zeikus 1983; Hamilton 1988).

Heterogeneities

Biofilms are dynamic structures. The pattern of their development can be divided into phases of transport of molecules and cells to the substratum, attachment, increase in biomass (that is, cells and extracellular polymeric substances), and detachment (Characklis & Marshall 1990). This holds true for each individual species, and a mature biofilm is considered to be in a steady state when the rate of growth is equal to the rate of detachment. In a multispecies biofilm this situation may be complicated by the requirement for organism A to establish the necessary conditions, for example of anaerobiosis, before organism B can become established within the biofilm. Other less well characterized features of changes with time, which may be of particular significance with respect to corrosion mechanisms, are the retention of corrosion products within the polymer matrix of the biofilm, and the random sloughing of whole areas of biofilm with the loss of cells, polymers and any inorganic inclusions.

Reference has already been made to the development of anaerobic conditions within biofilms. This is a common occurrence and it has been demonstrated even within thin biofilms of only 12 μm depth. Oxygen limitation results from the combination of mass transfer and diffusional resistances to the entry and passage of oxygen through the biofilm, and its utilization by aerobic organisms in the biofilm surface layers. Where an illuminated biofilm also contains diatoms and/or cyanobacteria, the effect may be partially alleviated by oxygenic photosynthesis. While the development of an aerobic/anaerobic interface is the most striking vertical heterogeneity noted in biofilms, the phenomenon is a general one and can equally give rise to carbon substrate or sulphate limitation in the deeper layers of particular biofilms. It is clear, therefore, that there is a close interdependence between these discontinuities throughout the depth of the biofilm and the essential character of the biofilm as a microbial consortium. Particular organisms exist within appropriate microenvironments which are created by their own and other organisms' activities and are maintained by virtue of the polymeric matrix within which they are imbedded.

It may appear surprising to stress the importance of anaerobic regions within a biofilm existing in an aerobic bulk phase, when one might expect the so-called anaerobic corrosion caused by the SRB to be more directly associated with anoxic and reducing environments. In fact, it is general experience that maximum levels of corrosion are most usually linked to environments where the necessary reducing conditions are in close proximity to an aerobic phase. In addition to any stimulation of the SRB themselves (Rosser & Hamilton 1983; Jørgensen 1988), there appears to be a direct effect of oxygen on the actual corrosion reactions. This important facet will be considered later in this chapter.

Biofilms also demonstrate a degree of heterogeneity in the plane horizontal to the underlying substratum. This so-called patchiness arises from localized or colonial growth of individual species with their associated exopolymers but can also occur when significant areas of the biofilm slough off, in what is a well documented but poorly understood phenomenon. Since the essential characteristic of corrosion processes is the establishment and operation of electrochemical cells with separately identifiable anodic and cathodic regions on the metal substratum, discontinuities across the surface of the biofilm are likely to be of direct and major importance.

Sulphate-reducing bacteria

The SRB are a relatively diverse taxonomic group of bacteria which share a common property that dominates both their physiology and their ecology. They are strict anaerobes with the capacity to reduce sulphate as the terminal electron acceptor in a respiratory mode of metabolism. Other oxidized forms of sulphur, such as thiosulphate, can also fulfil this role, and a few species reduce sulphur itself rather than sulphate. The group as a whole is sometimes referred to as the sulphidogens which has the merit of including the sulphur being applied extensively to the study of SRB and have given rise to a much more informed picture of phylogenetic relationships within the group (Devereux *et al.* 1989, 1990). Specific efforts are currently being made to extend such approaches with the objective of developing probes that will allow more accurate and detailed characterization of natural populations, without the bias introduced by cultural techniques based on a particular medium composition. Voordouw and his colleagues have found that a probe for the [NiFe] hydrogenase gene can identify species from the genus *Desulfovibrio* but not other SRB (Voordouw *et al.* 1990). (Three hydrogenase enzymes have been found variously distributed across the group and individually characterized by their metal cofactors Fe, NiFe and NiFeSe: Peck & Lissolo 1988). Recently a technique has been developed using whole genomic DNA to differentiate among at least 20 genotypically individual SRB from a range of oil field samples (Voordouw *et al.* 1991). Interestingly, there appears to be little or no relationship between this series of natural isolates and the supposedly representative species held in culture collections. Somewhat complementary to this observation is the fact that at no stage has any suggestion been made that corrosion might be the direct consequence of any one or more particular species of SRB, rather than a general property of the group as a whole. At a higher degree of resolution, it is at least theoretically possible that only the occurrence of different species at different locations within a sulphate reducing biofilm could give rise to the establishment of electrochemical corrosion cells. A first attempt to approach such a question has been made using polymerase chain reaction amplification and fluorescence microscopy to study the distribution within a multispecies biofilm of *Deslfovibrio vulgaris*-like and *Desulfuromonas*-like organisms (Amann *et al.* 1992).

A further significant development in our knowledge of sulphidogenic bacteria has been the increasing incidence of the isolation of both eubacterial and archaebacterial thermophilic species from a wide range of natural and manmade environments (Stetter *et al.* 1987; Cochrane *et al.* 1988; Rosnes *et al.* 1991). Although such organisms might have a significant effect on corrosion in oil field production systems, it is likely that their major economic impact

will arise from their potential to cause souring in petroleum reservoirs.

Microbially influenced corrosion

As our appreciation of the true nature of biological corrosion mechanisms has increased, it is interesting to note that the firm terminology of 'biocorrosion' or 'microbial corrosion' has been replaced by the rather more circumspect term of 'microbially influenced corrosion'. This is a direct reflection of there being a range of quite different microbial processes at least potentially involved, and the fact that they operate primarily by stimulating pre-existing standard electrochemical mechanisms of corrosion rather than by introducing any novel reactions schemes of their own.

Further, it must be pointed out that although the major thrust of this chapter is focused on the activities of SRB within multispecies biofilms, there are signifcant examples of microbially influenced corrosion that involve neither SRB nor biofilms (Tatnall 1991). Stainless steels, for example, are generally resistant to corrosion by virtue of a protective iron and chromium oxide-rich inorganic film. The integrity of this film can be disrupted by high levels of Cl^- or H_2S, or through colonial growth of *Gallionella* or *Siderocapsa* giving, respectively, iron-rich and iron and manganese-rich deposits. Separate instances of pitting corrosion in stainless steels have also been recorded in association with the growth of unidentified aerobic, slime-forming bacteria. In one case the pits had a tunnelling character and were associated with a grey-brown manganese-rich deposit, while in the other the biodeposits were chloride-rich, with no involvement of either manganese or iron. SRB were reported to be the cause of one case of crevice corrosion which developed under joint gaskets. This appeared, however, to be unique to type 304 stainless steel which is low in molybdenum and the problem was resolved by changing the specification to molybdenum-containing type 316 stainless.

The main instances of microbially influenced corrosion resulting from the action of the SRB are noted with cast iron and mild steel. In the case of buried cast iron pipes, the effect can be dramatic, with the complete removal of the iron leaving behind an apparently unaltered structure which is, in fact, composed solely of graphite and retains none of the strength of the original material. It is, however, with the mild steels, universally used in all construction work, that the SRB have their major impact. Here the corrosion is of a pitting character and is invariably linked to the production of black iron sulphide corrosion products. Although the pitting implies a localized corrosion mechanism, it is generally associated with a degree of confluent growth in the form of a biofilm. It is these aspects we can now explore further with reference to the various points already raised in our general considerations of biofilms and of the SRB.

Generalised mechanisms of corrosion

Metal loss during corrosion occurs by passage into solution of positively charged metal ions at the anode of an electrochemical cell. This must be balanced by transfer of the excess electrons to that part of the metal substratum which is cathodic and where a second reaction can absorb the electrons. The simple, non-stoichiometric, equations describing aerobic corrosion or the rusting of iron would be:

$$Fe \rightarrow Fe^{2+} + 2e \qquad \text{anode}$$
$$\tfrac{1}{2}O_2 + H_2O + 2e \rightarrow 2OH^- \qquad \text{cathode}$$

with the subsequent formation of a complex of iron oxides and hydroxides as corrosion products.

In the absence of oxygen, it has been hypothesized that protons (von Wolzogen Kuhr & van der Vlught 1934) or H_2S (Costello 1974) might serve as electron acceptor at the cathode:

$$\text{(a) } 2H^+ + 2e \rightarrow 2H \rightarrow H_2$$
$$\text{(b) } 2H_2S + 2e \rightarrow 2HS^- + H_2$$

with SRB subsequently oxidizing the H_2, thus both preventing polarization of the cathode by adsorbed H_2, and giving rise to S^{2-} and the potential for metal sulphide corrosion products:

$$4H_2 + SO_4^{2-} \rightarrow 4H_2O + S^{2-}$$
$$Fe^{2+} + S^{2-} \rightarrow FeS$$

That SRB can oxidize cathodic hydrogen and use it as a source of metabolic energy has now been demonstrated in several laboratories (Hardy 1983; Pankhania *et al.* 1986; Cord-Ruwisch *et al.* 1987). Additionally, it has been noted that the application of an impressed current, to confer protection against corrosion weight loss, also has the effect of stimulating sulphate reducer growth and activity in direct response to the increased tendency to produce hydrogen at the cathode (Guezennec *et al.* 1991).

Most discussions of microbially influenced corrosion involving the SRB have concluded, however, that the major factor determining the rate and extent of corrosion is the presence of the iron sulphide corrosion products themselves (King & Miller 1981). Even Bryant *et al.* (1991) noted that the highest rates of corrosion occurred after the period of microbial activity, and suggested that iron sulphide corrosion products may have a role to play in corrosion mechanisms in addition to that of hydrogenase. In purely electrochemical terms, iron sulphide is cathodic to unreacted iron. The deposition of corrosion product might also be expected to offer an increased surface area for the formation and subsequent oxidation of cathodic hydrogen. More detailed studies, however, have identified that thin adherent layers of the primary corrosion product, mackinawite, are generally protective but that time-dependent rupture, or high iron concentration-dependent bulky precipitation, can lead to high non-transient corrosion rates that are independent of microbial growth or hydrogenase activity (Mara & Williams 1972; King *et al.* 1973).

Aerobic/anaerobic interface

As was noted above, not only are SRB dependent upon the development of anaerobic regions within sediments or biofilms, but they also generally show their highest activities close to the interface between aerobic and anaerobic microenvironments. Almost certainly this reflects nutrient input from the metabolic activities of other aerobic and facultative organisms also present but with greater biodegradative capability than the SRB, as well as the possibility of biotic and abiotic oxidation of the produced sulphide, to which the sulphate reducers are themselves sensitive. It is clear, however, that oxygen can also play a much more direct role in microbially influenced corrosion, by at least two quite separate mechanisms.

Microbial growth at a metal surface may facilitate the establishment of separate anodic and cathodic regions by a number of specific reactions within the general mechanism of establishing concentration gradients. These may arise from the local production of protons or other metabolic products, or from differential binding of dissolved metal ions by the EPS produced by particular cellular species in colonial growth (Costerton *et al.* 1987). A particularly important example of this mechanism is the oxygen concentration or differential aeration cell arising from localized growth of aerobic organisms (Fig. 9.2). Here the region immediately below the area of active growth becomes anaerobic and hence anodic relative to the surrounding metallic surface exposed to the air, and is the site of metal dissolution. It is likely that this constitutes the most common and widespread mechanism of microbially influenced corrosion associated with generalized microbial growth, either as a discontinuous biofilm or in discrete colonies. The further microbial oxidation of iron from the ferrous to the ferric state can on occasion increase the extent of corrosion and lead to the formation of tubercles. The anaerobic regions created by such microbial activity serve, however, not only as anodes in electrochemical corrosion cells but also as ideal microenvironments for the growth of anaerobic species, including SRB. By this means, therefore, SRB microbially influenced corrosion generally occurs alongside more generalized mechanisms and may even be, to some extent, dependent upon them.

Most of the earlier studies of the so-called anaerobic corrosion caused by SRB were indeed carried out under strictly anaerobic conditions but, in field situations, it is widely recognized that the most extensive corrosion occurs where there is ready access of air. Under such condi-

tions, penetration rates of up to 5 mm per year have been recorded (Tatnall 1991). At least two studies suggest that this stimulation by oxygen is neither on the metabolic activity of the SRB nor even on the formation of the iron sulphide corrosion products themselves. Hardy & Bown (1984) showed that if a metal surface that had previously been subjected to growth of SRB was subsequently exposed to pulses of air, there was a marked increase in the instantaneous corrosion rate. A similar effect of oxygenation on preformed iron sulphides was noted by Braithwaite & Lichti (1980) in their examination of corrosion in geothermal power stations where the primary corrodent is H_2S in the high temperature steam, with no involvement of any microbial activity.

These findings have been confirmed and extended by some of our own work with corroding steel coupons in sea bed sediments associated with oil production platforms, and in a laboratory simulation of a stratified marine system (Moosavi *et al.* 1991; McKenzie & Hamilton 1992). Again, maximal rates of corrosion were noted in those bulk environments where conditions were aerobic. Two factors were considered to be of possible significance under these circumstances. First, visual observation revealed three discrete layers of corrosion product; a thin black adherent layer, surmounted by first a bulky layer of looser material, again black, and finally an overlying layer of brown, oxidized products. Preliminary attempts at chemical analysis served only to establish an extremely complex mixture of sulphides, oxides and carbonates. It should be noted at this point that iron sulphide exists in a wide range of forms, each with its characteristic iron/sulphur stoichiometry, and each with a unique physical form, some being crystalline and others amorphous. Much of the early seminal work on corrosion mechanisms suggested that the transition of one form of iron sulphide to another was the direct cause of a thin adherent layer of mackinawite (FeS_{1-x}) losing both its chemical and physical form, and consequently its protective character (King *et al.* 1976).

Secondly, it was noted that in the more oxidizing regions associated with the higher corrosion rates, up to 92% of the sulphides were in a form that did not give rise to H_2S on treatment with cold acid. The principal forms of non-acid volatile sulphur compounds have been identified as pyrite (FeS_2) and sulphur itself, the formation

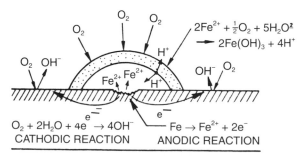

Fig. 9.2. Differential aeration cell. (From Schaschl 1980, with permission.)

of each of which is favoured by oxidizing conditions. Pyrite has been implicated previously in corrosion mechanisms, and sulphur is well known to be highly corrosive (Schaschl 1980; Schmitt 1991). These data confirm, therefore, that it is the effect of environmental variables, particularly oxygen, on the nature and form of the sulphide corrosion products that determines the true nature and extent of the corrosion which occurs as a result of the growth of SRB.

These analyses and conclusions have been further extended in an important series of studies carried out by Lee and Characklis and colleagues at the Center for Biofilm Engineering in Bozeman, Montana. The experimental system employed was a closed channel flow reactor which allowed measurement of the corrosion of test coupons by the techniques of corrosion potential, cathodic polarization, pitting potential and A.C. impedance, as well as by weight loss. It was first confirmed with a monoculture of *Desulfovibrio desulfuricans* that there is no direct correlation between corrosion and sulphate reduction activity, and that effectively no corrosion occurred anaerobically in the absence of added iron(II) (Lee & Characklis 1993). Only after the addition of 60 mg L^{-1} iron(II) to the wholly anaerobic system was there significant corrosion. Under these conditions they observed no adherent sulphide film but rather a bulky precipitate and, in accordance with the earlier findings of Mara & Williams (1972) and King *et al.* (1973), they concluded that the loose iron sulphide particles appear to play a more important role than the bacteria in the anaerobic corrosion process. They suggested that the role of the SRB is to maintain the production of H_2S which has the effect of keeping the iron sulphide particles permanently cathodically active.

These researchers then conducted a series of experiments with a mixed inoculum of *Pseudomonas aeruginosa*, *Klebsiella pneumonia* and *D. desulfuricans* under conditions of, respectively, low (1.5 mg L^{-1}) and high (7 mg L^{-1}) dissolved oxygen (DO) in the bulk phase, and with alternating (12 h/12 h) aerobic/anaerobic phases (Lee *et al.* 1993a, b; Nielsen *et al.* 1993). Associated with higher levels of oxygen were both thicker biofilms (up to 3 mm) and increased rates of corrosion (250 µm year^{-1} at 7 mg L^{-1} DO, and up to 4 mm year^{-1} under alternating aerobic/anaerobic phases). Even at the higher values of DO,

zero oxygen concentrations developed at the metal surface although there was also clear evidence of heterogeneities within the biofilm; SRB were present and active in advance of anaerobic conditions being identified by microelectrode analysis, and even after the development of such conditions DO values of 1.8 mg L^{-1} could be measured at specific sites on the metal surface. Perhaps most significantly, however, the corrosion products developed a three layered structure with pits first becoming evident beneath orange-coloured surface layers. Auger spectroscopy coupled with argon ion sputtering allowed elemental analysis of corrosion pits to the depth of 180 nm and showed a loose but definite spatial correlation between oxygen and sulphur signals. Although the specific mechanism remains uncertain, what these findings suggest is that the action of oxygen is to reoxidize reduced ferrous sulphides, with the SRB and their sulphide corrosion products thus effectively acting as electron carriers from the metal surface to the oxidized surface deposits.

Only reduced iron sulphides such as mackinawite and greigite are detected in a totally anaerobic system, and they are not permanently cathodic to unreacted mild steel. The role of the SRB in such systems is to maintain the electrochemical activity of the ferrous sulphide. Where oxygen is introduced into the system, however, oxidized iron sulphides such as pyrite, and elemental sulphur are also found to be present, and these are associated with greatly increased rates of corrosion. Where the data relating to the corrosivity of pyrite remain ambiguous, there is clear evidence that sulphur itself is highly corrosive. Furthermore, since this process is autocatalytic, the rate and extent of corrosion should be independent of the presence and activity of the SRB.

A full appreciation of the mechanisms of microbially influenced corrosion requires consideration of both the biological and the physical or engineering aspects of the phenomenon. With respect to the SRB this is especially so as the balance of evidence suggests that the true corrosion, as a mechanism that continues with time and can lead to extensive structural damage, is a process firmly in the realm of materials science. None the less, the reaction triggering these events is the microbial production of sulphide. In order to understand that reaction and the factors, biological and physical, that control it, one must examine the physiology of the SRB, and consider the nature of their growth as component members of microbial consortia within biofilms. In addition to their impact on the presence and activity of the SRB, biofilms also play a direct role in corrosion *per se* as they substantially change the local chemistry of the adjacent metal surface due to microbial transformation and mass transfer processes, and the physical, chemical and microbiological properties of the fouling deposits.

References

Aeckersberg, F., Bak, F. & Widdel, F. (1991). Anaerobic oxidation of saturated hydrocarbons to CO_2 by a new type of sulfate-reducing bacterium. *Archives of Microbiology*, **156**, 5–14.

Amann, R. I., Stromely, J., Devereux, R., Key, R. & Stahl, D. A. (1992). Molecular and microscopic identification of sulfate-reducing bacteria in multispecies biofilms. *Applied and Environmental Microbiology*, **58**, 614–23.

Braithwaite, W. R. & Lichti, K. A. (1980). Surface corrosion of metals in geothermal fluids at Broadlands, New Zealand. In *Geothermal scaling and corrosion*, ed. L. A. Casper & T. R. Pinchback, pp. 81–121. American Society for Testing of Materials.

Brandis-Heep, A., Gebhardt, N. A., Thauer, R. K., Widdel, F. & Pfennig, N. (1983). Anaerobic acetate oxidation to CO_2 by *Desulfobacter postgatei*. 1. Demonstration of all enzymes required for the operation of the citric acid cycle. *Archives of Microbiology*, **136**, 222–9.

Bryant, R. D., Jansen, W., Boivin, J., Laishley, E. J. & Costerton, J. W. (1991). Effect of hydrogenase and mixed sulfate-reducing bacterial populations on the corrosion of steel. *Applied and Environmental Microbiology*, **57**, 2804–9.

Characklis, W. G. & Marshall, K. C. (ed.) (1990). *Biofilms*. New York: Wiley.

Characklis, W. G. & Wilderer, P. A. (ed.) (1989). *Structure and function of biofilms*. Chichester: Wiley.

Cochrane, W. J., Jones, P. S., Sanders, P. F., Holt, D. M. & Moseley, M. J. (1988). Studies on the thermophilic sulfate-reducing bacteria from a souring North Sea oil field. *Society of Petroleum Engineering*, Paper 18368. 301–16.

Cord-Ruwisch, R., Kleinitz, W. & Widdel, F. (1987). Sulfate-reducing bacteria and their activities in oil production. *Journal of Petroleum Technology*, (January), 97–106.

Costello, J. A. (1974). Cathodic depolarisation by sulphate-reducing bacteria. *South African Journal of Science*, **70**, 202–4.

Costerton, J. W., Cheng, K.-J., Geesey, G. G. *et al.* (1987). Bacterial biofilms in nature and disease. *Annual Reviews of Microbiology*, **41**, 435–64.

Devereux, R., Delany, M., Widdel, F. & Stahl, D. A. (1989). Natural relationships among sulfate-reducing eubacteria. *Journal of Bacteriology*, **171**, 6689–95.

Devereux, R., He, S.-H., Doyle, C. L. *et al.* (1990). Diversity and origin of *Desulfovibrio* species: phylogenetic definition of a family. *Journal of Bacteriology*, **172**, 3609–19.

Dowling, N. J., Mittleman, M. W. & Danko, J. C. (ed.) (1991). *Microbially influenced corrosion and biodeterioration*. Washington, DC: National Association of Corrosion Engineers.

Flemming, H.-C. & Geesey, G. G. (ed.) (1991). *Biofouling and biocorrosion in industrial water systems*. Berlin: Springer-Verlag.

Fuchs, G. (1986). CO_2 fixation in acetogenic bacteria: variations on a theme. *FEMS Microbiology Reviews*, **39**, 181–213.

Gebhardt, N. A., Thauer, R. K., Linder, D., Kaulfers, P. M. & Pfennig, N. (1985). Mechanism of acetate oxidation to CO_2 with elemental sulfur in *Desulfuromonas acetoxidans*. *Archives of Microbiology*, **141**, 392–8.

Guezennec, J., Dowling, N. J., Conte, M., Antoine, E. & Fiksdal, L. (1991). Cathodic protection in marine sediments and the aerated seawater column. In *Microbially influencd corrosion and biodeterioration*, ed. N. J. Dowling, M. W. Mittleman & J. C. Danko, pp. 6.43–6.50. Washington, DC: National Association of Corrosion Engineers.

Hamilton, W. A. (1985). Sulphate-reducing bacteria and anaerobic corrosion. *Annual Review of Microbiology*, **39**, 195–217.

Hamilton, W. A. (1988). Energy transduction in anaerobic bacteria. In *Bacterial energy transduction*, ed. C. Anthony, pp. 83–149. London: Academic Press.

Hardy, J. A. (1983). Utilisation of cathodic hydrogen by sulphate-reducing bacteria. *British Corrosion Journal*, **18**, 190–3.

Hardy, J. A. & Bown, J. (1984). The corrosion of mild steel by biogenic sulfide films exposed to air. *Corrosion*, **40**, 650–4.

Jørgensen, B. B. (1982). Ecology of the bacteria of the sulphur cycle with special reference to anoxic–oxic interface environments. *Philosophical Transactions of the Royal Society of London, Series B*, **298**, 543–61.

Jørgensen, B. B. (1988). Ecology of the sulphur cycle: oxidative pathways in sediments. *Symposium of the Society for General Microbiology*, **42**, 31–63.

King, R. A., Dittmer, C. K. & Miller, J. D. A. (1976). Effect of ferrous iron concentration on the corrosion of iron in semicontinuous cultures of sulphate-reducing bacteria. *British Corrosion Journal*, **11**, 105–7.

King, R. A. & Miller, J. D. A. (1981). Corrosion by sulphate-reducing bacteria. *Nature*, **233**, 491–2.

King, R. A., Miller, J. D. A. & Wakerley, D. S. (1973). Corrosion of mild steel in cultures of sulphate-reducing bacteria: effect of changing the soluble iron concentration during growth. *British Corrosion Journal*, **8**, 89–93.

Kroger, A., Schroder, J., Paulsen, J. & Beilmann. (1988). Acetate oxidation with sulphur and sulphate as terminal electron acceptors. *Symposium of the Society for General Microbiology*, **42**, 133–45.

Lee, W. & Characklis, W. G. (1993). Corrosion of mild steel under anaerobic biofilm. *Corrosion*, **49**, 186–99.

Lee, W., Lewandowski, Z., Morrison, M., Characklis, W. G., Avci, R. & Nielsen, P. H. (1993b). Corrosion of mild steel underneath aerobic biofilms containing sulfate-reducing bacteria. Part 2. At high bulk oxygen concentration. *Biofouling* 7, 217–39.

Lee, W., Lewandowski, Z., Okabe, S., Characklis, W. G. & Avci, R. (1993a). Corrosion of mild steel underneath aerobic biofilms containing sulfate-reducing bacteria. Part 1. At low dissolved oxygen concentration. *Biofouling* 7, 197–216.

McKenzie, J. & Hamilton, W. A. (1992). The assay of *in-situ* activities of sulphate-reducing bacteria in a laboratory marine corrosion model. *International Biodeterioration and Biodegradation*, **29**, 285–97.

Mara, D. D. & Williams, D. J. A. (1972). The mechanism of sulphide corrosion by sulphate-reducing bacteria. In *Biodeterioration of materials*, Vol. 2, ed. A. M. Walters & E. H. Hueck van der Plas, pp. 103 13. London: Applied Science Publishers.

Moosavi, A. N., Pirrie, R. S. & Hamilton, W. A. (1991). Effect of sulphate-reducing bacteria activity on performance of sacrificial anodes. In *Microbially influenced corrosion and biodeterioration*, ed. N. J. Dowling, M. W. Mittleman & J. C. Danko, pp. 3.13–3.27. Washington, DC: National Association of Corrosion Engineers.

Nethe-Jaenchen, R. & Thauer, R. K. (1984). Growth yields and saturation constant of *Desulfovibrio vulgaris* in chemostat culture. *Archives of Microbiology*, **137**, 236–40.

Nielsen, P. H., Lee, W., Lewandowski, Z., Morrison, M. & Characklis, W. G. (1993). Corrosion of mild steel in an alternating oxic and anoxic biofilm system. *Biofouling* 7, 267–84.

Pankhania, I. P., Moosavi, A. N. & Hamilton, W. A. (1986). Utilization of cathodic hydrogen by *Desulfovibrio vulgaris* (Hildenborough). *Journal of General Microbiology*, **132**, 3357–65.

Parkes, R. J. (1987). Analysis of microbial communities within sediments using biomarkers. *Symposium of the Society for General Microbiology*, **41**, 147–77.

Peck, H. D. & Lissolo, T. (1988). Assimilatory and dissimilatory sulphate reduction: enzymology and bioenergetics. *Symposium of the Society for General Microbiology*, **42**, 99–132.

Postgate, J. R. (1984). *The sulphate-reducing bacteria*, 2nd edn. Cambridge: Cambridge University Press.

Rosnes, J. T., Torsvik, T. & Lien, T. (1991). Spore-forming thermophilic sulfate-reducing bacteria isolated from North Sea oil field waters. *Applied and Environmental Microbiology*, **57**, 2302–7.

Rosser, H. R. & Hamilton, W. A. (1983). Simple assay for accurate determination of [^{35}S] sulfate reduction activity. *Applied and Environmental Microbiology*, **45**, 1956–9.

Schaschl, E. (1980). Elemental sulfur as a corrodent in deaerated, neutral aqueous solutions. *Materials Performancem*, **19**, 9–12.

Schmitt, G. (1991). Effect of elemental sulfur on corrosion in sour gas systems. *Corrosion*, **47**, 285–308.

Stetter, K. O., Lauerer, G., Thomm, M. & Neuner, A. (1987). Isolation of extremely thermophilic sulfate reducers: evidence for a novel branch of Archaebacteria. *Science*, **236**, 822–4.

Tatnall, R. E. (1991). Case histories: biocorrosion. In *Biofouling and biocorrosion in industrial water systems*, ed. H.-C. Flemming & G. G. Geesey, pp. 165–85. Berlin: Springer-Verlag.

Tatnall, R. E. & Horacek, G. L. (1991). New perspectives on testing for sulfate-reducing bacteria. In *Microbially influenced corrosion and biodeterioration*, ed. N. J. Dowling, M. W. Mittleman & J. C. Danko, pp. 5.17–5.32. Washington, DC: National Association of Corrosion Engineers.

Videla, H. A. & Gaylarde, C. C. (ed.) (1992). Microbially influenced corrosion. *International Biodeterioration and Biodegradation*, **29**, 193–375.

von Wolzogen Kuhr, C. A. M. & van der Vlught, I. S. (1934). The graphitization of cast iron as an electrobiochemical process in anaerobic soils. *Water*, **18**, 147–65.

Voordouw, G., Niviere, V., Ferris, F. G., Fedorak, P. M. & Westlake, D. W. S. (1990). Distribution of hydrogenase genes in *Desulfovibrio* spp. and their use in identification of species from the oil field environment. *Applied and Environmental Microbiology*, **56**, 3748–54.

Voordouw, G., Voordouw, J. K., Karkhoff-Schweizer, R. R., Fedorak, P. M. & Westlake D. W. S. (1991). Reverse sample genome probing, a new technique for identification of bacteria in environmental samples by DNA hybridization, and its application to the identification of sulfate-reducing bacteria in oil field samples. *Applied and Environmental Microbiology*, **57**, 3070–8.

Widdel, F. (1988). Microbiology and ecology of

sulfate- and sulfur-reducing bacteria. In *Biology of anaerobic microorganisms*, ed. A. J. B. Zehnder, pp. 469–585. London: Wiley.

Wood, H. G., Ragsdale, S. W. & Pezacka, E. (1986). The acetyl-CoA pathway of autotrophic growth. *FEMS Microbiology Reviews*, **39**, 345–62.

Zeikus, J. G. (1983). Metabolic communication between biodegradative populations in nature. *Symposium of the Society for General Microbiology*, **34**, 423–62.

10

Microbial Consortia in Industrial Wastewater Treatment

R. Cam Wyndham and Kevin J. Kennedy

Introduction

Wastewater treatment is an essential component of our social order, for without it we soon find our communities suffering from waterborne disease, and our ecosystems suffering unwanted change. For over a century wastewater treatment systems have been designed to increase microbial growth in order to remove organic carbon and other nutrients, while limiting the release of suspended solids into receiving waters. Optimizing design parameters for these goals has been a successful strategy for many conventional wastewaters. However, in the past two decades we have confronted an increasing variety of non-conventional wastewaters in the form of chemical and industrial process effluents and landfill leachates. Legislation has been enacted in many countries that specifies the levels of toxic organic and inorganic chemicals that may be released into the environment. For example, in Canada the provinces of British Columbia, Ontario and Quebec have enacted strict regulations dealing with the discharge of chlorinated organics in pulp mill wastewaters. The long term cost to Canadian industry of this legislation plus proposed new regulations further limiting the chronic toxicity of wastewater discharges will be in the billions of dollars. These costs are already exerting pressure on the industries concerned to solve toxicity problems in the most efficient way possible.

There are three components to a successful toxic wastewater treatment process: knowledge of the nature of the toxic chemicals, an understanding of the concentrations of these chemicals that affect target organisms in receiving waters, and knowledge of the conditions required for growth of toxin degrading microorganisms in the treatment system. The first two of these have been placed on a sound scientific footing using the techniques of analytical chemistry and toxicology. The last component has traditionally been treated as a black box, manipulated by trial and error until optimal growth and detoxification are achieved. Unfortunately, in many cases this approach has failed because knowledge of the specific microbial populations needed to degrade the toxic components of the wastewater is lacking. The major barrier to industry is the knowledge of how to enhance toxicity reductions in wastewater treatment systems that were originally designed to achieve traditional goals: that is, the reduction of total organic carbon, nutrients and suspended solids. New techniques for characterizing microbial consortia are beginning to overcome this barrier and allow operators to enhance the growth of specific populations in wastewater treatment systems. These techniques include phenotype and genotype detection methods as well as *in situ* activity measurements. In the following chapter we will examine several types of wastewater treatment in both laboratory model systems and in full scale operating treatment systems. We will discuss the microbial consortia involved in these processes, their structures, and their genetics and physiology under different treatment conditions. The first two examples deal with aerobic processes; a laboratory scale system for polycyclic aromatic hydrocarbon (PAH) removal from dilute slurries, and a full scale groundwater bioremediation system. These

183

examples will be followed by a discussion of anaerobic treatment processes.

Aerobic consortia

Specialized aerobic treatment processes have been developed for a variety of organic wastes. Many of these are now being applied for the bioremediation of contaminated soils, sediments, groundwater and surface water. Of the 159 sites listed in the United States Environmental Protection Agency's Bioremediation Field Initiative (EPA 1993), 102 are petroleum and wood preserving waste contaminated. Most of these are being treated, or are in the design stage for treatment by aerobic slurry techniques. Many of the 60 groundwater and surface water sites are also candidates for pump and treat technologies using aeration, or are already being treated this way.

Bacterial biofilms adherent to crystalline asphaltenes

A fundamental limitation to bioremediation efforts in the field is the availability of the contaminant for microbial mineralization. PAHs and many chlorinated aromatic compounds partition onto soil or sediment particles, both organic and inorganic, and are only slowly released. The concentrations of these contaminants in the aqueous phase are often too low to support the growth of a biodegradative consortium. In a study undertaken to assess microorganisms for desulphurization of heavy crude oils, we developed a solid support for PAH degrading bacteria that shows promise (Wyndham & Costerton 1981). The surface consists of crystalline ashphaltenes, the high molecular weight component of crude oils, a fraction normally discarded during oil refining. This surface provides a hydrophobic site for the concentration of PAHs and also a site for the formation of a microbial biofilm active in PAH degradation.

The insoluble, high molecular weight asphaltene fraction of Cold Lake crude oil was used as an enrichment substrate under sulphur limited growth conditions to isolate a consortium of bacteria capable of growth on naphthalene and phenanthrene, with the asphaltenes providing the

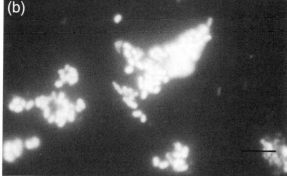

Fig. 10.1.   Epifluorescence photomicrographs of asphaltenes (black, crystalline particles) colonized by a PAH-degrading consortium. The sole source of sulphur for the culture was the asphaltene substrate (8% S by weight). The sole sources of carbon were naphthalene, phenanthrene and the asphaltene particles. Bars represent: (a) 50 μm; (b) 20 μm; (c) 20 μm.

sole source of sulphur. The consortium was characterized by epifluorescence and transmission electron microscopy, and the number of organisms capable of oxidation of polycyclic aromatic substrates was determined by the methods of Kiyohara *et al.* (1982) and Shiaris & Cooney (1983). Cluster analysis of phenotypic characters

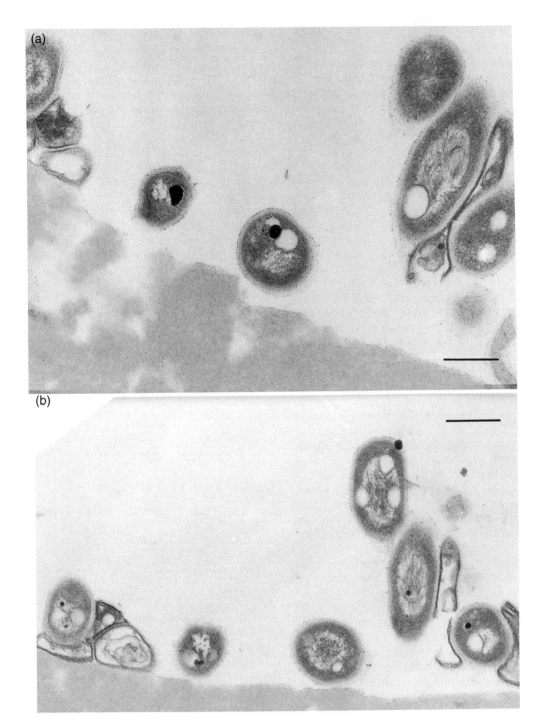

Fig. 10.2. Transmission electron micrographs of serial sections through an asphaltene particle showing the association of Gram negative bacteria with the asphaltene surface. The grey zone in the lower portion of each micrograph is the asphaltene surface. Lysed cells are evident. Whole cells do not penetrate the surface. Serial sections show the distribution of numerous electron dense and transparent inclusion bodies in whole cells. Bars = 0.5 μm.

of isolates from the biofilm revealed a complex community. Clusters were further characterized by plasmid screening and by their capacities for aromatic hydrocarbon metabolism and heterocyclic sulphur containing PAH co-metabolism.

A Versatec, 1.2 L fermentor was supplied with a dilute, S-limited, minimal medium at a dilution rate of 0.05 h^{-1}. The sole source of sulphur was the asphaltene substrate, containing approximately 8% sulphur by weight (Selucky *et al.* 1978). This substrate was supplied as fine particles recrystallized from 40 volumes of hexanes following dispersal of Cold Lake crude oil as a concentrated toluene solution. Recrystallized asphaltenes (3 g L^{-1}), plus 1.6 mM naphthalene and 1.6 mM phenanthrene were added as solids to the culture at the outset. A coarse screen prevented the loss of asphaltenes from the fermentor. At 4 week intervals the bulk of the asphaltenes was removed and replaced with fresh, recrystallized asphaltenes and PAHs. The fermentor was inoculated with heavy oil contaminated soils (a total of 1 g added) from an operating oil field near the town of Oil Springs, Ontario, the site of the first oil well in North America (1857). In addition, pure cultures isolated from Athabasca oil sands, Alberta (Wyndham & Costerton 1981) were inoculated into the mixed culture. For the first 2 months the culture was periodically reinoculated. Epifluorescence photomicrographs of colonized asphaltene particles from the continuous culture are shown in Fig. 10.1. The total number of bacteria was in the order of 10^8 ml^{-1}. Most cells were adherent and the mixed population size was dependent on the surface area of the crystalline asphaltene fraction. Transmission electron micrographs of the association between bacterial cells in the adherent biofilm and the crystalline asphaltene surface are shown in Fig. 10.2.

To characterize the culturable community of the biofilm, particles were washed with sterile minimal medium and sonicated to disperse the adherent cells. Isolations were made on non-selective NA and CPS media (Jones 1970) and on selective media (glucose, succinate, naphthalene, phenanthrene and dibenzothiophene). A subset of 45 of these isolates was characterized by Gram stain and oxidase reaction, then subjected to metabolic characterization using Biolog Microplates (Biolog Inc., Hayward, Calif.). These rapid diagnostic plates detect the ability of isolates to oxidize a test panel of organic carbon substrates. The results of the 96 tests were coded and the isolates grouped using a simple absolute value distance measure and single linkage cluster analysis. This approach allows us to pick out groups of isolates with similar metabolic patterns. The results are shown in Fig. 10.3 in the form of a dendrogram.

The cluster in Fig. 10.3 centred on isolate 80A, designated the '80 Cluster' and including isolates P4, P41, 80D and P1, was the most abundant polycyclic aromatic hydrocarbon oxidizing group of organisms in the biofilm. The group as a whole grows rapidly on phenanthrene and naphthalene as carbon sources and also forms large colonies on minimal medium plates in the presence of dibenzothiophene vapours. The isolates of the 80 Cluster co-metabolized

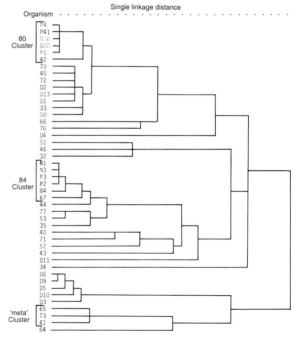

Fig. 10.3. Dendrogram of phenotypic characteristics of 45 isolates recovered from the asphaltene biofilm. The single linkage distance is shown on an arbitrary scale, with the greatest distance to the right representing approximately 31/96 phenotypic test differences. Isolates designated by initials N, P and D, followed by a number, were recovered from naphthalene, phenanthrene and dibenzothiophene enrichment plates, respectively. Isolates designated by a number alone were recovered from general CPS medium plates. The 80-, 84- and *meta*-clusters are discussed in the text.

indole to indigo, indicative of a non-specific dioxygenase activity, but failed to oxidize thianthrene and benzothiophene. Dibenzothiophene was metabolized by these isolates as a sole source of carbon. No accumulation of yellow ring fission metabolites occurred during the growth of the 80 Cluster isolates on dibenzothiophene, which is unlike the metabolism of this compound observed in naphthalene degrading *Pseudomonas* species (Monticello *et al.* 1985; Foght & Westlake 1988).

The '84 Cluster', centred on isolate 84 and including isolates N1, N3, P3, P2, 84 and 67, was also capable of rapid growth on the polycyclic hydrocarbons naphthalene and phenanthrene, but was not capable of forming colonies on dibenzothiophene. These isolates cometabolized the sulphur-heterocycle to yellow ring fission metabolites, including 4-[2-(3-hydroxy)-thianaphthenyl]-2-oxo-3-butenoic acid and 3-hydroxy-2-formyl-benzothiophene. These two dominant clusters of isolates are taxonomically distinct. The 80-Cluster has not been identified to genus and species; however, the 84 Cluster isolates belong to the fluorescent *Pseudomonas* genus.

As part of our characterization of these isolates, a plasmid screen was carried out. Plasmid encoded functions are known to be involved in the degradation of naphthalene and simpler aromatics such as toluene and xylenes (Singer & Finnerty 1984). In addition, the naphthalene plasmids are known to determine the co-metabolism of dibenzothiophene (Monticello *et al.* 1985). All of the isolates of both the 80 and 84 Clusters contained at least one high molecular weight plasmid in the 60 to 150 kilobase size range, with several isolates containing more than one plasmid.

The three isolates of the 'meta cluster' of Fig. 10.3 metabolized simple aromatics like benzoate, phenylacetate and toluene to meta ring fission products. This type of metabolism is normally associated with bacteria that carry catabolic plasmids for monoaromatic hydrocarbons like toluene and xylene. Their presence in the PAH degrading community growing on crystalline asphaltenes is interesting, as these compounds were not supplied for growth and the normal pathways for naphthalene and phenanthrene degradation do not produce these compounds as metabolites.

Crystalline asphaltenes colonized by PAH degrading bacteria may be an effective material for supplementing the natural activity of PAH degrading bacteria in contaminated soils and sediments. The asphaltene fraction itself provides both a surface for colonization, and also a hydrophobic matrix for the absorption of PAHs. We are currently investigating this material for its capacity to absorb PAHs from contaminated sediments and soil slurries, and for its capacity to provide a niche for PAH degrading bacteria. The goal of these studies is to use this material to increase the population density of PAH degrading bacteria, and hence the rate of PAH degradation, in contaminated soils and sediments.

Microbial consortia in an operating bioremediation system

There are few detailed analyses of the adherent microbial communities active in operating treatment systems used in the bioremediation of toxic wastes. Even fewer of these systems have been characterized genetically. In fact, few if any of the well characterized plasmids and transposons known to code for the biodegradation of chlorinated aromatic pollutants in laboratory strains have been shown to be functional during the bioremediation of groundwaters, industrial wastewaters or contaminated soils. There is no information on the frequency of plasmid transfer and transposition in these systems, processes that have long been suspected of being important in the adaptation of bacteria to toxic organics.

Others have studied the dynamics and activity of single inoculants in model wastewater treatment systems. For example, Dwyer *et al.* (1988) have followed the fate in laboratory activated sludge microcosms of *Pseudomonas* sp. strain B13 and genetically manipulated derivatives of this strain that degrade chloro- and methylbenzoates. Both of these strains survived at population densities of approximately 10^5 cells ml^{-1}, but biodegradative activity was poor. McClure *et al.* (1989, 1991) have demonstrated the survival of both naturally occurring (CDC Group IVC-2 strain AS2) and genetically engineered bacteria in a laboratory scale activated sludge unit. The engineered organism in this case was a *Pseudomonas putida* strain ASR2.8 isolated from activated sludge and carrying the cloned *clc* operon of plasmid pJP4 on plasmid pD10. Both strains maintained stable populations of greater than 10^6 CFU ml^{-1}, and in each case between 12 and 27% of the 3.2 mM 3-chlorobenzoate added to the activated sludge unit was removed.

In the following example we show that it is possible to track the distribution of specific genotypes in real wastewater treatment systems and to extract useful genetic and metabolic information from these studies.

The Hyde Park chemical landfill site is located in Niagara Falls, New York. The site was used between the 1950s and 1970s for waste organic chemicals from the manufacture of chlorinated pesticides, herbicides and other xenobiotics. There are an estimated 73 000 metric tons of waste at this site, buried in gravel with a high leachability. The groundwater at the site contains contaminants at the millimolar level including chlorinated benzoic acids, phenol and chlorophenols, plus a range of less concentrated organic and inorganic chemicals. Occidental Chemical Corporation operates a groundwater bioremediation system on site that treats groundwater collected from perimeter trenches. The site has been capped with clay. Groundwater collected from perimeter trenches is treated in the following steps: removal of non-aqueous phase liquids for incineration; adjustment of the aqueous phase pH to 7.5; venting and aeration to remove volatile organics to vapour phase absorbers and to precipitate iron hydroxides; removal of suspended solids; sand filtration; pretreatment sacrificial carbon adsorption of PCB and Dioxins; addition of urea and phosphoric acid supplements to bring the C/N/P ratio to 150/10/2; three sequencing batch reactors (SBRs) in series for biological treatment in 350 000 L working volumes with a 24 h cycle time; chlorination; sand filtration; post-treatment carbon adsorption; and finally, release of the treated groundwater aqueous phase to the municipal sewage system (Ying et al. 1992).

Bacterial growth in the sequencing batch reac-

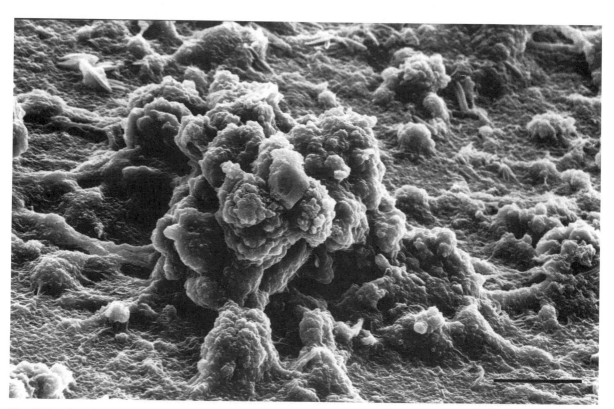

Fig. 10.4. Scanning electron micrograph of a large floc recovered from the Hyde Park sequencing batch reactor 2. The floc contains organic and inorganic material and microbial cells held together by a diffuse polysaccharide matrix. Bar = 20 µm.

tors occurs in macroscopic flocs that are of a size sufficient to settle during the resting stage of reactor operation. Smaller flocs and suspended bacteria that do not settle are removed from the reactors during the reactor draw stage and are lost from the active community. A scanning electron micrograph of flocs from the SBR2 tank is shown in Fig. 10.4. The active community of bacteria is structurally complex.

Our objective in this study was to determine the distribution of a particular catabolic genotype in the SBR community at Hyde Park. The genotype we selected encodes 3- and 3,4-dichlorobenzoate degradation. The genes involved, designated *cbaABC*, are located within a transposable element, designated Tn5271, on an IncPß plasmid in *Alcaligenes* sp. BR60 (Fig. 10.5a; Wyndham *et al.* 1988; Nakatsu *et al.* 1991; Nakatsu & Wyndham 1993). This organism was previously isolated from surface runoff waters near the Hyde Park site; however, it was not known whether the *cbaABC* genes it contains were significant in the ongoing bioremediation efforts at Hyde Park.

Two approaches were used to track the *cbaABC* genes in Hyde Park samples. The first involved polymerase chain reaction amplification of target sequences within the *cba* genes of Tn5271 (Fig. 10.5a). This approach revealed similar sequences in the groundwater community and in samples from the sequencing batch reactors treating that groundwater. Quantitative PCR indicated that *cba* sequences were carried in a large subset of the microbial community functional in the sequencing batch reactors (Wyndham *et al.* 1994). This result indicates that the metabolic pathway encoded by the *cba* genes of Tn5271 is important in the overall conversion of chlorobenzoates to biomass, Cl$^-$ and CO$_2$ in the bioremediation system at Hyde Park. This pathway has been described elsewhere (Nakatsu & Wyndham 1993) and is quite different from the pathway through chlorocatechol intermediates that is used by most fluorescent pseudomonads and *Alcaligenes eutrophus* strains. The *cba* encoded catabolic pathway proceeds through protocatechuate or chloroprotocatechuate intermediates.

The second approach, as for the PAH community described above, was to isolate and characterize culturable bacteria. Sequencing batch reactor SBR2 samples were sonicated to disperse

clumped cells and yield an optimum dilution plate count. The suspended cells were diluted in sterile water and spread on CPS agar for total counts (Jones 1970; Fulthorpe & Wyndham 1992). Selection for growth on various defined carbon sources was carried out on minimal medium supplemented with 4 mM of 2-, 3- or 4-chlorobenzoate, or in the presence of toluene saturated air. Isolates from these plates were characterized by Gram staining, oxidase reaction, catalase reaction and production of fluorescent pigments by established procedures (King *et al.* 1954; Smibert & Krieg 1981). Selected isolates were grown on CPS agar for 18 h at 32 °C and inoculated into Biolog GN Microplates (Biolog Inc., Hayward Calif.) for rapid identification according to the manufacturer's instructions. Microplate substrate utilization profiles were used to group isolates by single linkage cluster analysis as described for the PAH community above. Total genomic DNA was also prepared from each isolate and was probed with a digoxigenin labelled DNA probe prepared from the *cbaABC* genes of Tn5271 in *Alcaligenes* sp. BR60.

Phenotypic and genotypic characterization of isolates from the SBR2 reactor indicated that two clusters, *Alcaligenes faecalis* Type 2 and CDC Group IVC-2, contained all the *cbaABC* probe positive isolates from the community. These isolates were capable of growth on 3- and 4-chlorobenzoate, but not 2-chlorobenzoate or toluene. These two phylogenetic groups fall within the ß-subclass of the Proteobacteria, as shown in Fig. 10.5b. In contrast, most of the toluene degrading isolates from the SBR2 sample clustered within the fluorescent *Pseudomonas* group of the γ subclass of the Proteobacteria. Interestingly, the bacterial hosts of the *cbaABC* genes found in the SBR2 reactor were phenotypically distinct from the bacteria found to be effective recipients of these genes in freshwater microcosms (Fulthorpe & Wyndham 1992). The latter were characterized following horizontal transfer of pBRC60 (Tn5271) in pristine freshwater microcosms. These differences most likely reflect the different physical and chemical environments of the two systems.

The phenotypic characterization of isolates from the SBR2 reactor revealed some important properties of the community as a whole. For example, isolates capable of degrading

(a)

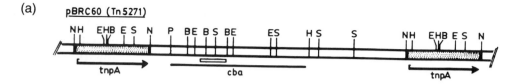

(b)

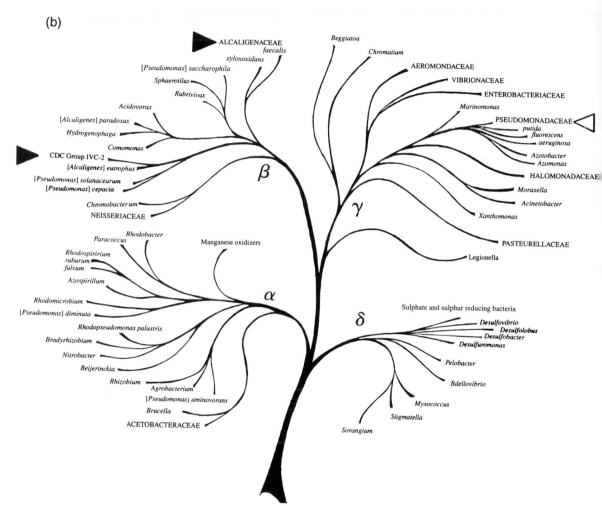

Fig. 10.5. Transposon Tn5271 distribution in the Hyde Park wastewater consortium. (a) A restriction enzyme digest map of the transposon as it exists on the IncPß plasmid pBRC60. The transposon is 17 kb in length and is flanked by direct repeats of the insertion sequence IS1071. The insertion sequences are indicated by the stippling and the *tnpA* (transposase) open reading frame marked by arrows. The non-repeated protion of the transposon carries *cba* genes in the region marked. These genes encode a dioxygenase activity that initiates the degradation of 3- and 3,4-dichlorobenzoate, contaminants found at the Hyde Park site. The open box within the *cba* region of Tn5271 shows the location of the PCR target.
(b) An illustration of the phylogenetic distribution of various isolates from the Hyde Park wastewater consortium within the class Proteobacteria. The α-, ß-, Γ- and ∂-subclasses within the Proteobacteria are shown, with a number of families, genera and species representing these subclasses shown for reference. The black triangles show the distribution of Tn5271-probe positive isolates (49 in all) recovered from the wastewater consortium. The open triangle shows the position of toluene degrading isolates from the consortium (4 in all).

2-chlorobenzoate did not degrade other isomers. These observations support previous work showing that isolates capable of degrading *ortho*-substituted aromatics like 2-chlorobenzoate usually cannot degrade *meta*- or *para*-substituted substrates (Sylvestre *et al.* 1989). An exception to this was recently reported in *Pseudomonas aeruginosa* JB2, which apparently carries two alternative chlorobenzoate dioxygenases (Hickey & Focht 1990). In our cluster analysis, 2-chlorobenzoate degraders were phenotypically diverse and distinct from the two clusters of 3- and 4-chlorobenzoate degraders.

The data for the structure of the SBR2 community also support the general observation that the metabolism of chloro- and methyl-substituted aromatics are mutually exclusive unless specific criteria for substituent positioning are met (Schmidt *et al.* 1985; Higson & Focht 1992). In our study, all 3- and 4-chlorobenzoate degraders were found to belong to the *Alcaligenes faecalis* Type II or CDC Group IVC-2 clusters, clearly differentiated from the toluene degrading fluorescent *Pseudomonas* cluster. Interestingly, in a study of chlorobenzoate degradation by natural and genetically engineered strains in a laboratory scale activated sludge unit, McClure *et al.* (1991) isolated an indigenous 3-chlorobenzoate degrading organism that was an effective survivor in their system. The isolate, designated AS2, was identified as a CDC Group IVC 2 strain. It contained a plasmid, pQM300, which conferred 3-chlorobenzoate degradation on a *Pseudomonas putida* transconjugant. The plasmid lacked homology to the *tfdC* and *tfdD* genes of pJP4 encoding the essential steps in chlorocatechol metabolism. The relationship of this CDC Group IVC-2 isolate and its plasmid to the cluster of similar isolates recovered from the Occidental sequencing batch reactors would be worth investigating.

This investigation has begun to show the detailed structure of a microbial community involved in bioremediation. It represents only a small subset of the genotypes likely to be involved in contaminant biodegradation, and yet it has already revealed some community organization. We are currently trying to determine why certain phylogenetic groups are effective at expressing specific genotypes like *cbaABC* and surviving in the SBR environment. We are also interested in the partitioning of resources by the ß- and Γ-subclasses of the Proteobacteria within a structured community like this. Recently, Wagner *et al.* (1993) have applied rRNA phylogenetic probes specific for the α-, ß- and Γ-subclasses of the Proteobacteria to activated sludge samples. This work showed that for activated sludge communities, phylogenetic subclasses as a whole respond differently to varying nutrient loadings. Nutrient enrichment strongly favoured the Γ-subclass at the expense of the ß-subclass, explaining why Γ-subclass bacteria tended to dominate in plate counts of culturable bacteria from the samples. The *in situ* hybridization approach, combined with analyses of the phylogenetic distribution of specific catabolic elements like Tn5271, will eventually provide a much clearer understanding of the factors that structure the microbial community in this wastewater treatment system.

Anaerobic biofilms in wastewater treatment

Nowhere are there more complex interactions between microbial populations than in anaerobic biofilms growing on complex organic wastes. These biofilms depend for continued energy transduction and nutrient cycling on the transfer of low molecular weight metabolites between populations. The most important of these metabolites, from the perspective of energy transduction, is hydrogen. However, acetate, formate and other volatile fatty acids are also critical intermediates.

In order to promote efficient growth and activity within anaerobic biofilms, various solid support materials and reactor designs have been evaluated. Achieving a large biomass concentration is the key to high rate performance of any biological system. A high solids retention time results in high biomass concentrations and so allows slowly growing microorganisms to maintain themselves within the reactor. Second generation, high rate anaerobic processes employ configurations and operational modes that embody these fundamental concepts of anaerobic biological treatment and inherently favour some significant secondary features of anaerobic metabolism.

In the following part of this chapter the start-up and operational phases of various anaerobic treatment systems for industrial wastes will be discussed. Many of the examples presented deal with a particular design, the downflow stationary fixed film reactor. The importance of structured biofilms in the efficient operation of these systems will be emphasized.

The start-up phase in anaerobic treatment systems

Start-up of anaerobic reactors is generally time consuming because of the slow growth rates of acetic acid converting methanogens. The establishment of these organisms and other methanogens in a fixed film, acclimatization of the community to the waste and finally, establishment of a biofilm of sufficient thickness to withstand shock loadings, each add to the time required to reach equilibrium.

There are few systematic studies on the start-up phase of anaerobic reactors. Four factors effect this start-up phase.

1. Quality of the inoculum in terms of the concentration of slow growing methanogenic bacteria acclimatized to a particular waste.
2. Rate of acclimatization of these microorganisms to the new waste.
3. Rate of growth of anaerobic microorganisms.
4. Rate of loss of anaerobic microorganisms, especially the slow growing methanogens.

The attachment and growth of the biofilm are the key factors for the start-up of most fixed film reactors. As an illustration, Fig. 10.6 is a plot of the distribution of acetoclastic methanogenic activity between the biofilm and the mixed liquor of a laboratory scale, downflow stationary fixed film reactor during start-up. As start-up progresses and hydraulic retention times become shorter the proportion of methanogenic activity associated with the biofilm increases while the portion associated with the mixed liquor decreases (Kennedy & Droste 1985). These data augment the importance of the reactor surface area to volume ratio and the surface characteristics of the biofilm support materials.

To enhance the start-up dynamics, Canovas-Diaz & Howell (1987) suggest promoting the growth of methanogenic and acetogenic H_2 producers by feeding a synthetic volatile fatty acid mixture supplemented with Ni, Co and Fe. Once

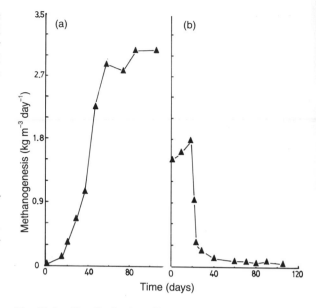

Fig. 10.6. The distribution of acetoclastic methanogenic acitivity during the start-up phase of an anaerobic, downflow stationary fixed film bioreactor.
(a) The activity in the biofilm. (b) The activity in the mixed liquor flowing through the reactor. The reactor was receiving a 10 g L^{-1} sucrose influent.

the methanogens have formed a biofilm, the reactor is then shifted to the more complex wastewater. Improved start up may be in part attributable to a higher rate of attachment and biofilm formation by methanogens. Attachment rates of mixed anaerobic cultures and enriched acetoclastic methanogens to similar support materials have been reported. Preliminary experiments conducted in the authors' laboratories indicate that methanogens alone have a higher attachment rate than mixed cultures, and that biofilm yields were higher and rates of biofilm accumulation faster (without loss in specific activity) for treatment of lower strength, nutritionally balanced wastewaters (Kennedy & Droste 1985). Start-up can best be accomplished by feeding a dilute wastewater (5000 mg organic carbon per L), maintaining volatile fatty acids between 200 and 1500 mg L^{-1} while increasing the flow rate to the reactor in 5–15% increments. High volatile fatty acid levels encourage the growth in the reactor of *Methanosarcina* species that are desirable for increasing the rate of acetoclastic methanogenesis that often limits carbon turnover rates in anaerobic biofilms.

Comparison of fixed film and suspended growth systems

Downflow stationary fixed film reactors maintain high rates of organic carbon removal along with high rates of methane production. These properties are also shared by various other anaerobic wastewater treatment system designs. For example, designs based on suspended growth such as the upflow anaerobic sludge blanket reactors are potentially capable of higher loading rates and sometimes higher organic carbon removal than the fixed film systems. The reason for this is that suspended growth systems, when operating efficiently, can retain a higher biomass concentration. Both systems depend upon the development of anaerobic biofilms, in the one case in the form of suspended pellets and in the other a fixed film. However, there are several advantages to fixed film systems.

1. Fixed film systems can treat concentrated wastes directly without dilution or recycling of the effluent.
2. A related observation is that fixed film systems can tolerate both hydraulic overloads and organic shockloads better than suspended growth systems.
3. Finally, downflow fixed film reactors are less sensitive to particulate matter in the waste stream than suspended growth systems.

To illustrate these points we will look at the treatment of two wastewaters by fixed film and suspended growth systems and contrast these. Richardson *et al.* (1991) examined the effects of the toxic components of chemi-thermomechanical pulp wastewaters on anaerobic bacteria. They found that up to 80% of the toxicity, in terms of inhibition of acetoclastic methanogenic activity, was localized in the suspended fines fraction, while only 20% was attributable to soluble components. During the treatment of the pulp wastewaters in two stage anaerobic sludge blanket reactors, the suspended fines caused a reduction in organic carbon removal, but this loss of performance was not as severe as predicted by batch tests. However, the fines caused a significant change in the appearance and composition of the biofilm in the reactors. If fines were removed by prefiltering the wastewater, the anaerobic sludge blanket developed normally as a granule suspension, black in colour, with good settling characteristics. In the presence of suspended particulate fines in the wastewater, initially these passed

through the sludge bed and accumulated on top of the granules, significantly increasing the bed volume and being washed out at intervals in the outflow. However, as the system continued to operate, fines accumulated within the granular sludge bed and began to coat the granules themselves. The appearance of the granular biofilm changed from black to pink and the granules increasingly washed out of the system.

Clearly, for this particular waste stream, design parameters must take into account not only the biological structure of the biofilm, but its tendency to bind and incorporate particulate matter. If sludge blanket granules increase in flotation due to changes in density brought on by the binding of suspended particulates, or by the entraining of gases, then the flow of wastewater must be reduced (or the hydraulic retention time increased) in order to save the sludge blanket from washout. These changes may offset any advantages due to the higher biomass density of these systems compared to fixed film systems.

Upflow anaerobic sludge blanket reactors subject to washout have been redesigned slightly to incorporate surface retainers of various types. For example, the upflow blanket filter reactor uses a large number of small 'biorings' as packing media that float in the upper portion of the sludge blanket. These serve to retain the sludge, act as gas solid separators, and reduce channelling or short-circuiting. In a direct comparison of the efficiency of downflow stationary fixed film and upflow blanket filter reactors for treating municipal landfill leachate, Kennedy *et al.* (1988) found that both systems successfully removed over 90% of the organic carbon from the waste stream at loading rates of 14.7 kg m^{-3} per day (hydraulic retention time, 1.5 days). Biomass yields for the two reactors were essentially the same (*c.* 0.05–0.08 kg biomass per kg organic carbon removed). However, an inorganic, heavy metal precipitate formed in both reactors. In the upflow blanket filter reactor, the precipitate was concentrated in the sludge layer and the top filter media, while in the downflow stationary fixed film system the metal precipitate coated the biofilm. In both systems the metal precipitates limited the organic loading rates and reduced the efficiency of the treatment processes.

These examples serve to illustrate limitations that can occur in the operation of anaerobic treatment systems, and the need to understand the chemical and biological processes that control biofilm development, structure, activity, and inactivation in real wastewater streams. In addition, this overview of both aerobic and anaerobic processes illustrates two quite different approaches to the study of wastewater treatment biofilms. The reductionist microbiological and genetic approach contrasts strongly with the operational, design approach. However, these two approaches are rapidly converging on a thorough understanding of the ecology and dynamics of industrial biofilms.

Acknowledgements

The research describes in this review was supported by the Natural Sciences and Engineering Research Council of Canada, the National Research Council of Canada and by a NSERC Cooperative Research and Development Grant with Esso Petroleum Canada.

References

Canovas-Diaz, M. & Howell, J. A. (1987). Stratified ecology techniques in the start up of an anaerobic downflow fixed film percolating reactor. *Biotechnology and Engineering*, **30**, 289–96.

Dwyer, D. F., Rojo, F. & Timmis, K. N. (1988). Fate and behaviour in an activated sludge microcosm of a genetically engineered microorganism designed to degrade substituted aromatic compounds. In *The release of genetically engineered microorganisms*, ed. M. Sussman, C. H. Collins, F. A. Skinner & D. E. Stewart-Tull, pp. 77–88. London: Academic Press.

EPA (1993). Update on the Bioremediation Field Initiative. In *Bioremediation in the field.* United States Environmental Protection Agency, Office of Research and Development EPA/540/N-93/001 No. 8, May 1993, pp. 1–39. Washington, DC.

Foght, J. M. & Westlake, D. W. S. (1988). Degradation of polycyclic aromatic hydrocarbons and aromatic heterocycles by a *Pseudomonas* species. *Canadian Journal of Microbiology*, **34**, 1135–41.

Fulthorpe, R. R. & Wyndham, R. C. (1992). Involvement of a chlorobenzoate catabolic transposon, Tn5271, in community adaptation to chlorobiphenyl, chloroaniline and 2,4-dichlorophenoxyacetic acid in a freshwater ecosystem. *Applied and Environmental Microbiology*, **58**, 314–25.

Hickey, W. J. & Focht, D. D. (1990). Degradation of mono-, di-, and trihalogenated benzoic acids by *Pseudomonas aeruginosa* JB2. *Applied and Environmental Microbiology*, **56**, 3842–50.

Higson, F. K. & Focht, D. D. (1992). Utilization of 3-chloro-2-methylbenzoic acid by *Pseudomonas cepacia* MB2 through the *meta* fission pathway. *Applied and Environmental Microbiology*, **58**, 2501–4.

Jones, J. G. (1970). Studies on freshwater bacteria: Effect of medium composition and method on estimates of bacterial population. *Journal of Applied Bacteriology*, **33**, 679–86.

Kennedy, K. J. & Droste, R. L. (1985). Start up of anaerobic downflow stationary fixed film (DSFF) reactors. *Biotechnology and Engineering*, **27**, 1152–65.

Kennedy, K. J., Hamoda, M. F. & Guiot, S. G. (1988). Anaerobic treatment of leachate using fixed film and sludge bed systems. *Journal of the Water Pollution Control Federation*, **60**, 1675–83.

King, E. O., Ward, M. K. & Raney, D. E. (1954). Two simple media for the demonstration of pyocin and fluorescin. *Journal of Laboratory and Clinical Medicine*, **44**, 301.

Kiyohara, H., Nagao, K. & Yana, K. (1982). Rapid screen for bacteria degrading water-insoluble solid hydrocarbons on agar plates. *Applied and Environmental Microbiology*, **43**, 454–7.

McClure, N. C., Fry, J. C. & Weightman, A. J. (1991). Survival and catabolic activity of natural and genetically engineered bacteria in a laboratory scale activated-sludge unit. *Applied and Environmental Microbiology*, **57**, 366–73.

McClure, N. C., Weightman, A. J. & Fry, J. C. (1989). Survival of *Pseudomonas putida* UWC1 containing cloned catabolic genes in a model activated sludge unit. *Applied and Environmental Microbiology*, **55**, 2627–34.

Monticello, D.J., Bakker, D. & Finnerty, W.R. (1985). Plasmid-mediated degradation of dibenzothiophene by *Pseudomonas* species. *Applied and Environmental Microbiology*, **49**, 756–60.

Nakatsu, C., Ng, J., Singh, R., Straus, N. & Wyndham, C. (1991). Chlorobenzoate catabolic transposon Tn5271 is a composite class I element with flanking class II insertion sequences. *Proceedings of the National Academy of Sciences USA*, **88**, 8312–16.

Nakatsu, C. H. & Wyndham, R. C. (1993). Cloning and expression of the transposable chlorobenzoate-3,4-dioxygenase genes of *Alcaligenes* sp. BR60. *Applied and Environmental Microbiology*, **59**, 3625–33.

Richardson, D. A., Andras, E. & Kennedy, K. J. (1991). Anaerobic toxicity of fines in chemi-thermomechanical pulp wastewaters: a batch assay reactor study comparison. *Water Science and Technology*, **24**, 102–12.

Schmidt, E., Bartels, I. & Knackmuss, H.-J. (1985).

Degradation of 3-chlorobenzoate by benzoate or 3-methylbenzoate-utilizing cultures. *FEMS Microbiology Ecology*, **31**, 381–9.

Selucky, M.L., Chu, Y., Ruo, T. C. S. & Strausz, O. P. (1978). Chemical composition of Cold Lake bitumen. *Fuel*, **57**, 9–16.

Shiaris, M.P. & Cooney, J. J. (1983). Replica plating method for estimating phenanthrene-utilizing and phenanthrene cometabolizing microorganisms. *Applied and Environmental Microbiology*, **45**, 706–10.

Singer, J.T. & Finnerty, W.R. (1984). Genetics of hydrocarbon utilizing microorganisms. In *Petroleum microbiology*, ed. R. M. Atlas, pp. 299–354. New York: Macmillan.

Smibert, R. M. & Krieg, N. R. (1981). General characterization. In *Manual of methods for general bacteriology*, ed. P. Gerhardt, pp. 409–43. Washington, DC: American Society for Microbiology.

Sylvestre, M., Mailhiot, K., Ahmad, D. & Massé, R. (1989). Isolation and preliminary characterization of a 2-chlorobenzoate degrading *Pseudomonas*. *Canadian Journal of Microbiology*, **35**, 439–43.

Wagner, M., Amann, R., Lemmer, H. & Schleifer, K.-H. (1993). Probing activated sludge with oligonucleotides specific for Proteobacteria: inadequacy of culture-dependent methods for describing microbial community structure. *Applied and Environmental Microbiology*, **59**, 1520–5.

Wyndham, R. C. & Costerton, J. W. (1981). *In vitro* microbial degradation of bituminous hydrocarbons and *in situ* colonization of bitumen surfaces within the Athabasca oil sands deposit. *Applied and Environmental Microbiology*, **41**, 791–800.

Wyndham, R. C., Nakatsu, C., Peel, M., Cashore, A. Ng, J. & Szilagyi, F. (1994). Distribution and structure of the catabolic transposon Tn5271 in a groundwater bioremediation system. *Applied and Environmental Microbiology*, **60**, 86–93.

Wyndham, R. C., Singh, R. K. & Straus, N. A. (1988). Catabolic instability, plasmid gene deletion and recombination in *Alcaligenes* sp. BR60. *Archives of Microbiology*, **150**, 237–43.

Ying, W.-C., Wnukowski, J., Wilde, D. & McLeod, D. (1992). Successful leachate treatment in SBR-adsorption system. In *Proceedings of the 47th Industrial Waste Conference*, pp. 501–18. Chelsea, Mich: Lewis Publishers.

11

Heterogeneous Mosaic Biofilm – A Haven for Waterborne Pathogens

James T. Walker, Craig W. Mackerness, Julie Rogers and
C. William Keevil

Waterborne microorganisms

The supply of good quality potable water has been of concern for many years and is of paramount importance for public health. Between 1911 and 1937 there were 20 outbreaks of waterborne disease in the United Kingdom, with 80% of these being due to enteric fever, and the remainder to dysentery and gastroenteritis. The implication that the water supply was the source of a large outbreak of typhoid fever in Croydon, Surrey in 1937, led to the chlorination of public water supplies. This treatment dramatically improved the quality of water supplied to the public and produced improvements in health. Despite chlorination there have been 21 incidents of disease associated with public water supplies since 1937, one of which was due to paratyphoid (George *et al*. 1972).

The importance of bacterial quality as a measure of water quality has been recognized for many years. Of 495 colonies identified from the water phase, 448 were Gram negative bacteria such as coliforms, *Alcaligenes* spp. and *Pseudomonas* spp. (Shannon & Wallace 1944). A diverse range of microorganisms, including *Pseudomonas*, *Flavobacterium*, *Achromobacter*, *Klebsiella*, *Bacillus*, *Corynebacterium*, *Mycobacterium*, *Spirillium*, *Clostridium*, *Arthrobacter*, *Gallionella* and *Leptothrix* spp., is present in potable drinking water (Geldreich *et al*. 1972), but it is generally considered that these bacteria are not harmful. However, *Flavobacterium* spp. and *Pseudomonas* spp. are known opportunistic pathogens, and *P. aeruginosa* is the major cause of hospital acquired infections (Favero *et al*. 1971). A total of 15 outbreaks of waterborne disease were identified in the UK between 1977 and 1986 (Galbraith *et al*. 1987), with *Campylobacter enteritis* and viral gastroenteritis accounting for 66% of outbreaks. More recently there has been a shift from enteritis to cryptosporidiosis, which accounted for 64% of the 11 outbreaks in the UK between 1987 and 1990 (Stanwell-Smith 1991). In the USA there were 224 outbreaks of waterborne disease, affecting 48 193 individuals, between 1971 and 1978 (Craun 1981).

More active surveillance in the UK has led to an apparent increase in the number of outbreaks in recent years; however, it is likely that there is still an underestimation of cases. In a report published in 1973, it was estimated that only 50% of waterborne outbreaks in municipal water systems were reported (Craun & McCabe 1973). This can be partially attributed to the criteria used to determine whether infections warrant investigation. At least two epidemiologically linked cases of disease must be reported before an outbreak is declared, a common source noted and further investigations undertaken. Since only 45% of cases had an identifiable aetiological agent, the criteria for investigation excluded 55% of all cases. The most commonly identified bacteria were *Salmonella* and *Shigella*, since the culture techniques were specific for these organisms and the most commonly identified pathogen was *Giardia lamblia*, a flagellated protozoan responsible for giardiasis.

Not all disease outbreaks are due to bacterial contamination of the water supply. Chemical poisonings accounted for 10% of outbreaks investigated in one study (Craun *et al.* 1976), four due to acute copper poisoning with copper concentrations of 4, 12.5, 38.5 and 80 mg L^{-1} in the water. The leaching of copper from water distribution systems was attributed to the low pH of the supply water which was naturally aggressive; an interruption of pH adjustment of this water and the release of CO_2 caused by a defective check valve on a drinks machine leading to increased dissolution of the copper. Fortunately, acute gastroenteritis due to chemical poisoning has also been rare in the UK. The cases reported included metallic poisoning due to copper and zinc leachate (Anon. 1986), poisonings due to phenol (Jarvis *et al.* 1985) and poisonings due to aluminium sulphate (Stanwell-Smith 1991), the latter two resulting from industrial accidents.

Potential pathogens in drinking water

Untreated water contains primary pathogens (e.g. *Vibrio fulvis*) and opportunistic pathogens such as *Aeromonas* sp., *Acinetobacter* spp., *Bacillus cereus*, *Flavobacteria* spp., *Moraxella* sp. and *Pseudomonas* spp. (Payment *et al.* 1988). Although water treatment is intended to remove coliforms, *Aeromonas* spp. and *P. aeruginosa* can be cultured from sampling sites downstream of the filtration plant (Payment *et al.* 1988). This can be explained by bacterial regrowth in distribution systems and problems encountered if indicator organisms cannot initially be cultured. The numbers of indicator organisms may be underestimated because of poor detection in the high background of heterotrophic bacteria or due to sublethal injury. This latter phenomenon was established when it was noted that the numbers of coliforms from water containing chlorine were consistently higher by the multiple tube fermentation–most probable number (MPN) method than by a membrane filtration procedure (Payment *et al.* 1988). It is thought that more than 90% of indicator bacteria present in water systems may become injured in less than 1 week (Bissonnette *et al.* 1975). Chlorination of water systems results in an inability to culture *Escherichia coli* from water samples, but by varying the diluent composition, temperature and time of exposure, and media increased recovery is possible (McFeters *et al.* 1984). The cell envelope is a major location of cell damage of water-injured coliforms, which may result in an increased sensitivity to any bacteriostatic or bactericidal compounds in the media. Of 16 different media normally used to culture bacteria from water, all were found to be suppressive and the 11 of those which contained deoxycholate or bile salts were also found to suppress injured cells (McFeters *et al.* 1986).

In the study of LeChevalier *et al.* (1987) it was concluded that competition between heterotrophic bacteria and coliform bacteria for assimilable organic carbon (AOC) was probably responsible for the decline of indicator organisms. It is generally accepted that the AOC plays a central role in the growth of microorganisms in distribution system biofilms. Previous experiments have shown that heterotrophic bacteria are well adapted to compete with coliforms for low levels of nutrients (LeChevallier & McFeters 1985a, b). Although the presence of heterotrophs has been hypothesized to predict the presence of total coliforms, there have been few documented reports, but there is a common belief that an increase in heterotrophic plate counts precedes an increase in total coliforms. Olson (1982) reported isolating *E. coli* from pipeline biofilms, and noted that it was not possible to assess the health significance of faecal coliforms originating from distribution system biofilms.

Generally, it is considered that microorganisms can be eradicated from water by approximately 0.3 mg L^{-1} residual chlorine, but even with remedial actions using 4.3 mg L^{-1} of free chlorine residual coliforms may persist (LeChevallier *et al.* 1984). Ridgeway & Olson (1982) reported that free residual chlorine is an extremely potent bactericidal agent at concentrations of less than 0.1 mg L^{-1} but found that some Gram positive spore forming bacteria were able to survive 2 min exposure to 10 mg L^{-1} of free chlorine. Such resistance is an advantageous survival mechanism to particular bacteria. Dead end pipe sections and air chambers within water systems make treatment difficult and also provide a source of microorganisms to reinoculate the treated section.

Microorganisms are often associated with the surfaces of the distribution systems. These bacteria can directly influence water quality by entering the bulk phase or may indirectly influence the quality by biodegrading the materials, possibly leading to taste or odour problems. Mains water may become discoloured and flow reduced through pipes due to the growth of microorganisms, including *Leptothrix* and *Gallionella* (Ridgeway *et al.* 1981). Another problem associated with biofilm formation is microbially induced corrosion, for example, pipelines in the marine environment (Hamilton 1985).

Pathogens in other man made water systems

Legionella pneumophila is the aetiological agent responsible for legionellosis. Although *Legionella* is ubiquitous in environmental water samples (Fliermans *et al.* 1979, 1981), it is often responsible for infection when present in man-made water systems. These systems, including potable water, storage tanks and cooling towers, have all been shown to contain legionellae (Dondero *et al.* 1980; Colbourne *et al.* 1984; Stout *et al.* 1985; Bhopal & Barr 1991). Infection results from the inhalation of aerosol particles of less than 5 μm in diameter which contain the bacterium (Baskerville *et al.* 1981). Legionellosis is manifest in two forms, Legionnaires' disease (which is the most serious and may be fatal) and Pontiac fever, which is a self-limiting illness with flu-like symptoms.

Many cases of legionellosis occur in hospitals, with hot water systems being implicated as the reservoir for *L. pneumophila* (Bartlett *et al.* 1983). Over the last decade hospital hot water systems have reduced operating temperatures to between 40 and 45 °C in order to reduce energy costs and to prevent scalding of patients. In laboratory experiments the numbers of *L. pneumophila* rapidly declined at temperatures greater than 42 °C indicating that the bacterium is unable to tolerate high operating temperatures (Wadowsky *et al.* 1982). However, *L. pneumophila* has been recovered from hot water circuits in hospitals and hotels with temperatures greater than 42 °C (Bartlett *et al.* 1983). Plouffe *et al.* (1983) were also able to recover *L. pneumophila* from water storage tanks at 43 to 45 °C but *L. pneumophila* was not detectable in systems operating between 58 and 60 °C.

The persistence of *L. pneumophila* in these systems can be aided by the presence of cool spots in hot water circuits (Ezzidine *et al.* 1989) or at the base of electric calorifiers where the temperatures may be lower (Ciesielski *et al.* 1984). In addition, *L. pneumophila* can survive within amoebae (Anand *et al.* 1983) which can protect the bacteria from chlorination (Kuchta *et al.* 1983). *L. pneumophila* has also been shown to occur in biofilms on the surfaces of materials in water systems (Colbourne *et al.* 1988), and biocides has been shown to be less effective against *L. pneumophila* in biofilms than against those in the planktonic phase (Wright *et al.* 1991).

Continuous culture biofilm model

Any laboratory model used to determine factors of importance within a natural ecosystem must have conditions which are representative of the environment intended to be modelled. Chemostat modelling of the aquatic ecosystem offers an holistic approach to the study, where environmental conditions can be closely controlled and reproducible. In our laboratory, continuous cultures have been used extensively to produce biofilms under defined conditions. The use of a two stage continuous culture model ensures that the initial inoculum is conserved and allows test conditions to be modified in the second vessel (Fig. 11.1).

The model systems have been inoculated with mixed populations of microorganisms which were concentrated by filtration from an appropriate source for the particular investigation. Hence, in modelling drinking water systems the inoculum was taken from drinking water in London (Mackerness *et al.* 1989); in modelling *Legionella* ecology the inoculum was taken from a calorifier associated with an outbreak of Legionnaires' disease (Rogers *et al.* 1989) and for investigations involving microbially induced corrosion the inoculum was from a building where the corrosion occurred (Walker *et al.* 1991).

Growth of the mixed population of microorganisms was achieved using unsupplemented, filter sterilized water as the sole source of nutrients. The models were assembled using inert materials such as glass, titanium and silicone to prevent metals or exogenous carbon sources from influencing microbial growth. Two continuous culture vessels (500 ml volume) were assembled in series, with the first vessel being used to provide a constant inoculum to the second vessel where biofilms were generated. The first vessel was maintained at a dilution rate of 0.05 h^{-1} and effluent from this vessel was supplied to the second vessel which received additional water

Table 11.1. Identifiable bacteria of the inoculum and the model

Inoculum	Biofilm
Flavobacterium spp.	*Flavobacterium* spp.
Alcaligenes spp.	*Alcaligenes latus*
Alcaligenes spp.	*A. xyloxidans*
Acinetobacter spp.	*Acinetobacter* spp.
Pseudomonas spp.	*Pseudomonas fluorescens*
	P. paucimoblis
	P. vesicularis
	P. stutzeri
	P. corrugata

giving an overall dilution rate of $0.2 \, \mathrm{h}^{-1}$. Temperature, oxygen tension and media addition were controlled using process control modules (Brighton Systems, UK). Biofilms were generated on $1 \, \mathrm{cm}^2$ tiles of suitable plumbing material suspended from silicone bungs by titanium wire, which were degreased in acetone before autoclaving.

Growth of microorganisms in the drinking water model

The diverse range of microorganisms present in the inoculum was maintained in a continuous culture in the model system, with the predominant microorganisms being identified to genus level (Table 11.1). This indicates that ordinary tap water contains sufficient nutrients to support the growth of heterotrophic microorganisms. Since microbial diversity was maintained in the model and since similar bacteria can be found in field investigations, the model system can be regarded as being representative of a distribution system.

Biofilms were developed on the surface of glass and bitumen painted mild steel surfaces. Glass tiles were used as control surfaces and bitumen painted mild steel was used since it is the approved material for use in distribution mains. These surfaces were found to be similarly colonized in terms of both numbers of microorganisms (Table 11.2) and microbial diversity. The majority of the bacterial population in the biofilms were pseudomonads, particularly *P. paucimobilis* and *P. fluorescens*.

The second chemostat vessel, which had a higher dilution rate, also had higher counts for planktonic and biofilm bacteria when compared with the first vessel, which had the lower dilution rate (Table 11.3). These higher counts in the second vessel are possibly due to increased nutrient availability at the high dilution rate, or due to

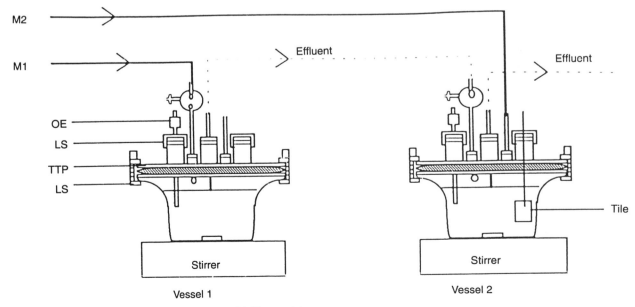

Fig. 11.1. Schematic diagram of the laboratory biofilm model.

Table 11.2. Effect of dilution rate on the heterotrophic bacteria of the biofilm and planktonic phases

Time (days)	Biofilm	Dilution rate 0.2 h⁻¹ Planktonic	0.05 h⁻¹ Biofilm	Planktonic
1	6.3	6.7	4.8	5.7
7	6.0	6.4	5.7	5.7
14	6.2	6.2	5.8	5.6
21	6.6	6.0	5.8	5.0

Viable counts are expressed as log10 CFU cm^{-2} for biofilm bacteria and log10 CFU ml^{-1} for planktonic bacteria. Cultures were grown aerobically at 25 °C, and the biofilm was developed on glass tiles.

Table 11.3. Comparison of the viable heterotrophic planktonic and biofilm bacteria when biofilm was developed on glass or bitumen painted mild steel

Time (days)	Glass Biofilm	Planktonic	Bitumen-painted mild steel Biofilm	Planktonic
1	6.3	6.7	6.9	5.7
7	6.0	6.4	5.1	5.8
14	6.2	6.2	6.1	6.0
21	6.6	6.0	6.1	5.7

Viable counts are expressed as log10 CFU cm^{-2} for biofilm bacteria and log10 CFU ml^{-1} for planktonic bacteria. Cultures were grown aerobically at 25 °C, with the dilution rate of 0.2 h^{-1}, and biofilm was developed on bitumen painted mild steel.

Table 11.4. Colonization of a heterotrophic biofilm by *Aeromonas hydrophila* and *Escherichia coli*

Time (days)	Heterotrophs	*A. hydrophila*	*E. coli*
1	6.1	4.8	5.6
4	5.8	3.3	5.0
7	4.9	3.7	3.4
14	5.8	3.8	3.3
21	5.6	3.8	3.1

Viable counts are expressed as log10 CFU cm^{-2}.

of microorganisms to form a stable heterotrophic biofilm community.

A biofilm was established on the bitumen painted mild steel surfaces for 7 days and then the model was challenged with *Aeromonas hydrophila* and *E. coli*. Both of these microorganisms became incorporated into the biofilm community, and remained viable after a further 7 days (Table 11.4). The initial fluctuations in numbers after addition of the two microorganisms were attributed to the re-establishment of steady state conditions. This work has demonstrated that both *E. coli* and *A. hydrophila* can become members of the biofilm and planktonic communities and growth can be achieved using tap water as the sole source of nutrient. The survival of coliforms in the distribution system is dependent on the presence of a stable multispecies heterotrophic biofilm.

Biofilm heterogeneity

Biofilms present in water systems do not necessarily occur as confluent film over the surface. Microscope studies have shown that microorgan-

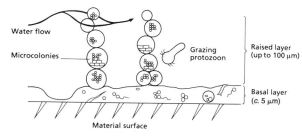

Fig. 11.2. Schematic diagram of the heterogeneic mosaic biofilm formed in the laboratory model.

the vessel being supplied by effluent from the first vessel which contains bacterial metabolites. Nonetheless, there were negligible fluctuations in the numbers of microorganisms in the planktonic and biofilm phases, which indicates that both phases are stable. The results also show that there was little difference between the numbers of microorganisms present in the planktonic and biofilm phases, indicating that no nutrient supplying or inhibitory materials were leaching from the bitumen paint. This would indicate that bitumen painted mild steel is a suitable material for use in distribution systems since it does not effect the overall quality of the water. However, the work has shown that bitumen painted mild steel surfaces are readily colonized by a diverse range

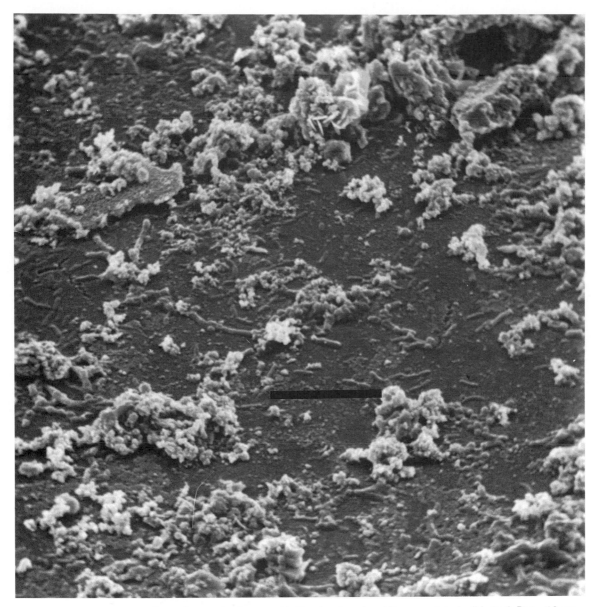

Fig. 11.3.   Scanning electron photomicrograph of mature biofilm developed on bitumen painted mild steel. Bar = 10 μm.

isms initially attach to surfaces and form small microcolonies (Fig. 11.2). After 14 days these microcolonies produce a 'film' over the surfaces (Fig. 11.3). The biofilm consists of a low background of microorganisms with tall stacks of microcolonies rising from the surfaces. This basal background is approximately 5 μm in depth and the stacks reach up to 100 μm in height (Rogers *et al.* 1989). When the biofilm is viewed unstained, under low power light microscopy it is evident that the microorganisms exist in microcolonies visible due to the colour of the biofilm (Fig. 11.4). The presence of microcolonies of *L. pneumophila* within the biofilm suggests that despite the complex nutritional requirements of the pathogen, the microbial consortium can sustain growth of this microorganism (Rogers & Keevil 1992). Additional evidence for the

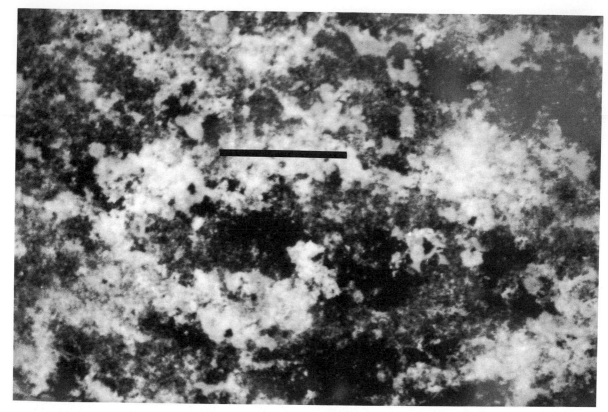

Fig. 11.4. Stereo light microscope of unstained heterogeneic mosaic biofilm. (From Julie Rogers (1993). The colonisation of potable water systems by *Legionella pneumophila*. PhD thesis.)

extracellular growth of *L. pneumophila* has been provided by operating the model systems at different temperatures, where *L. pneumophila* was demonstrated to grow in the absence of amoebae and that the biofilm protected the pathogen at high temperatures (Rogers *et al.* 1993).

References

Anand, C. M., Skinner, A. R., Malic, A. & Kurtz, J. B. (1983). Interaction of *Legionella pneumophilia* and a free living amoeba (*Acanthamoeba palestinansis*). *Journal of Hygiene*, **91**, 167–78.

Anonymous (1986). PHLS Communicable Disease Surveillance Centre. *Community Medicine*, **8**, 142.

Bartlett, C. L. R., Kurtz, J. B., Hutchinson, J. P. G., Turner, G. C. & Wright, A. E. (1983). Legionella in hospital and hotel water supplies. *Lancet*, **ii**, 1315.

Baskerville, A., Fitzgeorge, R. B., Broster, M., Hambleton, P. & Dennis, P. J. (1981). Experimental transmission of legionnaires' disease by exposure to aerosols of *Legionella pneumophila*. *Lancet*, **ii**, 1389–90.

Bhopal, R. S. & Barr, G. (1991). Are hot water and cold water systems in large premises maintained to prevent Legionnaires' disease? A preliminary answer. *Health and Hygiene*, **12**, 159–65.

Bissonette, G. K., Jezeski, J. J., McFeters, G. A. & Stuart, D. G. (1975). Influence of environmental stress on enumeration of indicator bacteria from natural waters. *Applied and Environmental Microbiology*, **29**, 186–94.

Ciesielski, C. A., Blaser, M. J., LaForce, M. & Lang, W.L. (1984). Role of stagnation and obstruction of water flow in isolation of *Legionella pneumophila* from hospital plumbing. In *Legionella, Proceedings of the 2nd International Symposium*, ed. C. Thornsberry, A. Balows, J. C. Feeley & W. Jakubowski, pp. 307–9. Washington, DC: American Society for Microbiology.

Colbourne, J. S., Pratt, D. J., Smith, M. G., Fisher-Hoch, S. P. & Harper, D. (1984). Water fittings as sources of *Legionella pneumophila* in a hospital plumbing system. *Lancet*, **i**, 210–13.

Colbourne, J. C., Trew, R. M. & Dennis, P. L. J. (1988). Treatment of water for aquatic bacterial growth studies. *Journal of Applied Bacteriology*, **65**, 79–85.

Craun, G. F. (1981). Outbreaks of waterborne disease in the United States: 1971–1978. *Journal of the American Water Works Association*, **70**, 360–9.

Craun, G. F. & McCabe, L. J. (1973). Review of the causes of waterborne disease outbreaks. *Journal of the American Water Works Association*, **65**, 74–84.

Craun, G. F., McCabe, L. J. & Hughes, J. M. (1976). Waterborne disease outbreaks in the US – 1971–1974. *Journal of the American Water Works Association*, **68**, 420–5.

Dondero, T. J., Roberts, M. D., Rentorff, C., Mallison, G. F., Weeks, R. M., Levy, J. S., Wong, E. W. & Schaffner, W. (1980). An outbreak of Legionnaires' disease associated with a contaminated air-conditioning cooling tower. *The New England Journal of Medicine*, **302**, 365–70.

Ezzedine, H., Van Ossel, C., Delmee, M. & Wauters, G. (1989). *Legionella* spp. in a hospital hot water system: effect of control measures. *Journal of Hospital Infection*, **13**, 121–31.

Favero, M. S., Carson, L. A., Bond, W. W. & Peterson, N. J. (1971). *Pseudomonas aeruginosa*: growth in distilled water from hospitals. *Science*, **173**, 836–7.

Fliermans, C. B., Cherry, B., Orrison, L. H., Smith, S. J., Tison, D. L. & Pope, D. H. (1981). Ecological niche of *Legionella pneumophila*. *Applied and Environmental Microbiology*, **41**, 9–16.

Fliermans, C. B., Cherry, W. B., Orrison, L. H. & Thacker, L. (1979). Isolation of *Legionella pneumophila* from non-epidemic related aquatic habitats. *Applied and Environmental Microbiology*, **37**, 1239–42.

Galbraith, N. S., Barrett, N. J. & Stanwell-Smith, R. (1987). Water and disease after Croydon: a review of water-borne and water-associated disease in the UK 1937–86. *Journal of the Institute of Water Environmental Management*, **1**, 7–21.

Geldreich, E. E., Nash, H. D., Reasoner, D. J. & Taylor, R. H. (1972). The necessity of controlling bacterial populations in potable water: community water supply. *Journal of the American Water Works Association*, **64**, 96.

George, J. T. A., Wallace, J. G., Morrison, H. R. & Harbourne, J. F. (1972). Paratyphoid in man and cattle. *British Medical Journal*, **3**, 208.

Hamilton, W. A. (1985). Sulphate-reducing bacteria and anaerobic corrosion. *Annual Reviews in Microbiology*, **39**, 195–217.

Jarvis, S. N., Straube, R. C., Williams, A. L. J. & Bartlett, C. L. R. (1985). Illness associated with contamination of drinking water supplies with phenol. *British Medical Journal*, **290**, 1800–2.

Kuchta, J. M., States, S. J., McNamara, A. M., Wadowsky, R. M. & Yee, R. M. (1983). Susceptability of *Legionella pneumophila* to chlorine in tap water. *Applied and Environmental Microbiology*, **46**, 1134–9.

LeChevallier, M. W., Babcock, T. M. & Lee, R. G. (1987). Examination and characterisation of distribution system biofilms. *Applied and Environmental Microbiology*, **53**, 2714–24.

LeChevallier, M. W., Hassenauer, T. S., Camper, A. L. & McFeters, G. A. (1984). Disinfection of bacteria attached to granular activated carbon. *Applied and Environmental Microbiology*, **48**, 918–23.

LeChevallier, M. W. & McFeters, G. A. (1985a). Enumerating injured coliforms in drinking water. *Journal of the American Water Works Association*, **77**, 81–7.

LeChevallier, M. W. & McFeters, G. A. (1985b). Interactions between heterotrophic plate count bacteria and coliform organisms. *Applied and Environmental Microbiology*, **49**, 1338–41.

Mackerness, C. W., Colbourne, J. C. & Keevil, C. W. (1991). Growth of *Aeromonas hydrophila* and *Escherichia coli* in a distribution system biofilm model. In *Proceedings of the UK Symposium on health related water microbiology*, ed. R. Morris, L. M. Alexander, P. Wyn-Jones & J. Sellwood, pp. 131–8. Glasgow: International Association on Water Pollution Research and Control.

McFeters, G. A., Kippin, J. S. & LeChevallier M. W. (1986). Injured coliforms in drinking water. *Applied and Environmental Microbiology*, **51**, 1–5.

McFeters, G. A., LeChevallier, M. W. & Domek, M. (1984). *Injury and improved recovery of coliform bacteria in drinking water*. EPA-600/2-84-166. Cincinnati: U.S. Environmental Protection Agency.

Olson, B. H. (1982). *Assessment and implications of bacterial regrowth in water distribution systems*. EPA-600/52-82-072. Cincinnati: US Environmental Protection Agency.

Payment, P., Gramade, F. & Paquette, G. (1988) Microbiological and virological analysis of water from two water filtration plants and their distribution systems. *Canadian Journal of Microbiology*, **34**, 1304–9.

Plouffe, J. F., Webster, L. R. & Hackman, B. (1983). Relationship between colonisation of hospital buildings with *Legionella pneumophila* and hot water temperatures. *Applied and Environmental Microbiology*, **46**, 769–70.

Ridgeway, H. F., Means, E. G. & Olson, B. H. (1981). Iron bacteria in drinking-water distribution systems: elemental analysis of *Gallionella* stalks, using X-ray energy-dispersive microanalysis. *Applied and Environmental Microbiology*, **41**, 288–97.

Ridgeway, H. F. & Olson, B. H. (1982). Chlorine resistance patterns of bacteria from two distribution systems. *Applied and Environmental Microbiology*, **44**, 972–87.

Rogers, J., Dennis, P. J., Lee, J. V. & Keevil, C. W. (1993). Effects of water chemistry and temperature on the survival and growth of *Legionella pheumophila* in potable water systems. In *Legionella: current status and emerging perspectives*, ed. J. M. Barbaree, R. F. Bieman & A. P. Dufour, pp. 248–50. Washington, DC: American Society of Microbiology.

Rogers, J. & Keevil, C. W. (1992). Immunogold and fluorescein immunolabelling of *Legionella pneumophila* within aquatic biofilms visualised using episcopic differential interference contrast microscopy. *Applied and Environmental Microbiology*, **58**, 2326–30.

Rogers, J., Lee, J. V., Dennis, P. J. & Keevil, C. W. (1989). Continuous culture biofilm model for the survival and growth of *Legionella pneumophila* and associated protozoa in potable water systems. In *Proceedings of the UK symposium on health related water microbiology*, ed. R. Morris, L. M. Alexander, P. Wyn-Jones & J. Sellwood, pp. 192–200. Glasgow: IAWPRC.

Shannon, A. M. & Wallace W. M. (1944). The bacteria in a distribution system. *Journal of American Water Works Association*, **36**, 1356–64.

Stanwell-Smith, R. (1991). Recent trends in the epidemiology of waterborne disease. In *Proceedings of the UK symposium on health related water microbes*, ed. R. Morris, L. M. Alexander, P. Wyn-Jones & J. Sellwood, pp. 44–52. Glasgow: IAWPRC.

Stout, J. E., Yu, V. L. & Best, M. G. (1985). Ecology of *Legionella pneumophila* within water distribution systems. *Applied and Environmental Microbiology*, **49**, 221–8.

Wadowsky, R. M., Yee, R. B., Mezmar, L., Wing, E. J. & Dowling, J. N. (1982). Hot water systems as sources of *Legionella pneumophila* in hospital and non-hospital plumbing fixtures. *Applied and Environmental Microbiology*, **43**, 1104–10.

Walker, J. T., Dowsett, A. B., Dennis, J. P. & Keevil, C. W. (1991). Continuous culture studies of biofilms associated with copper corrosion. *International Biodeterioration*, **27**, 121–134.

Wright, J. B., Ruseska, I. & Costerton, J. W. (1991). Decreased biocide susceptibility of adherent *Legionella pneumophila*. *Journal of Applied Bacteriology*, **71**, 531–8.

Part III BIOFILMS ON THE SURFACES OF LIVING CELLS

12

The Rhizosphere as a Biofilm

David Pearce , Michael J. Bazin and James M. Lynch

Introduction

The rhizosphere

The rhizosphere encloses the zone of soil around a plant root in which the plant root exerts an influence on the growth and distribution of microorganisms. An important source of microbial growth limiting nutrients are the products of rhizodeposition, which include exudates, secretions, lysates and gases (Whipps & Lynch 1985). The increase in the specific growth rate of microorganisms in response to increased organic carbon input has been shown to result in a 5 to 10-fold increase in the number of bacteria when compared with that of the population in the bulk soil (Rouatt et al. 1960; Rovira & Davey 1974). Rhizosphere microorganisms largely depend on root products for their carbon and energy supply (Merckx et al. 1986), and maximum microbial population density occurs at the root surface or rhizoplane (Clarke 1949). This is a consequence of the presence of the highest concentration of growth limiting nutrient at the rhizoplane. Rhizosphere microorganisms do not form a continuous layer on the root surface, but occur in microcolonies (Newman & Bowen 1974; Rovira & Campbell 1975). This microbial cover has been estimated to be below 10% (Rovira et al. 1974; Bowen & Rovira 1976; Bowen & Theodorou 1979). The concentrations of growth limiting nutrient and microbial numbers decrease as a function of radial distance from the root, and under optimal conditions, with adequate growth limiting nutrient supply, the total bacterial numbers would be limited by space. Chemotaxis towards plant root exudates and extracts is a well

established phenomenon (Morris et al. 1992), enabling motile bacteria to migrate towards the root through chemotaxis. Non-motile bacteria accumulate in response to an elevated specific growth rate. Additional bacteria may also enter the system suspended in percolating groundwater. This aggregation of bacteria and their metabolic products at the root surface leads to the formation of a biofilm.

Biofilm formation

Healthy root function is influenced by the presence of microorganisms, and the plant root in turn, induces changes in the rhizosphere microbiota (Merckx et al. 1986). Metabolite diffusion through a microbial film on a solid surface has been shown to act as a limiting factor for the attachment of microorganisms in soil columns (Saunders & Bazin 1973). However, Helal & Sauerbeck (1986) calculated that microbial biomass labelled with carbon-14 corresponded to 1.6% of the total photosynthetically fixed carbon-14; to about 15% of the organic carbon-14 input into the rhizosphere and to 58% of the plant carbon remaining in the soil after the removal of roots. Of this biomass carbon-14, 20% was found outside the immediate root zone, which would suggest that the biofilm is not carbon limited. Pirt (1973) has suggested that the most important consequence of the sessile mode of growth is that the biomass within the film that is formed can be active only within a surface layer whose thickness is restricted to the depth to which all of the necessary metabolites can permeate by diffusion. This theory has been modified by Saunders & Bazin (1973).

Experimental systems

Experimental work however, can be difficult to interpret, because factors which influence the behaviour of the bacterial population will also affect the behaviour of the plant root. Helal & Sauerbeck (1986) have suggested that direct measurements of root microbial interactions have been hampered mainly by the lack of convenient and reliable experimental methods. In any non-sterile system, it is not possible to study root exudates by chemical methods because release and degradation of the organic materials are under steady state conditions (Haller & Stolp 1985). The use of experimental systems which can mimic the behaviour of real roots, is beginning to untangle the complex web of interactions in the rhizosphere (for example, Theodorou *et al.* 1980; Bazin *et al.* 1982; Martens 1982; Thompson *et al.* 1983; Martin & Foster 1985; Odham *et al.* 1986; Streit *et al.* 1991; Pearce *et al.* 1992; Sun *et al.* 1992). Bowen (1980) has rightly cautioned about interpretations of rhizosphere biology from the results of such work; ultimately verification is necessary in the field. However, experimental systems allow the manipulation of independent parameters under precisely controlled conditions.

Chemical interactions

Rhizodeposition

A wide range of organic compounds are released as rhizodeposition from plant roots to the surrounding soil (after Rovira 1965);

sugars: arabinose, fructose, glucose, maltose, raffinose, rhamnose, sucrose, xylose;
amino acids: α-alanine, ß-alanine, γ -aminobutyric acid, cysteine/cystine, glutamic acid, glutamine, glycine, homoserine, leucine/*iso*-leucine, lysine, methionine, phenylalanine, proline, serine, threonine, tryptophan, tyrosine, valine;
vitamins: *p*-aminobenzoic acid, biotin, choline, inositol, *n*-methylnicotinic acid, niacin, pantothenate, pyridoxine, thiamin;
organic acids: acetic, butyric, citric, fumaric, glycolic, malic, propionic, oxalic, succinic, tartaric, valeric;
nucleotides: adenine, guanine, uridine/cytidine;
enzymes: amylase, invertase, phosphatases, polygalacturonase, proteases;
miscellaneous: auxins, flavonone, glycosides, saponin, scopolotin (6–methoxy–7–hydroxy-coumarin).

Rhizodeposition provides a diffusion source and is affected by the presence of a biofilm. Microorganisms stimulate the loss of soluble organic materials from plant roots (Barber & Lynch 1977; Warembourg & Billes 1979; Pricryl & Vancura 1980). Lee & Gaskins (1982) demonstrated increased root exudation of carbon-14 containing organic compounds by *Sorghum* seedlings inoculated with nitrogen fixing bacteria. Under sterile conditions between 5 and 10% of photosynthetically fixed carbon is lost by plant roots, compared with 12 to 18% from unsterilized roots (Barber & Martin 1976). The quantity of material released by the roots will presumably be much less than would occur in the soil, since there is evidence that both contact with surfaces and the presence of microorganisms greatly enhance exudation (Barber & Gunn 1974; Barber & Martin 1976). Helal & Sauerbeck (1986) have demonstrated that plant roots induced a 197% increase in microbial biomass and a 5.4% decrease in soil organic carbon compared with a 1.2% decrease in the unplanted control soil. The contributions of plant and soil carbon to this increased microbial growth amounted to 68% and 32% respectively.

Rhizodeposits can be divided into a number of categories according to their origin and chemistry.

Exudates are water soluble compounds such as glucose, which diffuse from the plant without the expenditure of metabolic energy. Exudation rate increases in response to glucose metabolism in the biofilm, increasing the glucose concentration differential between the root and the soil solution. Modification of the nature of the organic compounds released could lead to a reduction in the growth rate. Roberts & Sheets (1990) demonstrated that *Enterobacter cloacae*, which is a common rhizosphere organism, would not grow on carbohydrate polymers supplied as energy sources, but did grow on all of the mono- and oligosaccharides including most ß-glucosides supplied.

Secretions need not be water soluble and are products of metabolic processes which are actively pumped from the root. For example, mucigel used for lubrication, is composed of highly hydrated, complex polysaccharides, probably secreted from root cap cells. Photomicrographs have indicated that the mucilaginous layer varies from 1 to 10 µm in thickness (Estermann & McLaren 1961; Jenny & Grossenbacher 1963; Dart & Mercer 1964;

Pickett-Heaps 1968). O'Brien (1972) stated that mucilage layers on maize, oats, wheat, peas and willow were about 10 µm thick. Greaves & Derbyshire (1972) reported that the thickness of the mucilage on a range of plants was influenced by the presence of bacteria. Foster (1986) has reviewed the spatial distribution of the microorganisms colonizing the mucilage. The biofilm may lead to metabolism of mucigel or reduction in lubrication by trapping mucigel around the plant root surface.

Lysates, including nucleic acids, are released when cells autolyse. These cells are sloughed off through physical abrasive action or may be an integral part of damaged root tissue. The biofilm could contribute to root damage, increasing infection potential, or it may protect the root tissue from physical abrasion through lubrication.

Gaseous compounds are released into soil pore spaces in small amounts. Examples include ethylene, a plant growth hormone, and carbon dioxide. A bacterial biofilm could trap or recycle ethylene, preventing its removal from the system, increasing the local concentration. Microbial respiration would increase carbon dioxide concentration in the soil, reducing oxygen tension and reducing soil pH.

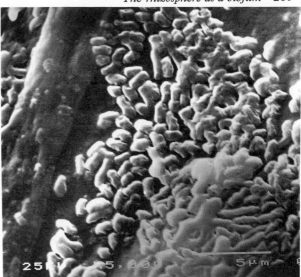

Fig. 12.1. *Enterobacter cloacae* on the surface of lettuce (*Latuca sativa*). Note the slimy bacterial cell morphology. Bar = 5 µm.

Chemical diffusion gradients

Chemical diffusion gradients are established in the rhizosphere as a result of the concentration differential between plant root tissues and the surrounding medium. Substances leaving the root diffuse beyond the surface layer to form a diffusion gradient. Caldwell & Lawrence (1986) demonstrated the occurrence of a glucose diffusion limited growth of adhered cells, even at glucose concentrations of 0.1 g L^{-1}. The decrease in growth at high surface population density observed by Ellwood *et al.* (1982) might also be caused by substrate diffusion limitation. Both substrates and microbial cells are distributed in a concentration gradient (Kunc 1988). Chemical diffusion gradients are established between all chemical substances that are water soluble and interact with the plant. Examples include most materials released by the plant. The biofilm disrupts such gradient formation through extraction or addition of nutrients, or through reduction in the surface area available for diffusion.

Extracellular polysaccharides

The presence of a polysaccharide sheath leads to the reduction in the soil solution polysaccharide concentration, increasing the water potential of the soil solution relative to the root tissue and encouraging water uptake by the plant. Extracellular polysaccharide serves to bind soil particles together, improving soil aggregation and increasing soil stability (Lynch 1987). *Enterobacter cloacae* produces extracellular polysaccharides in response to excess growth limiting nutrient (Fig. 12.1). The polysaccharides produced form a barrier to diffusion, reducing the diffusive surface area. The barrier increases the hydrodynamic residence time, preventing the removal of bacteria by soil water flow. The barrier may protect root tissue from abrasion, predation and pathogenic infection. In this symbiotic relationship, the plant provides raw materials through rhizodeposition and the biofilm produces a buffer zone protecting the root from adverse changes in the soil solution. Conversely Delaquis *et al.* (1989) have demonstrated that depletion of either glucose or nitrogen led to the active detachment of cells from the biofilm in a batch culture. They observed an increase in the hydrophobicity of unattached *Pseudomonas fluorescens* on depletion of carbon. This increase was the result of emigration of cells from the surface into the bulk phase.

Carbohydrate sink

The rhizosphere biofilm will increase the net carbohydrate requirement by the soil microbiota compared to the bulk soil requirement. The biofilm is a larger carbohydrate sink and maintenance of a larger diffusion gradient would stimulate rhizodeposition. Increased carbohydrate loss through rhizodeposition would stimulate further translocation of photosynthate to the roots. Greater allocation of primary production to cover this increased loss would make the plant less efficient at converting solar energy and less competitive, as plant competitive ability governs the extent of competitive exclusion. The biofilm may convert bioavailable polysaccharides into insoluble residues, no longer utilized by pathogenic microorganisms, or vice versa, so a knowledge of the growth of microorganisms in the soil and rhizosphere in relation to the supply of organic nutrients is necessary for the evaluation of the significance of microbial processes which are important to plants (Barber & Lynch 1977).

Modification of pH

Modification of pH takes place in the rhizosphere biofilm in response to proton fluxes towards the root tip during growth. Nye (1986) and Hadley et al. (1982), have demonstrated that imbalance between nutrient cation and anion uptake across the root–soil interface results in a decrease in rhizosphere pH by as much as two units. Increased carbon dioxide production, which dissolves in situ, produces carbonic acid. Carbonic acid production will depend on the buffering capacity of the soil and on the local carbon dioxide concentration. The removal of oxygen through respiration and the resultant reduction in oxygen tension could lead to anaerobic respiration. By-products of respiration, particularly those of anaerobic respiration, are often acidic and can lead to a further reduction in soil pH if released. A change in pH may alter the competitive advantage of a colonizing microorganism. For example, Gauemannomyces graminis, a soil borne pathogen, is sensitive to pH whereas pseudomonads can be more tolerant. Reducing pH for a species which grows well under acidic conditions would establish a positive feedback loop, greater metabolic activity producing greater carbon dioxide concentrations and then more growth. pH is, however, only one type of ionic modification. Uptake of mineral nutrients by both plant and microorganisms may result in the production of an electrochemical gradient.

Physical interactions

Physical gradients

Physical diffusion gradients are produced in the rhizosphere in the presence of a plant root and its associated microbiota. The two key physical diffusion gradients established are thermal and osmotic diffusion gradients. Insoluble polysaccharides reduce the rate of heat transfer, as organic molecules have a lower thermal conductivity than water. The additional layer of cells at the root surface would afford some degree of protection from frost damage in the surface layers of the soil through insulation. This effect would diminish rapidly with depth, as soil is an extremely good thermal insulator and internal temperatures rarely fluctutate to ecologically significant levels. The biochemical reactions taking place within microbial cells liberate heat energy, increasing the local temperature. Meharg & Killham (1989) showed that temperature affects the nature of organic compounds released from roots. It will also influence rhizodeposition, underlining the need to study the two systems both independently in model systems and together in the field.

The biofilm may also influence the plant soil water balance. The osmotic diffusion gradient could be increased through the removal of solutes by bacteria. Such removal of solutes would result in a higher or lower water potential relative to the surrounding soil solution and encourage water uptake by the plant. There is a strong influence of soil moisture on the release of carbon-14 containing organic material into soil from cereal roots (Barber & Martin 1976). Water availability is in itself a major constraint on microbial activity in the soil (Bleakley & Crawford 1989). Water movement through the soil is restricted by the presence of a biofilm. The hydrodynamic residence time is increased by trapping a thin layer of water between the rhizoplane and the surface of the biofilm. Nutrient loss from the system through leaching is reduced by tight internal recycling.

Mechanical stress

Mechanical stress in the plant root arises through root extension, wind action on aerial parts of the plant and through the movement of microfauna such as earthworms and nematodes through the soil. Mechanical stress is largely absorbed by the plasticity of the plant root cells and by sloughing of small amounts of cellular material. Barber & Gunn (1974) have shown that the exudation of both amino acids and carbohydrates increases when roots are grown through a solid medium, subjecting them to mechanical stress. Abrasion occurs as a result of frictional forces against sharp soil aggregates. This can puncture plasma membranes, leading to leakage of cell contents. The biofilm provides some degree of protection from frictional resistance through lubrication. The effectiveness of the biofilm would consequently depend on its thickness and the presence of extracellular lubricants such as polysaccharides.

Charge interactions

Charge interactions occur in the biofilm as a result of electrostatic imbalance. The surface charge properties of bacterial and plant cells play a role in rapid firm cell adhesion. This binding is strongly promoted by cation mediated events involving physical masking of negative charges at the cell surface (James *et al.* 1985). The biofilm will influence surface charges on both the root surface and the soil particles. This would change the balance of root–nutrient electrostatic interactions. The growing cells of eukaryotic microorganisms generate electrical currents that are detectable with extracellular vibrating microelectrodes (Gow 1989; Nuccitelli 1990). These currents are circulating ions that result from the spatial segregation of electronic ion transport proteins in the plasma membrane. Studies using a vibrating microelectrode and ion substitution experiments have suggested that the root-generated electrical current of *Zea mays* is due primarily to this circulation of protons.

Detoxification

Bacteria within the biofilm may absorb and degrade harmful ions and toxins, mask pathogen receptors on the surface of the root by possibly 'disguising' the plant and convert insoluble ions into bioavailable species. The bacteria would compete for essential trace elements and nutrients. However, as bacterial growth appears to be carbon limited, bacteria would compete with the plant where the requirement for micronutrients was the same as that required by the plant.

Soil structuring

Soil aggregation around the root is assisted by the presence of polysaccharides, although growth of a biofilm would tend to block solid pores and reduce water movement. Soil binding may be prevented or encouraged by biofilm interaction with charge on soil colloids. The biofilm may encourage disruption and mixing of the soil around the root by microfauna, through the attraction of bacterial predators.

Nutrient interactions

Release and activation of minerals

Changes in nutrient concentration and flux through the soil profile can affect the physiological status of microorganisms (Pickett & Bazin 1980; Parke *et al.* 1986). The biofilm may activate chelated minerals and liberate nutrients into the soil solution, either as metabolic by-products from the bacteria, or through pH modification, altering the ionic balance. Liberation of nutrient can be a beneficial or detrimental process. Cakmak & Marschner (1988) have demonstrated a distinct role of zinc ions in membrane integrity and hence root extension. They also postulate ecological implications with respect to nutrient mobilization and microbial activity in the rhizosphere. Bhat (1983) compared factors which influence nutrient inflows into apple roots. More specifically, reductions in rhizosphere pH could liberate toxic ions reducing or even terminating root growth. Aluminium ions can be extremely toxic in solution, depending on speciation, but require low pH in order to become solubilized from soil particles and enter the soil solution (Wild 1988). Microorganisms may reduce ion influx into the plant or reduce the rate of ion loss by recycling nutrients.

Nutrient recycling

Cheng & Coleman (1990) have shown that living roots have a stimulatory effect on soil organic matter decomposition due to the higher microbial activity induced by the roots. Humus degradation is an integral part of the nutrient recycling process occurring in soils and requires the presence of specialized enzymes capable of degrading complex plant polymers. The rhizosphere biofilm could act as a source of such enzymes, for example cellulases. This would increase the efficiency of nutrient recycling, allowing humus degradation to take place in the soil around the root, facilitating the reabsorption of ions by the same plant. The presence of a crop enhances the ability of the microfauna to compete for soil nitrogen (Cortez & Hameed 1992). Siderophores are released by microorganisms and sequester iron from the soil solution, preventing its loss through leaching. The production of such extracellular organic molecules serves to increase the efficiency of tight nutrient cycling around the root (Lynch 1990).

Ionic modification

Potassium supply modifies the composition of soluble exudates of maize (Kraffczyk *et al.* 1984) and, therefore, any microbial change in the potassium supply will influence the composition of exudates. The overall flux of carbon compounds from the system is reduced as some are converted to insoluble polysaccharides. Such carbon compounds are unavailable to other microorganisms, which could prove either competitive or pathogenic. The biofilm may provide prey for predators such as nematodes. Such movement would encourage soil solution movement and encourage transport of minerals both towards and away from the root.

Pilet & Senn (1980) demonstrated the pattern of endogenous ion currents measured in the medium surrounding horizontally growing tips of *Zea mays*, and showed that current consistently flowed into meristematic and actively elongating tissues of the root. Miller & Gow (1989a) demonstrated the extension rate of the root correlates with the density of the proton current entering actively dividing and elongating cells for *Zea mays*; it may also exist for species of *Trifolium* and *Hordeum*. Factors which promoted or restricted root growth, enhanced or reduced the root generated current respectively (Miller *et al.* 1988). Toko *et al.* (1987) showed that electrical isolation of the elongating and mature root tis-

sues led to a decrease in root growth rate, further emphasizing the importance of the current to root growth. However, there is no good correlation between either the extension rate or the length of the growing peripheral cell and the current density (Gow 1989). The electrical current is incidental to the process of tip growth (De Silva *et al.* 1992), and the influence of the biofilm in the electrical current is, therefore, also incidental to the process of tip growth.

Plant roots generate electrical fields in the rhizosphere as a consequence of their ion transport activities. Zoospores are chemotactic (Dukes & Apple 1961; Katsura & Hosomi 1963; Hickman & Ho 1966; Bimpong & Clerk 1970), accumulating at sites of maximum exudation of low molecular weight solutes. Zoospores of the plant pathogen *Phytophthora palmivora* exhibit anodal electrotaxis in electrical fields >0.5 V m^{-1}, comparable in size to the physiological fields around roots (Morris *et al.* 1992). Motile zoospores of the fungal pathogen *P. parasitica* var. *nicotianae* have been reported to associate preferentially with induced wound sites in *Nicotiana tabacum*, where a large positive current (70 mA m^{-2}) develops at the root surface. This might suggest that electrotaxis may be part of the mechanism by which zoospores locate regions susceptible to fungal infection (Miller *et al.* 1988). Growing fungal hyphae normally generate an electrical current so that positive charge enters the hyphal tip and exits from the rear (Gow 1984; Harold *et al.* 1985; Miller & Gow 1989a, b). For a range of mycelial fungi, electrical fields cause orientation of germ tubes, hyphae, branches and rhizoids and polarization of sites of germination (De Vries & Wessels 1982; McGillivray & Gow 1986; Van Laere 1988; Youatt *et al.* 1988). The masking or changing of charges on the root surface, and the metabolism of exudates by the biofilm, will consequently influence the attraction of zoospores.

Nitrogen dynamics

Dinitrogen fixation and denitrification are critical soil processes in terms of soil fertility. Both processes are mediated by rhizosphere bacteria, for example, *Rhizobium* species. Legumes appear to support considerably larger rhizosphere populations than non-leguminous plants (Katznelson 1965). Ardakani *et al.* (1973) have demonstrated that both *Nitrobacter* and *Nitrosomonas* species show higher density near the soil surface where the concentrations of nitrogenous substrates are always the highest. Crop plants are frequently cited as being either nitrogen or phosphorus limited. Nitrogen fixation capacity will therefore have a direct effect on plant productivity. The presence of a nitrogen fixing colony in a biofilm in the rhizosphere may increase the amount of nitrogen available for plant growth, stimulating herbivory, and resulting in nutrient addition through urine and faeces.

For example, inoculation with *Azotobacter* and *Bacillus megaterium* has been shown to intensify the activity of rhizosphere bacteria involved in ammonification, anaerobic nitrogen fixation, nitrification, phosphate mineralization and cellulose decomposition (Samtsevich 1962; Berezova 1963; Mosiashuih *et al.* 1963; Marendiak 1964). Conversely, denitrification and the presence of denitrifying microorganisms would be detrimental by removing essential nitrates. Drury *et al.* (1991) have shown that nitrogen immobilization could be restricted to a relatively limited location of microbial biomass in microsites containing readily decomposable residues with high carbon: nitrogen ratios.

Biotic interactions

With a large variety of different species living in close proximity, direct biotic interactions invariably occur in the rhizosphere. Such interactions include symbiosis, synergy, competition, antagonism, pathogenesis, predation and mutualism.

Symbiosis

Mycorrhizal fungi and rhizosphere microorganisms are involved in a symbiosis which is of benefit to mutual development (Linderman 1988). Azcon (1989) showed that bacterial inoculation generally increased the growth of mycorrhizal plants, and that there is a selective interaction between the free-living bacteria of the rhizosphere and vesicular-arbuscular mycorrhizal fungi. Phosphorus is a very immobile element in the soil, so any root rapidly uses the phosphorus in the surrounding soil. Many different plant species use the symbiotic relationship associated with mycorrhizal infection to increase the uptake of phosphorus (Hartley & Smith 1983). The external hyphae increase the volume of soil which can be exploited. In sheathing mycorrhizas, those with a tight biofilm, the phosphorus may accumulate in the sheath, partly as polyphosphate, especially when soil phosphorus levels are low. Other substances such as nitrogen and potassium also accumulate in the sheath, either by active absorption or by passive diffusion along or through the sheath.

Synergy

Beneficial rhizobacteria have been credited with having positive effects on the growth of plants (Burr & Ceasar 1983). The presence of a microbial community in the rhizosphere could influence minerals in the soil solution. It may also produce and decompose plant growth hormones and plant toxins, stimulating and retarding plant growth, respectively. For example, maize roots show a growth response to *Azospirillum* inoculation resulting from the effect of soil organic matter content, number of rhizosphere bacteria and timing of inoculation (Fallik *et al.* 1988). Rhishitin, a substance produced by potato tubers rotted in air, caused a 50% reduction in viability of *Erwinia atroseptica* (Lyon & Bayliss 1975). Bacterial populations in the rhizosphere and phylloplane vary with plant age (Rovira 1965), and their composition may be greatly influenced by the changes in root exudation patterns that occur during plant growth (Matsumoto *et al.* 1979). Patel (1969), examining bacteria and root surface fungi from wheat, found that, compared with uninoculated plants, seed inoculated with *Azotobacter* significantly depressed colonization of emerging roots by bacteria and actinomycetes. Bacterial growth is stimulated by nematode grazing (Sundin *et al.* 1990).

Competition

Dorofeev & Panikov (1989) studied growth efficiency for microorganisms with different ecological strategies. They found that microorganisms with r and K growth strategies had different growth efficiencies in batch culture. Root respiration results in competition between the root and the biofilm for oxygen, with an attendant increase in the carbon dioxide concentration and its associated decrease in pH. This increase in carbon dioxide concentration has been shown to affect growth, carbon distribution within the plant and exudation rate in maize (Whipps 1985). Ethylene is a plant hormone which induces lateral branching in roots, stimulates fruit ripening, contributes to leaf abscission and other changes characterizing senescence, stimulates radial growth of stems and roots, can aid in breaking dormancy in the buds and seeds, and can help initiate flowering in some plants, for example, pineapple. Moreover, some effects attributed to auxin (for example, lateral bud inhibition and root geotropism) may in some cases be due to an auxin induced increase in ethylene production. Such plant hormones, being relatively simple organic molecules, may act as substrates for bacterial growth, reducing their concentration. This would influence the effect of the hormone on its target cells. Sinclair & Alexander (1984) suggested that starvation susceptible bacteria will not persist in environments that are nutrient poor or in which they fail to compete for organic nutrients and that starvation resistance is a necessary but not sufficient condition for persistence in environments that are nutrient poor or that support intense interspecific competition. *Streptococcus faecalis*, *Staphylococcus aureus*, *Bacillus subtilis* and *Streptococcus* sp. readily lost viability in the absence of organic nutrients. *Rhizobium meliloti* and two pseudomonads were resistant to starvation. The greater survival of the lake water isolates than the skin and mouth isolates supports the view that bacteria from a nutrient poor environment are better able to withstand starvation than those from a nutrient rich habitat, in which starvation resistance is not a necessary trait.

Antagonism

Potential antibiotic and hormone production by both root tissue and biofilm could alter the root community structure. Antibiotic resistant microorganisms and those organisms which use antibiotics as an energy source would be favoured. Microorganisms which adapted to fast growth (r strategists), with a consequent inability to deal with antibiotics, would be reduced. Hormone production is encountered in many soil organisms which produce indolyl-3–acetic acid (IAA) with or without added tryptophan (Brown 1972; Scott 1972). Root exudates contain tryptophan and related compounds that could act as precursors for IAA synthesis (Scott 1972).

Examples of antagonism which lead to potential control of root diseases are given by Lynch (1990).

Pathogenesis

Infection by many root pathogens is known to require the prior stimulation by exudates from host plants. The physical biofilm barrier provides direct competition with pathogenic species for nutrient and indirect competition for space. For example, infection of *Medicago lupulina* by *Ascochyta imperfecta* was decreased, an effect directly related to the amount of pathogen on the seed (Brown & Burlingham 1963). Van Vuurde & Schippers (1980) showed that the interaction between bacteria and actinomycetes may be involved in the reduction in the number of bacteria present. Microbial activity has been linked to the suppression of damping-off caused by *Pythium ultimum* (Chen *et al.* 1988; Lumsden *et al.* 1990). If the biofilm is not composed of pathogenic bacteria, it could lead to a reduction in the materials being released and amenable to chemotaxis, and the concentration of substrate available for pathogen growth and proliferation. This would hinder attainment of a minimum infection potential, reducing the chance of infection, or reducing the extent of final infection.

Pathogens which may use microorganisms as indicator species, or for access through damage to the root epidermis, may use a biofilm as an infection route. Natural senescence of the root cortex of spring wheat has been related to susceptibility to common root rot (*Cochliobolus sativus*) and growth of a free living, nitrogen fixing bacterium (Deacon & Lewis 1982). Hawes *et al.* (1988) refer to the considerable literature which describes bacterial attraction to roots and root exudates. In most such attractions, simple metabolites seem to be involved. Bacterial metabolites or substances released from the plant root through damaged tissue may attract pathogenic microorganisms. Phytoalexins are a chemically heterogeneous group of low molecular weight compounds with antimicrobial properties (Lyon & Wood 1975). Phytoalexins are not present in healthy plant tissues, but they appear at the site of an infection (Mansfield *et al.* 1974; Bell *et al.* 1984). The microfauna attracted to this rich source of bacteria may transport pathogenic microorganisms to the root.

The environmental release of genetically manipulated microorganisms

The release of genetically engineered microorganisms is a technology which is currently under development for use in agriculture. Genetically engineered microorganisms might be used in plant nutrition, for example, nutrient release and nitrogen fixation; in the control of pests and diseases; in the control of frost damage; and in the disposal of waste materials. The rhizosphere is a useful target area for such organisms due to the abundance of growth limiting carbon nutrient. In the assessment of the risks associated with such releases through genetic exchange, the rhizosphere has been a useful study area with high growth rates and high incidence of cell-to-cell contact. Van Elsas & Trevors (1990) have shown that plant roots enhance the frequency of bacterial matings, not only between organisms present in the same niches but also between organisms from different niches, or in different conditions of stress. Biofilm formation in the environment often results in colonies of single species that might provide the prerequisite numbers of recipient cells and the concentration of homologous DNA necessary for transformation (Paul *et al.* 1991). This could occur through stimulation of bacterial migration and/or growth, or by providing additional surfaces for cell–cell contact. Van Elsas *et al.* (1986) showed that the detection of transconjugants in nutrient amended soil and their absence from unamended soil suggested that the physiological activity of the cells is essential for effective plasmid transfer between bacilli in soil. In this respect, it may be necessary for pili or pilus-like structures to form on the bacterial cells, to promote conjugation, and it is well established that surfaces can have a role in this (Berkeley *et al.* 1980). Although conjugal plasmid transfer does not depend on cell reproduction, it depends on the metabolic state of the donors and recipients (Trevors *et al.* 1987). The metabolic rate of organisms in their natural habitats is reduced in response to the scarcity of growth substrates (Reanney *et al.* 1982). Starvation has been shown to affect plasmid expression and maintenance and stimulate protein degradation in bacteria (Reeve *et al.* 1984; Caldwell *et al.* 1989).

Colonization

The biofilm would be a rapid colonization source as microbes leached from the edges of the colony would reach new areas of root on the plant rapidly. Different bacteria have different root colonization capacities (Bennett & Lynch 1981; Turner & Newman 1984; Scher *et al.* 1984). Growth along the root itself would be rapid due to the spatial constraints of the biofilm and an elevated rhizodeposition rate at the root tip. Hozore & Alexander (1991) have shown that there is no significant difference between rhizosphere and soil bacteria in chemotactic response to exudates. Most organisms grew faster in exudates than in an inorganic salts solution. Rapid growth did not seem necessary for migration along the root, neither did it ensure that this migration occurred. By occupying colonization sites, it seems that ectomycorrhizas may protect plants against pathogenic colonization from species such as *Phytophthora cinnamomi*, *Rhizoctonia solani*, *Pythium* spp. and *Fusarium oxysporum*. However, there is little experimental evidence to support this.

Rhizosphere biofilms can have a significant effect on soil ecology. Effects are determined by the species richness, diversity and the relative population densities. The biofilm can alter the physicochemical properties of the system. Chemical diffusion gradients are altered through the increase in exudation rate, polysaccharide production, increased carbohydrate utilization and through pH modification. Physical diffusion gradients are altered through interference with the water balance, reduction of mechanical stress, charge interactions, reduction in leaching rate, protection from frost damage and abrasion, soil structuring and reduced heat exchange. The biofilm can alter the rate of liberation of nutrients, including aluminium toxicity, recycling of nutrients, nitrogen and phosphorus dynamics, activation of chelated minerals, humus degradation and siderophore production. Plant growth can be inhibited directly by the presence of the biofilm by hormone and toxin production and degradation. Biotic interactions are developed such as mycorrhizal infection, competition, predation, pathogenesis, colonization potential, genetic exchange and antibiotic production.

The rhizosphere biofilm plays an integral part in the complex web of interactions which govern plant growth and microbial population dynamics in the rhizosphere. Even chemical and physical factors in the rhizosphere biofilm may become interactive. For example, the presence of NH_4^+ or NO_3^- influences the current density and the position of the current crossover point, although the overall pattern of the circulating current is maintained in the presence of either form of combined nitrogen (Miller *et al.* 1991). With such a diverse range of interactions in such a heterogeneous habitat, further work will need to be undertaken in both experimental systems and in the field, before it is possible to make accurate predictions about the behaviour of this important ecological system. The biofilm established on the surface of a plant root is an integral part and influential factor in the ecology of the plant and the rhizosphere.

References

Ardakani, M. S., Rehbock, J. T. & McLaren, A. D. (1973). Oxidation of nitrite to nitrate in a soil column. *Soil Science Society of America Proceedings*, **37**, 53–6.

Azcon, R. (1989). Selective interaction between free-living rhizosphere bacteria and vesicular-arbuscular mycorrhizal fungi. *Soil Biology and Biochemistry*, **21**, 639–44.

Barber, D. A. & Gunn, K. B. (1974). The effect of mechanical forces on the exudation of organic substances by roots of cereal plants grown under sterile conditions. *New Phytologist*, **73**, 39–45.

Barber, D. A. & Lynch, J. M. (1977). Microbial growth in the rhizosphere. *Soil Biology and Biochemistry*, **9**, 305–8.

Barber, D. A. & Martin, J. K. (1976). The release of organic substances by cereal roots into soil. *New Phytologist*, **76**, 69–80.

Bazin, M. J., Cox, D. J. & Scott, R. I. (1982). Nitrification in a column reactor: limitations, transient behaviour and effect of growth on a solid substrate. *Soil Biology and Biochemistry.*, **14**, 477–87.

Bell, J. N., Dixon, R. A., Bailey, J. A., Rowell, P. M. & Lamb, C. J. (1984). Differential induction of chalcone synthase mRNA activity at the onset of phytoalexin accumulation in compatible and incompatible plant pathogen interactions. *Proceedings of the National Academy of Sciences, USA*, **81**, 3384–8.

Bennett, R. A. & Lynch, J. M. (1981). Colonization potential of bacteria in the rhizosphere. *Current Microbiology*, **6**, 137–8.

Berezova, E. P. (1963). The effectiveness of bacterial fertilizers. *Mikrobiologiya*, **32**, 358–61.

Berkeley, R. C. W., Lynch, J. M., Melling, J., Rutter, P. R., & Vincent, B. (ed) (1980). *Microbial adhesion to surfaces*. Chichester: Ellis Horwood.

Bhat, K. K. S. (1983). Nutrient inflows into apple roots. *Plant and Soil*, **71**, 371–80.

Bimpong, C. E. & Clerk, G. C. (1970). Motility and chemotaxis in zoospores of *Phytophthora palmivora* (Butl.) Butl. *Annals of Botany*, **34**, 617–24.

Bleakley, B. H. & Crawford, D. L. (1989). The effects of varying moisture and nutrient levels on the transfer of a conjugative plasmid between *Streptomyces* species in soil. *Canadian Journal of Microbiology*, **25**, 544–9.

Bowen, G. D. (1980). Misconceptions, concepts and approaches in rhizosphere biology. In *Contempory microbial ecology*, ed. D. C. Ellwood, J. N. Hedger, M. J. Latham, J. M. Lynch & J. H. Slater, pp. 283–304. London: Academic Press.

Bowen, G. D. & Rovira, A. D. (1976). Microbial colonization of plant roots. *Annual Reviews in Phytopathology*, **14**, 121–44.

Bowen, G. D. & Theodorou, C. (1979). Interactions between bacteria and ectomycorrhizal fungi. *Soil Biology and Biochemistry*, **11**, 119–26.

Brown, M. E. (1972). Seed and root bacterization. *Annual Review of Phytopathology*, **12**, 181–97.

Brown, M. E. & Burlingham, S. K. (1963). Azotobacter and plant disease. *Annual Report, Rothamsted Experimental Station*, 73.

Burr, T. J. & Ceasar, A. (1983). Beneficial plant bacteria. In *CRC Critical reviews in Plant Science*, Vol. 2, ed. B. V. Conger, pp. 1–20. Boca Raton, Fla: CRC Press.

Cakmak, I. & Marschner, H. (1988). Increase in membrane permeability and exudation in roots of zinc deficient plants. *Journal of Plant Physiology*, **132**, 356–61.

Caldwell, D. E. & Lawrence, J. R. (1986). Growth kinetics of *Pseudomonas fluorescens* microcolonies within the hydrodynamic boundary layers of surface microenvironments. *Microbial Ecology*, **12**, 299–312.

Caldwell, B. A., Ye, C., Griffiths, R. P., Moyer, C. L. & Morita, R. Y. (1989). Plasmid expression and maintenance during long-term starvation-survival of bacteria in well water. *Applied and Environmental Microbiology*, **55**, 1860–4.

Chen, W., Hoitink, H. A. J., Schmitthenner, A. F. & Tuovinen, O. H. (1988). The role of microbial activity in suppression of damping-off caused by *Pythium ultimum*. *Phytopathology*, **78**, 314–22.

Cheng, W. & Coleman, D. C. (1990). Effect of living roots on soil organic matter decomposition. *Soil Biology and Biochemistry*, **22**, 781–7.

Clarke, M. (1949). Soil microorganisms and plant roots. *Advances in Agronomy*, **1**, 241–88.

Cortez, J. & Hameed, R. (1992). Mineralization of 15–N labelled organic compounds adsorbed on soil size fractions: effect of successive wheat cropping. *Soil Biology and Biochemistry*, **24**, 113–9.

Dart, P. J. & Mercer, F. V. (1964). The legume rhizosphere. *Archives for Microbiology*, **47**, 344–8.

Deacon, J. W. & Lewis, S. J. (1982). Natural senescence of the root cortex of spring wheat to common root rot (*Cochliobolus sativus*) and growth of a free-living nitrogen fixing bacterium. *Plant and Soil*, **66**, 13–20.

Delaquis, P. J., Caldwell, D. E., Lawrence, J. R. & McCurdy, A. R. (1989). Detachment of *Pseudomonas fluorescens* from biofilms on glass surfaces in response to nutrient stress. *Microbial Ecology*, **18**, 199–210.

De Silva, L. R., Youatt, J., Gooday, G. W. & Gow, N. A. R. (1992). Inwardly directed ionic currents of *Allomyces macrogynus* and other water moulds indicate sites of proton-driven nutrient transport but are incidental to tip growth. *Mycological Research*, **96**, 925–31.

De Vries, S. C. & Wessels, J. G. H. (1982). Polarized outgrowth of hyphae by constant electrical fields during reversion of *Schizophyllum commune* protoplasts. *Experimental Mycology*, **6**, 95–8.

Dorofeev, A. G. & Panikov, N. S. (1989). Dynamics of the dying off of starving microorganisms as a function of the initial growth rate. *Mikrobiologiya*, **60**, 814–22.

Drury, C. F., Voroney, R. P. & Beauchamp, E. G. (1991). Availability of NH_4^+–N to microorganisms and the soil internal N cycle. *Soil Biology and Biochemistry*, **23**, 165–9.

Dukes, P. D. & Apple, J. L. (1961). Chemotaxis of zoospores of *Phytophthora parasitica* var. *nicotianae* by plant roots and certain chemical solutions. *Phytopathology*, **51**, 195–7.

Ellwood, D. C., Keevil, C. W., Marsh, P. D., Brown, C. M. & Wardell, J. N. (1982). Surface-associated growth. *Philosophical Transactions of the Royal Society of London, Series B*, **297**, 517–32.

Estermann, E. F. & McLaren, A. D. (1961). Contribution of rhizoplane organisms to the total capacity of plants to utilize organic nutrients. *Plant and Soil*, **15**, 243–60.

Fallik, E., Okon, Y. & Fischer, M. (1988). Growth response of maize roots to *Azospirillum* inoculation., effect of soil organic matter content, number of rhizosphere bacteria and timing of inoculation. *Soil Biology and Biochemistry*, **20**, 45–9.

Foster R. C. (1986). 'The ultrastructive of the rhizophere and rhizophere *Annual Review of Phytopathology*, **24**, 211–34.

Gow, N. A. R. (1984). Transhyphal electrical currents in fungi. *Journal of General Microbiology*, **130**, 3313–18.

Gow, N. A. R. (1989). Relationship between growth and the electrical current of fungal hyphae. *Biological Bulletin*, **176**, 31–5.

Greaves, M. P. & Derbyshire, J. F. (1972). The ultrastructure of the mucilaginous layer on plant roots. *Soil Biology and Biochemistry*, **4**, 443–9.

Hadley, M. J., Nye, P. H. & White, R. E. (1982). Plant-induced changes in the rhizosphere of rape (*Brassica napus* var. emerald) seedlings. II. Origin of the pH change. *New Phytologist*, **91**, 31–44.

Haller, T. & Stolp, H. (1985). Quantitative estimation of root exudation of maize plants. *Plant and Soil*, **86**, 207–16.

Hartley, J. L. & Smith, S. E. (1983). *Microbial symbiosis*. London: Academic Press.

Harold, F. M., Schreurs, W. J., Harold, R. L. & Caldwell, J. H. (1985). Electrobiology of fungal hyphae. *Microbiological Sciences*, **2**, 363–6.

Hawes, M. C., Smith, L. Y. & Howarth, A. J. (1988). *Agrobacterium tumefaciens* mutants deficient in chemotaxis to root exudates. *Molecular Plant–Microbe Interactions*, **1**, 182–6.

Helal, H. M. & Sauerbeck, D. (1986). Effect of plant roots on carbon metabolism of soil microbial biomass. *Z. Pflanzenernaehr. Bodenk*, **149**, 181–8.

Hickman, C. J. & Ho, H. H. (1966). Behaviour of zoospores in plantpathogenic phycomycetes. *Annual Review of Phytopathology*, **4**, 195–220.

Hozore, E. & Alexander, M. (1991). Bacterial charactcristics important to rhizosphere competence. *Soil Biology and Biochemistry*, **23**, 717–23.

James, D. W. J., Suslow, V. & Steinback, K. E. (1985). Relationship between rapid, firm adhesion and long-term colonization of roots by bacteria. *Applied and Environmental Microbiology*, **50**, 392–7.

Jenny, H. & Grossenbacher, K. (1963). Root–soil boundary zones as seen in the electron microscope. *Soil Science Society of American Proceedings*, **27**, 273–7.

Katsura, K. & Hosomi, T. (1963). Chemotaxis of zoospores for plant roots in relation to infection by *Phytophthora capsici* var. *Leonian. Science Report from Kyoto Prefectural University*, **15**, 27–32.

Katznelson, H. (1965). Nature and importance of the rhizosphere. In *Ecology of soil borne pathogens*, ed. K. F. Baker & W. G. Snyder, pp. 187–209. London: John Murray.

Kraffczyk, I., Trolldenier, G. & Beringer, H. (1984). Soluble root exudates of maize: influence of potassium supply and rhizosphere microorganisms. *Soil Biology and Biochemistry*, **16**, 315–22.

Kunc, F. (1988). Three decades of heterocontinuous flow cultivation method in soil microbiology. In *Continuous culture*, ed. P. Kyslik, V. Krumphaiz, M. Novak & E. A. Dawes, pp. 43–55. London: Academic Press.

Lee, K. J. & Gaskins, M. H. (1982). Increased root

exudation of 14-C compounds inoculated with nitrogen-fixing bacteria. *Plant and Soil*, **69**, 391–9.

Linderman, R. G. (1988). Mycorrhizal interactions with the rhizosphere microflora: the mycorrhizosphere effect. *Phytopathology*, **78**, 366–71.

Lumsden, R. D., Carter, J. P., Whipps, J. M. & Lynch, J. M. (1990). Comparison of biomass and viable propagule measurements in the antagonism of *Trichoderma harzianum* against *Pythium ultimum*. *Soil Biology and Biochemistry*, **22**, 187–94.

Lynch, J. M. (1987). Microbial interactions in the rhizosphere. *Soil Microorganisms*, **30**, 33–41.

Lynch, J. M. (1990). Microbial metabolites. In *The rhizosphere*, ed. J. M. Lynch, pp. 177–206. Chichester: Wiley.

Lyon, G. D. & Bayliss, C. E. (1975). The effect of rishitin on *Erwinia carotovora* var. *atroseptica* and other bacteria. *Physiological Plant Pathology*, **6**, 177–86.

Lyon, F. M. & Wood, R. K. S. (1975). Production of phaseollin, coumesterol and related compounds in bean leaves inoculated with *Pseudomonas* species. *Physiology and Plant Pathology*, **70**, 406–9.

MacLeod, W. J., Robson, A. D. & L. K. Abbott (1986). Effects of phosphate supply and inoculation with a vesicular-arbuscular mycorrhizal fungus on the death of the root cortex of wheat, rape and subterranean clover. *New Phytologist*, **103**, 349–57.

McGillivray, A. M. & Gow, N. A. R. (1986). Applied electrical fields polarize the growth of mycelial fungi. *Journal of General Microbiology*, **113**, 2515–25.

Malajczuk, N. (1983). Microbial antagonism to *Phytophthora*. In Phytophthora: *its biology, taxonomy, ecology and pathology*, ed. D. C. Erwin, S. Bartnicki-Garcia & P. H. Tsao, pp. 197–218. St. Paul, Minn: The American Phytopathological Society.

Mansfield, J. W., Hargreaves, J. A. & Boyle, F. C. (1974). Phytoalexin production by live cells in broad bean leaves infected with *Botrytis cinerea*. *Nature*, **252**, 316–17.

Marendiak, D. (1964). Contribution to the question of dynamism and quantitative relations of some groups of microorganisms in the rhizosphere of maize. *Rostlinna Vyroba*, **10**, 137–44.

Martens, R. (1982). Apparatus to study the quantitative relationships between root exudates and microbial populations in the rhizosphere. *Soil Biology and Biochemistry*, **14**, 315–17.

Martin, J. K. & Foster, R. C. (1985). A model system for studying the biochemistry and biology of the root–soil interface. *Soil Biology and Biochemistry*, **17**, 261–9.

Matsumoto, H., Okada, K. & Takahashi, E. (1979). Excretion products of maize roots from seedlings to seed development stage. *Plant and Soil*, **53**, 17–26.

Meharg, A. A. & Killham, K. (1989). Distribution of assimilated carbon within the plant and rhizosphere of *Lolium perenne*: influence of temperature. *Soil Biology and Biochemistry*, **21**, 487–9.

Merckx, R., van Ginkel, J. H., Sinnaeve, J. & Cremers, A. (1986). Plant-induced changes in the rhizosphere of maize and wheat. I. Production and turnover of root-derived material in the rhizosphere of maize and wheat. *Plant and Soil*, **96**, 85–93.

Miller, A. L. & Gow, N. A. R. (1989a). Correlation between root-generated ionic currents, pH, fusicoccin, IAA and the growth of the primary root tip of *Zea mays*. *Plant Physiology*, **89**, 1198–1206.

Miller, A. L. & Gow, N. A. R. (1989b). Correlation between profile of ion-current circulation and root development. *Physiologia Plantarum*, **75**, 102–8.

Miller, A. L., Shand, E. & Gow, N. A. R. (1988). Ion currents associated with root tips, emerging laterals and induced wound sites in *Nicotiana tabacum*: spatial relationship proposed between resulting electrical fields and phytophthoran zoospore infection. *Plant, Cell and Environment*, **11**, 21–5.

Miller, A. L., Smith, G., Raven, J. A. & Gow, N. A. R. (1991). Ion currents and the nitrogen status of roots of *Hordeum vulgare* and *Trifolium repens*. *Plant, Cell and Environment*, **14**, 559–68.

Morris, B. M., Reid, B. & Gow, N. A. R. (1992). Electrotaxis of zoospores of *Phytophthora palmivora* at physiologically relevant field strengths. *Plant, Cell and Environment*, **15**, 645–53.

Mosiashvili, G. I., Topuridze, K. V. & Kiriakova, N. G. (1963). Effectiveness of azotobacterin in vineyard soils. *Mikrobiologiya*, **32**, 835–7.

Newman, E. I. & Bowen, H. J. (1974). Patterns of distribution of bacteria on root surfaces. *Soil Biology and Biochemistry*, **6**, 205–9.

Nuccitelli, R. (1990). Vibrating probe technique for studies of ion transport. In *Non-invasive techniques in cell biology*, ed. J. K. Foskett & S. Grinstein, pp. 273–310. New York: Wiley-Liss.

Nye, P. H. (1986). Acid–base changes in the rhizosphere. *Advances in Plant Nutrition*, **2**, 129–53.

O'Brien, T. P. (1972). Cytology of cell-wall formation in some eukaryotic cells. *Botanical Review*, **38**, 87–118.

Odham, G., Tunlid, A., Valeur, A., Sundin, P. & White, D. C. (1986). Model system for studies of microbial dynamics at exuding surfaces such as the rhizosphere. *Applied and Environmental Microbiology*, **52**, 191–6.

Parke, J. L., Moen, R., Rovira, A. D. & Bowen, G. D. (1986). Soil water flow affects the rhizosphere distribution of a seed-borne biological control agent *Pseudomonas fluorescens*. *Soil Biology and Biochemistry*, **18**, 583–8.

Patel, J. J. (1969). Microorganisms in the rhizosphere of plants inoculated with *Azotobacter chroococcum*. *Plant and Soil*, **31**, 209–23.

Paul, J. H., Frischer, M. E. & Thurmond, J. M.

(1991). Gene transfer in marine water column and sediment microcosms by natural plasmid transformation. *Applied and Environmental Microbiology*, **57**, 1509–15.

Pearce, D., Lynch, J. M. & Bazin, M. J. (1992). An artificial microcosm to study microbial interactions in the rhizosphere. In *The release of genetically modified microorganisms-REGEM2*, ed. E. S. Duncan & M. S. Stewart-Tull, pp. 229–30. New York: Plenum Press.

Pickett, A. M. & Bazin, M. J. (1980). Growth and composition of *Escherichia coli* subjected to square-wave perturbations in nutrient supply: effect of varying amplitudes. *Biotechnology and Bioengineering*, **22**, 1213–24.

Pickett-Heaps, J. O. (1968). Further ultrastructural observations on polysaccharides. *Journal of Cell Science*, **3**, 55–64.

Pilet, P. E. & Senn, A. (1980). Root growth gradients: a critical analysis. *Zeitschrift für Pflanzenphysiologie*, **99**, 121–30.

Pirt, S. J. (1973). A quantitative theory of the action of microbes attached to a packed column: relevant to trickling filter effluent purification and to microbial action in the soil. *Journal of Applied Chemistry and Biotechnology*, **23**, 389–400.

Pricryl, Z. & Vancura, V. (1980). Root exudates of plants. VI. Wheat root exudation as dependent on growth, concentration gradient of exudates and the presence of bacteria. *Plant and Soil*, **57**, 69–83.

Reanney, D. C., Roberts, W. P. & Kelly, W. J. (1982). Genetic interactions among microbial communities. In *Microbial interactions and communities*, ed. A. T. Bull & J. H. Slater, pp. 287–322. London: Academic Press.

Reeve, C. A., Bockman, A. T. & Matin, A. (1984). Role of protein degradation in the survival of carbon-starved *Escherichia coli* and *Salmonella typhimurium*. *Journal of Bacteriology*, **157**, 758–63.

Roberts, D. P. & Sheets, C. J. (1990). Carbohydrate nutrition of *Enterobacter cloacae* ATCC 39978. *Canadian Journal of Microbiology*, **37**, 168–71.

Rouatt, J. W., Katznelson, H. & Payne, T. M. B. (1960). Statistical evaluation of the rhizosphere effect. *Soil Science Society of America Proceedings*, **24**, 271–3.

Rovira, A. D. (1965). Plant root exudates and their influence upon soil microorganisms. In *Ecology of soil-borne plant pathogens – prelude to biological control*, ed. K. F. Baker & W. C. Snyder, pp. 170–86. London: John Murray.

Rovira, A. D. & Campbell, R. (1975). A scanning electron microscope study of interactions between microorganisms and *Gaemannomyces graminis* (syn. *Ophiobolus graminis*) on wheat roots. *Microbial Ecology*, **2**, 177–85.

Rovira, A. D. & Davey, H. (1974). Biology of the rhizosphere. In *The plant root and its environment*, ed. E. W. Carson, pp. 153–204. Charlottesville, Va: University of Virginia Press.

Rovira, A. D., Newman, E. I., Bowen, H. J. & Campbell, R. (1974). Quantitative assessment of the rhizoplane microflora by direct microscopy. *Soil Biology and Biochemistry*, **6**, 211–6.

Samtsevich, S. A. (1962). Preparation, use and effectiveness of bacterial fertilizers in the Ukrainian SSR. *Mikrobiologiya*, **31**, 923–33.

Saunders, P. T. & Bazin, M. J. (1973). Attachment of microorganisms in a packed column: metabolite diffusion through the microbail film as a limiting factor. *Applied Chemistry and Biotechnology*, **23**, 847–53.

Scher, F. M., Ziegla, J. S. & Kloepper, J. W. (1984). A method for assessing the root-colonizing capacity of bacteria on maize. *Canadian Journal of Microbiology*, **30**, 151–7.

Scott, T. K. (1972). Auxins and roots. *Annual Review of Plant Physiology*, **23**, 235–58.

Sinclair, J. L. & Alexander, M. (1984). Role of resistance to starvation in bacterial survival in sewage and lake water. *Applied and Environmental Microbiology*, **48**, 410–5.

Streit, W., Kipe-Nolt, J. & Werner, D. (1991). Competitive growth of *Rhizobium leguminosarum* bv. *phaseoli* strains under oligotrophic conditions. *Current Microbiology*, **23**, 159–63.

Sun, L., Bazin, M. J. & Lynch, J. M. (1993). Plasmid dynamics in a model soil column. *Molecular Ecology*, **2**, 9–15.

Sundin, P., Valeur, A., Olsson, S. & Odham, G. (1990). Interactions between bacteria feeding nematodes and bacteria in the rape rhizosphere: effects on root exudation and distribution in bacteria. *FEMS Microbiology Ecology*, **73**, 13–22.

Theodorou, M. K., Bazin, M. J. & Trinci, A. P. J. (1980). Cellulose degradation in a structured ecosystem which is analogous to soil. *Transactions of the British Mycological Society*, **75**, 451–4.

Thompson, L. A., Nedwell, D. B., Balba, M. T., Banat, I. M. & Senior, E. (1983). The use of multiple-vessel, open flow systems to investigate carbon flow in anaerobic microbial communities. *Microbial Ecology*, **9**, 189–99.

Toko, K., Iiyama, S., Tanaka, C., Hayashi, K., Yamafugi, K. & Yamafugi, K. (1987). Relation of plant growth processes to spatial patterns of electric potential and enzyme activity in bean roots. *Biophysical Chemistry*, **27**, 39–58.

Trevors, J. T., Barkay, T. & Bourquin, A. W. (1987). Gene transfer among bacteria in soil and aquatic environments: a review. *Canadian Journal of Microbiology*, **33**, 191–8.

Turner, S. M. & Newman, E. I. (1984). Fungal abundance on *Lolium perenne* roots: the influence of nitrogen and phosphorus. *Transactions of the British Mycological Society*, **82**, 315–22.

Van Elsas, J. D., Dijkstra, A. F., Govaert, J. M. & Van Veen, J. A. (1986). Survival of *Pseudomonas fluorescens* and *Bacillus subtilis* introduced into two soils of different texture in field microplots. *FEMS Microbiology Ecology*, **38**, 151–60.

Van Elsas, J. D. & Trevors, J. T. (1990). Plasmid transfer to indigenous bacteria in soil and rhizosphere: problems and perspectives. In *Bacterial genetics in natural environments*, ed. J. C. Fry & M. J. Day, pp. 188–99. London: Chapman & Hall.

Van Laere, A. (1988). Effect of electrical fields on polar growth of *Phycomyces blakesleeanus*. *FEMS Microbiology Letters*, **49**, 111–16.

Van Vuurde, J. W. L. & Schippers, B. (1980). Bacterial colonization of seminal wheat roots. *Soil Biology and Biochemistry*, **12**, 559–65.

Warembourg, F. R. & Billes, G. (1979). Estimating carbon transfers in the plant rhizosphere. In *The soil-root interface*, ed. J. L. Hartley & R. S. Russell, pp. 183–96. London: Academic Press.

Whipps, J. M. (1985). Effect of CO_2 concentration on growth, carbon distribution and loss of carbon from the roots of maize. *Journal of Experimental Botany*, **36**, 644–51.

Whipps, J. M. & Lynch, J. M. (1985). Energy losses by the plant in rhizodeposition. *Annual Proceedings of the Phytochemical Society of Europe*, **26**, 59–71.

Wild, A. (ed.) (1988). *Russell's Soil Conditions and Plant Growth*, 11th edn. London: Longman.

Youatt, J., Gow, N. A. R. & Gooday, G. W. (1988). Bioelectric and biosynthetic aspects of cell polarity in *Allomyces macrogynus*. *Protoplasma*, **146**, 118–126.

13

Biofilms of the Ruminant Digestive Tract

K.-J. Cheng, Tim A. McAllister and J. William Costerton

Introduction

Direct observations of microbial growth in a large number of natural ecosystems have shown that the predominant populations are attached to surfaces where they grow in glycocalyx enclosed microcolonies that develop into adherent biofilms (Costerton *et al.* 1987; Cheng *et al.* 1991; Lappin-Scott *et al.* 1992). In various digestive tracts, this adhesion takes the form of very specific associations with insoluble nutrients (Cheng *et al.* 1977; Akin 1979; Cheng *et al.* 1991) to form particle associated microbial populations, and with tissue surfaces to form tissue associated populations (Cheng & Costerton 1980; Cheng *et al.* 1981a, b). A variation of this latter mode of growth is the association of a major microbial population with the mucous blanket of secretory intestinal tissue (Rozee *et al.* 1982). Bacteria of the ruminant digestive tract can be divided into three distinct subpopulations, (i) those associated with the digesta, (ii) those associated with gastrointestinal tissue and (iii) those associated with gastrointestinal fluid (Cheng & Costerton 1980; Cheng *et al.* 1981a; Czerkawski & Cheng 1988).

General principles of microbial ecology dictate that any organism will establish itself in a favourable nutrient niche and proliferate within a physiologically integrated community. Our observations of the growth of bacteria associated with feed particles indicate that each species adheres to its own particular insoluble substrate (e.g. cellulose, protein, starch) and produces enzymes to degrade insoluble substrates to soluble nutrients (Cheng *et al.* 1984; McAllister *et al.* 1990b). Because most digestive processes involve the coordination of physiologically related activities, we often see the development of structured consortia (Fig. 13.1) within which the metabolic products of the primary digestive organisms are the substrates for sequential members of a layered biofilm (Wolin & Miller 1988). The digestible dead epithelial cells of some tissues concentrate bacterial nutrients (for example urea) at their surfaces (Wallace *et al.* 1979; Cheng & Wallace 1979; Dinsdale *et al.* 1980) and constitute well defined niches in which taxonomically distinct microbial populations develop. These populations often play an important role in the overall function of the digestive tract (Cheng &

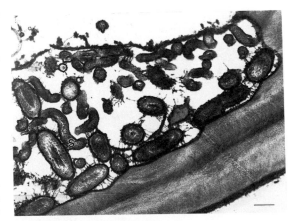

Fig. 13.1. Transmission electron micrograph (TEM) of a ruthenium red stained preparation of barley straw incubated in a continuous culture of a mixed natural population of bacteria from the bovine rumen. Note the extensive pit formation by rod-shaped Gram negative bacteria resembling *Fibrobacter succinogenes* and the presence of spirochaetes (s) in the same adherent consortium. Bar = 1 μm. (From Cheng & Costerton 1986.)

Costerton 1980). The free microbial cells in gastrointestinal fluid can be largely available for attachment to new nutrients or freshly exposed tissue and this population contains some cells from the adherent population that are in transit between surfaces.

Generally, the presence of well adapted microbial biofilms on a tissue surface is beneficial in that these cells cooperate physiologically and the biofilm precludes the adhesion of pathogenic bacteria and viruses. Thus, an objective of successful management should be to encourage the development of beneficial biofilms on feed particles and on tissues at all levels of the digestive tract (Cheng *et al.* 1991). Modification of the physiological activity of the digestive tract populations of bacteria in mature animals can be achieved either through chemical agents or processing of the feed.

Biofilms on the ruminant tongue

The tongue is exposed to air, food, saliva and contents of the reticulo-rumen during regurgitation. The dorsum of the tongue, where particulate feed is chewed and compressed, is very highly abraded during regurgitation. Accordingly, the adherent biofilm of this tissue surface must persist in spite of massive shear forces. Resistance to mechanical displacement is exemplified by the tightly adherent growth of trichome forming bacteria and members of the *Alysiella* and *Simonsiella* genera (Fig. 13.2a, b) which predominate in particularly high abrasion areas (McCowan *et al.* 1979). However, many other chain forming bacteria and rod shaped organisms are present in less abraded areas. *Alysiella* and *Simonsiella* genera adhere firmly to the tongue tissues by fibrillar glycoprotein structures formed on only one side

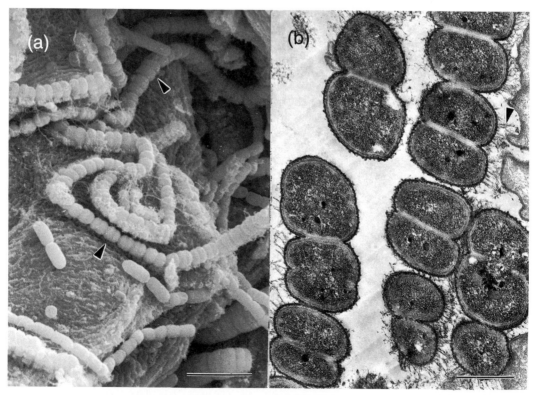

Fig. 13.2. (a) Scanning electron micrograph (SEM) of a mixed population on the tissue of the tongue. Almost all the bacteria are either filamentous or chain forming, and many follow the contours of the tissue cells, often appearing wedged between them (arrows). (From McCowan *et al.* 1979). (b) TEM showing the adhesion of these trichome forming cells to tissues by means of fibres (arrow) found only on one side of the bacterial cells. Bars = 5 μm. (From McCowan *et al.* 1979.)

Table 13.1. Proportion of bacteria associated with plant cell wall material and liquor in rumen digesta of cows fed straw based diets[a]

Genera	Composition (%) Plant cell wall material	Rumen liquor
Butyrivibrio	32	7
Selenomonas	14	10
Unnamed spirochaete	8	0
Lachnospira	8	1
Megasphaera	0	11
Streptococcus	3	12
Ruminococcus	16	6
Bacteroides	11	38
Others	8	15
Total number of isolates	368	292
Viable count (CFU[b] ml-1×107)	230 ± 38	201 ± 13

[a] Data from these animals each sampled twice (Latham 1980).
[b] cfu, colony forming units.

of these bacterial cells. Adhesion is so avid that disruption of the host tissue is usually necessary to detach these cells (McCowan *et al.* 1978, 1979). The adherent biofilm of the bovine tongue remains remarkably consistent in both morphological balance and numbers throughout the life of the animal, and it is notably resistant to the influences of different diets.

Biofilms within the reticulo rumen

Approximately 20–30% of the bacteria in the reticulo-rumen are found in the rumen fluid, while 70–80% are associated with feed particles (Forsberg & Lam 1977; Craig *et al.* 1987), and approximately 1% are associated with the epithelial surface. The fluid population contains detached representatives from the surfaces of feed particles and tissue, in addition to organisms that are not normally adherent and subsist on soluble nutrients in the rumen fluid (McAllister *et al.* 1994). The rumen fluid population has a distinct composition (Table 13.1; Latham 1980) with some shared genera in both the particle and tissue associated populations (Table 13.2). Within minutes of ingestion, feed particles (e.g. legumes, grass, cereal grains) are colonized by organisms associated with the rumen fluid (Bonhomme 1990). Starch granules (Minato & Suto 1979; McAllister *et al.* 1990b) and cellulose (Akin 1986; Bowman & Firkins 1993) rapidly

Table 13.2. Genera recovered from epithelial tissue at 13 sites in the bovine rumen

Genus	Number of sites yielding this genus[a]
Fusobacterium	12
Propionibacterium	12
Eubacterium	8
Bacteroides	6
Selenomonas	5
Butyrivibrio	4
Streptococcus	4
Peptococcus	3
Bifidobacterium	3
Lactobacillus	3
Micrococcus	2
Flavobacterium	2
Hemophilus	2

[a] Blocks of tissue were removed, washed, homogenized, and cultured to recover these adherent organisms.

become heavily colonized by amylolytic and cellulolytic bacteria, respectively. These initial colonizing bacteria multiply and are joined by additional bacteria from the ruminal fluid to form the mature biofilms that are frequently observed on the surfaces of feed particles and epithelial tissue (McAllister *et al.* 1994).

Direct examination of the leaves of forage legumes exposed to rumen fluid shows that certain bacterial species invade the intercellular spaces and colonize pectin rich surfaces (Cheng

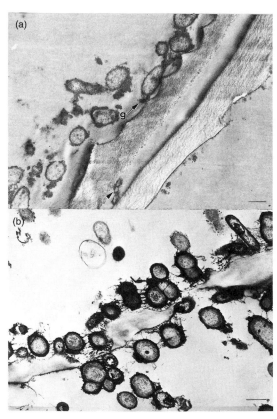

Fig. 13.3. (a) TEM of a ruthenium red stained preparation of sterilized barley straw incubated with a pure culture of strain BL-2 of *Bacteroides succinogenes*. Note the 'pits' and discontinuities of the cellulose where these bacterial cells have adhered, the vesicular residues of bacterial cell envelope (arrow) which are still adherent to the plant cell wall, and the fibrous bacterial product (g) that mediates bacterial adhesion. (b) TEM of a ruthenium red stained preparation of barley straw incubated with a continuous culture of a mixed natural population of bacteria from the bovine rumen. Note the extensive 'pit' formation by cells resembling Ruminococcus species (arrows). Bars = 1 μm. (Authors' unpublished data.)

& Costerton 1980). When a particular bacterium has adhered to its specific nutrient substrate (e.g. cellulose) and finds itself in a favourable nutrient niche, it proliferates to form a functional microcolony in which specific digestive enzymes are produced in especially high concentrations. These enzymes may remain cell associated, but they remain active even after bacterial cell breakdown (Stewart *et al.* 1979; Forsberg *et al.* 1981; Groleau & Forsberg 1981). Cell envelope fragments of *Fibrobacter succinogenes* concentrate sufficient cellulase activity to form deep pits in the surface of cellulose fibres (Fig. 13.3a). The adherent bacterial biofilm can be thought of as constituting a local 'critical mass' of enzyme activity that focuses its attack on cellulose to produce distinct localized pits (Fig. 13.3b). This process is characteristic of the bacterial digestion

of feeds such as legumes, grass and straw (Cheng *et al.* 1991).

The population of bacteria associated with the epithelial tissues of the reticulo-rumen is relatively small (i.e. less than 1%: Fig. 13.4a, b, c). Taxonomically, this tissue associated microbial population is unique in that it contains large numbers of *Fusobacterium* species and Gram positive cocci (Table 13.2). These latter organisms are ideally suited to the epithelial environment, because many of them are facultative anaerobes and produce urease, which hydrolyses urea and generates the ammonia required for bacterial metabolism within the rumen (Cheng & Wallace 1979). Bacteria contained within this biofilm are unique in producing an essential enzyme (i.e. urease) lacking in the animal tissue (Cheng & Wallace 1979; Wallace *et al.* 1979) and this

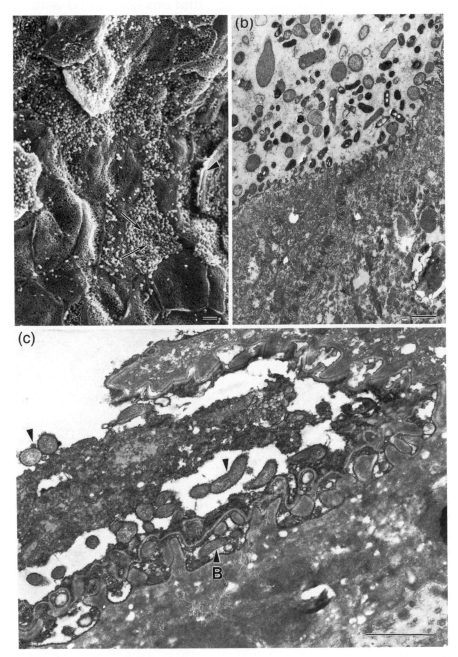

Fig. 13.4.   (a) SEM of the bacterial population of the rumen epithelium from a milk-fed Holstein calf. The number of adherent bacteria varies from one tissue cell to the next. One tissue cell (small arrows) may be completely covered by adherent bacteria while adjacent cells carry very low numbers. Bacteria can be seen colonizing the undersurface of a shedding epithelial cell (large arrow). (From Cheng *et al.* 1981b.) (b) TEM of a ruthenium red stained population of the rumen tissue showing the thick and complex bacterial biofilm that develops on this organ wall. These cells are enclosed in a delicate fibrous network within the biofilm. (Cheng *et al.* 1979). (c) TEM of a section of a ruthenium red stained preparation of the tissue of the bovine rumen showing the digestion of a dead tissue cell by adherent bacteria (arrows). Note that one of these bacteria (B) appears to have penetrated the intercellular space under the cell undergoing digestion. Bars represent: (a), (b) 5 µm; (c) 1 µm. (From Cheng *et al.* 1981b.)

association between animal and ruminal bacteria is an excellent example of the mutualistic benefits of biofilms.

The tissue associated biofilm is acquired early (within 2 days of birth) in the life of the ruminant (Fonty *et al.* 1987, 1988; Stewart *et al.* 1988; Cheng *et al.* 1991). Whole animal infusion studies have shown that this population is maintained even if no feed is ingested (Wallace *et al.* 1979; Dinsdale *et al.* 1980). High numbers of tissue associated bacteria have been found at 14 sites on the epithelia of cattle fed hay and concentrates. Milk-fed animals (up to 200 kg) also possess microbial cells on their eptihelial surfaces, indicating that the development of mature biofilms does not require the presence of solid feed in the digestive tract (Wallace *et al.* 1979).

Studies with Svalbard reindeer suggest that the formation of tissue adherent biofilm may be related to energy intake. In the severe Norwegian winter, these animals are frequently starved and the epithelial tissue of the rumen is sparsely colonized by small bacteria (Cheng *et al.* 1993). In contrast, when feed is readily available in the spring and summer, the rumen epithelium is completely covered by a thick bacterial biofilm of several component species.

Biofilms and ruminant digestion and physiology

In addition to the primary digestion of food materials in the rumen, the three major microbial populations participate in a myriad of metabolic processes that help sustain the animal by the production of volatile fatty acids, vitamins and the microbial biomass that is itself digested in the lower digestive tract. These microbial activities can sometimes be located in the rumen by direct cytological identification of the enzymes concerned (Fay *et al.* 1979, 1983). It has been estimated that 75% of the protease activity (Brock *et al.* 1982), 70% of the amylase activity and 80% of the endoglucanase activity (Minato *et al.* 1966) is associated with the particulate fraction of rumen contents. These data indicate that the biofilms associated with feed particles are responsible for the majority of feed digestion in the rumen.

Biofilms associated with the omasum

The epithelium of the omasum is colonized by a biofilm that resembles that of the reticulo-rumen in both its density and its distribution (Fig. 13.5a). A large proportion of the surface of this tissue is occupied by a varied population of bacteria (Fig. 13.5a), but a mycelial fungus (Fig. 13.5b) and a bacterium that grows in especially coherent glycocalyx enclosed microcolonies (Fig. 13.6) are also observed. The bacteria of this adherent tissue associated population have a taxonomic composition similar to that on the epithelium of the reticulo-rumen, but some of the Gram positive species that colonize the rumen wall appear to be absent in the omasum (Table 13.3).

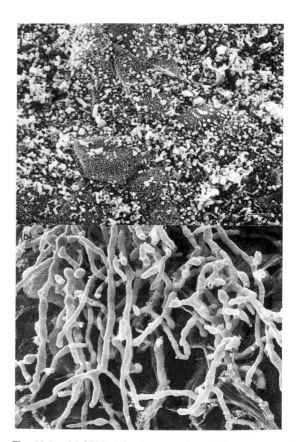

Fig. 13.5. (a) SEM of the tissue surface of the omasum of a 6 month old Holstein calf fed a grain diet. This tissue surface is colonized by a varied bacterial population similar to that of the reticulo-rumen. (b) SEM of the surface of the omasum of a calf fed as in Fig.13.7 showing one of the areas of the tissue colonized by a fungus that produces elongated cells. Bars = 5 μm.

Table 13.3. Bacteria isolated from the lower digestive tract of cattle

Omasum	Abomasum	Duodenum, small intestine and caecum
Propionibacterium acnes	*Propionibacterium acnes*	*Staphylococcus* sp.
Propionibacterium sp.	*Fusobacterium prausnitzii*	*Propionibacterium acnes*
Fusobacterium prausnitzii	*Fusobacterium* species	*Propionibacterium* sp.
Butyrivibrio fibrisolvens	*Eubacterium aerofaciens*	*Clostridium* sp.
Eubacterium aerofaciens	*Eubacterium* species	
Clostridum clostridiiforme	*Clostridium* species	
Peptostreptococcus intermedius	*Peptostreptococcus intermedius*	
	Staphylococcus species	
	Gram positive cocci	
	Gram positive bacillus	

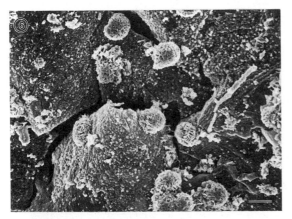

Fig. 13.6. SEM of one of the rare areas of the tissue of the omasum which is colonized by bacteria that form discrete ball-like adherent microcolonies. Bar = 5 μm. (From Cheng & Costerton 1986.)

The microbiology of the contents of the omasum has not been examined to date, but the direct flow of digesta from the reticulo-rumen into this organ leads us to anticipate no major differences in the particle associated and fluid populations.

Biofilms within the abomasum

A major change in the ruminant digestive system occurs at the abomasum, where the epithelium becomes secretory and the pH of the lumen is reduced to between 2 and 4. These conditions are not as conducive to the formation of the complex biofilms that are observed in other regions of the digestive tract. In cattle and sheep, direct observation shows relatively few tissue associated bacteria in the mucous blanket at the epithelial surface of

Table 13.4. Bacterial urease production

	Urease activity[a]		
Area sampled	Holstein	Hereford	Sheep
Reticulum	8.08	2.10	1.15
Rumen	4.72	1.04	0.96
Omasum	0.52	0.71	0.35
Abomasum	0.01	3.15	0.02
Duodenum	0.07	-	0.02

[a]μmoles N min^{-1} in^{-2} of epithelial tissue.

the abomasum. In addition, the low level of urease associated with tissue from the abomasum suggests that this organ is sparsely colonized by ureolytic bacteria in Holstein cattle and sheep (Table 13.4). Bacteria are seen within the mucous blanket at the tissue surface and particular types appear to predominate in certain areas (Fig. 13.7). The species of bacteria within this limited biofilm are not markedly different from those of the upper tract, except for the presence of relatively high numbers of Gram positive cocci and of an unidentified group of Gram positive bacilli (Table 13.3). An exception to this sparse bacterial colonization of the abomasum is seen in certain breeds of beef cattle (Angus, Hereford) in which greater colonization and high levels of bacterial urease are detected on the tissue surface (Table 13.4).

Biofilms of the lower digestive tract

A thick and coherent mucous blanket on the tissue surfaces of the small intestine, caecum and

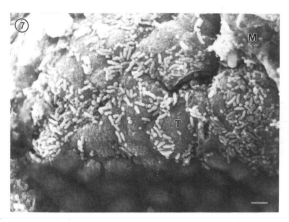

Fig. 13.7.   SEM of the tissue of the bovine abomasum showing 'buried' bacteria where the copious mucous layer (M) is intact and clear outlines of adherent native bacteria where the tissue surface (T) is exposed by loss of the covering mucus. The diversity of bacterial forms seen in the rumen and omasum is sharply reduced in the abomasum. Bar = 5 µm. (From Cheng *et al.* 1981b).

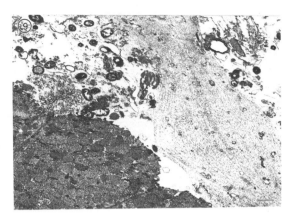

Fig. 13.9.   TEM of a section of a ruthenium red-stained preparation of the bovine ileum showing the presence of bacteria associated with the mucous layer that is retained on the tissue surface in this preparation. Bar = 1 µm. (From Cheng *et al.* 1981b).

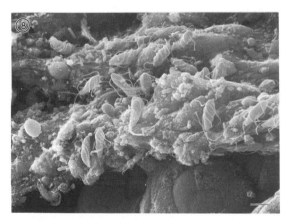

Fig. 13.8.   SEM preparation of the mouse ileum in which the mucous blanket has been retained. Note the association of very large numbers of bacteria and protozoa with the mucous blanket and very few with the actual tissue surface (arrow). Bar = 5 µm. (From Costerton *et al.* 1985.)

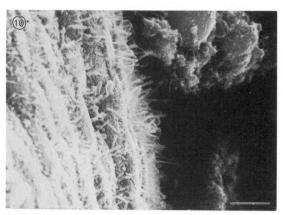

Fig. 13.10.   SEM of a colon of a milk-fed Holstein calf showing the end-on association of rod shaped bacteria with the surface of this tissue. Bar = 1 µm. (From Cheng & Costerton 1986).

large intestine of ruminants presents considerable technical problems in digestive tract microbiology. Routine preparation of tissues of monogastric animals for electron microscopy leads to the loss of the mucous blanket and, with it, the greater part of the biofilm on the tissue surface (Rozee *et al.* 1982). Most of the autochthonous bacteria and protozoa that occupy the tissue surfaces of the intestine live in the mucous layer

(Fig. 13.8) and only a very small number of organisms grow immediately on the tissue surface (Savage 1977; Savage *et al.* 1968). Examinations of the lower digestive tract of ruminants have not yet been made using immunological methods for mucous retention, but some micrographs of these tissues show numerous bacteria in association with remnants of the mucous blanket (Fig. 13.9). In these preparations, small numbers of organisms are sufficiently adherent to, or embedded within, the tissue surface to remain associated with duodenal and colonic epithelia even after removal of the mucous blanket (Fig. 13.10).

Bacterial recovery from these tissues has been plagued by the same loss of the mucous blanket and its epithelial embedded population, a technical problem that may account for the relatively low numbers and few species isolated from this region (Table 13.3).

Manipulation of biofilms to benefit ruminants

The digestive tract of the newborn ruminant constitutes a microbial 'vacuum' which may be colonized by beneficial autochthonous bacteria or by pathogens, such as the pathogenic *Escherichia coli* that cause neonatal scours (Fig. 13.11). Studies in our laboratory have shown that oral inoculation of newborn lambs with 36 autochthonous organisms, at 16 h and 14 days, resulted in complete bacterial colonization of the ruminal epithelium (Fig. 13.12). Inoculated animals exhibited a greater average weight gain (at 135 days), accelerated development of the upper digestive tract, and were more resistant to enterotoxigenic *E. coli* than uninoculated control animals reared under identical vivarium conditions (T. A. McAllister, K.-J. Cheng & J. W. Costerton, unpublished data). However, this laboratory inoculum failed to improve the formation of biofilms within the digestive tracts of inoculated lambs, compared with those in lambs that were allowed to develop

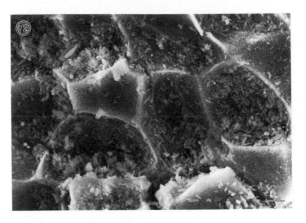

Fig. 13.12. SEM of the rumen tissue surface of a newborn lamb reared for 4 months under vivarium conditions and inoculated orally at 16 h and 14 days after birth with a mixture of 36 strains of bacteria isolated from all regions of the bovine digestive tract. Bar = 1 μm (From Cheng & Costerton 1986.)

natural microbial populations while suckling their mothers. Apparently these laboratory grown bacterial strains lose their ability to compete effectively with microorganisms that are adapted to forming functional biofilms under the more rigorous conditions of the natural environment.

Established biofilms are difficult to change by biological manipulation, but they can be altered by chemical manipulations that favour the induction of particular enzyme activities or the establishment of certain species within the digestive system. The production of nitrotoxin degrading enzymes may be induced by feeding nitrate and nitro-compounds (Majak *et al.* 1982), while methane production can be inhibited by halogenated compounds (Chalupa 1977; Demeyer & van Nevel 1975; Prins 1978). Ionophores are now routinely used to alter rumen fermentation to favour the production of propionate (van Nevel & Demeyer 1988).

Once a stable microbial population has been established, feed efficiency can be improved and digestive disturbances prevented by the manipulation of specific plant materials for digestion by existing microbial populations. For decades, physical and chemical processing have been used to improve the utilization of feed by ruminants. Many of these techniques were developed prior to an appreciation for the importance of microbial biofilms in feed digestion. When processing is examined in detail, it is evident that the

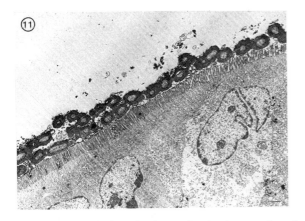

Fig. 13.11. TEM of a ruthenium red-stained preparation of the ileum of a newborn calf that had been infected with an enterotoxigenic strain of *E. coli* (ETEC). Note the heavy colonization of the microvillar (M) border of this tissue and the extensive matrix (X) surrounding the adherent bacterial cells (arrow).

effectiveness of many processes is due to their ability to enhance the formation of biofilms on the surface of feed. Grinding and alkali treatment increase the number of adhesion sites, enhance the availability of nutrients for the biofilm (Latham *et al.* 1979; Kolankaya *et al.* 1985; Bowman & Firkins 1993) and increase the digestion of these feeds in the rumen. Biological delignification via white rot basidiomycetes has been shown to increase the *in vitro* digestibility of oat straw, presumably by breaking down microbial barriers and stimulating biofilm formation (Jung *et al.* 1992). In contrast, fat and formaldehyde inhibit the formation of biofilms and limit the microbial digestion of feeds (Stewart 1977; McAllister *et al.* 1990a). Many of the advances possible in physical and chemical processing of feed have probably already been made and future strategies to enhance the formation of digestive biofilms on plant tissues are likely to result from technologies in genetic engineering and plant breeding.

The too rapid release of nutrients that results from the bacterial digestion of bloat-causing legumes and finely ground cereal grains can be reduced by selecting plants for a slow initial rate of ruminal digestion and by processing grains to a coarse particle size. Conversely, the slow initial rate of ruminal digestion of grasses can be increased by breeding plants with cell walls that are more susceptible to digestion by rumen bacteria (Cherney *et al.* 1986). Selection for brown midrib mutants of maize, sorghum and pearl millet has resulted in plants that contain substantially lower (25–50%) lignin than normal genotypes and these plants have increased digestibilities *in vitro* and in dairy cows and sheep (Porter *et al.* 1978; Wedig *et al.* 1988). Recently, Watson (1990) has advocated using antisense RNA genes or catalytic RNA molecules as anti-lignin 'knock-out' genes to perturb the biosynthesis of lignin in alfalfa. The improved ruminal digestibility of these selected or genetically altered forages is most likely due to a reduction in the barriers to biofilm formation. Further characterization of natural plant compounds (e.g. condensed tannins) that inhibit the formation of digestive biofilms could provide additional selection criteria for plant breeding programmes or potential targets for genetic manipulation.

References

Akin, D. E. (1979). Microscopic evaluation of forage digestion by rumen microorganisms – a review. *Journal of Animal Science*, **48**, 701–10.

Akin, D. E. (1986). Chemical and biological structure in plants as related to microbial degradation of forage cell walls. In *Control of digestion and metabolism in ruminants*, ed. L. P. Milligan, W. L. Grovum & A. Dobson, pp. 139–57. New Jersey: Prentice-Hall.

Bonhomme, A. (1990). Rumen ciliates: their metabolism and relationships with bacteria and their hosts. *Animal Feed Science and Technology*, **30**, 203–66.

Bowman, J. G. P. & Firkins, J. L. (1993). Effects of forage species and particle size on bacterial cellulolytic activity and colonization *in situ*. *Journal of Animal Science*, **71**, 1623–33.

Brock, F. M. Forsberg, C. W. & Buchanan-Smith, J. G. (1982). Proteolytic activity of rumen microorganisms and effects of proteinase inhibitors. *Applied and Environmental Microbiology*, **44**, 561–9.

Chalupa, W. (1977). Manipulating rumen fermentation. *Journal of Animal Science*, **45**, 585–99.

Cheng, K.-J., Akin, D. E. & Costerton, J. W. (1977). Rumen bacteria: interaction with dietary components and response to dietary variation. *Federation Proceedings of the American Society of Experimental Biology*, **36**, 193–7.

Cheng, K.-J. & Costerton, J. W. (1980). Adherent rumen bacteria: their role in the digestion of plant material, urea, and epithelial cells. In *Digestive physiology and metabolism in ruminants*, ed. Y. Ruckebush & P. Thivend, pp. 227–50. Lancaster, UK: MTP Press.

Cheng, K.-J. & Costerton, J. W. (1986). Microbial adhesion and colonization within the digestive tract. In *Anaerobic bacteria in habitats other than man*, ed. E. M. Barnes and G. C. Mead, pp. 239–61. London: Blackwells.

Cheng, K.-J., Fay, J. P., Coleman, R. N., Milligan, L. P. & Costerton, J. W. (1981a). Formation of bacterial microcolonies on feed particles in the rumen. *Applied and Environmental Microbiology*, **41**, 298–305.

Cheng, K.-J., Fay, J. P., Howarth, R. E. & Costerton, J. W. (1980). Sequence of events in the digestion of fresh legume leaves by rumen bacteria. *Applied and Environmental Microbiology*, **40**, 613–25.

Cheng, K.-J., Forsberg, C. W., Minato, H. & Costerton, J. W. (1991). Microbial ecology and physiology of feed degradation within the rumen. In *Physiological aspects of digestion and metabolism in ruminants*, ed. T. Tsuda, Y. Sasaki & R. Kawashima, pp. 595–624. Toronto: Academic Press.

Cheng, K.-J., Irvine, R. T. & Costerton, J. W.

(1981b). Autochthonous and pathogenic colonization of animal tissues by bacteria. *Canadian Journal of Microbiology*, **27**, 461–90.

Cheng, K.-J., McAllister, T. A., Mathiesen, S. D., Blix, A. S., Orpin, C. G. & Costerton, J. W. (1993). Seasonal changes in the adherent microflora of the rumen in high-arctic Svalbard reindeer. *Canadian Journal Microbiology*, **39**, 101–8.

Cheng, K.-J., McCowan, R. P. & Costerton, J. W. (1979). Adherent epithelial bacteria in ruminants and their roles in digestive tract function. *American Journal of Clinical Nutrition*, **32**, 139–48.

Cheng, K.-J., Stewart, C. S., Dinsdale, D. & Costerton, J. W. (1984). Electron microscopy of bacteria involved in the digestion of plant cell walls. *Animal Feed Science and Technology*, **10**, 93–120.

Cheng, K.-J. & Wallace, R. J. (1979). The mechanism of passage of endogenous urea through the rumen wall and the role of ureolytic epithelial bacteria in the urea flux. *British Journal of Nutrition*, **42**, 533–57.

Cherney, J. H., Moore, K. J., Volenec, J. J., & Axtell, J. D. (1986). Rate and extent of digestion of cell wall components of brown-midrib sorghum species. *Crop Science*, **26**, 1055–9.

Costerton, J. W., Cheng, K.-J., Geesey, G. G. *et al.* (1987). Bacterial biofilms in nature and disease. *Annual Reviews of Microbiology*, **41**, 435–64.

Costerton, J. W., Marrie, T. J. & Cheng, K.-J. (1985). Phenomena of bacterial adhesion. In *Bacterial adhesion*, ed. D. C. Savage & M. Fletcher, pp. 3–43. New York: Plenum Press.

Craig, W. M., Broderick, G. A. & Ricker, D. B. (1987). Quantitation of microorganisms associated with the particulate phase of ruminal ingesta. *Journal of Nutrition*, **117**, 56–62.

Czerkawski, J. W. & Cheng, K.-J. (1988). Compartmentation in the rumen. In *The rumen microbial ecosystem*, ed. P. N. Hobson, p. 361. New York: Elsevier.

Demeyer, D. I. & van Nevel, C. J. (1975). Methanogenesis, an integrated part of carbohydrate fermentation, and its control. In *Digestion and metabolism in the ruminant*, ed. I. W. McDonald & A. C. I. Warner, pp. 366–82. Armidale: University of New England Publishers.

Dinsdale, D., Cheng, K.-J., Wallace, R. J. & Goodlad, R. A. (1980). Digestion of epithelial tissue of the rumen wall by adherent bacteria in infused and conventionally fed sheep. *Applied and Environmental Microbiology*, **39**, 1059–66.

Fay, J. P., Cheng, K.-J. & Costerton, J. W. (1979). Production of alkaline phosphatase by epithelial cells and adherent bacteria of the bovine rumen and abomasum. *Canadian Journal of Microbiology*, **25**, 932–36.

Fay, J. P., Cheng, K.-J. & Costerton, J. W. (1983).

Effects of breed, diet and sex on the alkaline phosphatase activity in walls of the bovine rumen and abomasum. *Journal of Animal Science*, **56**, 1427–33.

Fonty, G., Gouet, P., Jouany, J.-P. & Senaud, J. (1987). Establishment of the microflora and anaerobic fungi in the rumen of lambs. *Journal of General Microbiology*, **133**, 1835–43.

Fonty, G., Senaud, J., Jouany, J.-P. & Gouet, P. (1988). Establishment of ciliate protozoa in the rumen of conventional and conventionalized lambs: influence of diet and management conditions. *Canadian Journal of Microbiology*, **34**, 235–41.

Forsberg, C. W. & Lam, K. (1977). Use of adenosine 5'-triphosphate as an indicator of the microbiota biomass in rumen contents. *Applied and Environmental Microbiology*, **33**, 528–37.

Forsberg, C. W., Beveridge, T. J. & Hellstrom, A. (1981). Cellulase and xylanase release from *Bacteroides succinogenes* and its importance in the rumen environment. *Applied and Environmental Microbiology*, **42**, 886–96.

Groleau, D. & Forsberg, C. W. (1981). Cellulolytic activity of the rumen bacterium *Bacteroides succinogenes*. *Canadian Journal of Microbiology*, **27**, 517–30.

Jung, H. G., Valdez, F. R., Abad, A. R., Blanchette, R. A. & Hatfield, R. D. (1992). Effect of white rot basidiomycetes on chemical composition and *in vitro* digestibility of oat straw and alfalfa stems. *Journal of Animal Science*, **70**, 1928–35.

Kolankaya, N., Stewart, C. S., Duncan, S. H., Cheng, K.-J. & Costerton, J. W. (1985). The effect of ammonia treatment on the solubilization of straw and the growth of cellulolytic bacteria. *Journal of Applied Bacteriology*, **58**, 371–79.

Lappin-Scott, H. M., Costerton, J. W. & Marrie, T. J. (1992). Biofilms and biofouling. In *Encyclopedia of microbiology*, Vol. 1, ed. Alexandra, D. A. Hopwood, B. H. Iglewski & A. I. Laskin, pp. 277–84. Toronto: Academic Press.

Latham, M. J. (1980). Adhesion of rumen bacteria to plant cell walls. In *Microbial adhesion to surfaces*, ed. R. C. W. Berkeley, J. M. Lynch, J. Melling, P. R. Rutter & B. Vincent, pp. 339–50. Chichester: Ellis Horwood.

Latham, M. J., Hobbs, D. G. & Harris, P. J. (1979). Adhesion of rumen bacteria to alkali-treated plant stems. *Ann. Rech. Vet.*, **10**, 244–5.

Majak, W., Cheng, K.-J. & Hall, J. W. (1982). The effect of cattle diet on the metabolism of 3–nitropropanol by ruminal microorganisms. *Canadian Journal of Animal Science*, **62**, 855–60.

McAllister, T. A., Bae, H. D., Jones, G. A. & Cheng, K.-J. (1994). Microbial attachment and feed digestion in the rumen. *Journal of Animal Science* (in press).

McAllister, T. A., Cheng, K.-J., Rode, L. M. & Buchanan-Smith, J. G. (1990a). Use of formaldehyde to regulate the digestion of barley starch. *Canadian Journal of Animal Science*, **70**, 581–9.

McAllister, T. A., Rode, L. M., Major, D. J., Cheng, K.-J. & Buchanan-Smith, J. G. (1990b). The effect of ruminal microbial colonization on cereal grain digestion. *Canadian Journal of Animal Science*, **70**, 571–9.

McCowan, R. P., Cheng, K.-J., Bailey, C. B. M. & Costerton, J. W. (1978). Adhesion of bacteria to epithelial cell surfaces within the reticulo-rumen of cattle. *Applied and Environmental Microbiology*, **35**, 149–55.

McCowan, R. P., Cheng, K.-J. & Costerton, J. W. (1979). Colonization of a portion of the bovine tongue by unusual filamentous bacteria. *Applied and Environmental Microbiology*, **37**, 1224–9.

Minato, H., Endo, A., Ootomo, Y. & Uemura, T. (1966). Ecological treatise on the rumen fermentation. II. The amylolytic and cellulolytic activities of fractionated bacterial portions attached to the rumen solids. *Journal of General Microbiology*, **12**, 53–69.

Minato, H. & Suto, T. (1979). Technique for fractionation of bacteria in rumen microbial ecosystem. III. Attachment of bacteria isolated from bovine rumen to starch granules *in vitro* and elution of bacteria attached therefrom. *Journal of General Applied Microbiology*, **25**, 71–93.

Porter, K. S., Axtell, J. D., Lechtenberg, V. L. & Colenbrander, V. F. (1978). Phenotype, fiber composition, and *in vitro* dry matter disappearance of chemically induced brown midrib (bmr) mutants of sorghum. *Crop Science*, **18**, 205–8.

Prins, R. A. (1978). Nutritional impact of intestinal drug microbe interactions. In *Nutrition and drug interrelations*, ed. J. N. Hatcock & J. Coon, pp. 189–251. New York: Academic Press.

Rozee, K. R., Cooper, D., Lam, K. & Costerton, J. W. (1982). Microbial flora of the mouse ileum mucous layer and epithelial surface. *Applied and Environmental Microbiology*, **43**, 1451–63.

Savage, D. C. (1977). Microbial ecology of the gastrointestinal tract. *Annual Reviews of Microbiology*, **31**, 107–33.

Savage, D. C., Dubos, R. & Schaedler, R. W. (1968). The gastrointestinal epithelium and its autochthonous bacterial flora. *Journal of Experimental Medicine*, **127**, 67–76.

Stewart, C. S. (1977). Factors affecting the cellulolytic activity of rumen contents. *Applied and Environmental Microbiology*, **33**, 497–502.

Stewart, C. S., Dinsdale, D., Cheng, K.-J. & Paniagua, C. (1979). The digestion of straw in the rumen. In *Straw decay and its effect on disposal and utilization*, ed. E. Grossbard, pp. 123–30. Chichester: Wiley.

Stewart, C. S., Fonty, G. & Gouet, P. (1988). The establishment of rumen microbial communities. *Animal Feed Science and Technology*, **21**, 69–97.

van Nevel, C. J. & Demeyer, D. I. (1988). Manipulation of rumen fermentation. In *The rumen microbial ecosystem*, ed. P. N. Hobson, pp. 387–443. New York: Elsevier.

Wallace, R. J., Cheng, K.-J., Dinsdale, D. & Orskov, E. R. (1979). An independent microbial flora of the epithelium and its role in the ecomicrobiology of the rumen. *Nature*, **279**, 424–6.

Watson, J. M. (1990). Genetic engineering of low-lignin pasture plants. In *Microbial and plant opportunities to improve lignocellulose utilization by ruminants*, ed. D. E. Akin, L. G. Ljungdahl, J. R. Wilson & P. J. Harris, pp. 215–26. New York: Elsevier.

Wedig, C. L., Jaster, E. H. & Moore, K. J. (1988). Effect of brown midrib and normal genotypes of sorghum X sudangrass on ruminal fluid and particulate rate of passage from the rumen and extent of digestion at various sites along the gastrointestinal tract of sheep. *Journal of Animal Science*, **66**, 559–65.

Wolin, M. J. & Miller, T. L. (1988). Microbe–microbe interactions. In *The rumen microbial ecosystem*, ed. P. N. Hobson pp. 343–59. New York: Elsevier.

14

The Immune Response to Bacterial Biofilms

Niels Høiby, Anders Fomsgaard, Elsebeth Tvenstrup Jensen, Helle Krogh Johansen, Gitte Kronborg, Svend S. Pedersen, Tacjana Pressler and Arsalan Kharazmi

Introduction

The immune response to bacterial biofilms has been extensively studied in patients suffering from cystic fibrosis (CF), since nearly all of these patients eventually contract chronic *Pseudomonas aeruginosa* lung infection (Pedersen 1992; Koch & Høiby 1993). This infection has been shown to be a typical biofilm (cryptic) infection, where the bacteria colonize the smaller airways (endo-

bronchiolitis) without invading the deeper tissues (Fig.14.1, 14.2) (Baltimore *et al.* 1989). The tissue damage has been shown to be generated by immune complexes between the pronounced antibody response and *P. aeruginosa* antigens (Pedersen 1992). The chronic *P. aeruginosa* lung infection in CF patients can therefore be regarded as a model for other chronic infections caused by biofilm forming bacteria. The understanding of the pathogenesis of this infection has, however, been facilitated by *in vitro* experiments of the interaction of *P. aeruginosa* biofilms and the components of the humoral and cellular defence systems, and by animal models of this chronic infection. The most important results generated by these experimental models will therefore be described in addition to the results obtained from studies in CF patients.

In vitro interaction between *P. aeruginosa* biofilm and humoral and cellular components of the defence system

Persistence of bacteria in spite of a normal host immune system and relevant antibiotic treatment is a key problem in many chronic infections, such as the bronchopulmonary *P. aeruginosa* infection in CF patients. In recent years the persistence of the bacteria has been increasingly attributed to the concept of bacterial biofilms (Costerton *et al.* 1987; Brown *et al.* 1990). The capability of

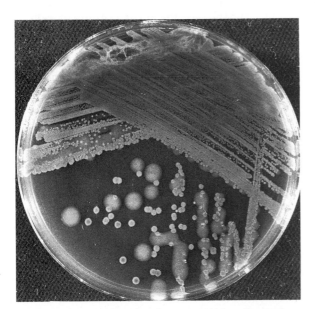

Fig. 14.1. Mucoid (large) and non-mucoid (small) colonies of P. *aeruginosa* isolated from a cystic fibrosis patient.

233

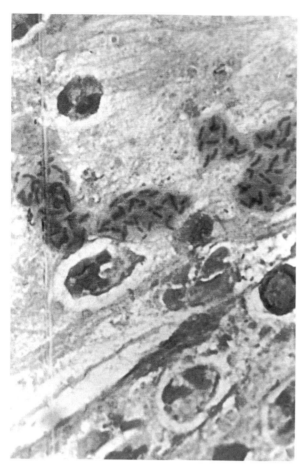

Fig. 14.2.   Gram stained smear of cystic fibrosis sputum with mucoid *P. aeruginosa* accumulated in microcolonies. A few PMNs can also be seen. Magnification × 700.

bacteria to establish themselves in microcolonies or biofilms, where they are enmeshed in glycocalyx and escape elimination by host defences as well as by antibiotics, has been documented in clinical cases and under experimental conditions (Bergamini *et al.* 1992; Giwercman *et al.* 1991; Kharazmi *et al.* 1986; Lee *et al.* 1983; Marrie *et al.* 1990). One of the approaches to study the immune response to bacterial biofilms has been the establishment of bacterial biofilms *in vitro* and the study of the response of each component of the immune system to the bacteria grown as biofilm.

Establishment and composition of biofilm

Bacterial biofilms have been established *in vitro* by using several types of devices. The most commonly used is the Robbins device (Costerton *et al.* 1987). Robbins device biofilms are produced on silicone discs (2 mm thick, 7 mm diameter, washed for 3 days in distilled water at 4 °C before sterilization by autoclaving or ethylene oxide) using the log phase broth culture of a given bacterium such as *P. aeruginosa*. The bacterial culture is pumped through the device at 40–60 ml h^{-1} at room temperature. After 1–5 days the studs protruding into the flow channel are taken out, the discs are aseptically removed and rinsed with physiological saline. Bacterial counts are made by scraping off the discs into 2 ml of saline, mildly sonicating for 10 min, and counts being performed as colony forming units (CFU) or by microscopy. The *in vitro* prepared biofilm usually consists of bacteria enmeshed in a glycocalyx matrix containing lipopolysaccharide (LPS), alginate and proteins (Jensen *et al.* 1993). The LPS of biofilm grown *P. aeruginosa* appears predominantly rough as compared with the smooth, planktonically grown bacteria on agar plates or in suspension (Giwercman *et al.* 1992).

Interaction of biofilm with polymorphonuclear leucocytes

It has been shown that polymorphonuclear neutrophils (PMN) are the most prominent cells in many chronic bacterial infections where the bacteria grow as biofilm. An excellent example of such an inflammatory process is the lungs of CF patients with chronic *P. aeruginosa* infection (Høiby & Koch 1990; Kharazmi *et al.* 1986; Suter 1989). Therefore, the study of the PMN response to biofilm seems appropriate. Studies using purified human peripheral blood PMN and *P. aeruginosa* biofilm have shown a lower oxidative burst response (31–48% of that of the planktonic cells) in PMN as compared with similar numbers of planktonic bacteria (Jensen *et al.* 1990, 1992; Table 14.1). Moreover, it was shown that the response was weaker to both opsonized and non-opsonized bacteria.

Interaction of biofilm with complement system

Complement activation by bacteria is not a new phenomenon. It is known that both Gram

Table 14.1. PMN chemiluminescence response to *Pseudomonas aeruginosa* in biofilm and planktonic form

Percentiles	Peak (kcpm)			Peak in % of response to planktonic bacteria 50th
	25th	50th	75th	
A. Planktonic bacteria non-opsonized	158	253	852	100
B. Biofilm bacteria non-opsonized	89	176	343	47.5
C. Planktonic bacteria opsonized	406	499	985	100
D. Biofilm bacteria opsonized	128	249	329	30.5

Results are median and 25th–75th percentiles of 22 (A + B) and 9 (C + D) experiments. A vs B: $p < 0.04$; C vs D: $p < 0.03$. From Jensen *et al.* 1992, with permission.

positive and Gram negative bacteria are able to activate the complement system (Inada 1980). Peptidoglycan appears to be the major component of Gram positive bacteria to activate complement (Riber *et al.* 1990), whereas LPS in Gram negative bacteria is the factor responsible for their activation of complement (Inada 1980; Riber *et al.* 1990). Complement activation by the bacteria could have positive as well as negative implications for the chronically infected patient. A positive consequence would be eradication of the typically serum sensitive bacteria, namely *P. aeruginosa,* from the lungs of CF patients (Høiby & Olling 1977). On the other hand, live and dead bacteria or bacterial fragments would activate complement by themselves, and also the concomitant active production of specific antibodies and immune complex formation (Pedersen 1992; Döring *et al.* 1984; Høiby *et al.* 1986; Kronborg *et al.* 1992) would add to a complement mediated inflammation. This has been shown by demonstration of activation of non specific host components such as PMN, complement split products, basophils and platelets (Høiby *et al.* 1986; Kharazmi *et al.* 1986) resulting in a constant inflammatory destruction of pulmonary tissue (Döring *et al.* 1986; Berger 1991).

Most studies on complement activation by bacteria have dealt with the activation by planktonic bacteria (Schiller 1988). However, some recent studies also show complement activation by bacteria grown as biofilms (Pier *et al.* 1991; Anwar *et al.* 1992). Anwar *et al.* (1992) actually found that biofilmgrowing mucoid *P. aeruginosa* resisted and survived the lytic action of normal human serum in contrast to planktonic growing bacteria. In a recent study from our group complement activation by *P. aeruginosa* biofilm was

assessed by complement consumption assay, production of C3 and factor B conversion products by crossed immunoelectrophoresis, C5a generation by a PMN chemotaxis assay, and finally formation of terminal complement complex using an ELISA (Jensen *et al.* 1993). It was found that *P. aeruginosa* grown in biofilm activated complement less than planktonic bacteria, and that activation was submaximal. Addition of EGTA indicated that complement activation was mainly mediated via the classical pathway and it was inhibited by polymyxin B, indicating that lipid A of LPS was the main mediator of this complement activation. It has also been shown that some fragments of activated complement are deposited on the biofilm (Pier *et al.* 1991). Thus, the results from these studies (Jensen *et al.* 1990, 1992, 1993) indicate that for the patients with chronic *P. aeruginosa* lung infection the bacteria persisting in biofilms mediate a constant low grade complement and PMN activation and thereby contribute to the chronic inflammation seen in these patients.

Chronic *P. aeruginosa* lung infection in laboratory animals

Chronic *P. aeruginosa* lung infection (for at least 35 days) in experimental animals was first produced by Cash *et al.* (1979), by intratracheal inoculation of bacteria enmeshed in agar beads. The histopathology surrounding the beads containing bacteria mimicked the lesions seen in CF patients with an abundance of PMNs. According to Nacucchio *et al.* (1984), however, the same effect was produced if the inoculum was given together with sterile agar beads instead of encased in agar beads, perhaps by providing an appropriate surface for bacterial adherence and growth, and obstruction of small airways.

Other techniques have also been used to establish lung infections, for example Döring & Dauner (1988) used (i) reserpine, (ii) PMN elastase, and (iii) a pulmonary type III hypersensitivity reaction in preimmunized rats, and Boyd *et al.* (1983) used oral hexamethylphosphoramide to denude the ciliated respiratory epithelia and pave the way for chronic *P. aeruginosa* infection. Woods *et al.* (1982), using the same rat model, showed that within the 30 day period after challenge, toxin A and elastase producing strains induced more severe histopathologic changes than mutants negative for these toxins, thereby indicating the importance of these virulence factors. The model of Cash *et al.* (1979) was established in cats by Winnie *et al.* (1982), who showed that in serum of all chronically infected cats, just as in CF serum, a substance (most probably antibodies) developed which inhibited the phagocytic activity of normal cat alveolar macrophages. Cochrane *et al.* (1988) showed that *P. aeruginosa* in the agar bead model produced iron regulated membrane proteins (IRMP) and that antibodies produced by the rats reacted with these IRMPs and other major outer membrane proteins. These results indicate that the bacteria grow in iron limited conditions in the rat lungs just as in the CF lungs.

Iwata & Sato (1991) employed the Cash *et al.* (1979) model to investigate the importance of the bronchus associated lymphoid tissue (BALT) and provided evidence that BALT regulates the local immune response against chronic lung infection due to *P. aeruginosa*. The rat model of chronic lung infection with *P. aeruginosa* was further developed by Pedersen *et al.* (1990) who introduced seaweed alginate (which resembles *P. aeruginosa* alginate, although some differences exist) beads containing *P. aeruginosa*. The antibody response to *P. aeruginosa* antigen (including IgA and IgG against alginate, flagella, outer membrane proteins), measured by crossed immuno-electrophoresis, immunoblotting and ELISA, was significantly stronger in rats challenged with alginate containing bacteria compared with that induced by agar beads containing the bacteria. The histopathology, however, was similar in both models consisting of an inflammatory response with mainly PMNs.

Johansen *et al.* (1993) have recently established the alginate bead model of chronic *P. aeruginosa* lung infection in normal and athymic rats which lack T-cells and measured the antibody response to a *P. aeruginosa* sonicate and to purified alginate from *P. aeruginosa* by the ELISA technique. Interestingly, the histopathology of the lung lesions in the athymic rats was similar – although more severe – to that seen in the normal rats, abundance of PMNs dominating the inflammatory reaction. Early transitory IgM titres were demonstrated in both normal and athymic rats,

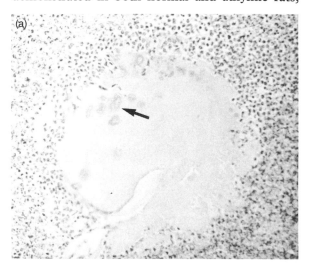

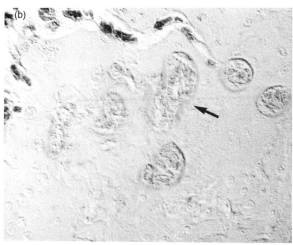

Fig. 14.3. Photomicrographs of a section from the base of a rat lung infected with a large inoculum of *P. aeruginosa* encased in alginate beads (4.7 × 10^8 CFU ml^{-1}). The alginate beads are surrounded by a dense accumulation of PMNs. Arrows indicate microcolonies of *P. aeruginosa* within the beads. (a) Normal rat, × 12. (b) Normal rat, × 200.

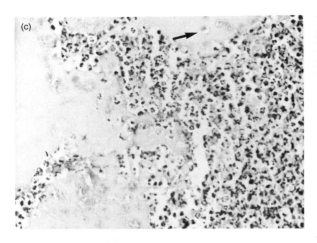

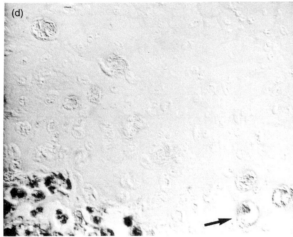

Fig. 14.3 *(cont.)*. (c) Athymic rat, × 12. (d) Athymic rat, × 200. (From Johansen *et al.* 1993, with permission.)

whereas the IgG titres were much lower in athymic rats, presumably due to the absence of CD4+ cells, but higher primary IgA titres were obtained in the athymic rats. In spite of the antibody response no clearance of the infection was seen. Compared to similar challenges with free living *P. aeruginosa*, the rats challenged with alginate beads containing the same number of bacteria experienced a more severe lung pathology and a more uniform antibody response (Figs. 14.3, 14.4).

Effect of preimmunization

Several authors have studied the effect of preimmunization on the chronic *P. aeruginosa* infec-

tion in laboratory animals. The Cash *et al.* (1979) model was employed in guinea pigs by Pennington *et al.* (1981) who found that preimmunizing with a *P. aeruginosa* LPS vaccine resulted in lower numbers of bacteria and significantly milder histopathology of the lungs compared with non-immunized animals. Furthermore, they found no evidence that preimmunization increased the frequency of circulating immune complexes in infected animals. Likewise, Klinger *et al.* (1983) used the PEV-01 vaccine consisting of LPS and other cell wall antigens from 16 O–groups of *P. aeruginosa* in the Cash *et al.* (1979) model. They found a significantly milder lung pathology but no decrease of *P. aeruginosa* count in the immunized rats investigated 8–10 days after challenge. Similar results of protection were obtained by Gilleland *et al.* (1988, 1993), using outer membrane protein F from *P. aeruginosa* for immunization prior to challenge of rats with agar beads containing *P. aeruginosa*; and by Pier *et al.* (1990), who used purified alginate to immunize rats and mice prior to challenge with the bacteria encased in agar beads. The amount of alginate used for immunization was, however, crucial since 10 µg alginate induced protective opsonizing antibodies, whereas 100 µg alginate induced non-protective, low opsonizing antibodies.

In contrast Freihorst *et al.* (1989), by oral immunization with *P. aeruginosa* which induced both secretory IgA and systemic IgG antibodies in the rats, could not detect any protective effects against subsequent challenge with the bacteria encased in agar beads. Buret *et al.* (1993), however, by priming of the gut associated lymphoid tissues with mucoid *P. aeruginosa* (paraformaldehyde-killed *P. aeruginosa* cells emulsified in Freund's incomplete adjuvant injected subserosally in Peyer's plaques), observed enhanced recruitment, chemotaxis, chemokinesis, and phagocytic activity of PMNs following pulmonary challenge infection with free living *P. aeruginosa* and these changes correlated with enhanced bacterial clearance from the lungs. Woods & Bryan (1985) immunized rats with purified *P. aeruginosa* alginate prior to challenge with agar beads, but instead of protection by the high titres of anti-alginate antibodies they found evidence of immune complex formation and deposits in the lung tissue containing IgM, IgG and IgA.

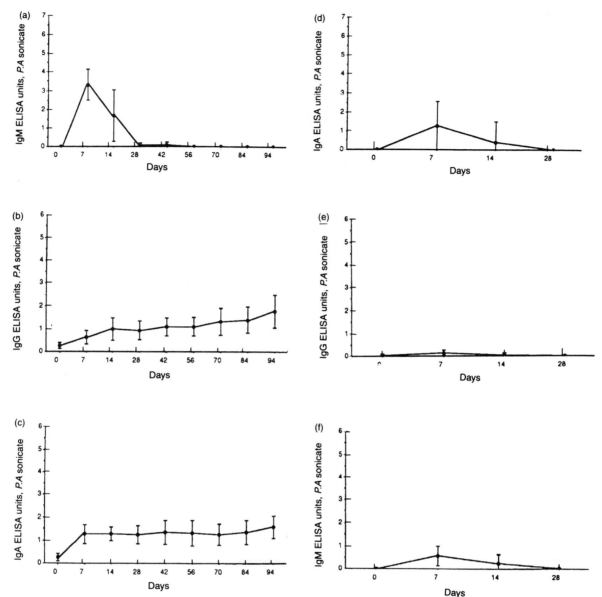

Fig. 14.4. Development of IgM, IgG and IgA antibodies in normal rats (a, b, c) and athymic rats (d, e, f) in serum against *P. aeruginosa* sonicate when given 10^8 CFU ml^{-1} *P. aeruginosa* in alginate beads intratracheally on day 0. ●, mean values; vertical bars = 95% confidence limits. (From Johansen et al. 1993, with permission).

According to the published data, the chronic lung infection established in rats or other animals by encasing the bacteria in alginate or agar beads can be designed to mimic very closely the chronic *P. aeruginosa* infection in CF patients: development of a pronounced antibody response including blocking antibodies, histopathological lesions including immune complex formation and PMN infiltration, no protective or curative effect of antibodies and no major importance of T-cells. These models are therefore well suited for preimmunization and other prophylactic and therapeutic experiments and, as detailed above, they have actually been used for that purpose in

Table 14.2. The pathogenesis of chronic *P. aeruginosa* infection in cystic fibrosis: immune complex mediated tissue damage

Stage of infection	Mechanisms and pathogenesis (Reference)	Clinical signs
Aquisition	Cross-infection. Environmental reservoir. Airway damage by other bacteria. Concomitant virus infection. (Høiby & Pedersen 1989; Johansen & Høiby 1992b; Pedersen 1992)	None Acute exacerbations
Attachment	Pili, haemagglutinin, exotoxin S, alginate. (Baker & Svanborg-Edén 1989; Baker *et al.* 1991)	None
Initial persistent colonization	Bacterial toxins: Elastase, alkaline protease, exotoxins A and S, fosfolipase, lipase etc. (Kharazmi 1991)	None or minimal
Chronic infection	Persistence: Microcolonies embedded in alginate. PMN-pseudomonas mismatch. Tissue damage: immune complexes. PMN* elastase, cytokines. (Høiby & Koch 1990; Tosi *et al.* 1990)	Chronic suppurative lung inflammation, progressive loss of lung function
Modifying mechanisms	PMN* elastase cleaves immune complexes. IRAP inhibits IL-1. Increase of antibodies to *P. aeruginosa*, especially of IgG2 and IgG3 subclasses. Δ508 homozygotes vs. other mutations. (Høiby *et al.* 1977; Döring *et al.* 1986) (Pressler *et al.* 1990; Johansen *et al.* 1991b; Kronborg *et al.* 1993)	Individual clinical course of the infection

From Høliby *et al.* 1986.

the past. The use of the CF mouse (Dorin *et al.* 1992) has yet to be reported and may offer some new insights.

Chronic *P. aeruginosa* lung infection in cystic fibrosis patients

The CF gene product, which is the membrane bound protein called the CF transmembrane conductance regulator (CFTR) protein, has been shown to be the chloride ion channel regulating the transportation of chloride ions across fluid transporting epithelial cells such as exocrine glands. The CF defect of the CFTR protein leads to altered secretions (salty sweat, thick mucus), blocked ducts and thereby reduced mucosal defence. CF patients have no detectable immune deficiency and, except for the respiratory tract, CF patients are no more susceptible to infections than normal children of the same age and bacteraemia is rarely recorded in CF patients. The altered secretions of the respiratory tract, leading to thick dehydrated mucus, are thought to be the reason why CF patients suffer from recurrent and chronic respiratory tract infections (Collins 1992; Pedersen 1992; Koch & Høiby 1993).

The non-specific defence of the lungs consists of the *primary non-inflammatory defence mechanisms* such as the mucociliary escalator, coughing, alveolar macrophages and surfactant, and the secondary *non-inflammatory defence mechanisms* such as s-IgA. The action of these defence mechanisms is silent and very efficient in normal people and their activity does not give rise to any symptoms with the exception of occasional coughing. No tissue damage is mediated through these defence mechanisms. Defects of the primary non-inflammatory defence mechanisms compromise CF patients. Such defects lead to secondary acute or chronic bacterial infections and recruitment of the *inflammatory defence mechanisms* of the lungs such as IgG and PMNs. The activity of the inflammatory defence mechanisms may lead to successful killing of the offending pathogens but in addition gives rise to local (and systemic) symptoms of inflammation such as fever, tissue damage and impaired function. If the infection is not eradicated (persistent or chronic infection) then immunopathologic tissue lesions occur, such as immune complex mediated tissue damage (Table 14.2).

P. aeruginosa in CF

This Gram negative, motile rod is an environmental species, which is found in numerous natural habitats including fresh water and soil, but also in man-made habitats such as poorly chlorinated swimming pools and whirlpool baths (Botzenhart & Döring 1993). It is rarely found in the stools of normal humans and then only in low numbers. The most prevalent and severe chronic lung infection (≥6 months) in CF patients is caused by *P. aeruginosa* (Høiby 1982; Govan & Glass 1990; Gilligan 1991) which has become endemic in CF patients in all countries (Figs. 14.1, 14.2). In addition to the lungs, CF patients are often colonized in the sinuses by this bacterium (Taylor *et al.* 1992) and they may be present in the stools probably originating from swallowed sputum (Agnarsson *et al.* 1989; Speert *et al.* 1993). Whether colonization of the upper respiratory tract precedes establishment of bronchial infection with *P. aeruginosa* is not known. *P. aeruginosa* shows chemotaxis towards mucin rich mucosal surfaces (Nelson *et al.* 1990). In animal studies *P. aeruginosa* adheres to buccal, nasal turbinate, and tracheobronchial epithelial cells and to mucus (Baker & Svanborg Edén 1989; Ramphal *et al.* 1987). Four types of adhesion factors have been identified on *P. aeruginosa* – pili, alginate and possibly haemagglutinin and exotoxin S – which bind to corresponding receptors on the host cells: glycolipids, glycosphingolipids and glycoproteins containing lactosyl and sialosyl residues (Baker & Svanborg-Edén 1989; Baker *et al.* 1991; Ramphal *et al.* 1987; Lee *et al.* 1989; Plotkowski *et al.* 1989; Hata & Fick 1991). Injury to epithelial cells of mucous membranes by trypsin or human leucocyte elastase exposed new receptors for pili, and increased bacterial adhesion (Baker & Svanborg Edén 1989; Plotkowski *et al.* 1989).

Chronic *P. aeruginosa* infection

In most (82%) CF patients a period (median 12 months) of intermittent colonization precedes the persistent colonization (Johansen & Høiby 1992b). The factors which in addition to virus infection (Petersen *et al.* 1981) determine the

Table 14.3. Antibodies to *P. aeruginosa* antigens in cystic fibrosis patients with chronic P. aeruginosa infection

Antigen	
Alkaline protease	Alginate
Elastase	Lipopolysaccharide
Exotoxin A	Outer membrane proteins
Phospholipase C	Flagella
Exotoxin S	Common antigen (GroEL)

From Pedersen 1992.

transition to persistent colonization are probably the toxins produced by *P. aeruginosa*, although this point of view is supported only by circumstantial evidence. *P. aeruginosa* produces many toxins and other virulence factors with potential effect on the lungs of CF patents (Table 14.3; Pedersen 1992). Some of these toxins are thought to play a role during establishment of the initial persistent colonization of the CF respiratory tract, notably elastase and alkaline protease. These have been shown to interfere with the non-specific (phagocytes) and immunological specific (T-cells, NK-cells, immunoglobulins) defence mechanisms (Döring *et al.* 1985; Kharazmi 1989), but also LPS and sometimes alginate, since antibodies against LPS and alginate can be detected early before the infection becomes chronic (Brett *et al.* 1992; Pedersen 1992; Fomsgaard 1990). Later in the course of infection, the significance of the action of these toxins becomes doubtful, since specific antibodies are produced by the CF patients. In particular, free elastase and alkaline protease can only be detected in bronchial secretions during the first few months of the infection before neutralizing antibodies develop (Table 14.2; Döring *et al.* 1983, 1984; Döring & Høiby 1983; Pedersen 1992).

Alginate is the only antigen which clinically correlates to poor prognosis in CF patients (Pedersen *et al.* 1992a; Henry *et al.* 1992). The most characteristic feature of the persistent *P. aeruginosa* infection is the production of mucoid alginate and the formation of microcolonies in the lungs of the patients (Figs. 14.1, 14.2; Lam *et al.* 1980; Høiby *et al.* 1986; Baltimore *et al.* 1989; Govan & Glass 1990; May *et al.* 1991; Deretic *et al.* 1991; Roychoudhury *et al.* 1991). The microcolony form of growth (biofilm) is the survival

Table 14.4. Some biological functions of human immunoglobulins

Function	Immunoglobulin							Inflammation
	IgG1	IgG2	IgG3	IgG4	IgM	IgA/s IgA	IgE	
Complement fixation	++	+	+++	−	+++	−	−	++
Basophil binding	−	−	−	?	−	−	+++	++
Platelet binding	+	+	+	+	−	−	?	++
Neutrophil binding	+	−	+	+	−	+	−	++
Monocyte binding	+	−	+	−	−	−	?	+
T-B lymphocyte binding	+	+	+	+	+	+	+	+

+, present; ++, strong; +++, very strong; −, absent.

strategy of environmental bacteria (Costerton *et al.* 1987) and the major component of the matrix of the microcolony is alginate (Jensen *et al.* 1990; Figs. 14.1, 14.2). The median concentration of the mucoid exopolysaccharide (alginate) in sputum from CF patients is 35.5 μg ml^{-1} (Pedersen *et al.* 1990). Although mucoid strains are also found in other chronically colonized patients, such strains are characteristic of CF (Høiby 1975).

It has been shown *in vitro*, that *P. aeruginosa* growing in alginate biofilms are protected against phagocytes and complement (Jensen *et al.* 1990; Anwar *et al.* 1992). In most patients non-mucoid strains initiate the infection, and the transition to the mucoid variant correlates with the development of a pronounced antibody response against virtually all antigens and toxins of *P. aeruginosa*, with different biological properties depending on isotype and allotype of the antibodies (Fig. 14.5, Table 14.3, 14.4; Pedersen *et al.* 1992a, b; Cordon *et al.* 1992) and the occurrence of mucoid variants correlates also to a poor prognosis (Høiby 1974; Høiby *et al.* 1986; Pedersen *et al.* 1989, 1992; MacDougall *et al.* 1990; Henry *et al.* 1992).

Complement deposition on the surface of mucoid microcolonies may be deficient in CF patients (Pier *et al.* 1991), favouring the survival of such colonies, although the embedded bacteria are unusual in several other aspects: they are often serum sensitive to normal human serum (Høiby & Olling 1977; Penketh *et al.* 1983), polyagglutinable (Pitt *et al.* 1986; Ojeniyi *et al.* 1991), lacking the LPS O-side chain (Hancock *et al.* 1983; Ojeniyi *et al.* 1985), non-motile, and expressing iron regulated outer membrane pro-

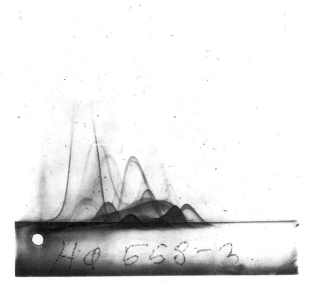

Fig. 14.5. Many precipitating antibodies against *P. aeruginosa* demonstrated by crossed immunoelectrophoresis of *P. aeruginosa* sonicated antigens against serum from a cystic fibrosis patient with chronic *P. aeruginosa* lung infection.

teins, indicating that the bacterium grows under iron restricted conditions in the CF lungs (Brown *et al.* 1984; Høiby & Koch 1990; Shand *et al.* 1991). The polyagglutinability is due to the semi-rough nature of the LPS and the exposure of antigenetically common determinants in the LPS core/lipid A part (Fomsgaard *et al.* 1988), or the presence of a common A band of LPS, and it seems to be related to bacteriophages, which are present in sputum of CF patients (Ojeniyi *et al.* 1985; 1987; Ojeniyi 1988; Lam *et al.* 1989). The semi-rough nature of LPS in strains from CF patients is also in accordance with *in vitro* results from *P. aeruginosa* growing in biofilm

Table 14.5. Clinical condition and antibody response to *P. aeruginosa* antigens

	Low FVC (<59%)	High FVC (>88%)	Low FEV 1 (<36 %)	High FEV1 (>72%)
Age (years)	18	16.9	19.7	15.7
Duration of infection (years)	6.7	6.7	7.6	6.4
Precipitins (see Fig. 14.5)[a]	25	18*	28	17*
Anti-*P. aeruginosa* IgA[a]	771	129*	787	148*
Anti-*P. aeruginosa* IgG[a]	1502	341*	1645	444*
Anti-alginate IgA	163	78*	181	82*
Anti-alginate IgG	400	223*	433	194*

*$P < 0.05$. [a] sonicated extract of *P. aeruginosa* serotype 1 to 17.
FVC: forced vital capacity; FEV1, forced expiratory volume in 1 sec. Values given as % of predicted for normal persons of the same height and sex.
The lung functions of 105 CF patients with chronic *P. aeruginosa* infection were recorded and the antibody response (in arbitrary ELISA units) of the patients with lung function in the upper quartile range were compared with that of those in the lower quartile range. The actual cut-off lung function values (in % of predicted) are shown in parentheses. From Pedersen 1992, with permission.

(Giwercman *et al.* 1992) and with the change of their hydrophobicity (Allison *et al.* 1990).

LPS has been found to be the major antigen component of immune complexes in sputum of CF patients with chronic *P. aeruginosa* infection (Kronborg *et al.* 1992; Fomsgaard *et al.* 1993) but elastase, alkaline phosphatase and the common antigen of *P. aeruginosa* (= GroEL, a heat shock protein which is important for protein folding) have also been detected in some immune complexes (Döring *et al.* 1984; Kronborg *et al.* 1992).

The most remarkable host response to the infection is the pronounced antibody response, which continues to increase during several years, and which is correlated with poor prognosis (Table 14.5). These antibodies are eventually directed against most, if not all, antigens of *P. aeruginosa* including alginate, and they belong to all classes and subclasses of immunoglobulins (Stahl Skov *et al.* 1980; Pedersen 1992) (Tables 14.3, 14.5, 14.6, 14.7). Genetically individual differences between the IgG subclass antibody response – high IgG2 and IgG3 response was associated to Gm (3;5) and Gm (23) genotypes (chromosome no. 14) – and correlated with a more severe course of the lung infection (Table 14.7; Pressler *et al.* 1990, 1992a, b, c; Pedersen *et al.* 1989).

The correlation between the antibody response and poor prognosis has been shown to be due to immune complex mediated chronic inflammation

Table 14.6. Local production of specific immunoglobulins in patients with cystic fibrosis. Values shown are median (range)

	Sputum concentration as % of serum concentration		
	IgG	IgA	IgM
Total immunoglobulin	3	16	2
Anti-St-Ag[a]	131 (18–243)	587 (188–1030)	5 (0–18)
Anti-alginate	27 (3–126)	636 (19–1538)	4 (1–41)

[a] Sonicated extract of *P. aeruginosa* serotype 1–17.
Modified from Pedersen *et al.* 1992, with permission.

in the lungs of CF patients (Høiby *et al.* 1977; Schiøtz 1981; Döring *et al.* 1984; Hodson *et al.* 1985; Wisnieski *et al.* 1985; Dasgupta *et al.* 1987a, b; Kronborg *et al.* 1992) and the IgG3 subclass is the strongest activator of complement and thereby inflammation (Table 14.4). This inflammatory reaction is dominated by PMNs, and released PMN proteases, myeloperoxidase and oxygen radicals are the main mechanisms of lung tissue damage; in this process the high levels of inflammatory cytokines in sputum (IL-1, IL-6, TNF and IRAP) are probably also important (Table 14.8; Høiby *et al.* 1986; Goldstein & Döring 1986; Ammitzbøll *et al.* 1988; Suter

Table 14.7. Correlation between IgG subclass antibody levels to *P. aeruginosa* alginate and lung function parameters in 51 CF patients with chronic *P. aeruginosa* infection expressed as Spearman rho and p-values

IgG subclass	IgG subclass concentration (ELISA units, median) (N = 46[a])	Lung function	
		FVC (N = 51)	FEVI (N = 51)
IgG1, rho	36.5	− 0.16, ns	− 0.25, ns
IgG2, rho	9	− 0.04, ns	− 0.21, ns
IgG3, rho	14	− 0.31, P < 0.03	− 0.30, P < 0.03
IgG4, rho	4	− 0.29, P < 0.03	− 0.23, ns

[a] Duration of infection > 2 years. ns, not significant (P > 0.05)
From Pressler *et al.* 1992c, with permission.

1989; Zach 1991; Kronborg *et al.* 1993).

Based on the concept of inflammation mediated tissue damage several clinical trials of the use of anti-inflammatory drugs in CF patients have been started. The first, whose results were published some years ago, showed a significant beneficial effect (Auerbach *et al.* 1985), but side effects have subsequently developed in some of these patients (Donati *et al.* 1990). Use of non-steroid anti-inflammatory drugs is currently being tried and hopefully their side effects are less serious (Sordelli *et al.* 1990; Konstan *et al.* 1991). Another approach is based on the imbalance between PMN elastase and proteinase inhibitors in the respiratory tract of CF patients (Goldstein & Döring 1986; Suter 1989; Meyer *et al.* 1991). McElvaney *et al.* (1991) showed recently that aerosol α-l-antitrypsin treatment could suppress PMN elastase in the respiratory secretion of CF patients and reverse the inhibitory effect of that fluid on *P. aeruginosa* killing by PMNs. This is thought to be at least partly due to cleavage of IgG and complement C3b receptors on phagocytic cells and of C3b on *P. aeruginosa* by PMN elastase (Tosi *et al.* 1990). The results of long term controlled trials are needed to show whether such treatment regimes with α-1-antitrypsin will improve the prognosis of CF patients.

Immune prophylaxis and therapy

Some theoretical immuno-prophylactic strategies, according to the pathophysiology of the chronic *P. aeruginosa* infection (Table 14.2) are given in Table 14.9. Vaccination of non-infected CF patients with a polyvalent *P. aeruginosa* vaccine had been tried previously without success; there was actually a trend to more rapid deterioration in the vaccinated patients (Langford & Hiller 1984). This is hardly surprising considering the immune complex mediated tissue damage that occurs during chronic *P. aeruginosa* lung infection (Pedersen 1992). Other vaccines are now being evaluated in animals and CF patients, hopefully providing beneficial effects and not immunological side effects if colonization with *P. aeruginosa* is not prevented (Cryz *et al.* 1991; Schaad *et al.* 1991; Pier 1991; Johansen *et al.* 1991a; Johansen & Høiby 1992a). Naturally acquired anti-*P. aeruginosa* antibodies fail to afford protection against *P. aeruginosa* bronchopulmonary infection and exacerbations of the

Table 14.8. Plasma and sputum cytokines (pg ml^{-1}) in CF patients with chronic *P. aeruginosa* infection and controls

	Controls, serum (N = 15)			CF patients, serum (N = 28)			CF patients, sputum (N = 24)		
	% Pos.	median	(range)	% Pos.	median	(range)	% Pos.	median	(range)
TNF	0	u.d.[a]		11		(600–1440)	71	410	(300–4000)
IL-6	68	15	(15–40)	68	38	(25–400)*	85	146	(66–6600)
IL-1α	0	u.d.[a]		0			94	400	(250–4000)
IL-1ß	87	25	(10–68)	89	92	(40–1280)*	100	2520	(256–4000)
IRAP[b]	100	280	(180–710)*	100	75	(40–1300)	100	7000	(1280–140 000)

*P < 0.05.
[a] u.d., undetectable.
[b] IRAP, Interleukin-1 receptor antagonist, which blocks effects of IL-1{alpha} and IL-1 (beta).
Pos, positive.
Modified from Kronborg *et al.* 1993, with permission.

Table 14.9. Prophylactic strategies to enhance the secondary non-inflammatory and inflammatory defence mechanisms of the lungs

Enhance s-IgA = *non-inflammatory defence mechanism* (no risk?)
and/or
Enhance opsono-phagocytic *inflammatory defence mechanism* (risk: tissue damage leading to persistent infection)

Both strategies *may* lead to effective elimination of inhaled *P. aeruginosa* and thus prevent persistent colonization

infection in CF patients (Fomsgaard 1990). Bruderer *et al.* (1992) have investigated the titre and affinity constants of serum anti-LPS) IgG in CF patients and healthy adults before and after immunization with a polyvalent LPS-based vaccine (O-polysaccharide-toxin A conjugate vaccine containing O-polysaccharide of eight serotypes). Immunization elicited a significant rise in total anti-LPS IgG levels (mostly IgG1 and IgG2) and affinity constants in both healthy adults and CF patients. Although chronically colonized patients had elevated levels of total anti-LPS antibodies, they possessed affinities at least 100–fold less than those of vaccine induced antibodies. The vaccine induced high affinity antibodies possessed higher opsonophagocytic activity than antibodies from non-immunized chronically colonized patients. There was no systemic and only mild to moderate local reactions at the injection site associated with the vaccinations even in the two recently colonized CF patients. The seven healthy non-colonized CF patients remained non-colonized and without any significant change in their clinical score during a 2 year follow-up period. This study is interesting because it shows significant functional differences between vaccine induced and infection induced antibodies. However, the possibility of immune complex mediated tissue damage in the lungs which may be aggravated by vaccination is still a possible side effect of vaccination. Further studies on larger groups of CF patients followed for several years are necessary to establish the role of immunoprophylaxis against *P. aeruginosa*.

Immunotherapy is currently being evaluated in CF patients as an adjunctive treatment based on the work of Moss (1990), who advocates that a shift from non-opsonizing to opsonizing anti-

bodies can be obtained by giving passive antibody therapy. Some effect has been found in short term trials (Moss 1990). The risk is, however, similar (although of a shorter period of time due to the elimination of the administred immunoglobulins) to that of vaccine trials, that some patients may experience aggravation of the immune complex disease.

The chronic *P. aeruginosa* lung infection in CF is a prototype of how biofilm producing bacteria may survive for decades in spite of a strong immune mediated inflammatory response and also in spite of intensive chemotherapy (Koch & Høiby 1993). The consequence of the chronic inflammation for the CF patient is a gradual destruction of the lung tissue, the production of copious amounts of viscid sputum and eventually respiratory failure and death. The increased knowledge of the pathogenesis of biofilm infection will hopefully facilitate improved rational prophylaxis and therapy in the near future to the benefit of the patients.

References

Agnarsson, U., Glass S. & Govan, J. R. W. (1989). Fecal isolation of *Pseudomonas aeruginosa* from patients with cystic fibrosis. *Journal of Clinical Microbiology*, **27**, 96–8.

Allison, D. G., Brown, M. R. W., Evans, D. E. & Gilbert, P. (1990). Surface hydrophobicity and dispersal of *Pseudomonas aeruginosa* from biofilms. *FEMS Microbiology Letters*, **71**, 101–4.

Ammitzbøll, T., Pedersen, S. S., Espersen, F. & Schieøler, H., (1988). Excretion of urinary collagen meabolites correlates to severity of pulmonary disease in cystic fibrosis. *Acta Paediatrica*, 77, 824–6.

Anwar, H., Strap, J. L. & Costerton J. W. (1992). Susceptibility of biofilm cells of *Pseudomonas aeruginosa* to bactericidal actions of whole blood and serum. *FEMS Microbiology Letters*, **92**, 235–42.

Auerbach, H. S., Kirkpatrick, J. A., Williams, M. & Colten, H. R. (1985). Alternate-day prednisone reduces morbidity and improves pulmonary function in cystic fibrosis. *Lancet*, **ii**, 686–8.

Baker, N. R., Minor, V., Deal, C., Sharhrabadi, M. S., Simpson, D. A. & Woods, D. E. (1991). *Pseudomonas aeruginosa* exoenzyme-S is an adhesin. *Infection and Immunity*, **59**, 2859–63.

Baker, N. R. & Svanborg-Edén, C. (1989). Role of alginate in the adherence of *Pseudomonas aeruginosa*. In *Pseudomonas aeruginosa* infection, ed. N. Høiby,

S. S. Pedersen, G. H. Shand, G. Döring & I. A. Holder, pp. 72–9. Basel: Karger.

Baltimore, R. S., Christie, C. D. C. & Smith, G. J. W. (1989). Immunohistopathologic localization of *Pseudomonas aeruginosa* in lungs from patients with cystic fibrosis – implications for the pathogenesis of progressive lung deterioration. *American Review of Respiratory Disease*, **140**, 1650–61.

Bergamini, T. M., Peyton, J. C. & Cheadle, W. G. (1991). Prophylactic antibiotics prevent bacterial biofilm graft infection. *Journal of Surgical Research*, **52**, 101–5.

Berger, M. (1991). Inflammation in the lung in cystic fibrosis – a vicious cycle that does more harm than good. *Clinical Reviews in Allergy*, **9**, 119–42.

Botzenhart, K. & Döring, G. (1993). Ecology and epidemiology of *Pseudomonas aeruginosa*. In Pseudomonas aeruginosa *as an opportunistic pathogen*, ed. M. Campa, M. Bertinelli & H. Friedman, pp. 1–18. New York: Plenum Press.

Boyd, R. L., Ramphal, R. & Mangos, J. A. (1983). Chronic colonization of rat airways with *Pseudomonas aeruginosa*. *Infection and Immunity*, **39**, 1403–10.

Brett, M. M., Simmonds, E. J., Goneim, A. T. M. & Littlewood, J. M. (1992). The value of serum IgG titres against *Pseudomonas aeruginosa* in the management of early Pseudomonal infection in cystic fibrosis. *Archives of Disease in Childhood*, **69**, 1086–8.

Brown, M. R. W., Anwar, H. & Lambert, P. A. (1984). Evidence that mucoid *Pseudomonas aeruginosa* in the cystic fibrosis lung grows under iron restricted conditions. *FEMS Microbiology Letters*, **21**, 113–17.

Brown, M. R. W., Collier, P, J, & Gilbert, P. (1990). Influence of growth rate on susceptibility to antimicrobial agents – modification of the cell envelope and batch and continuous culture studies. *Antimicrobial Agents and Chemotherapy*, **34**, 1623–8.

Bruderer, U., Cryz, S. J., Jr, Schaad, U. B., Deusinger, M., Que, J. U. & Lang, A. B. (1992). Affinity constants of naturally acquired and vaccine-induced anti-*Pseudomonas aeruginosa* antibodies in healthy adults and cystic fibrosis patients. *Journal of Infectious Diseases*, **166**, 344–9.

Buret, A., Dunkley, M., Clancy, R. L. & Cripps, A. W. (1993). Effector mechanisms of intestinal induced immunity to *Pseudomonas aeruginosa* in the rat lung: role of neutrophils and leukotriene B4. *Infection and Immunity*, **61**, 671–9.

Cash, H. A., Woods, D. E., McCullough, B., Johanson, W. G. & Bass, J. A. (1979). A rat model of chronic respiratory infection with *Pseudomonas aeruginosa*. *American Review of Respiratory Disease*, **119**, 453–9.

Cochrane, D. M. G., Brown, M. R. W., Anwar, H.,

Lam, K. & Costerton, J. W. (1988). Antibody response to *Pseudomonas aeruginosa* surface protein antigens in a rat model of chronic lung infection. *Journal of Medical Microbiology*, **27**, 255–61.

Collins, F. S. (1992). Cystic fibrosis – molecular biology and therapeutic implications. *Science*, **256**, 774–9.

Cordon, S. M., Elborn, J. S., Rayner, R. J., Hiller, E. J. & Shale, D. J. (1992). Antibodies in early *Pseudomonas aeruginosa* infection in cystic fibrosis. *Archives of Childhood Disease*, **67**, 737–40.

Costerton, J. W., Cheng, K.-J., Geesey, G. G. et al. (1987). Bacterial biofilms in nature and disease. In *Annual Reviews of Microbiology*, ed. L. N. Ornston, A. Balows & P. Baumann, **41**, 435–64. Paolo Alto: Annual Reviews Inc.

Cryz Jr. , S. J., Fürer, E., Que, J. U., Sadoff, J. C., Brenner, M. & Schaad, U. B. (1991). Clinical evaluation of an octavalent *Pseudomonas aeruginosa* conjugate vaccine in plasma donors and in bone marrow transplant and cystic fibrosis patients. In Pseudomonas aeruginosa *in human diseases*, ed. J. Y. Homma, H. Tanimota, I. A. Holder, N. Høiby & G. Döring, pp. 157–62. Basel: Karger.

Dasgupta, M. K., Lam, J., Döring, G. et al. (1987a). Prognostic implications of circulating immune complexes and *Pseudomonas aeruginosa* specific antibodies in cystic fibrosis. *Journal of Clinical and Laboratory Immunology*, **23**, 25–30.

Dasgupta, M. K., Zuberbuhler, P., Abbi, A. et al. (1987b). Combined evaluation of circulating immune complexes and antibodies to *Pseudomonas aeruginosa* as an immunological profile in relation to pulmonary function in cystic fibrosis. *Journal of Clinical Immunology*, **7**, 51–7.

Derectic, V., Mohr, C. D. & Martin, D. W. (1991). Mucoid *Pseudomonas aeruginosa* in cystic fibrosis – signal transduction and histone like elements in the regulation of bacterial virulence. *Molecular Microbiology*, **5**, 1577–83.

Donati, M. A., Haver, K., Gerson, W., Klein, M., McLaughlin, F. J. & Wohl, M. E. B. (1990). Long term alternate day prednisone therapy in cystic fibrosis. *Pediatric Pulmonology* Suppl., **5**, 277 (abstract).

Dorin, J. R., Dickinson, P., Alton, E. W. F. W. et al. (1992). Cystic fibrosis in the mouse by targeted insertional mutagenesis. *Nature*, **359**, 211–15.

Döring, G., Buhl, V., Høiby, N., Schiøtz, P. O. & Botzenhart, K. (1984). Detection of proteases of *Pseudomonas aeruginosa* in immune complexes isolated from sputum of cystic fibrosis patients. *Acta Pathologica et Microbiologica Scandinavica*, Section C, **92**, 307–11.

Döring, G. & Dauner, H.-M. (1988). Clearence of *Pseudomonas aeruginosa* in different rat lung models. *American Review of Respiratory Disease*, **138**, 1249–53.

Döring, G., Goldstein, W., Röll, A., Schiøtz, P. O., Høiby, N. & Botzenhart, K. (1985). The role of *Pseudomonas aeruginosa* exoenzymes in lung infections of patients with cystic fibrosis. *Infection and Immunity*, **49**, 557–62.

Döring, G., Goldstein, W., Schiøtz, P. O., Høiby, N., Dasgupta, M. & Botzenhart, K. (1986). Elastase from polymorphonuclear leukocytes – a regulatory enzyme in immune complex disease. *Clinical and Experimental Immunology*, **64**, 597–605.

Döring, G. & Høiby, N. (1983). Longitudinal study of immune response to *Pseudomonas aeruginosa* antigens in cystic fibrosis. *Infection and Immunity*, **42**, 197–201.

Döring, G., Obernesser, H.-J., Botzenhart, K., Flehmig, B., Høiby, N. & Hofman, A. (1983). Proteases of *Pseudomonas aeruginosa* in patients with cystic fibrosis. *Journal of Infectious Diseases*, **147**, 744–50.

Fomsgaard, A. (1990). Antibodies to lipopolysaccharides: some protective and diagnostic aspects. *Acta Pathologica, Microbiologica et Immunologica Scandinavica*, Suppl. **19**, 3–38.

Fomsgaard, A., Conrad, R. S., Galanos, C., Shand, G. H. & Høiby, N. (1988). Comparative immunochemistry of lipopolysaccharides from typable and polyagglutinable *Pseudomonas aeruginosa* strains isolated from cystic fibrosis patients. *Journal of Clinical Microbiology*, **26**, 821–6.

Fomsgaard, A., Shand, G. H., Freudenberg, M. A. *et al.* (1993). Antibodies from chronically infected cystic fibrosis patients react with lipopolysaccharides extracted by micromethods from all serotypes of *Pseudomonas aeruginosa*. *Acta Pathologica, Microbiologica et Immunologica Scandinavica*, **101**, 101–12.

Freihorst, J., Merrick, J. M. & Ogra, P. L. (1989). Effect of oral immunization with *Pseudomonas aeruginosa* on the development of specific antibacterial immunity in the lungs. *Infection and Immunity*, **57**, 235–8.

Gilleland, H. E., Gilleland, L. B. & Fowler, M. R. (1993). Vaccine efficacies of elastase, exotoxin A, and outer membrane protein F in preventing chronic pulmonary infection by *Pseudomonas aeruginosa* in a rat model. *Journal of Medical Microbiology*, **38**, 79–86.

Gilleland, H. E., Gilleland, L. B. & Matthews-Greer, J. M. (1988). Outer membrane protein F preparation of *Pseudomonas aeruginosa* as a vaccine against chronic pulmonary infection with heterologous immunotype strains in a rat model. *Infection and Immunity*, **56**, 1017–22.

Gilligan, P. H. (1991). Microbiology of airway disease in patients with cystic fibrosis. *Clinical Microbiology Review*, **4**, 35–51.

Giwercman, B., Fomsgaard, A., Mansa, B. & Høiby, N. (1992). Polyacryamide gel electrophoresis analysis of lipopolysaccharide from *Pseudomonas aeruginosa* growing planktonically and as biofilm. *FEMS Microbiology Immunology*, **89**, 225–9.

Giwercman, B., Jensen, E. T., Høiby, N., Kharazmi, A. & Costerton, J. W. (1991). Induction of beta-lactamase production in *Pseudomonas aeruginosa* biofilm. *Antimicrobial Agents and Chemotherapy*, **35**, 1008–10.

Goldstein, W. & Döring, G. (1986). Lysosomal enzymes from polymorphonuclear leukocytes and proteinase inhibitors in patients with cystic fibrosis. *American Review of Respiratory Disease*, **134**, 49–56.

Govan, J. R. W. & Glass, S. (1990). The microbiology and therapy of cystic fibrosis lung infections. *Reviews in Medical Microbiology*, 19–28.

Hancock, R. E. W., Mutharia, L. M., Chan, L., Darveau, R. P., Speert, D. P. & Pier, G. B. (1983). *Pseudomonas aeruginosa* isolates from patients with cystic fibrosis: a class of serum-sensitive, nontypable strains deficient in lipopolysaccaride *O* side chains. *Infection and Immunity*, **42**, 170–7.

Hata, J. S. & Fick, R. B, (1991). Airway adherence of *Pseudomonas aeruginosa* – mucoexopolysaccharide binding to human and bovine airway proteins. *Journal of Laboratory and Clinical Medicine*, **117**, 410–22.

Henry, R. L., Mellis, C. M. & Petrovic, L. (1992). Mucoid *Pseudomonas aeruginosa* is a marker of poor survival in cystic fibrosis. *Pediatric Pulmonology* **12**, 158–61.

Hodson, M. E., Beldon, I. & Batten, J. C. (1985). Circulating immune complexes in patients with cystic fibrosis in relation to clinical features. *Clinical Allergy*, **15**, 363–70.

Høiby, N. (1974). *Pseudomonas aeruginosa* infection in cystic fibrosis. Relationship between mucoid strains of *Pseudomonas aeruginosa* and the humoral immune response. *Acta Pathologica et Microbiologica Scandinavica*, Section B, **82**, 551–8.

Høiby, N. (1975). Prevalence of mucoid strains of *Pseudomonas aeruginosa* in bacteriological specimens from patients with cystic fibrosis and patients with other diseases. *Acta Pathologica et Microbiologica Scandinavica*, Section B, **83**, 549–52.

Høiby, N. (1982). Microbiology of lung infections in cystic fibrosis patients. *Acta Paediatrica Scandinavica* (Suppl.), **301**, 33–54.

Høiby, N., Döring, G. & Schiøtz, P. O. (1986). The role of immune complexes in the pathogenesis of bacterial infections. *Annual Review of Microbiology*, **40**, 29–53.

Høiby, N., Flensborg, E. W., Beck, B., Friis, B., Jacobsen, L. & Jacobsen, S. V. (1977). *Pseudomonas aeruginosa* infection in cystic fibrosis. Diagnostic and prognostic significance of *Pseudomonas aeruginosa* precipitins determined by means of crossed immunoelectrophoresis. *Scandinavian Journal of Respiratory Diseases*, **58**, 65–79.

Høiby, N., & Koch, C. (1990). *Pseudomonas aeruginosa* infection in cystic fibrosis and its management. *Thorax*, **45**, 881–4.

Høiby, N. & Olling, S. (1977). *Pseudomonas aeruginosa* infection in cystic fibrosis. Bactericidal effect of serum from normals and patients with cystic fibrosis on *P. aeruginosa* strains from patients with cystic fibrosis or other diseases. *Acta Pathologica et Microbiologica Scandinavica*, Section C, **85**, 107–14.

Høiby, N. & Pedersen, S. S. (1989). Estimated risk of cross-infection with *Pseudomonas aeruginosa* in Danish cystic fibrosis patients. *Acta Paediatrica Scandinavica*, **78**, 395–404.

Inada, K. (1980). Complement activating property of the protein-rich endotoxin (OEP) of *Pseudomonas aeruginosa*. 2) Complement activating property of the lipopolysaccharide portion and the inhibition by polymyxin B. *Japanese Journal of Experimental Medicine*, **50**, 197–215.

Iwata, M. & Sato, A. (1991). Morphological and immunohistochemical studies of the lungs and bronchus associated lymphoid tissue in a rat model of chronic pulmonary infection with *Pseudomonas aeruginosa*. *Infection and Immunity*, **59**, 15514–20.

Jensen, E. T., Kharazmi, A., Garred, P. *et al.* (1993). Complement activation by *Pseudomonas aeruginosa* biofilms. *Microbial Pathogenesis*, **15**, 377–88.

Jensen, E. T., Kharazmi, A., Høiby, N. & Costerton, J. W. (1992). Some bacterial parameters influencing the neutrophil oxidative burst response to *Pseudomonas aeruginosa* biofilms. *Acta Pathologica, Microbiologica et Immunologica Scandinavica*, **100**, 727–33.

Jensen, E. T., Kharazmi, A., Lam, K., Costerton, J. W. & Høiby, N. (1990). Human polymorphonuclear leukocyte response to *Pseudomonas aeruginosa* grown in biofilm. *Infection and Immunity*, **58**, 2383–5.

Johansen, H. K., Espersen, F., Pedersen, S. S., Hougen, H. P., Rygaard, J. & Høiby, N. (1993). Chronic *Pseudomonas aeruginosa* lung infection in normal and athymic rats. *Acta Pathologica, Microbiologica et Immunologica Scandinavica*, **101**, 207–25.

Johansen, H. K. & Høiby, N. (1992a). Local IgA and IgG response to intratracheal immunization with *Pseudomonas aeruginosa* antigens. *Acta Pathologica, Microbiologica et Immunologica Scandinavica*, **100**, 87–90.

Johansen, H. K. & Høiby, N. (1992b). Seasonal onset of initial colonisation and chronic infection with *Pseudomonas aeruginosa* in patients with cystic fibrosis in Denmark. *Thorax*, **47**, 109–11.

Johansen, H. K., Høiby, N. & Pedersen, S. S. (1991a). Experimental immunization with *Pseudomonas aeruginosa* alginate induces IgA and IgG antibody responses. *Acta Pathologica, Microbiologica et Immunologica Scandinavica*, **99**, 1061–8.

Johansen, H. K., Nir, M., Høiby, N., Koch, C. & Schwartz, M. (1991b). Severity of cystic fibrosis in patients homozygous and heterozygous for DeltaF508 mutation. *Lancet*, **337**, 631–4.

Kharazmi, A. (1989). Interactions of *Pseudomonas aeruginosa* proteases with the cells of the immune system. In Pseudomonas aeruginosa *infection*, ed. N. Høiby, S. S. Pedersen, G. H. Shand, G. Döring & I. A. Holder, pp. 42–9. Basel: Karger.

Kharazmi, A. (1991). Mechanisms involved in the evasion of the host defence by *Pseudomonas aeruginosa*. *Immunology Letters*, **30**, 201–6.

Kharazmi, A., Schiøtz, P. O., Høiby, N., Bæk, L. & Döring, G. (1986). Demonstration of neutrophil chemotactic activity in the sputum of cystic fibrosis patients with *Pseudomonas aeruginosa* infection. *European Journal of Clinical Investigation*, **16**, 143–8.

Klinger, J. D., Cash, H. A., Wood, R. E. & Miller, J. J. (1983). Protective immunization against chronic *Pseudomonas aeruginosa* pulmonary infection in rats. *Infection and Immunity*, **39**, 1377–84.

Koch, C. & Høiby, N. (1993). Pathogenesis of cystic fibrosis. *Lancet*, **341**, 1065–9.

Konstan, M. W., Hoppel, C. L., Chai, B.-L. & Davis, P. B. (1991). Ibuprofen in children with cystic fibrosis: pharmacokinetics and adverse effects. *Journal of Pediatrics*, **118**, 956–64.

Kronborg, G., Hansen, M., Svenson, M., Fomsgaard, A., Høiby, N. & Bendtzen, K. (1993). Cytokines in sputum and serum from patients with cystic fibrosis and chronic *Pseudomonas aeruginosa* infection as markers of destructive inflammation in the lungs. *Pediatric Pulmonology*, **15**, 292–7.

Kronborg, G., Shand, G. H., Fomsgaard, A. & Høiby, N. (1992). Lipopolysaccharide is present in immune complexes isolated from sputum in patients with cystic fibrosis and chronic *Pseudomonas aeruginosa* lung infection. *Acta Pathologica, Microbiologica et Immunologica Scandinavica*, **100**, 75–180.

Lam, J., Chan, R., Lam, K. & Costerton, J. W. (1980). Production of mucoid microcolonies by *Pseudomonas aeruginosa* within infected lungs in cystic fibrosis. *Infection and Immunity*, **28**, 546–56.

Lam, M. Y. C., McGroarty, E. J., Kropinski, A. M. *et al.* (1989). The occurrence of a common lipopolysaccharide antigen in standard and clinical strains of *Pseudomonas aeruginosa*. *Journal of Clinical Microbiology*, **27**, 962–7.

Langford, D. T. & Hiller, J. (1984). Prospective, controlled study of a polyvalent *pseudomonas* vaccine in cystic fibrosis – three year results. *Archives of Disease in Childhood*, **59**, 1131–4.

Lee, D. A., Hoidal, J. R., Clawson, C. C., Quie, P. G. & Peterson, P. K. (1983). Phagocytosis by polymorphonuclear leukocytes of *Staphylococcus auerus* and *Pseudomonas aeruginosa* adherent to plastic, agar, or glass. *Journal of Immunological Methods*, **63**, 103–14.

Lee, K. K., Doig, P., Paranchych, W. & Hodges, R. S. (1989). Mapping the surface regions of *Pseudomonas aeruginosa* PAK pilin: the importance of the C-terminal region for adherence to human buccal epithelial cells. *Molecular Microbiology*, **11**, 1493–9.

MacDougall, J., Hodson, M. E. & Pitt, T. L. (1990). Antibody response of fibrocystic patients to homologous *O*-typable and *O*-defective isolates of *Pseudomonas aeruginosa*. *Journal of Clinical Microbiology*, **43**, 567–71.

Marrie, T. J., Sung, J. Y. & Costerton, J. W. (1990). Bacterial biofilm formation on nasogastric tubes. *Journal of Gastroenterology and Hepatology*, **5**, 503–6.

May, T. B., Shinabarger, D., Maharaj, R. *et al.* (1991). Alginate synthesis by *Pseudomonas aeruginosa* – a key pathogenic factor in chronic pulmonary infections of cystic fibrosis patients. *Clinical Microbiology Reviews*, **4**, 191–206.

McElvaney, N. G., Hubbard, R. C., Birrer, P. *et al.* (1991). Aerosol alpha 1–antitrypsin treatment for cystic fibrosis. *Lancet*, **337**, 392–4.

Meyer, K. C., Lewandoski, J. R., Zimmerman, J. J., Nunley, D., Calhoun, W. J. & Dopico, G. A. (1991). Human neutrophil elastase and elastase/alpha 1–antiprotease complex in cystic fibrosis – Comparison with interstitial lung disease and evaluation of the effect of intravenously administered antibiotic therapy. *American Review of Respiratory Disease*, **144**, 580–5.

Moss, R. B. (1990). Antibody production in CF and possibilities for immunotherapy. *Pediatric Pulmonology* (Suppl.) **5**, 66–7.

Nacucchio, M., Cerquetti, M. C., Meiss, R. P. & Sordelli, D. O. (1984). Role of agar beads in the pathogenicity of *Pseudomonas aeruginosa* in the rat respiratory tract. *Pediatric Research*, **18**, 295–6.

Nelson, J. W., Tredgett, M. W., Sheehan, J. K., Thornton, D. J., Notman, D. & Govan, J. R. W. (1990). Mucinophilic and chemotactic properties of *Pseudomonas aeruginosa* in relation to pulmonary colonization in cystic fibrosis. *Infection and Immunity*, **58**, 1489–95.

Ojeniyi, B. (1988). Bacteriophages in sputum of cystic fibrosis patients as a possible cause of *in vivo* changes in serotypes of *Pseudomonas aeruginosa*. *Acta Pathologica, Microbiologica et Immunologica Scandinavica*, **96**, 294–8.

Ojeniyi, B., Bæk, L. & Høiby, N. (1985). Polyagglutinability due to loss of *O*-antigenic derminants in *Pseudomonas aeruginosa* strains isolated from cystic fibrosis patients. *Acta Pathologica et Microbiologica Scandinavica*, Section B, **93**, 7–13.

Ojeniyi, B., Høiby, N. & Rosdal, V. T. (1991). Prevalence and persistence of polyagglutinable *Pseudomonas aeruginosa* in cystic fibrosis patients. *Acta Pathologica, Microbiologica et Immunologica Scandinavica*, **99**, 187–95.

Ojeniyi, B., Rosdal, V. T. & Høiby, N. (1987). Changes in serotype caused by cell to cell contact between different *Pseudomonas aeruginosa* strains from cystic fibrosis patients. *Acta Pathologica et Microbiologica Scandinavica*, Section B, **95**, 23–7.

Pedersen, S. S. (1992). Lung infection with alginate-producing, mucoid *Pseudomonas aeruginosa* in cystic fibrosis. *Acta Pathologica, Microbiologica et Immunologica Scandinavica*, **100**, (Suppl. 28), 5–79.

Pedersen, S. S., Høiby, N., Espersen, F. & Koch, C. (1992a). Role of alginate in infection with mucoid *Pseudomonas aeruginosa* in cystic fibrosis. *Thorax*, **47**, 6–13.

Pedersen, S. S., Høiby, N., Shand, G. H. & Pressler, T. (1989). Antibody response to *Pseudomonas aeruginosa* antigens in cystic fibrosis. In *Pseudomonas aeruginosa (infection)*, ed. N. Høiby, S. S. Pedersen, G. H. Shand, G. Döring & I. A Holder, pp. 130–53. Basel: Karger.

Pedersen, S. S., Kharazmi, A., Espersen, F. & Høiby, N. (1990). *Pseudomonas aeruginosa* alginate in cystic fibrosis sputum and the inflammatory response. *Infection and Immunity*, **50**, 3363–8.

Pedersen S. S., Møller, H., Espersen, F., Sørensen, C. H., Jensen, T. & Høiby, N. (1992b). Mucosal immunity to *Pseudomonas aeruginosa* alginate in cystic fibrosis. *Acta Pathologica, Microbiologica et Immunologica Scandinavica*, **11**, 326–34.

Pedersen, S. S., Shand, G. H., Hansen, B. L. & Hansen, G. N. (1990). Induction of experimental chronic *Pseudomonas aeruginosa* lung infection with *P. aeruginosa* entrapped in alginate microspheres. *Acta Pathologica, Microbiologica et Immunologica Scandinavica*, **93**, 203–11.

Penketh, A. R. L., Pitt, T. L., Hodson, M. E. & Battern, J. C. (1983). Bactericidal activity of serum from cystic fibrosis patients for *Pseudomonas aeruginosa*. *Journal of Medical Microbiology*, **16**, 401–8.

Pennington, J. E., Hickey, W. F., Blackwood, L. L. & Arnaut, M. A. (1981). Active immunization with lipopolysaccharide *Pseudomonas* antigen for chronic *Pseudomonas* bronchopneumonia in guinea pigs. *Journal of Clinical Investigation*, **68**, 1140–8.

Petersen, N. T., Høiby, N., Mordhorst, C.-H., Lind, K., Flensborg, E. W. & Bruun, B. (1981). Respiratory infections in cystic fibrosis caused by virus, chlamydia and mycoplasma – possible synergism with *Pseudomonas aeruginosa*. *Acta Paediatrica Scandinavica*, **70**, 623–8.

Pier, G. B. (1991). Vaccine potential of *Pseudomonas aeruginosa* mucoid exopolysaccharide (alginate). In *Pseudomonas aeruginosa in Human Diseases*, ed. J. Y.

Homma, H. Tanimoto, I. A. Holder, N. Høiby & G. Döring, pp. 136–42. Basel: Karger.

Pier, G., Grout, M. & DesJardins, D. (1991). Complement deposition by antibodies to *Pseudomonas aeruginosa* mucoid exopolysaccharide (MEP) and by non-MEP specific opsonins. *Journal of Immunology*, **147**, 1369–76.

Pier, G. B., Small, G. J. & Warren, H. B. (1990). Protection against mucoid *Pseudomonas aeruginosa* in rodent models of endobronchial infections. *Science*, **249**, 537–40.

Pitt, T. L., MacDougall, J., Penketh, A. R. L. & Cooke, E. M. (1986). Polyagglutinating and non-typable strains of *Pseudomonas aeruginosa* in cystic fibrosis. *Journal of Medical Microbiology*, **21**, 179–86.

Plotkowski, M. C., Beck, G., Tournier, J. M., Bernardo, M., Marques, E. A. & Puchelle, E. (1989). Adherence of *Pseudomonas aeruginosa* to respiratory epithelium and the effect of leukocyte elastase. *Journal of Medical Microbiology*, **30**, 285–93.

Pressler, T., Kronborg, G., Shand, G. H., Mansa, B. & Høiby, N. (1992a). Determination of IgG subclass antibodies to *Pseudomonas aeruginosa* outer membrane proteins in cystic fibrosis lung infection using immunoblotting and enzyme linked immunosorbent assay. *Medical Microbiology and Immunology*, **181**, 339–49.

Pressler, T., Pandey, J. P., Espersen, F. *et al.* (1992b). Immunoglobulin allotypes and IgG subclass antibody response to *Pseudomonas aeruginosa* antigens in chronically infected cystic fibrosis patients. *Clinical and Experimental Microbiology*, **90**, 209–14.

Pressler, T., Pedersen, S. S., Espersen, F., Høiby N. & Koch, C. (1990). IgG subclass antibodies to *Pseudomonas aeruginosa* in sera from patients with chronic *P. aeruginosa* infection investigated by enzyme linked immunosorbent assay. *Clinical and Experimental Immunology*, **81**, 428–34.

Pressler, T., Pedersen, S. S., Espersen, F., Høiby, N. & Koch, C. (1992c). IgG subclass antibody responses to alginate from *Pseudomonas aeruginosa* in patients with cystic fibrosis and chronic *P. aeruginosa* infection. *Pediatric Pulmonology*, **14**, 44–51.

Ramphal, R., Guay, C. & Pier, G. B. (1987). *Pseudomonas aeruginosa* adhesins for tracheobronchial mucin. *Infection and Immunity*, **55**, 600–3.

Riber, U., Espersen, F., Wilkinson, B. J. & Kharazmi, A. (1990). Neutrophil chemotactic activity of peptidoglycan: a comparison between *Staphylococcus aureus* and *Staphylococcus epidermidis*. *Acta Pathologica, Microbiologica et Immunologica Scandinavica*, **98**, 881–6.

Roychoudhury, S., Zielinski, N. A., Devault, J. D. *et al.* (1991). *Pseudomonas aeruginosa* infection in cystic fibrosis – biosynthesis of alginate as a virulence factor. In *Pseudomonas aeruginosa in human diseases*, ed. J. Y. Homma, H. Tanimoto, I. A. Holder, N. Høiby & G. Döring, pp. 63–7. Basel: Karger.

Schaad, U. B., Lang, A. B., Wedgewood, J., Ruedeberg, A., Que, J. U., Fürer, E. & Cryz, S. J., Jr (1991). Safety and immunogenicity of *Pseudomonas aeruginosa* conjugate-A vaccine in cystic fibrosis. *Lancet*, **338**, 1236–7.

Schiller, N. L. (1988). Characterization of the susceptibility of *Pseudomonas aeruginosa* to complement mediated killing: role of antibodies to the rough lipopolysaccharide on serum-sensitive strains. *Infection and Immunity*, **56**, 632–9.

Schiøtz, P. O. (1981). Local humoral immunity and immune reactions in the lungs of patients with cystic fibrosis. *Acta Pediatrica Scandinavica* (Suppl. 276), 3–25.

Shand, G. H., Pedersen, S. S., Brown, M. R. W. & Høiby, N. (1991). Serum antibodies to *Pseudomonas aeruginosa* outer membrane proteins and iron regulated membrane proteins at different stages of chronic cystic fibrosis lung infection. *Journal of Medical Microbiology*, **34**, 203–12.

Sordelli, D. O., Macri, C. N. & Maillie, A. J. (1990). A study on the effect of piroxicam (PIR) treatment to prevent lung damage in CF patients with *Pseudomonas aeruginosa* (Psa) pneumonia. *Pediatric Pulmonology* (Suppl.) **5**, 247–8.

Speert, D. P., Campbell, M. E., Davidson, G. F. & Wong, L. T. K. (1993). *Pseudomonas aeruginosa* colonization of the gastrointestinal tract in patients with cystic fibrosis. *Journal of Infectious Diseases*, **167**, 226–9.

Stahl, P. S., Norn, S., Schiøtz, P.O., Permin, H. & Høiby, N. (1980). *Pseudomonas aeruginosa* allergy in cystic fibrosis. Involvement of histamine release in the pathogenesis of the lung tissue damage. *Allergy*, **35**, 23–9.

Suter, S. (1989). The imbalance between granulocyte neutral proteases and antiproteases in bronchial secretions from patients with cystic fibrosis. In *Pseudomonas aeruginosa infection*, ed. N. Høiby, S. S. Pedersen, G. H. Shand, G. Döring & I. A. Holder, pp. 158–68. Basel: Karger.

Suter, S. & Chevallier, I. (1991). Proteolytic inactivation of alpha-1–proteinase inhibitor in infected bronchial secretions from patients with cystic fibrosis. *European Respiratory Journal*, **4**, 40–9.

Taylor, R. F. H., Morgan, D. W., Nicholson, P. S., Mackay, I. S., Hodson, M. E. & Pitt, T. L. (1992). Extrapulmonary sites of *Pseudomonas aeruginosa* in adults with cystic fibrosis. *Thorax*, **47**, 426–8.

Tosi, M. F., Zakem, H. & Berger, M. (1990). Neutrophil elastase cleaves C3Bi on opsonized *Pseudomonas* as well as Crl on neutrophils to create a functionally important opsonin receptor mismatch. *Journal of Clinical Investigation*, **86**, 300–8.

Winnie, G. B., Klinger, J. D., Sherman, J. M. & Thomassen, M. J. (1982). Induction of phagocytic inhibatory activity in cats with chronic *Pseudomonas aeruginosa* pulmonary infection. *Infection and Immunology*, **32**, 1088–93.

Wisnieski, J. J., Todd, E. W., Fuller, R. *et al.* (1985). Immune complexes and complement abnormalities in patients with cystic fibrosis. *American Review of Respiratory Disease*, **132**, 770–6.

Woods, D. E. & Bryan, L. E. (1985). Studies on the ability of alginate to act as protective immunogen against infection with *Pseudomonas aeruginosa* in animals. *Journal of Infectious Diseases*, **151**, 581–8.

Woods, D. E., Cryz, S. J., Jr, Friedman, R. L. & Iglewski, B. H. (1982). Contribution of toxin A and elastase to virulence of *Pseudomonas aeruginosa* in chronic lung infections of rats. *Infection and Immunity*, **36**, 1223–8.

Zach, M. S. (1991). Pathogenesis and management of lung disease in cystic fibrosis. *Journal of the Royal Society of Medicine*, **84** (Suppl. 18), 10–17.

15

Bacterial Biofilms in the Biliary System

Joseph J. Y. Sung and Joseph W. C. Leung

Introduction

Cholangitis consists of bacterial infection of bile in the biliary system. The syndrome of acute cholangitis has been well recognized since Charcot (1877) described the classical triad of pain, fever and jaundice in these patients. It is an important cause of abdominal emergency cases and septicaemia with a high rate of morbidity and mortality (Li *et al.* 1985; French *et al.* 1990). Biliary obstruction, due to gallstones obstructing the bile ducts, or to benign or malignant stricture of the biliary tract, is an essential element in the development of cholangitis. In the past decade, endoscopic drainage by biliary stenting has become a standard procedure in palliation for inoperable biliary malignancies and some cases of large biliary stones causing obstructive jaundice. Unfortunately, there is an increased incidence of cholangitis related to the blockage of the biliary stents with the use of this technique (Huibregtse *et al.* 1986; Cotton 1990). Some studies have revealed that the pathogenesis of pigment gall-stones (Stewart *et al.* 1987; Leung *et al.* 1988), and the blockage of the biliary stents (Leung *et al.* 1988; Speer *et al.* 1988) are closely related to the formation of bacterial biofilms in which the glycocalyx enclosed microcolonies coalesce to form an adherent structure (Jacques *et al.* 1987). This chapter reviews the present knowledge of the microbial ecology of the biliary system, formation of bacterial biofilm from bacterial infection within this system, the pathogenesis of brown biliary pigment stone and blockage of the biliary stent.

Microbial ecology of the biliary tract

Information of the normal microbial ecology of the biliary system is limited. Studies based on culturing gall bladder bile from patients with no evidence of biliary disease suggest that few bacteria normally exist in the biliary tract (Csendes *et al.* 1975; Dye *et al.* 1978). These reports are based on a small number of patients and single samples of bile aspirated during laparotomies, so such studies could, for example, only identify free living (planktonic) bacteria existing in the bile, whereas the surface associated (sessile) bacterial populations embedded within the mucus layer on the mucosal epithelia would not be detected. Information on a possible sessile bacterial population in the 'normal' biliary tract of humans is not available. Animal studies have shown that the feline biliary tract is sterile (Sung *et al.* 1990). The sphincter of Oddi, located at the junction between the sterile biliary tract and the colonized gastrointestinal tract, may harbour a few bacteria on its mucosal surface and probably served as a functional barrier to bacterial colonization between these two systems. The sphincter of Oddi is a distinct high pressure zone, which in the human is about 4 mm Hg higher than in the common bile duct and about 16 mm Hg greater than in the duodenum (Geenen *et al.* 1980). Superimposed on this basal sphincter zone are peristaltic contractions which aid the trans-sphincteric flow and prevent regurgitation (Helm *et al.* 1985; Dodds & Hogan 1989). Retrograde peristalsis of the sphincter occurs and perhaps transient relaxation allows retrograde contamination even on a temporary basis. A continuous bile flow and the shedding of mucus from the biliary mucosa ensure that the biliary system is free of bacteria under normal conditions.

Despite the fact that an intact biliary tract is sterile, the transient appearance of bacteria in bile has been demonstrated (Sung et al. 1991b). When a sterile foreign body is surgically implanted into the gall bladder, a bacterial biofilm forms on the surface of these implants in 6–12 weeks (Sung et al. 1991b). These foreign material implants provide physical surfaces for bacterial adhesion and proliferation, thus trapping the planktonic bacteria passing through the biliary tree and verifying their periodic appearance. The formation of bacterial biofilm on surfaces allows the microorganisms to stay in the hostile environment of the biliary tract.

Although the route of entry of bacteria into the biliary tract is still controversial, current observations suggest that the potential source of biliary pathogens originates in the gastrointestinal tract and infection descends via the hematogenous route in the portal circulation to invade the biliary tract (Schatten et al. 1955; Dineen 1964). In animal experiments, infusion of a genetically labelled Escherichia coli into the portal circulation results in isolation of the same organism from the common bile duct (Sung et al. 1991c). The entry of bacteria via the portal blood is facilitated by a raised intra-biliary pressure associated with obstructive jaundice (Sung et al. 1991c). Relatively fewer bacteria in the portovenous blood are needed to saturate the clearance mechanism in the liver in biliary obstruction and this results in bacterobilia. Disruption of tight junctions between hepatocytes and impairment of Kupffer cell functions in chronic biliary obstruction are likely to contribute to portal venous entry of bacteria.

The sphincter of Oddi is an important barrier to guard against duodenal bacteria. When the sphincteric mechanism is disrupted, either by surgical sphincterotomy or biliary endoscopy, duodenal biliary reflux is inevitable, allowing bacterial entry from the gastrointestinal tract into the normally sterile biliary tract: the so-called ascending route of infection (Feretis et al. 1984; Gregg et al. 1985; Sung et al. 1992a). There is some evidence that an abnormal sphincter function in patients with choledocholithiasis (Toouli et al. 1982; De Masi et al. 1984) and periampullary diverticulum (McSherry & Gleen

1970) contributes to the development of cholangitis. Whether this is a cause or an effect of the disease is still disputable. Nevertheless, both the hematogenous route via the portovenous blood and the ascending route in duodenal–biliary reflux are possible in the development of cholangitis.

Bacteriology in cholangitis

Bacteria commonly isolated from bile cultured in biliary tract infections included the coliforms (E. coli, Klebsiella sp.), faecal streptococci and anaerobic organisms such as Bacteroides fragilis and Clostridium perfringes (Keighley 1977; Borriello 1986). It is not uncommon to find more than one organism colonizing the biliary tract in cholangitis. Single species isolation accounts for over 70% of cases (Ong 1962). Synergism between E. coli and B. fragilis has been demonstrated in a biliary tract infection and it is postulated that the lowering of oxygen tension and pH achieved by aerobic bacteria in the acute inflammatory process facilitates the proliferation of the anaerobes in cholangitis.

Biofilms in sludge and brown pigment stone formation

Brown pigment stones represent a major health problem in Orientals and are most commonly associated with biliary infection (Ong 1962; Li et al. 1985). Calcium salts of bilirubinate, carbonate, fatty acid, and to a lesser extent phosphate constitute a major part of the biliary sludge which, with time, consolidates to form pigment stones (Fig. 15.1). The bile of these patients is almost always contaminated with sludge (Fig. 15.2). Calcium ions enter the hepatobiliary system by passive convection and diffusion from the sinusoidal blood (Rege et al. 1990). The entry of Ca^{2+} is found to be tightly linked to bile salt secretion and bile flow (Moore et al. 1985). In the bile, Ca^{2+} complexes with bicarbonate, carbonate and bilirubin diglucuronide in significant amounts leaving only about 20% of the total biliary calcium as free Ca^{2+} (Moore 1990). Acidification of bile by the gall bladder greatly enhances the solubility of these calcium salts. Defective acidification of bile, which usually

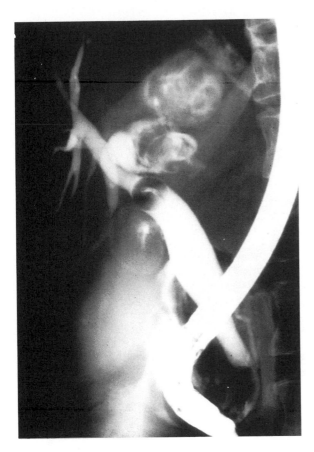

Fig. 15.1. Occlusion cholangiogram by endoscopic retrograd (ERC) showing intrahepatic duct stones impacting the left intrahepatic bile ducts.

Fig. 15.2. Sludge deposited when bile collected from a patient suffering from suppurative cholangitis was allowed to settle.

reflects increased buffering of H^+ by excessive mucin production, reduces the solubility of calcium bilirubinate, calcium carbonate and calcium salt of fatty acids and thus precipitates these calcium salts in bile mucin, forming biliary sludge and brown pigment stones (Lee 1990).

Besides the inorganic components, electron microscopy (EM) studies reveal that bacterial microcolonies are embedded in the matrix in up to 80% of brown pigment stones forming an integral part of the calculi (Stewart *et al.* 1987; Leung *et al.* 1989). In places where the bacteria were washed away in the EM processing, a honeycomb pattern appears representing the negative imprints of the bacteria. Bacteria contribute to the pathogenesis of brown pigment stones in several ways. First, bacterial ß-glucuronidase deconjugates bilirubin diglucuronide to form amorphous calcium bilirubinate (Maki 1962). Secondly, bacteria such as *E. coli* also produce phospholipase A_1 (lecithinase), which hydrolyses lecithin releasing fatty acids (mainly palmitic and stearic acids), forming calcium soaps of palmitate and stearate (Trotman 1991).

A further important role may be played by the bacteria forming brown pigment stones. The bacterial glycocalyx, like mucin, promotes the agglomeration of bile sediments and bacterial microcolonies in the formation of bacterial biofilm. With trapping of more bacteria, further deconjugation and precipitation of calcium

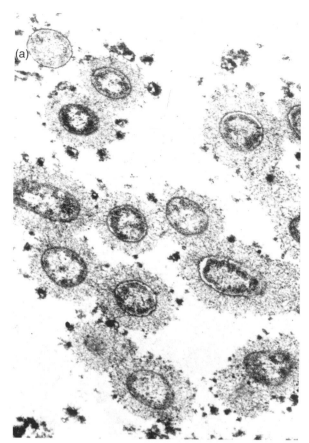

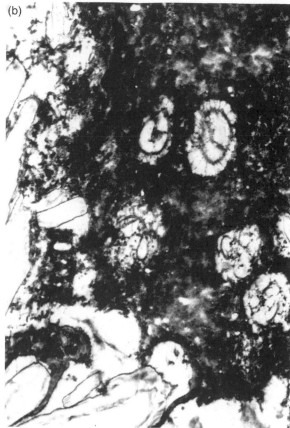

Fig. 15.3. (a) A microcolony of Gram negative bacteria from the centre of a brown pigment stone extracted from the common bile duct of a patient suffering from acute cholangitis. The microorganisms are surrounded by a ruthenium red stained material compatible with exopolysaccharide. They are protected from the hostile environment in the biliary system by the glycocalyx.

(b) Transmission electron micrograph of the same pigment stone where degenerated bacterial cells were seen surrounded by calcified amorphous material and crystalline structures. The bacterial glycocalyx may help the agglomeration of the pigment particles and biliary sediments together in forming pigment stones.

bilirubinate, the biofilm consolidates to form pigment stone (Stewart *et al.* 1987; Leung *et al.* 1989). Figure 15.3a shows a transmission electron micrograph of a brown pigment stone extracted from the common bile duct of a patient. Bacterial microcolonies surrounded by bacterial glycocalyx are seen inside the stone, forming a protective cover for the bacteria to proliferate despite the hostile environment of the biliary system. The glycocalyx forms an adherent mesh to bring together the amorphous bile pigment and bile sediment. Even after bacteria are dead, precipitation and agglomeration of bile sediments continue (Fig. 15.3b) and, with time, the consolidate to become the brown pigment

stones. Brown pigment gallstones are, therefore, in many ways, similar to the struvite stones in the urinary system (Nickel *et al.* 1986; see McLean *et al.*, Chapter 16). Bacteria play a crucial part in both situations by producing an enzyme and bacterial glycocalyx which promotes precipitation of sediments in bile and urine forming pigment stones and renal stones respectively.

In vitro studies on bacterial biofilms have shown that Ca^{2+} confers protection for biofilm bacteria to antibiotics (Marrie *et al.* 1982; Nickel *et al.* 1985; Hoyle *et al.* 1990). Condensation of the polyanionic bacterial glycocalyx in the presence of Ca^{2+} may explain the inability of antibiotics to penetrate the biofilm (Hoyle &

Costerton 1989). Brown pigment stones can be viewed as thick and calcified bacterial biofilms that predispose to obstruction and recurrent infection within the biliary system. As viable bacterial cells are often present within the biofilm, any treatment that involves crushing the stone *in vivo* by endoscopic manoeuvre has the risk of releasing these bacteria and reactivating the infection. Such patients should receive prophylactic antibiotics prior to lithotripsy.

Biofilms in the blockage of biliary stents

Endoscopic insertion of biliary stents has become a standard procedure in the palliation of biliary malignancy and some cases of large biliary stones causing obstructive jaundice (Csendes *et al.* 1975; Jacques *et al.* 1987; Fig. 15.4). The success rate for endoscopic stenting exceeds 90% and immediate complications are uncommon.

Unfortunately, the plastic stents in current use have a strong tendency to plug, resulting in recurrence of jaundice, cholangitis and septicaemia. Stent blockage leading to chills, fever and deteriorating liver function requires removal and replacement of the stent. The median patency interval for a 10 French gauge stent is only about 6 months (Leung *et al.* 1983; Huibregtse *et al.* 1986; Jacques *et al.* 1987). The blockage of biliary stents thus poses a limitation to their use in the treatment of biliary obstruction, especially in patients with benign conditions.

Microcolonies of bacteria, together with amorphous material form biofilms and are commonly found in the blocked stents (Speer *et al.* 1988; Leung *et al.* 1988). The microscopic appearance and chemical composition of the material scraped from these stents is very similar to that of biliary sludge (Wosiewitz *et al.* 1985). Studies on infection associated with medical devices have shown

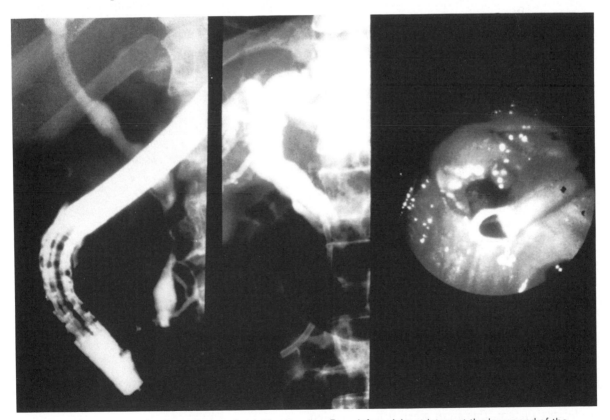

Fig. 15.4. Endoscopic stenting for malignant obstructive jaundice. From left to right, stricture at the lower end of the common bile duct, stent inserted to bypass the stricture and drainage of bile from the obstructed system achieved after stenting.

that once implanted, the surfaces of such medical implants are rapidly covered by a layer of host proteins, such as fibronectin, laminin, fibrin, collagen or immunoglobulin; these condition the surface and facilitate the subsequent adhesion of bacteria (Christensen *et al.* 1989; Dickson & Bisno 1989). This is probably also true for biliary stent blockage and infection. A layer of sticky material with a high protein content has been reported on the walls of stents (Groen *et al.* 1987). Further analysis using SDS-polyacrylamide gel electrophoresis (SDS-PAGE) revealed two major proteins of 13 and 16 kDa (Groen *et al.* 1987). The nature of these proteins is still not confirmed and their role in the blockage the biliary stent remains to be investigated.

In summary, the pathogenesis of stent blockage is initiated by bacteria attaching to the stent surface with their pili and/or glycocalyx to form a biofilm (Costerton *et al.* 1987). The elaboration of bacterial ß-glucuronidase and phospholipase subsequently form amorphous calcium bilirubinate and calcium salts of fatty acids. With time, the growth of biofilm and progressive agglomeration of bile sediments form biliary sludge which finally results in occlusion of the lumen.

Attempts to prolong the survival of the stents by using different materials and coatings have so far not been successful. The microorganisms in bacterial biofilm, when well covered with glycocalyx matrix, are resistant to antimicrobial agents in normal dosages (Anwar *et al.* 1990). Oxidizing biocides, such as chlorine, are effective in removing biofilm on the surface of these implants but they are too toxic to be used inside the human body (Costerton *et al.* 1987). Administration of aspirin, a mucolytic agent, and an antibiotic, doxycycline, have recently been shown to reduce the amount of encrusted material on the stent but have not eliminated the problem (Smit *et al.* 1989).

The current solutions are to use biliary stents made of different design, such as stents of a larger internal diameter that may delay complete blockage of the lumen (Speer *et al.* 1987; Seigel *et al.* 1988), or undertaking a regular exchange of the stent. A new self-expandable metal stent has been advocated recently (Huibregtse *et al.* 1989). Delivered in a compressed form using a special applicator system, this stent when fully expanded has a length of 6.7 cm and a diameter of 1 cm. With such an enlarged lumen, the problem of

stent blockage should be minimized, but longer follow-up studies are required to establish the benefit of these metal stents. Coene *et al.* (1990) have demonstrated that the formation of bacterial biofilm was maximal around the side holes, probably in relation to the rough surface and turbulent flow at these sites. Use of a straight stent without side holes has shown some improved drainage.

Currently, biliary stents are usually positioned in the common bile duct transversing the sphincter of Oddi, creating free communication between the biliary tract and the duodenum. Duodenal–biliary reflux is inevitable. Animal studies have shown that stents placed entirely within the common bile duct proximal to the sphincter remain patent for a longer time (De Masi *et al.* 1984; Geoghegan *et al.* 1991). Although advisable to leave the biliary stents above the sphincter, this poses problems with removal when blockage develops. A controlled clinical trial is needed to confirm any benefit to a more proximally placed biliary stent. Thus the pathogenesis of brown pigment stones and the clogging of biliary stents is very similar with biofilm formation playing a central role (Fig. 15.5). The possible ways to control these include methods to control biofilm formation.

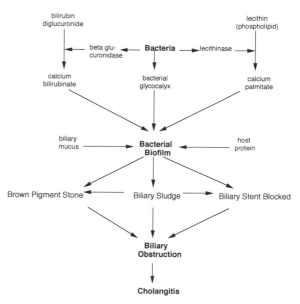

Fig. 15.5. Diagram showing bacterial biofilms in the pathogenesis of brown pigment stones, biliary sludge and the blockage of biliary stents.

Effects of bile salts on bacterial biofilms

Bile salts have antibacterial properties and this contributes to the sterility of the biliary system. They possess both cytotoxic and bacteriostatic properties *in vitro*. Velardi *et al.* (1991) have shown that the hydrophobic bile salts have more potent cytopathic effects on mammalian cells than the hydrophilic bile salts. Lecithin, the predominant phospholipid in bile, reduces the cytotoxic effect of bile salts (Coleman *et al.* 1980; Velardi *et al.* 1991). When the antibacterial activities of bile salts possessing different hydrophobicities were tested on common biliary pathogens (Sung, Costerton & Schaffer 1991a), the more hydrophobic bile salts (such as sodium taurodeoxycholate and sodium deoxycholate) have more significant inhibition on the growth of *E. coli* (O21:H25) when compared with the hydrophilic bile salts (that is, sodium taurocholate, sodium chenodeoxycholate and sodium tauroursodeoxycholate). However, with the addition of lecithin creating a mixed micellar solution, which mimics the *in vivo* conditions, the antibacterial activity of even the more potent bacteriostatic bile salt was significantly reduced. It is likely that in the mixed micellar solution, the hydrophobic components of the bile salts were engaged with the phospholipid in the mixed micellar aggregates, hence reducing their bacteriostatic property. These findings raise questions about the clinical significance of such a bacteriostatic effect *in vivo*, as bile salts in the biliary system exist in mixed micellar solution with lecithin. Free monomers/dimers of bile salts might have a minor protective influence. However, once the bacterial biofilm is established, in the form of pigment stone and sludge which occlude the biliary stent, it is extremely difficult to eradicate the microorgansims even with very high concentrations of hydrophobic bile salt (Sung *et al.* 1991a).

It is interesting to consider whether there may be a role for hydrophobic bile salt in the prevention of stent blockage and recurrent pigment stone formation following the establishment of bacterial biofilms. Electrostatic interactions and surface hydrophobicity are known to be the most crucial factors affecting bacterial adhesion to plastics (Klotz 1990). Microbial adhesion correlates directly with increasing substrate hydrophobicity (Fletcher & Loeb 1979; Klotz *et al.* 1985). Surface active agents (surfactants), which interfere with hydrophobic interactions, might reduce microbial adhesion to plastics by reducing the surface tension of the suspending medium (Absolom *et al.* 1983). An example of applying non-ionic surfactant on the surface of plastic is the use of alcohol propoxylate in coating the surface of polystyrene (Humphries *et al.* 1987). It has been demonstrated that the surfactant dramatically inhibits the adhesion of *Pseudomonas* sp. and *Serratia* sp. on plastics.

The enterohepatic recycling of bile salts provides a natural source of detergent in the bile. Our preliminary data shows that the hydrophilic bile salt (sodium taurocholate) did not affect the adhesion of *E. coli* to silastic. Conversely, the more hydrophobic bile salt (sodium taurodeoxycholate) at 25–50 mmol L^{-1} reduced adhesion by 100 to 1000-fold and the surface active property of this amphipathic molecule is likely to account for the difference (Sung *et al.* 1992b). The clinical utility of hydrophobic interaction inhibitors has yet to be demonstrated. It is possible that the use of such agents may prove advantageous in attempts to reduce adhesion and thus formation of brown pigment stone and blockage of biliary stents. A major problem with the use of hydrophobic bile salts is the cytopathic effect of these compounds causing gastrointestinal upset. Possible solutions include the use of urosdeoxycholic acid in conjunction with hydrophobic bile salt as the former can minimize the cytopathic effect of bile salts on the gastrointestinal mucosa. Conversely, incorporation of hydrophobic bile salts or other novel agents into biliary stents may prevent their self-defecting nature if the sludge brown stone sequence is aborted.

Bacterial pathogens play a key role in the pathogenesis of biliary sludge, brown pigment stones and the blockage of biliary stents. In the adherent bacterial biofilm, the microorganisms are dwelling in a favourable environment protected from antimicrobial agents and phagocytic leucocytes. Two major factors contributing to the protective environment in the biofilm are bacterial and mucous glycoprotein, and calcification of the structure. To date, recurrent pigment stone formation in the bile duct is still an unsolved problem. The prevention of stent blockage is, so far, confined to physical methods using larger stents and stents without side holes. As bile salts are abundant in the biliary system and preliminary results show that they interfere with the initial step of biofilm formation, namely bacterial adhesion on surfaces, it may be worth investigating the usefulness of these surfactants in the prevention of sludge and brown pigment stone formation and the blockage of biliary stents.

References

Absolom, D. R., Lamberti, F. V., Policova, Z., Zigg, W., van Oss, C. J. & Neumann, A. W. (1983). Surface thermodynamics of bacterial adhesion. *Applied and Environmental Microbiology*, **46**, 90–7.

Anwar, H, Dasgupta, M. K. & Costerton, J. W. (1990). Testing the susceptibility of bacteria in biofilms to antimicrobial agents. *Antimicrobial Agents and Chemotherapy*, **34**, 2043–6.

Borriello, S. P. (1986). Microbial flora of the gastrointestinal tract. In *Microbial metabolism in the digestive tract*, ed. M. J. Hill. Boca Raton, Fla: CRC Press.

Charcot, J. M. (1877). *Leçons sur les maladies du foie des voies filiares et des reins*. Paris: Faculté de Médécine de Paris.

Christensen, G. D., Baddour, L. M., Hasty, D. L., Lawrence, J. H. & Andrew Simpson, W. (1989). Microbial and foreign body factors in the pathogenesis of medical device infection. In I*nfection associated with indwelling medical devices*, ed. A. L. Bisno & F. A. Waldvogel, pp. 4–27. Washington, DC: American Society for Microbiology.

Coene, P. P., Groen, A. K., Cheng, J., Out, T., Tytgat, G. N. J. & Huibregtse, K. (1990). Clogging of biliary endoprothesis: a new perspective. *Gut*, **31**, 913–17.

Coleman, R., Lowe, P. J. & Billington, D. (1980). Membrane lipid composition and susceptibility to bile salt demage. *Biochimica et Biophysica Acta*, **242**, 825–34.

Costerton, J. W., Cheng, K. J. & Geesey, G. G. (1987). Bacterial biofilms in nature and disease. *Annual of Review of Microbiology*, **41**, 435–64.

Cotton, P. B. (1990). Critical appraisal of therapeutic endoscopy in biliary tract disease. *Annual Review of Medicine*, **41**, 211–22.

Csendes, A., Fernandez, M. & Uribe, P. (1975) Bacteriology of the gallbladder bile in normal subjects. *American Journal of Surgery*, **129**, 629–31.

De Masi, E., Corazziari, E., Habib, F. I. *et al.* (1984). Manometric study of the sphincter of Oddi in patients with and without common bile duct stones. *Gut*, **25**, 275–8.

Dickson, G. M. & Bisno, A. L. (1989). Infection associated with indwelling devices: concepts of pathogenesis of infection associated with extravascular devices. *Antimicrobial Agents and Chemotherapy*, **33**, 602–7.

Dineen, P. (1964). The importance of the route of infection in experimental biliary tract infection. *Surgery Gynecology and Obstetrics*, **119**, 1001–8.

Dodds, W. J. & Hogan, W. J. (1989). Motility of the biliary system. In *Handbook of physiology: the gastrointestinal system*, vol. 1, section 6, ed. S. G. Schultz, pp. 1055–1102. Washington, DC: American Physiological Society.

Dye, M., MacDonald, A. & Smith, G. (1978) The bacterial flora of the biliary tract and liver in man. *British Journal of Surgery*, **65**, 285–7.

Feretis, C. B., Contou, C. T., Manouras, A. J., Apostolidis, N. S. & Golematis, B. C. (1984). Long term consequences of bacterial colonization of the biliary tract after choledochotomy. *Surgery Gynecology and Obstetrics*, **159**, 363–6.

Fletcher, M. & Loeb, G. I. (1979). Influence of substratum characteristics on the attachment of a marine pseudomonad to solid surfaces. *Applied Environmental Microbiology*, **37**, 67–72.

French, G. L., Cheng, A. F. B, Duthie, R. & Cockram, C. S. (1990). Septicemia in Hong Kong. *Journal of Antimicrobial Chemotherapy*, **25**, 137–45.

Geenen, J. E., Hogan, W. I., Dodds, W. J., Steward, E. T. & Ardirfer, R. C. (1980). Intraluminal pressure recording of the human sphincter of Oddi. *Gastroenterology*, **78**, 317–23.

Geoghegan, J. G., Branch, M. S., Costerton, J. W., Pappas, T. N. & Cotton, P. B. (1991). Biliary stent occludes earlier if the distal tip is in the duodenum in dogs. *Gastroenterology Endoscopy*, **37**, 257 (abstract).

Gregg, J.A., Girolami, P. D. & Carr-Lock, D. L. (1985) Effects of sphincteroplasty and endoscopic sphincterotomy on the bacteriologic characteristics of the common bile duct. *American Journal of Surgery*, **149**, 668–71.

Groen, A. K., Out, T., Huibregtse, K., Belzenne, B., Hoek, F. J. & Tytgat, G. N. J. (1987). Characterization of the content of occluded biliary endoprothesis. *Endoscopy*, **19**, 57–9.

Helm, J. F., Dodds, W. J., Christensen, J. & Sarna, S. K. (1985). Control mechanism of spontaneous *in vitro* contractions of the opossum sphincter of Oddi. *American Journal of Physiology*, **249**, G572–9.

Hoyle, B. D. & Costerton, J. W. (1989). Transient exposure to a physiologically-relevant concentration of calcium confers tobramycin resistance upon sessile cells of *Pseudomonas aeruginosa*. *FEMS Microbiology Letters*, **60**, 339–42.

Hoyle, B. D., Jass, J. & Costerton, J. W. (1990). The biofilm glycocalyx as a resistant factor. *Journal of Antimicrobial Chemotherapy*, **26**, 1–6.

Huibregtse, K., Cheng, J., Coene, P. P., Fockens, P. & Tytgat, G. N. J. (1989). Endoscopic placement of expandable metal stent for biliary stricture: a preliminary report on experience with 33 patients. *Endoscopy*, **21**, 280–2

Huibregtse, K., Katon, R. M., Coene, P. P. & Tytgat, G. N. J. (1986). Endoscopic palliative treatment of pancreatic cancer. *Gastrointestinal Endoscopy*, **32**, 334–8.

Humphries, M., Jaworzyn, J. F., Cantwell, J. B. & Eakin, A. (1987). The use of non-ionic ethoxylated and propoxylated surfactants to prevent the adhesion of bacteria to solid surface. *FEMS Microbiology Letters*, **42**, 91–101.

Jacques, M., Marrie, T. J & Costerton, J. W. (1987). Microbial colonization of prosthetic devices. *Microbial Ecology*, **13**, 173–91.

Keighley, M. R. B. (1977). Microorganisms in the bile. *Annals of the Royal College of Surgery*, **59**, 329–34.

Klotz, S. A. (1990). Role of hydrophobic interactions in microbial adhesion to plastics used in medical devices. In *Microbial cell surface hydrophobicity*, ed. R. J. Doule & M. Rosenberg, pp. 107–36. Washington, DC: American Society for Microbiology.

Klotz, S. A., Drutz, D. J. & Zajic, J. E. (1985). Factors governing the adherence of *Candida* species to plastic surfaces. *Infection and Immunity*, **50**, 97–101.

Lee, S. P. (1990). Pathogenesis of biliary sludge. *Hepatology*, **12**, 200S–205S.

Leung, J. W. C., Emery, R. & Cotton, P. B. (1983). Management of malignant obstructive jaundice at the Middlesex Hospital. *British Journal of Surgery*, **70**, 584–6.

Leung, J. W. C., Ling, T. K. W., Kung, J. L. S. & Vallence-Owen, J. (1988). The role of bacteria in blockage of biliary stent. *Gastrointestinal Endoscopy*, **34**, 19–22.

Leung, J. W. C., Sung, J. Y. & Costerton, J. W. (1989). Bacteriological and electron microscopy examination of brown pigment stones. *Journal of Clinical Microbiology*, **27**, 915–21.

Li, A. C. K., Chung, S. C. S., Leung, J. W. C. & Mok, S. D. (1985). Recurrent pyogenic cholangitis: an update. *Tropical Gastroenterology*, **6**, 119–31.

Maki, T. (1962). Pathogenesis of calcium bilirubinate gallstone: role of *E. coli* ß-glucuronidase and coagulation by inorganic ions, polyelectrolytes and agitation. *Annals Of Surgery*, **164**, 90–100.

Marrie, T. J., Nelligan, J. & Costerton, J. W. (1982). A scanning and transmission electron microscopic study of an infected endocardial pacemaker lead. *Circulation*, **66**, 1339–43.

McSherry, C. K. & Gleen, F. (1970). Biliary tract obstruction and duodenal diverticula. *Surgery Gynecology and Obstetrics*, **130**, 829–36.

Moore, E. W. (1990). Biliary calcium and gallstone formation. *Hepatology*, **12**, 206S–218S.

Moore, E. W., Gleeson, D., Murphy, G. M. & Dowling, R. H. (1985). Biliary calcium secretion in man: a test of the computer model in bile salt-depleted subject. *Hepatology*, **5**, 994 (abstract).

Nickel, J. C., Reid, G., Bruce, A. W. & Costerton, J. W. (1986). Ultrastructural microbiology of infected urinary stone. *Investigational Urology*, **6**, 512–15.

Nickel, J. C., Ruseska, I., Wright, J. B. & Costerton, J. W. (1985). Tobramycin resistance of *Pseudomonas aeruginosa* cells growing as a biofilm on urinary catheter material. *Antimicrobial Agents and Chemotherapy*, **27**, 19–24.

Ong, G. B. (1962). A study of recurrent pyogenic cholangitis. *Archives of Surgery*, **84**, 199–255.

Rege, R. V., Dawes, L. G. & Moore, E. W. (1990). Biliary calcium secretion in the dog occurs primarily by passive convection and diffusion and is linked to bile flow. *Journal of Laboratory and Clinical Medicine*, **115**, 593–602.

Schatten, W. E., Desprez, J. D. & Holden, W. D. (1955). A bacteriological study of portal venous blood in man. *Archives of Surgery*, **71**, 404–9.

Seigel, J. H., Pullano, W., Kodsi, B., Cooperman, A. & Ramsey, W. (1988). Optimal palliation of malignant bile duct obstruction: experience with 12 French prostheses. *Endoscopy*, **20**, 137–41.

Smit, J. M., Out, T., Groen, A. K. *et al.* (1989). A placebo controlled study on the efficacy of aspirin and doxycycline in preventing clogging of biliary endoprostheses. *Gastrointestinal Endoscopy*, **35**, 485–9.

Speer, A., Cotton, P. B., Rode, J. *et al.* (1988). Biliary stent blockage with bacterial biofilm: a light and electron microscopy. *Annals of Internal Medicine*, **108**, 546–53.

Speer, A. G., Cotton, P. B., Russell, R. C. G. *et al.* (1987). Randomized trial of endoscopic versus percutaneous stent insertion in malignant obstructive jaundice. *Lancet*, **ii**, 57–62.

Stewart, L., Smith, A. L., Pellegrini, C. A., Roger, M. W. & Way, L. W. (1987). Pigment gallstones form as a composite of bacterial microcolonies and pigment solid. *Annals Of Surgery*, **206**, 242–50.

Sung, J. Y., Costerton, J. W. & Shaffer, E. A. (1991a). Bacteriostatic activities of bile salts of different hydrophobicity and in the presence of phospholipids. *Hepatology*, **14**, 262A (abstract).

Sung, J. Y., Leung, J. W. C., Shaffer, E. A., Lam, K. & Costerton, J.W. (1992a). Ascending of the biliary tract after surgical sphincgterotomy and biliary stent. *Journal of Gastroenterology and Hepatology*, **7**, 240–5.

Sung, J. Y., Olson, M. E., Leung, J. W. C., Lundburg, M. S. & Costerton, J. W. (1990) The sphincter of Oddi is the boundary of bacterial colonization of the gastrointestinal and the biliary system. *Microbial Ecology of Health and Disease*, **3**, 199–207.

Sung, J. Y., Olson, M. E., Leung, J. W. C., Lundberg, M. S. & Costerton, J.W. (1991b). The demonstration of transient bactebilia by foreign body implantation. *Digestive Disease and Science*, **36**, 943–8.

Sung, J. Y., Shaffer, E. A., Lam, K. & Costerton, J. W. (1992b). Hydrophobic bile salts inhibit the adhesion of bacteria on the surface of biliary stent. *Gastrointestinal Endoscopy*, **38**, 263 (abstract).

Sung, J. Y., Shaffer, E. A., Olson, M. E., Leung, J. W. C., Lam, K. & Costerton, J. W. (1991c). Bacterial invasion of the biliary system by way of the portal venous system. *Hepatology*, **14**, 313–17.

Toouli, J., Geenen, J. E., Hogan, W. J., Dodds, W. J. & Arndorfer, R. C. (1982). Sphincter of Oddi motor activity: a comparison between patients with common bile duct stones and controls. *Gastroenterology*, **82**, 111–17.

Trotman, B. W. (1991). Pigment gallstone disease. *Gastroenterology Clinics of North America*, **20**, 111–26.

Velardi, A. L. M., Groen, A. K., Oute Elferink, R. P. J., Van der Meer, R., Palascino, R. & Tytgat, G. N. J. (1991). Cell type-dependent effect of phospholipid and cholesterol on bile salt cytotoxicity. *Gastroenterology*, **101**, 457–64.

Wosiewitz, U., Schrameyer, B. & Safrany, L. (1985). Biliary sludge: its role during bile drainage with an endoprosthesis. *Gastroenterology*, **88**, 1076.

16

Biofilm Associated Urinary Tract Infections

Robert J. C. McLean, J. Curtis Nickel and Merle E. Olson

Introduction

Urinary tract infections (UTI) of the lower urinary tract are a common problem causing significant morbidity in females and males (Nickel 1990). Most of the causal organisms are Gram negative bacilli such as *Escherichia coli* and *Proteus mirabilis*, or Gram positive *Enterococcus* and *Streptococcus* spp., which have their reservoir in the gastrointestinal tract. The other major source of UTI pathogens is direct transmission through sexual activity. These organisms first colonize the introitus and the periurethral area before entering the bladder or prostate. There is an indigenous population of Gram positive, acid producing lactobacilli in this environment which under normal circumstances appears to enhance the urinary defence mechanisms and inhibit the successful progression of the enterics into the urethra and bladder (Chan *et al.* 1985; Reid *et al.* 1987, 1990a). When the balance between enteric uropathogens and host defences is upset, uropathogens are able to ascend through the urethra into the bladder, the prostate (in males) and less often into the kidneys where they colonize and cause infection. Usually only the bladder is infected giving rise to simple, uncomplicated acute cystitis. This condition is readily treated by several standard antibiotic regimes (Nickel 1990). Problems arise when this infection spreads from the bladder into other organs such as the kidneys (pyelonephritis) or prostate (prostatitis), or induces the formation of calculi (struvite urolithiasis: McLean *et al.* 1988, 1992). In these cases, significant tissue damage may occur, possibly causing permanent damage to renal

function, which can be life threatening. Numerous studies have shown the importance of uropathogen adhesion to the tissue surface as a necessary prerequisite for infection (Sandberg *et al.* 1988; Stamm *et al.* 1989; Svanborg Edén *et al.* 1989, 1990, Nickel 1990). As outlined below, a major aspect of the host defence strategy in this environment is geared towards prevention and control of initial bacterial adhesion and subsequent growth. The major rationale behind this strategy is that once the organisms have established themselves on urinary tissue, they adopt a microcolony and biofilm mode of growth (Nickel *et al.* 1992b) that enables them to develop chronic infections (cystitis, prostatitis) and to persist in spite of host defences and high levels of antibiotics (Elliott *et al.* 1984; Nickel *et al.* 1990). Experimental manipulation of host defences, and appropriate selection of uropathogens, are essential for the duplication of these various infections in animal models. In this chapter we outline the host defence mechanisms of the lower urinary tract, and summarize recent work on biofilm associated UTI.

Host defence mechanisms in the lower urinary tract

The bladder resists infection through several means (Kaye 1975; Nickel 1990; Svanborg Edén *et al.* 1990). Mechanical defence mechanisms are of primary importance in this environment. They include the periodic voiding of urine and washing out of unattached pathogens (McLean *et al.* 1988), sloughing of pathogen colonized uroepithelial cells (Orikasa & Hinman, 1977), the glycosaminoglycan (GAG) mucous layer (Parsons 1982; Cornish *et al.* 1987, 1988, 1990; Parsons *et al.* 1988) and the cellular and humoral immune responses (Fukushi & Orikasa 1981; Gillon *et al.* 1984; Hopkins *et al.* 1987; Cornish *et al.* 1988). The overall defence strategy of the bladder is to prevent adhesion and growth of incoming pathogens to the epithelial tissue through the production of a mucous layer, secretory IgA and urine components such as the Tamm–Horsfall glycoprotein (Fukushi & Orikasa 1981; Gillon *et al.* 1984; Hopkins *et al.* 1987; Cornish *et al.* 1987, 1988, 1990; Dulawa *et al.* 1988; Parkkinen *et al.* 1988; Hawthorn & Reid 1990; Reinhart *et al.* 1990; Hawthorn *et al.* 1991; Sobel 1991). Adherent organisms are removed through the sloughing of colonized epithelial cells (Orikasa & Hinman 1977) and cell mediated immunity (Fukushi & Orikasa 1981; Gillon *et al.* 1984). Incoming pathogens must also compete with the normal urogenital *Lactobacillus* flora for nutrients and attachment sites (Chan *et al.* 1985; Reid *et al.* 1987; 1988; 1990a). Inhibition of bacterial growth in urine is mediated through the low availability of iron (Shand *et al.* 1985) and osmotic stress caused by high concentrations of urea (Kaye 1975). Unattached organisms unable to overcome these defences are invariably removed through voiding (McLean *et al.* 1988).

Although the immune response is certainly important, there is evidence that IgG and sIgA production in themselves do not prevent recurrent cystitis or prostatitis (Rene *et al.* 1982, Schmidt *et al.* 1993). One other bladder defence mechanism makes a significant contribution to bladder defence, that is the protective GAG mucous layer (J. C. Nickel, J. Cornish & R. J. C. McLean, unpublished data; Hurst *et al.* 1987; Cornish *et al.* 1988, 1990; Holm-Bentzen & Ammitzboll 1989). This GAG layer is a very thin cover on the transitional cell epithelium of the bladder. It is thought to act as an important mechanism in physically shielding the bladder surface from pathogens, microcrystals, proteins and even carcinogenic molecules (Cornish *et al.* 1988; 1990; Parsons *et al.* 1988, 1990). In these studies, bladder mucus was removed or disrupted (often by administering acid to the tissue), and an increase in inflammation or infection was noted. In some manner, the degradation or deterioration of the GAG layer allowed the organisms

to infect the bladder tissue cells and recently we confirmed these findings (R. J. C. McLean, J. C. Nickel & M. E. Olson, unpublished data). However, ultrastructural observations showed that two techniques used for GAG layer disruption, notably mild acid washing and especially detergent washing (Cornish *et al.* 1988; Parsons *et al.* 1988, 1990), themselves cause considerable damage to rat bladder tissue in the absence of experimental infection (J. C. Nickel, J. Cornish & R. J. C. McLean, unpublished). Nevertheless, protamine sulphate removes bladder mucus (Parsons *et al.* 1990) with little damage to underlying tissues (J. C. Nickel, J. Cornish & R. J. C. McLean, unpublished data). These animals were then quite susceptible to *E. coli* cystitis.

There are still several unanswered questions regarding the role of the GAG layer. It is not clear whether inflammation is due to the mucus disruption treatment itself or whether it is due to the selective interaction between bacteria and bladder tissue in the localized microenvironments that have been deprived of mucus. It is also unclear whether this layer becomes disrupted prior to or during the course of infection, or what effect normal and abnormal urine chemistry has on the GAG layer. This knowledge is crucial before any protective role can be unequivocally assigned to the GAG layer.

Acute and chronic cystitis

Several animal models of cystitis are reported in the literature. In the majority of cases, these involve ascending models of infection in various animals (mice, rats, dogs, cats, and non-human primates). In these cases the uropathogens are administered to the periurethral area in anaesthetized animals or even administered into the bladder through urethral catheterization. The animals are allowed to recover and are then observed and evaluated for overt behavioural signs of urinary infection (such as lethargy, frequency of urination, painful voiding), bacterial concentrations in the urine, antibody concentrations in the blood or urine, or upon sacrifice, gross morphological or histological changes to the various components of the urinary tract. Ultrastructural observations of urinary tissue experimentally infected with a pyelonephritogenic strain of *E. coli* show that uropathogen adhesion to tissues, even in acute infections, is quickly followed (within 24 h) by microcolony formation (Hagberg *et al.* 1986). This confirms clinical observations by Elliot *et al.* (1984). It has been our experience that these experimental models of cystitis mimic acute cystitis. The infectious dose required is generally very high (*c.* 10^8–10^9 *E. coli* per ml in Sprague Dawley rats) and the success rate of infection in our test animals is in the order of 50–70%. In the absence of any other experimental manipulation, these infections are rapidly cleared usually within 2 or 3 days, and further complications such as prostatitis, pyelonephritis, and calculus formation are extremely rare.

Much higher rates of chronic bladder colonization can be obtained if the bladder environment is experimentally altered. One such approach is to insert a foreign object such as a zinc disc (Satoh *et al.* 1984; Nickel *et al.* 1987; Olson *et al.* 1989) or polyurethane sponge (Miller *et al.* 1987) into the bladder – a process which mimics foreign object implantation leading to UTI. In these cases, the best experimental approach is surgically to implant the sterile foreign object into the lumen of the bladder and to allow the animal (usually rat), to recover from surgery for several days before introducing the pathogens via urinary catheterization. This protocol of surgical, sterile foreign object-implantation → convalescence → introduction of infection is very successful in that the onset of experimental chronic UTI approaches 100%, yet the stress and discomfort on the test animals is minimalized (Satoh *et al.* 1984; Nickel *et al.* 1987).

In our experience, chronic UTI associated with foreign object implants generally lasts for a month or more. The foreign object acts as an 'immunologically inert' surface in that incoming organisms have a place on which to colonize where they do not have to combat the host defence mechanisms present in the mucous layer and urothelium. Colonization and biofilm formation occur rapidly (<24 h) (McLean 1986; Nickel *et al.* 1987), and the infected foreign object then acts as a reservoir for continuous infection of the surrounding tissues. Figure 16.1 illustrates a bacterial biofilm on the bladder of a rat that had been implanted with a foreign body. If uropathogenic *E. coli* strains are employed in this model, chronic cystitis always results. Complications which often arise from chronic *E.*

coli cystitis include pyelonephritis (due in part through urine reflux into the kidneys) and prostatitis when male rats are used (Nickel 1990). We have also noted that experimental infection by *E. coli* in this model is often accompanied by *Morganella morganii* infections (J. C. Nickel, J. Cornish & R. J. C. McLean, unpublished data). Normally, without predisposing conditions, *M. morganii* cannot readily cause infections in the urinary tract (Senior 1983).

As mentioned above, the experimental evidence to date suggests that the normal intact urinary tract of animals is well equipped to clear certain uropathogenic organisms, unless foreign objects are present. While this helps explain the onset of foreign object associated UTI, other host and bacterial mechanisms play a role in the onset

Fig. 16.1. Scanning electron micrograph of a sponge implanted in the bladder of a rat. Note the thick biofilm (B) and the numerous bacteria and small crystals (C). Bar = 5 μm.

and development of UTI not associated with prostheses (McTaggart *et al.* 1990). The expression of type 1 and P fimbriae, for example, provide bacteria with a capacity to colonize the bladder (Reid & Sobel 1987; Svanborg Edén *et al.* 1989, 1990, 1991). Also when host defences arc compromised or deficient, infection can evolve. The importance of host defences has been the subject of more recent studies by Svanborg Edén and her colleagues. Animal models such as C3H/HeJ and C57BL/10ScCr mice (Hagberg *et al.* 1984, 1985) can be employed with genetically defined LPS immunological response defects to illustrate that differences in infection rates partly reflect host properties. Alternatively the inflammatory response can be pharmacologically suppressed in animals having otherwise normal immune responses (Linder *et al.* 1988, 1990). Both genetic and pharmacological suppresion of the immune system similarly enhance the onset and development of experimental UTI.

Struvite urolithiasis

Struvite (infection) stones account for only 10% of all urinary calculi, yet represent a significant health problem and more often a greater danger to the integrity of the urinary tract than do conventional metabolic stones (Griffith 1978; Griffith & Klein 1983; Lerner *et al.* 1989). In spite of the most modern treatments, including extra-corporeal shockwave lithotripsy (Pode *et al.* 1988), or surgical removal coupled with the use of antibiotics, these calculi recur and persist in approximately 40–50% of patients (Griffith & Klein 1983). Left untreated, this condition can result in the loss of the kidney and mortality within 5–10 years (Wojewski & Zajaczkowski 1973). The need to study the pathogenesis, treatment and prevention of this debilitating infection with an animal model therefore becomes apparent.

When a urease producing organism, such as *Proteus mirabilis*, is instilled into an animal with a foreign implant, struvite urolithiasis results (Nickel *et al.* 1987; McLean *et al.* 1988; Olson *et al.* 1989). The urease activity of these organisms cause the liberation of ammonia from urea, elevating urine pH which in turn causes the deposition of Mg^{2+} and Ca^{2+} from urine as struvite ($NH_4MgPO_4 \cdot 6H_2O$) and carbonate apatite

($Ca_{10}(PO_4)_6 \cdot CO_3$). The biofilm nature of *P. mirabilis* colonization (McLean *et al.* 1988, 1991; McLean & Nickel 1991) and its anionic capsule polymer (Beynon *et al.* 1992) provide microenvironments that enhance struvite mineral growth and calculus formation (Clapham *et al.* 1990; Dumanski *et al.* 1994; McLean *et al.* 1991). This mode of growth also ensures that these organisms and their associated struvite crystals are protected against antibiotic therapy (Rocha & Santos 1969; McLean *et al.* 1988), acid dissolution (McLean *et al.* 1991), and physical disruption by extra-corporeal shockwave lithotripsy (Reid *et al.* 1990b; Stoller & Workman 1990). As in experimental *E. coli* chronic cystitis, spreading of *Proteus* into the kidneys induces pyelonephritis. However, *P. mirabilis* associated prostatitis is relatively rare.

During animal studies of struvite urolithiasis, the infecting organisms should be allowed to

Fig. 16.2. Scanning electron micrograph of a struvite crystal with adherent bacteria encased in a thick biofilm. Bar = 5 μm.

grow as biofilms. This is accomplished by inserting a foreign object into the bladder, prior to experimental infection. Once administered, the pathogens colonize the foreign object, grow as biofilms, and thus mimic their growth *in situ* (Nickel *et al.* 1985a, 1986; McLean *et al.* 1988, 1989). Also the biofilm structure may be important in binding the struvite mineral components, providing the foundation for the organic matrix component, and determining the final architecture of the calculus. Figures 16.2 and 16.3 illustrate struvite crystals with a bacterial biofilm that was experimentally induced in a rat. These studies will be crucial in identifying strategies for the prevention and ultimate eradication of this disorder.

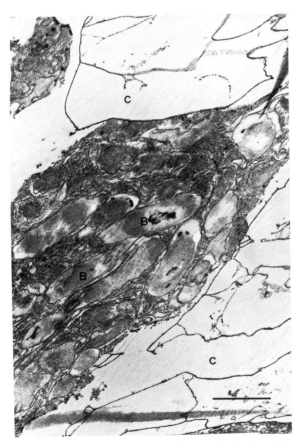

Fig. 16.3. Transmission electron micrograph of a bacterium (B) encased within the crystal matrix (C). Bar = 1 μm.

Chronic prostatitis

Prostatitis occurs mainly in adult males when bacteria enter the prostatic ducts from the urethra, and when the natural immunological defence mechanisms and prostatic antibacterial secretions are defective (Blacklock 1974; Blacklock & Beavis 1974; Fair *et al.* 1976). It is a particularly frustrating clinical problem in that traditional expression of prostatic fluid (Meares & Stamey 1968) may fail to yield positive bacterial cultures, particularly after a course of antibiotic therapy. In addition, it is quite resistant to conventional antibiotic regimes (Eykyn *et al.* 1974) and often results in a chronic infection. Using a rat model (Nickel *et al.* 1991a), we recently showed that the pathogens responsible for this infection sequester themselves in the interstices of the prostate as encapsulated microcolonies (Nickel *et al.* 1990). These biofilm encased microcolonies are quite resistant to antibiotics (See Korber *et al.*, Chapter 1; Gilbert & Brown, Chapter 6) and are not easily dislodged for microbiological culturing during expression of prostatic fluid.

Concentrations of antibiotics, only slightly higher than their *in vitro* minimum inhibitory concentration levels, reach the bacteria in the prostatic ducts. No significant differences in antibiotic concentrations were found between inflamed and non-inflamed glands (J. C. Nickel & J. Downey, unpublished data). It is likely that the small sporadic microcolonies adherent to the ductal walls are relatively resistant to the low concentration of antibiotics that reach them, because of their biofilm mode of growth.

Our animal model for prostatitis (Nickel *et al.* 1990, 1991a) employs male Sprague Dawley rats (*c*.150–175 g) and a uropatheogenic bacterial strain (Fig. 16.4). Induction of prostatitis involves the insertion of a urethral catheter (PE10) into the prostatic urethra on an anaesthetized animal. After insertion of the catheter into the mid-urethra, bacteria are injected so that some organisms are forced into the prostatic gland. This protocol has a very high success rate in that it ensures that sufficient organisms enter the prostate so as to induce prostatitis within 3 days of infection. Development of prostatitis varies among the animals. Those able to clear the infection generally do so within the acute phase (14 days). In other animals, bacterial micro-

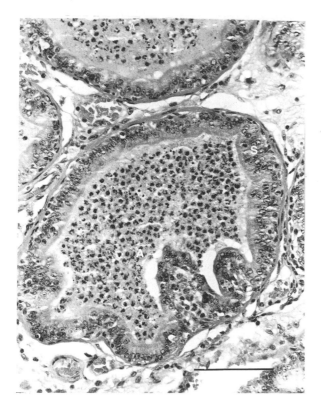

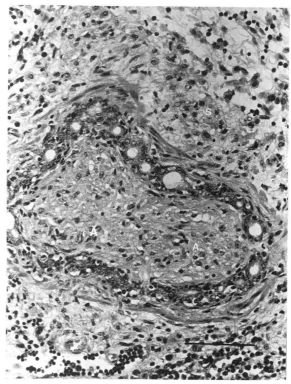

Fig. 16.4.   Light micrograph of a rat with acute prostatitis. The acinus (A) is dilated with numerous leucocytes and there is extensive stromal (S) oedema. Bar =100 µm.

Fig. 16.5.   Light micrograph of a rat with chronic prostatitis. The acinus (A) has collapsed and there is extensive acinar and stromal fibrosis and mononuclear infilitation. Bar = 100 µm.

colony and biofilm formation lead to chronic prostatitis which may last up to 56 days. Figure 16.5 illustrates the histological changes associated with chronic prostatitis in a rat model and Fig. 16.6 demonstrates a bacterial microcolony in a rat with chronic prostatis. This model is very useful for the study of the pathogenesis of prostatitis and in the evaluation of therapeutic regimes to treat this disease.

Catheter associated infections

Catheter associated infections represent the most common nosocomial infection in North America (Breitenbucher 1984; Nickel *et al.* 1985b, 1989, 1992a; Cox *et al.* 1989; Hawthorn & Reid 1990; Johnson *et al.* 1990; Harding *et al.* 1991). Catheterization of adult patients is commonly

performed during most surgical procedures. The risk of catheter associated infection increases by approximately 10% for each day a catheter is in place (Johnson *et al.* 1990; Nickel *et al.* 1989). These infections are especially threatening to patients with voiding dysfunction who undergo long-term catheterization (Kunin *et al.* 1987). The infecting organisms initially colonize external surfaces of the catheter such as the drainage spout, form a biofilm, and then ascend into the bladder. This ascent of bacteria along catheter surfaces into the bladder as a biofilm occurs within 1–3 days and has been referred to as 'creeping crud' (Nickel *et al.* 1992b), a phrase which vividly describes its motility and appearance. Once the biofilm has reached the catheter tip in the bladder, the organisms act as a chronic source of infection to the bladder and kidneys (Norl'n *et al.* 1988; Nickel *et al.* 1991b; 1992b).

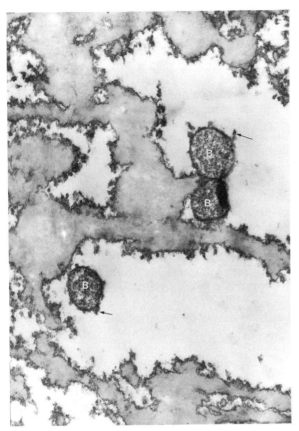

Fig. 16.6. TEM of a microcolony of bacteria in the prostate of a rat with chronic prostatitis. The glycocalyx (arrows) has condensed on the surface of the bacteria (B) in the fixation processing.

By virtue of their biofilm mode of growth, they are very resistant to antibiotics and host defences (Nickel *et al.* 1985c, 1992b; Ladd *et al.* 1987) by a variety of mechanisms as discussed by Gilbert & Brown (Chapter 6).

Mineralization is a frequent complication of catheter associated infections. Generally these minerals consist of crystalline struvite along with poorly crystalline apatite (Hukins *et al.* 1983, 1989; Cox *et al.* 1987b, 1989; Cox & Hukins 1989). While the underlying catheter material certainly influences the mineralization processes (Cox *et al.* 1987a; Hedelin *et al.* 1991), biofilms of urease producing pathogens are now believed to be the chief culprit (Cox *et al.* 1989). These encrustations reduce the effective diameter of, or even block, the catheter lumen and cause significant irritation to urinary tissue.

Frequent replacement of urinary catheters, when practical, would certainly be a practical approach to reduce the incidence of infection. However, for a number of medical reasons, this is not always a practical option (Norl'n *et al.* 1988). The high cost and incidence of catheter associated nosocomial infections has resulted in significant research efforts to develop materials or compounds which block bacterial adhesion and/or kill adherent organisms. Strategies used in this regard include altering the hydrophobicity of polymers to interfere with the adhesion processes (Hüttinger *et al.* 1987; Liedberg *et al.* 1990), and the incorporation of antimicrobial compounds such as disinfectants, antibiotics (Olson *et al.* 1988b; Jansen *et al.* 1992), and metals (Schaeffer *et al.* 1988; Johnson *et al.* 1990; Liedberg *et al.* 1990; Farrah & Erdos 1991; McLean *et al.* 1993). The catheter size has been shown to influence the rate of catheter colonization, but this factor is eliminated when the catheter is coated with a hydromer material (A. E. Khoury *et al.*, unpublished data). Hydromer coatings presumably act as a physical barrier to the bacteria from migrating from the meatus toward the bladder.

While the development of 'non-sticky' or poisonous surface materials for use in medical devices is appealing, this approach has one major drawback. Any surface, when placed into a liquid medium, will interact with chemical and biological components in that liquid. The net result of this interaction is that the physical and chemical characteristics of the original surface rapidly become obscured. This has been well documented in the dental field in that tooth enamel becomes rapidly coated with a pellicle of salivary proteins (see Marsh, Chapter 18). During plaque formation, initial bacterial adhesion by organisms such as *Streptococcus sanguis* and *Actinomyces viscosus* is directed to this pellicle layer (Rosan *et al.* 1985). Other plaque bacteria such as *S. mutans* adhere preferentially to the primary colonizers, *S. sanguis* and *A. viscosus* (Kolenbrander 1988). The microbial ecology of the urinary tract is similar to the oral cavity in some aspects. Alteration of catheter surfaces by urine components has also been investigated (Hawthorn & Reid 1990; G. Reid, personal communication). In addition, bacterial growth and adhesion to urinary surfaces may be negatively influenced by other organisms present, such as *Lactobacillus* sp. (Reid *et al.* 1988).

Catheter infections can be easily modelled both *in vitro* and in animal models. *In vitro* models are very useful in studying the progression of biofilm organisms along catheter surfaces. They involve dipping the distal end of the catheter into a suspension of the test organism and flushing the device continuously with sterile urine (Nickel *et al.* 1992a). The organisms colonize the tip of the catheter, grow as a biofilm, and creep along the lumen surface against the urine flow. Progression of the organisms along the catheter can be monitored by stopping the experiment at set times, cutting the catheter into sections and culturing for organisms present in each section. These *in vitro* experiments are also very useful for testing different catheter materials as to their ability to inhibit bacterial adherence and progression along the device.

Catheter infections are modelled in rabbits using paediatric catheters (Nickel *et al.* 1985b, 1991b; Olson *et al.* 1988a; Morck *et al.* 1993). The model is illustrated in Fig. 16.7. In this

manner, one can test biocompatability, toxicity, new catheter designs, catheter coatings and biofilm growth (Nickel *et al.* 1985d, 1991b; Olson *et al.* 1988a; Morck *et al.* 1993). Animals may be challenged with a uropathogen by applying bacteria to the catheter lumen or the meatus. The progression of infections is monitored by culturing urine from different sites and antibiotic levels may be determined from the blood and urine. At the end of the experimental period the catheter is removed, cut into sections, and cultured for the test organisms. If accurate localization of the progress of bacterial biofilms is required, the animal should be sacrificed, and the catheter carefully dissected from the urinary tissues to minimize disruption of organisms adherent to the catheter. Tissues are also collected for histopathology and culture. Test bacterial strains may have some readily distinguishable phenotypic marker (such as antibiotic resistance) so as

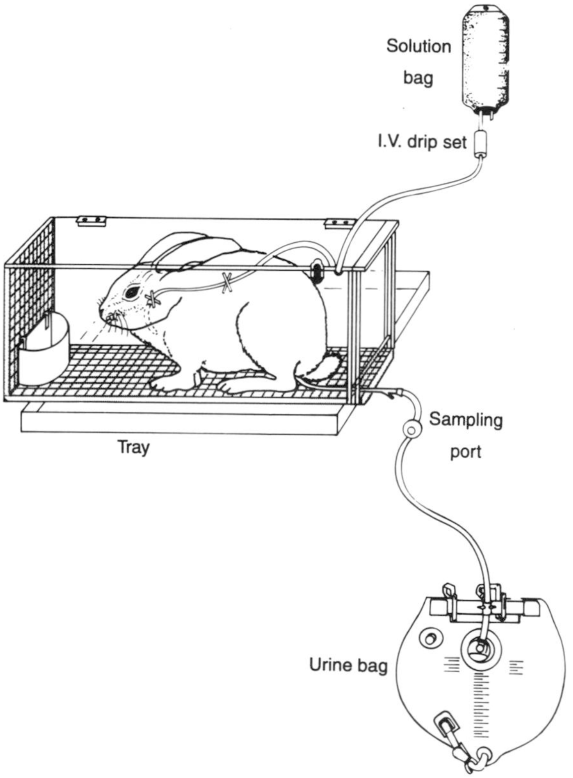

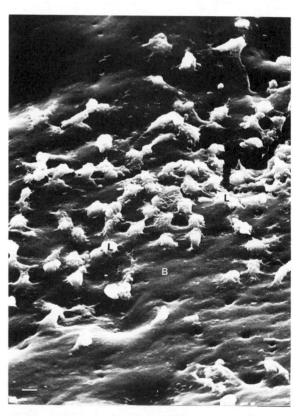

Fig. 16.8. Scanning electron micrograph of the luminal surface of a catheter from a rabbit showing large numbers of phagocytic leucocytes (L) adherent to the biofilm (B) on the catheter surface. Bar = 5 μm.

Solution bag

I.V. drip set

Tray

Sampling port

Urine bag

Fig. 16.7. A drawing of the rabbit model for evaluation of urinary catheters, drainage systems and antibiotics.

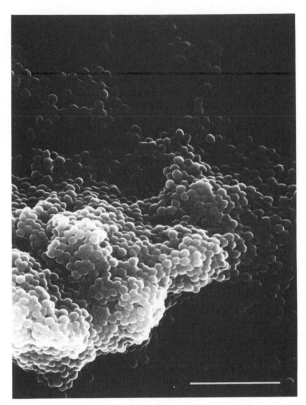

Fig. 16.9.   Scanning electron micrograph of the lumen surface of a catheter from a rabbit with bacteruria showing a thick biofilm of coccoid bacteria. Bar = 5 μm.

to distinguish them from normal flora. Figures 16.8 and 16.9 illustrate bacteria growing on the luminal surface of an experimental catheter of a test rabbit. This model continues to prove valuable in designing catheter systems that resist colonization and therapeutics for elimination of catheter associated urinary tract infections.

The pivotal role of biofilms in the pathogenesis of UTI is becoming increasingly recognized as a significant aspect of urinary pathogenicity. By recognizing this natural, biofilm mode of pathogen growth in the urinary enviroment, one can design models of infection that more accurately reflect conditions *in vivo*. Our basic rationale is that these urinary diseases are very complex phenomena. Yet through well designed experiments coupled with comprehensive data collection, one can identify the mechanisms and risk factors which contribute to the onset and persistence of urinary infections.

Acknowledgements

The model experiments have been made possible by generous grants from the Kidney Foundation of Canada (R. J. C. M. and J. C. N.), the Alberta Heritage Foundation for Medical Research (R. J. C. M., M. E. O. & J. C. N.), the Natural Sciences and Engineering Research Council of Canada (R. J. C. M. & M. E. O.), the Physicians' Services Foundation Inc. and the Department of Urology, Queens University (J. C. N.). R. J. C. M. was funded by a Career Scientist Fellowhip from the Ontario Ministry of Health. We express our gratitude to our collaborators, J. W. Costerton, K. Lam and J. Cornish, for their keen scientific insights and help through the years. We fondly dedicate this chapter to Ivanka Ruseska.

References

Beynon, L. M., Dumanski, A. J., McLean, R. J. C., MacLean, L. L., Richards, J. C. & Perry, M. B. (1992). Capsule structure of *Proteus mirabilis* (ATCC 49565). *Journal of Bacteriology*, **174**, 2172–7.

Blacklock, N. J. (1974). Anatomical factors in prostatitis. *British Journal of Urology*, **46**, 47–54.

Blacklock, N. J. & Beavis J. P. (1974). The response of prostatic fluid pH in inflammation. *British Journal of Urology*, **46**, 537–42.

Breitenbucher, R. B. (1984). Bacterial changes in the urine samples of patients with long term indwelling catheters. *Archives of Internal Medicine*, **144**, 1585–8.

Chan, R. C. Y., Reid, G., Irvin, R. T., Bruce, A. W. & Costerton, J. W. (1985). Competitive exclusion of uropathogens from human uroepithelial cells by *Lactobacillus* whole cells and cell wall fragments. *Infection and Immunity*, **47**, 84–9.

Clapham, L., McLean, R. J. C., Nickel, J. C., Downey, J. & Costerton, J. W. (1990). The influence of bacteria on struvite crystal habit and its importance in urinary stone formation. *Journal of Crystal Growth*, **104**, 475–84.

Cornish J., Lecamwasam, J. P., Harrison, G., Vanderwee, M. A. & Miller, T. E. (1988). Host defence mechanisms in the bladder. **II.** Disruption of the layer of mucus. *British Journal of Experimental Pathology*, **69**, 759–70.

Cornish, J., Nickel, J. C., Vanderwee, M. & Costerton, J. W. (1990). Ultrastructural visualization of human bladder mucous. *Urological Research*, **18**, 263–6.

Cornish J., Vanderwee, M. A. & Miller, T. ((1987). Mucus stabilization in the urinary bladder. *British*

Journal of Experimental Pathology, **68**, 369–75.

Cox, A. J., Harries, J. E., Hukins, D. W. L., Kennedy, A. P. & Sutton, T. M. (1987b). Calcium phosphate in catheter encrustation. *British Journal of Urology*, **59**, 159–63.

Cox, A. J., & Hukins, D. W. L. (1989). Morphology of mineral deposits on encrusted urinary catheters investigated by scanning electron microscopy. *Journal of Urology*, **142**, 1347–50.

Cox, A. J., Hukins, D. W. L., Davies, K. E., Irlam, J. C. & Sutton, T. M. (1987a). An automated technique for *in vitro* assessment of the susceptibility or urinary catheter materials to encrustation. *Engineering in Medicine*, **16**, 37–41.

Cox, A. J., & Hukins, D. W. & Sutton, T. M. (1989). Infection of catheterised patients: bacterial colonisation of encrusted Foley catheters shown by scanning electron microscopy. *Urological Research*, **17**, 349–52.

Dulawa, J., Jann, K., Thomsen, M., Rambausek, M. & Ritz, E. (1988). Tamm Horsfall glycoprotein interferes with bacterial adherence to human kidney cells. *European Journal of Clinical Investigation*, **18**, 87–91.

Dumanski, A. J., Hedelin, H., Edin-Liljegren, A., Beauchemin, D. & McLean, R. J. C. (1994). Unique ability of the *Proteus mirabilis* capsule to enhance mineral growth in infectious urinary calculi. *Infection and Immunity* **62**, 2998–3003.

Elliott, T. S., Slack, R. C. & Bishop, M. C. (1984). Scanning electron microscopy of human bladder mucosa in acute and chronic urinary tract infection. *British Journal of Urology*, **56**, 38–43.

Eykyn, S., Bultitude, M. I., Mayo, M. E. & Lloyd-Davies, R. W. (1974). Prostatic calculi as a source of recurrent bacteriuria in the male. *British Journal of Urology*, **46**, 527–32.

Fair, W. R., Couch, J. & Wehner, N. (1976). Prostatic antibacterial factor identity and significance. *Urology*, **7**, 169–77.

Farrah, S. R. & Erdos, G. W. (1991). The production of antibacterial tubing, sutures, and bandages by *in situ* precepitation of metallic salts. *Canadian Journal of Microbiology*, **37**, 445–9.

Fukushi, Y. & Orikasa, S. (1981). The role of intravesical polymorphonuclear leukocytes in experimental cystitis. *Investigative Urology*, **18**, 471–4.

Gillon, G., Small, M., Medalia, O. & Aronson, M. (1984). Sequential study of bacterial clearance in expermental cystitis. *Journal of Medical Microbiology*, **18**, 319–26.

Griffith, D. P. (1978). Struvite stones. *Kidney International*, **13**, 372–82.

Griffith, D. P. & Klein, A. S. (1983). Infection induced urinary stones. In *Stones: clinical management of urolithiasis*, ed. R. A. Roth & B. Finlayson,

pp. 210–27. Baltimore, Md: Williams & Wilkins.

Hagberg, L., Briles, D. E. & Svanberg Edén, C. (1985). Evidence for separate genetic defects in C3H/HeJ and C3HeB/FeJ mice that affect susceptibility to Gram negative infections. *Journal of Immunology*, **134**, 4118–22.

Hagberg, L., Hull, R., Hull, S., McGee, J. R., Michalek, S. M. & Svanberg Edén, C. (1984). Difference in susceptibility to Gram negative urinary tract infection between C3H/HeJ and C3H/HeN mice. *Infection and Immunity*, **46**, 839–44.

Hagberg, L., Lam, J., Svanberg Edén, C. & Costerton, J. W. (1986). Interaction of a pyelonephritogenic *Escherichia coli* strain with the tissue components of the mouse urinary tract. *Jounal of Urology*, **136**, 165–72.

Harding, G. K. M., Nicolle, L. E., Ronald, A. R. *et al.* (1991). How long should catheter acquired urinary tract infection in women be treated? A randomized controlled study. *Annals of Internal Medicine*, **114**, 713–19.

Hawthorn, L. A., Bruce, A. W. & Reid, G. (1991). Ability of uropathogens to bind to Tamm Horsfall protein coated renal tubular cells. *Urological Research*, **19**, 301–4.

Hawthorn, L. A. & Reid, G. (1990). The effect of protein and urine on uropathogen adhesion to polymer substrata. *Journal of Biomedical Materials Research*, **24**, 1325–32.

Hedelin, H., Grenabo, L. & Pettersson, S. (1991). Urease induced precipitation of phosphate salts *in vitro* on indwelling catheters made of different materials. *Urological Research*, **19**, 297–300.

Holm-Bentzen, M. & Ammitzboll, T. (1989). Structure and function of glycosaminoglycans in the bladder. *Annales d'Urologie*, **23**, 167–8.

Hopkins, W. J., Uehling, D. T. & Balish, E. (1987). Local and systematic antibody responses accompany spontaneous resolution of experimental cystitis in cynomolgus monkeys. *Infection and Immunity*, **55**, 1951–6.

Hukins, D. W. L., Hickey, D. S. & Kennedy, A. P. (1983). Catheter encrustation by struvite. *British Journal of Urology*, **55**, 304–5.

Hukins, D. W. L., Nelson, L. S., Harries, J. E., Cox, A. J. & Holt, C. (1989). Calcium environment in encrusting deposits from urinary catheters investigated by interpretation of EXAFS spectra. *Journal of Inorganic Biochemistry*, **36**, 141–8.

Hurst, R. E., Rhodes, S. W., Adamson, P. B., Parsons, C. L. & Roy, J. B. (1987). Functional and structural characteristics of the glycosaminoglycans of the bladder luminal surface. *Journal of Urology*, **138**, 433–7.

Hüttinger, K. J., Rudi, H. & Bomar, M. T. (1987). Influence of surface chemistry of the substrate on the absorption of *Escherichia coli*. *Zentralblatt für*

Bakteriologie, Mikrobiologie, und Hygiene, Series B, **184,** 538–47.

Jansen, B., Kristinsson, K. G., Jansen, S., Peters, G. & Pulverer, G. (1992). *In vitro* efficacy of a central venous catheter complexed with iodine to prevent bacterial colonization. *Journal of Antimicrobial Chemotherapy,* **30,** 135–9.

Johnson, J. R., Roberts, P. L., Olsen, R. J., Moyer, K. A. & Stamm, W. E. (1990). Prevention of catheter associated urinary tract infection with a silver oxide coated urinary catheter: clinical and microbiologic correlates. *Journal of Infectious Diseases,* **162,** 1145–50.

Kaye, D. (1975). Host defence mechanisms in the urinary tract. *Urological Clinics of North America,* **2,** 407–22.

Kolenbrander, P. E. (1988). Intergeneric coaggregation among human oral bacteria and ecology of dental plaque. *Annual Reviews of Microbiology,* **42,** 627–56.

Kunin, C. M., Chin, Q. F. & Chambers, S. (1987). Indwelling urinary catheters in the elderly. *American Journal of Medicine,* **82,** 405–11.

Ladd, T. I., Schmiel, D., Nickel, J. C. & Costerton, J. W. (1987). The use of a radiorespirometric assay for testing the antibiotic sensitivty of catheter assocaiated bacteria. *Journal of Urology,* **138,** 1451–6.

Lerner, S. P., Gleeson, M. J. & Griffith, D. P. (1989). Infection stones. *Journal of Urology,* **141,** 753–8.

Liedberg, H., Ekman, P. & Lundeberg, T. (1990). *Pseudomonas aeruginosa:* adherence to and growth on different urinary catheter coatings. *International Urology and Nephrology,* **22,** 487–92.

Linder, H., Engberg, I., Mattsby-Baltzer, I., Jann, K. & Svanborg Edén, C. (1988). Natural resistance by urinary tract infection determined by endotoxin induced inflammation. *FEMS Microbiology Letters,* **49,** 219–22.

Linder, H., Engberg, I., van Kooten, C., de Man, P. & Svanborg Edén, C. (1990). Effects of anti inflammatory agents on mucosal inflammation induced by infection with Gram negative bacteria. *Infection and Immunity,* **58,** 2056–60.

McLean, R. J. C. (1986). *The role of ureolytic bacteria in infectious urinary stone production.* Ph. D. thesis, University of Calgary, Canada.

McLean, R. J. C., Downey, J. A., Lablans, A. L., Clark, J. M., Dumanski, A. J. & Nickel, J. C. (1992). Modelling biofilm associated urinary tract infections. *International Biodeterioration,* **30,** 201–16.

McLean, R. J. C., Hussain, A. A., Sayer, M., Vincent, P. J., Hughes, D. J. & Smith, T. J. N. (1993). Antibacterial activity of multilayer silver copper surface films on catheter material. *Canadian Journal of Microbiology,* **39,** 895–9.

McLean, R. J. C., Lawrence, J. R., Korber, D. R. & Caldwell, D. E. (1991). *Proteus mirabilis* biofilm

protection against struvite crystal dissolution and its implications in struvite urolithiasis. *Journal of Urology,* **146,** 1138–42.

McLean, R. J. C. & Nickel, J. C. (1991). Bacterial colonization behaviour: a new virulence strategy in urinary infections? *Medical Hypotheses,* **36,** 269–72.

McLean, R. J. C., Nickel, J. C., Beveridge, T. J. & Costerton, J. W. (1989). Observations of the ultrstructure of infected kidney stones. *Journal of Medical Microbiology,* **29,** 1–7.

McLean, R. J. C., Nickel, J. C., Cheng, K. J. & Costerton, J. W. (1988). The ecology and pathogenicity of urease producing bacteria in the urinary tract. *Critical Reviews in Microbiology,* **16,** 37–79.

McTaggart, L. A., Rigby, R. C. & Elliott, T. S. J. (1990). The pathogenesis of urinary tract infections associated with *Escherichia coli, Staphylococcus saprophyticus* and *S. epidermidis. Journal of Medical Microbiology,* **32,** 135–41.

Meares, E. M. & Stamey, T. A. (1968). Bacteriological localization patterns in bacterial prostatitis and urethritis. *Investigative Urology,* **5,** 492–518.

Miller, T. E., Lecamwasam, J. P., Ormrod, D. J., Findon, G. & Cornish, J. (1987). An animal model for chronic infection of the unobstructed urinary tract. *British Journal of Experimental Pathology,* **68,** 575–83.

Morck, D. W., Olson, M. E., McKay, S. G. *et al.* (1993). Therapeutic efficacy of fleroxacin for eliminating catheter associated urinary tract infection in a rabbit model. *American Journal of Medicine,* **94** (3A), 23–37.

Nickel, J. C. (1990). The battle of the bladder: the pathogenesis and treatment of uncomplicated cystitis. *International Urogynecology Journal,* **1,** 218–22.

Nickel, J. C., Downey, J. A. & Costerton, J. W. (1989). Ultrastructural study of microbiologic colonization of urinary catheters. *Urology,* **34,** 284–91.

Nickel, J. C., Downey, J. A. & Costerton, J. W. (1992a). Movement of *Pseudomonas aeruginosa* along catheter surfaces: a mechanism in pathogenesis of catheter associated infections. *Urology,* **39,** 93–8.

Nickel, J. C., Downey, J. & Costerton, J. W. (1992b). Movement of *Pseudomonas aeruginosa* along catheter surfaces. *Urology,* **39,** 93–8.

Nickel, J. C., Emtage, J. & Costerton, J. W. (1985a). Ultrastructural microbial ecology of infection induced urinary stones. *Journal of Urology,* **133,** 622–7.

Nickel, J. C., Grant, S. K. & Costerton, J. W. (1985d). Catheter associated bacteriuria. An experimental study. *Investigative Urology,* **26,** 369–75.

Nickel, J. C., Grant, S. K., Lam, K., Olson, M. E. &

Costerton, J. W. (1991b). Bacteriologically stressed animal model of new closed catheter drainage system with microbicidal outlet tube. *Urology*, **38**, 280–9.

Nickel, J. C., Gristina, A. G. & Costerton, J. W. (1985b). Electron microscopic study of an infected Foley catheter. *Canadian Journal of Surgery*, **28**, 50–4.

Nickel, J. C., Olson, M. E., Barabas, A., Benediktsson, H., Dasgupta, M. K. & Costerton, J. W. (1990). Pathogenesis of chronic bacterial prostatitis in an animal model. *British Journal of Urology*, **66**, 47–54.

Nickel, J. C., Olson, M. E. & Costerton, J. W. (1991a). Rat model of experimental prostatitis, *Infection*, **19**, S126–30.

Nickel, J. C., Olson, M. E., McLean, R. J. C., Grant, S. K. & Costerton, J. W. (1987). An ecological study of infected urinary stone genesis in an animal model. *British Journal of Urology*, **59**, 21–30.

Nickel, J. C., Reid, G., Bruce, A. W. & Costerton, J. W. (1986). Ultrastructural microbiology of infected urinary stone. *Urology*, **28**, 512–15.

Nickel, J. C., Ruseka, I., Wright, J. B. & Costerton, J. W. (1985c). Tobramycin resistance of *Pseudomonas aeruginosa* cells growing as a biofilm on urinary catheter material. *Antimicrobial Agents and Chemotherapy*, **27**, 619–24.

Norl'n, L. J., Ekelund, P., Hedelin, H. & Johansson, S. L. (1988). Effects of indwelling catheters on the urethral mucosa (polypoid urethritis). *Scandinavian Journal of Urology and Nephrology*, **22**, 81–6.

Olson, M. E., Nickel, J. C. & Costerton, J. W. (1989). Infection induced struvite urolithiasis in rats. *American Journal of Pathology*, **135**, 581–3.

Olson, M. E., Nickel, J. C., Khoury, A. E., Morck, D. W., Cleeland, R. & Costerton J. W. (1988a). Amdinocillin treatment of catheter associated bacteriurea in rabbits. *The Journal of Infectious Diseases*, **159**, 1065–72.

Olson, M. E., Ruseka, I. & Costerton, J. W. (1988b). Colonization of *n*-butyl-2-cyanoacrylate tissue adhesive by *Staphylococcus epidermidis. Journal of Biomedical Materials Research*, **22**, 485–95.

Orikasa, S. & Hinman, F. Jr (1977). Reaction of the vesical wall to bacterial penetration. Resistance to attachment, desquamation, and leukocytic activity. *Investigative Urology*, **15**, 185–93.

Parkkinen, J., Virkola, R. & Korhonen, T. K. (1988). Identification of factors in human urine that inhibit the binding of *Escherichia coli* resins. *Infection and Immunity*, **56**, 2623–30.

Parsons, C. L. (1982). Prevention of urinary tract infection by the exogenous glycosaminoglycan sodium pentosanpolysulphate. *Journal of Urology*, **127**, 167–9.

Parsons, C. L., Boychuk, D., Jones, S., Hurst, R. & Callahan, H. (1990). Bladder surface glycosaminoglycans: an epithelial permeability barrier. *Journal of Urology*, **143**, 139–42.

Parsons, C. L., Stauffer, C. W. & Schmidt, J. D. (1988). Reversable inactivation of bladder surface glycosaminoglycan antibacterial activity by protamine sulphate. *Infection and Immunity*, **56**, 1341–3.

Pode, D., Lenkovsky, Z., Shapiro, A. & Pfau, A. (1988). Can extracorporeal shock wave lithotripsy eradicate persistent urinary infection associated with infected stones? *Journal of Urology*, **140**, 257–9.

Reid, G., Bruce, A. W., McGroaty, J. A., Cheng, K. J. & Costerton, J. W. (1990a). Is there a role for lactobacilli in prevention of urogenital and intestinal infections? *Clinical Microbiology Reviews*, **3**, 335–44.

Reid, G., Cook, R. L. & Bruce, A. W. (1987). Examination of strains of lactobacilli for properties that may influence bacterial interfernce in the urinary tract. *Journal of Urology*, **138**, 330–5.

Reid, G., McGroaty, J. A., Angotti, R. & Cook, R. L. (1988). *Lactobacillus* inhibitor production against *Escherichia coli* and coaggregation ability with uropathogens. *Canadian Journal of Microbiology*, **34**, 344–51.

Reid, G., Jewett, M. A. S., Nickel, J. C., McLean, R. J. C. & Bruce, A. W. (1990b). Effect of extra corporeal shock wave lithotripsy on bacterial viability: relationship to the teatment of struvite stones. *Urological Research*, **18**, 425–7.

Reid, G. & Sobel, J. D. (1987). Bacterial adherence in the pathogenesis of urinary tract infection: a review. *Reviews of Infectious Diseases*, **9**, 470–87.

Reinhart, H. H., Obedeanu, N. & Sobel, J. D. (1990). Quantitation of Tamm Horsfall protein binding to uropathogenic *Escherichia coli* and lectins. *Journal of Infectious Diseases*, **162**, 1335–40.

Rene, P., Dinolfo, M. & Silverblatt, F. J. (1982). Serum and urogenital antibody responses to *Escherichia coli* pili in cystitis. *Infection and Immunity*, **38**, 542–7.

Rocha, H. & Santos, L. C. S. (1969). Relapse of urinary tract infection in the presence of urinary tract calculi: the role of bacteria within the calculi. *Journal of Medical Microbiology*, **2**, 372–6.

Rosan, B., Eifert, R. & Golub, E. (1985). Bacterial surfaces, salivary pellicles, and plaque formation. In *Molecular basis of oral microbial adhesion*, ed. S. E. Mergenhagen & B. Rosan, pp. 69–76. Washington, DC: American Society for Microbiology.

Sandberg, T., Kaijser, B., Lidin-Janson, G. *et al.* (1988). Virulence of *Escherichia coli* in relation to host factors in women with symptomatic urinary tract infection. *Journal of Clinical Microbiology*, **26**, 1471–6.

Satoh, M., Munakata, K., Kitoh, K., Takeuchi, H. & Yoshida, O. (1984). A newly designed model for infection induced bladder stone formation in the rat. *Journal of Urology*, **132**, 1247–9.

Schaeffer, A. J., Story, K. O. & Johnson, S. M. (1988). Effect of silver oxide/trichloroisocyanuric acid antimicrobial urinary drainage system on cathether associated bacteriuria. *Journal of Urology*, **139**, 69–73.

Senior, B. W. (1983). *Proteus morgani* is less frequently associated with urinary tract infections than *Proteus mirabilis* – an explanation. *Journal of Medical Microbiology*, **16**, 317–22.

Shand, G. H., Anwar, H., Kadurugamuwa, J., Brown, M. R. W., Silverman, S. H. & Melling, J. (1985). *In vivo* evidence that bacteria in urinary tract infection grow under iron restricted conditions. *Infection and Immunity*, **48**, 35–9.

Sobel, J. D. (1991). Bacterial etiological agents in the pathogenesis of urinary tract infection. *Medical Clinics of North America*, **75**, 253–73.

Stamm, W. E., Hooton, T. M., Johnson, J. R. *et al.* (1989). Urinary tract infections: from pathogenesis to treatment. *Journal of Infectious Diseases*, **159**, 400–6.

Stoller, M. L. & Workman, S. J. (1990). The effect of extracorporeal shock wave lithotripsy on the microbiolgical flora of urinary calculi. *Journal of Urology*, **144**, 619–21.

Svanborg Edén, C., de Man, P. & Sandberg, T. (1991). Renal involvement in urinary tact infection. *Kidney International*, **39**, 541–9.

Svanborg Edén, C., Engberg, I., Hedges, S. *et al.* (1989). Consequences of bacterial attachment in the urinary tract. *Biochemical Society Transactions*, **17**, 464–6.

Svanborg Edén, C., Andersson, B., Aniansson, G., Lindstedt, R., de Man, P., Nielsen, A., Leffler, H & Wold, A. (1990). Inhibition of bacterial attachment: examples from the urinary and respiratory tracts. *Current Topics in Microbiology and Immunology*, **151**, 167–84.

Wojewski, A. & Zajaczkowski, T. (1973). The treatment of bilateral staghorn calculi of the kidneys. *International Urology and Nephrology*, **5**, 249–60.

17

The Role of the Urogenital Flora in Probiotics

Gregor Reid and Andrew W. Bruce

Introduction

In order to examine whether or not the flora of the healthy adult female urogenital tract has any role in protecting a host from infection, and thereby performing a probiotic function, we must first outline the formation, composition and fluctuations of the flora. This is not a simple task as factors such as age and hormonal status influence the type and quantity of organisms present. In simple terms, the establishment of the flora can be seen to follow the path outlined in Fig. 17.1. This figure is based upon epidemiological studies of the urogenital flora (Reid *et al.* 1990b, c; Sadhu *et al.* 1989) and a theory for maintenance and causation of infection.

The primary colonizers comprise organisms such as lactobacilli, Gram positive cocci and diphtheroids which have dominated the flora from puberty. In general, the secondary colonizers can comprise a number of species, including potential pathogenic coliforms, *Escherichia coli*, coagulase negative staphylococci, *Klebsiella*, *Proteus* sp. and other Gram positive and Gram negative bacteria. Depending upon the virulence of these secondary colonizers, the host may be able to maintain an infection free state or succumb to the pathogens which then infect the bladder or vagina.

Morphological and structural analyses of the urogenital flora adherent to the epithelia have shown the presence of many distinct organisms, often interacting and coaggregating, in microcolonies or diffuse patterns on the cells (Sadhu *et al.* 1989; Reid *et al.* 1990c). Figure 17.2 illustrates this adherence, primarily dominated by *Lactobacillus* species. The presence of glycocalyx material intertwined between the cells is evident.

In vitro studies have to some extent mimicked this coaggregation or cooperativity between lactobacilli and other urogenital organisms, including potential pathogens. Of particular interest is the fact that type 1 fimbriated *E. coli*, the most common cause of symptomatic, lower urinary tract infection (UTI), were found to coaggregate with some strains of lactobacilli. In addition, P fimbriated *E. coli*, *Pseudomonas aeruginosa*, diphtheroids, *Klebsiella pneumoniae*, *Staphylococcus saprophyticus* and *Enterococcus faecalis* have been found to coaggregate with lactobacilli (Reid *et al.*

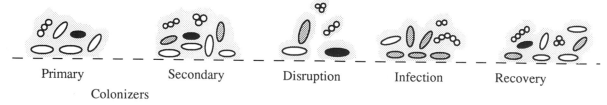

Primary   Secondary   Disruption   Infection   Recovery

Colonizers

Fig. 17.1. Primary colonization with mainly glycocalyx producing lactobacilli, disruption for example due to spermicide, infection and dominance with fimbriated uropathogens, then recovery after treatment leaving the microbial community made up of potential pathogens, lactobacilli and other species. (See text for discussion.)

Table 17.1. Coaggregating pairs found after *in* vitro assay

Lactobacillus	*E. coli* ATCC	2239	917	*Ps.aerug* P(R)	P(S)	*K.pn* 3a	*S.sap* Ya	*S.epi* 1938	*Ent* 4b
L. casei GR-1	3	2	0	1	0	1	0	2	1
L. acidophilus T-13	2	0	0	1	1	0	1	NT	NT
L. fermentum A-60	4	4	1	1	0	0	0	NT	NT
L. brevis 189	1	1	0	0	0	0	0	NT	NT
L. casei RC-9	2	1	0	0	0	0	0	NT	NT

E. coli: ATCC 25922, type 1 fimbriated 2239, P fimbriated 917; *Ps.aerug, Pseudomonas aeruginosa*: P(R) = rough, P(S) = smooth; *K.pn, Klebsiella pneumoniae*; *S.sap, Staphylococcus saprophyticus*; *S.epi, Staphylococcus epidermidis*; *Ent, Enterococcus faecalis*. NT, not tested; coaggregation score increases from 0 to 4.

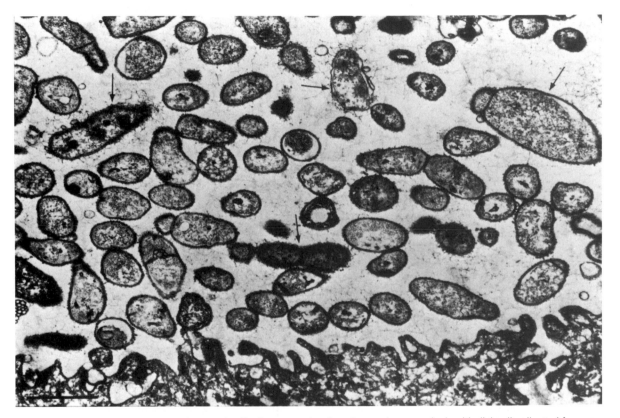

Fig. 17.2. Transmission electron micrograph of indigenous microflora (arrows) on a vaginal epithelial cell collected from a healthy adult female. (From Reid *et al.* 1990c, with permission.)

1988, 1990c; McGroarty *et al.* 1992) as demonstrated in Table 17.1.

The ability of lactobacilli to bind to other organisms is not apparently restricted to one species, and any given *Lactobacillus* strain may or may not be able to coaggregate with one or more of the other species. The mechanisms of coaggregation have not been elucidated, but preliminary evidence suggests the presence of pH and heat sensitive adhesins on the surfaces of the lactobacilli and other flora (Reid *et al.* 1988). The pH effect also influences bacterial autoaggregation and adhesion to cells. It is not known if *Lactobacillus* by-products, such as lactic acid, which are presumed to lower the urogenital pH and help protect the host against infection, actually alter the adhesiveness and colonization ability of the potential pathogens.

Based upon *in vitro* and *in vivo* studies, this coaggregation phenomenon seems to be real and it supports the hypothesis of a dynamic and interactive flora, proposed in Fig. 17.1 as the secondary colonization state.

Disruption of the flora

The understanding of the process, to date, suggests that infection occurs following (i) the domination of the flora by a virulent organism, (ii) the transient passage of a virulent uropathogen from the intestine via the vagina and introitus into the bladder, and (iii) the use of antibiotic agents, or spermicides (containing nonoxynol-9) which disrupt the flora and permit the development of a uropathogen dominated flora (Reid & Sobel 1987; McGroarty *et al.* 1990; Reid *et al.* 1990a). In a patient with recurrent bladder infection, the uropathogens can often be found as the dominant member of the genital tract flora, adhering well to uroepithelial cells (Bruce *et al.* 1983). However, it is still not understood how this alteration in domination occurs, although studies have shown that micturition alone can greatly alter the composition of the urogenital flora (Seddon *et al.* 1976).

Various studies of women with recurrent UTI have shown that the urethra and vagina of infected women are dominated by the infecting organism (Marrie *et al.* 1980; Pfau & Sacks 1981). It is assumed, but not verified fully, that domination of the flora occurs prior to infection; however, it may possible for progeny organisms already colonizing the bladder and growing in the urine to seed the external sites and cause domination upon micturition. The competitive mechanisms used by the pathogens to dominate the surfaces have not, to our knowledge, been examined. It is assumed that the properties which make a bacterium virulent to the host (for example haemolysin and toxin production) also confer a competitive edge in their microbial niche. The adherence of the bacteria to specific receptor sites also appears to be an important component. Factors such as bacteriocins and higher growth rates may also contribute to species domination.

Having stated that, many secondary colonizers in healthy women express these characteristics; the question is, why do these bacteria not always overgrow their microflora and cause infection? The answer may well lie in host factors and in the ability of indigenous organisms to keep the pathogens in check. Evidence for this latter point will be described in the section on protective factors.

Infection

The sequences illustrated in Fig. 17.1 need not necessarily hold true in all cases. For example, it is possible that a transient infecting agent may not disrupt or alter the flora, but may simply induce infection in the bladder or on the vaginal epithelium. An example could be the sexually transmitted pathogen, *Chlamydia trachomatis* which infects the genital tract or urethra by penetrating and multiplying inside cells (Schachter 1978).

The aim of this chapter is not to examine the pathogenesis of various urogenital infections. However, brief mention should be given to develop an appreciation for the process by which pathogens infect the host.

In the case of acute, uncomplicated UTI, it is known that *E. coli* enter the bladder, attach to the mucosa and multiply. The presentation of an infection is primarily reflected in the detection of large numbers of planktonic progeny bacteria in the urine, along with white and possibly red blood cells indicative of an inflammatory response, and the resultant signs and symptoms such as frequency, dysuria, urgency, and suprapubic pain. How many bacteria are required to invade the bladder and how many organisms have to attach to the cells for this process to occur? There are no data available on numbers of bacteria entering the bladder. In fact, whilst it is assumed that entry is via sequential attachment along the urethral mucosa, this has not been verified, nor has the trigger point been established where infection and symptoms occur. One study has shown that interleukin-6 production can be detected within 30 min of bacteria being inoculated into the bladder (Hedges *et al.* 1991), perhaps identifying part of the physiological mechanism. *In vitro* studies had suggested that large numbers of adherent bacteria were required for symptomatic UTI (Svanborg Edén *et al.* 1976), but *in vivo* findings indicate that small numbers per cell can also induce infection (Reid & Brooks 1985).

In vaginal infections, specific organisms such as *Gardnerella*, *Mobiluncus*, *Fusobacterium*, *Bacteroides*, *Chlamydia* sp., viruses and *Candida* sp. can be the infecting agents (Schachter 1978; Cook *et al.* 1989; Sobel 1989). In general, the presence of uropathogens in the vagina does not lead to localized infection. This leads us to question why this is the case and also what is the difference between vaginal and bladder cells and the surrounding microenvironment of the vagina compared to the bladder. In relation to uropathogenic *E. coli*, receptor sites have been found on both cell types, antibodies against the organisms can be recovered from both sites, nutritional components are available for the organisms to survive and multiply, cell sloughing occurs for transitional and squamous epithelia, and hormones and inflammatory cells gain access to organisms on bladder and vaginal cells. One important difference is that the cells of the vagina are covered by a diverse range of microorganisms, unlike bladder cells which, in general, are free of adherent organisms. It seems highly feasible then, that the normal vaginal flora is an important determinant of whether or not *E. coli* infect the host.

In the case of organisms such as *Gardnerella*, which infect the vagina but not the bladder, it seems to be their ability to survive in the former and not the latter site, and to utilize their virulence properties, which determines their occurrence as vaginal and not bladder infectants.

The consequence of treating the infection

The use of antimicrobial agents allows infections of the urogenital tract to be eradicated. Depending upon the agent(s) used, the genital flora will be depleted to some extent, but various organisms will still be retained on the mucosal surfaces. Studies have shown that antibiotic therapy targeted for other sites can disrupt the vaginal flora (Hertelius *et al.* 1989; Reid *et al.* 1990a). The recovery of the genital tract's so-called normal flora after antibiotic exposure can take several weeks.

Recovery of the flora

The flora which results from antimicrobial therapy will not resemble the primary indigenous colonizers unless all the organisms are eradicated by treatment and the same primary colonizers reattach. Thus, the emerging flora will comprise a consortium which may or may not be dominated by pathogenic bacteria, as illustrated in Fig. 17.1. In this scenario, the patient is at particular risk of becoming reinfected, and this phenomenon is probably the explanation for recurrent infections occurring in patients. In such cases, the concept of artificial implantation of indigenous organisms has been proposed initially by our group (Chan *et al.* 1984; Reid *et al.* 1987, 1990b, 1992; Bruce *et al.* 1992) and by others (Hertelius *et al.* 1989). In theory, such an approach would provide a more balanced flora less likely to infect the host.

For patients who do not suffer a recurrence and who do acquire a flora dominated by indigenous organisms, how does this occur, what determines the composition and how can it be maintained? These are all questions which face microbial ecologists and clinicians over the next decade. If we are to utilize probiotics to our advantage, it will be essential to understand the intricacies of the urogenital microbial community.

Protective aspects of normal flora

If the assumption is made that the maintenance and/or restoration of a normal flora is important in disease prevention (and some might argue that this remains to be verified), what components confer protection?

It has long been our contention that the indigenous flora of the urogenital tract have a protective function. In that regard, we have sought to define properties of *Lactobacillus* which could play a role in this disease prevention. The rationale has been fairly straightforward.

1. For the organisms to compete with adherent pathogens on a surface, they must surely themselves adhere. This ability to adhere to uroepithelial cells has been documented *in vitro* (Fig. 17.3) and *in vivo*, and found to vary from isolate to isolate (Reid *et al.* 1987, 1989; Bruce *et al.* 1992). The mechanism of adherence appears to involve lipoteichoic acid (Chan *et al.* 1985) and other pro-

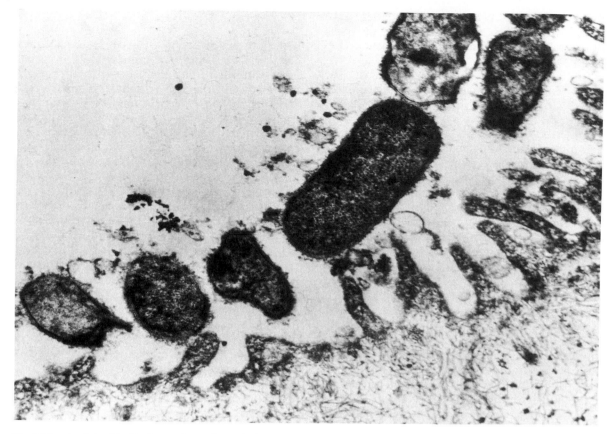

Fig. 17.3. Lactobacilli adhering to the surface of a human uroepithelial cell. (From Chan *et al.* 1985, with permission.)

teinaceous, polysaccharide and trypsin insensitive components present on the cell wall and secreted into the microenvironment (Cuperus *et al.* 1993a, 1993b; G. Reid *et al.*, unpublished data).

2. If the lactobacilli are able to prevent or reduce the adhesion and colonization of uropathogens, then the resultant flora will be less likely to be dominated by infecting pathogens. This concept of competitive exclusion was first shown *in vitro* in 1984 (Chan *et al.* 1984) and further demonstrated in animals (Reid *et al.* 1985, 1989) and again *in vitro* (Reid *et al.* 1987).

3. Once present in a microenvironment with potential pathogens, a useful trait which would aid the ability of lactobacilli to survive and compete would be the production of substances which adversely affect the growth and survival of other organisms. The ability of lactobacilli to produce acidic inhibitors and more specific bacteriocin-like compounds which inhibit pathogenic bacterial growth or kill these organisms has been well documented *in vitro*, but not verified *in vivo* (Reeves 1965; Tagg *et al.* 1976; Skarin & Sylwan 1986; Reid *et al.* 1987;

McGroarty & Reid 1988a, b). While a few such by-products have been subject to patent allowance, they have not, to our knowledge, been used in human trials or in the make-up of any commercial product. Perhaps one reason is that they are simply antimicrobial agents, and pharmaceutical companies already have a number of well defined agents to choose from. Nevertheless, within the context of a microcommunity, the production of such compounds could be vital to the maintenance of a normal flora. The administration of these substances by oral or even vaginal route would probably be less effective than the localized application of lactobacilli, although they could have an effect in killing pathogens.

4. If a balanced flora is to exist, the lactobacilli will have to be able to cope with the insurgence of many different types of bacteria. By coaggregating with such organisms, it is believed that a microbial balance can be achieved. As indicated earlier, examples of coaggregating lactobacilli have been found *in vitro* and *in vivo*.

5. Given that the urogenital tract can be exposed to

antimicrobial agents and spermicides that may kill many lactobacilli (McGroarty *et al.* 1990), an ability to resist the action of such compounds could provide strains with a survival advantage.

In summary, our present knowledge of urogenital probiotics suggests that these five characteristics, at least, assist with the protective function of lactobacilli in the urogenital tract.

Confirmation of protective function of probiotic regimens

The confirmation that lactobacilli can protect the host against infection is more difficult to prove conclusively. The relative success of the method in rats (Reid *et al.* 1985) and mice (Reid *et al.* 1989) has paved the way for human studies. A preliminary attempt to implant organisms into the bladder failed, due to an apparent inability of the bacteria to maintain colonization of this mucosa, and a somewhat poor patient selection (Hagberg *et al.* 1989). The problem did not appear to be in the choice of the bacterial strains. Further, it was felt by the authors that careful selection of bacteria based upon *in vitro* and *in vivo* studies should provide a more reliable product than over-the-counter preparations. An examination of a number of these products has shown how unreliable their contents can be (Hughes & Hilliar, 1990). It was thus decided that the bladder was not the best ecological niche for the lactobacilli (perhaps also explaining why they are not pathogenic to this area). Accordingly, the genital tract was chosen as the site for preferred implantation, and two small studies were undertaken in humans. The first demonstrated a level of success with *L. casei* GR-1 using a douche method of delivery (Bruce & Reid 1988). An improved method of preparation and delivery was then developed using two strains freeze-dried and placed in a gelatin suppository. This led to a 66.3% reduction in UTI over one year (Bruce *et al.* 1992). A more recent and larger human study (A. W. Bruce & G. Reid, unpublished data) has confirmed the previous findings and shown a reduction of over 70% in infection rate over one year. In addition, a method was devised to stimulate the patient's own *Lactobacillus* flora on a weekly basis using innocuous natural substances, and this too resulted in significantly decreased infection rates

during the subsequent year. Recently we have shown that *L. casei* GR-1 does colonize and survive upon implantation into the vagina (Reid *et al.* 1994).

Other researchers have used combinations of non-pathogenic bacteria, including lactobacilli, to prevent colonization by uropathogens. This has been successful in limited studies using monkeys (Hertelius *et al.* 1989), but no human data, to our knowledge, are available.

The definitive test will be to compare probiotic therapy with a suitable control, possibly long term antibiotics or an inactive placebo. The problem of interpretation may still be a matter of some controversy, as the recurrences of UTI often have a sporadic pattern (Kraft & Stamey 1977); thus some will argue that the rate in any given year is coincidental and not changed by probiotics. Nevertheless, having now treated over 100 patients, it would seem to us that there is definitely merit in probiotics; most patients are more compliant and favour an approach which does not involve synthetic chemicals. The advent of molecular technologies and improved understanding of microbial ecology, especially in defining the role of Gram positive bacteria, will provide many opportunities for future research investigation.

Acknowledgements

This work has been supported by the Medical Research Council of Canada. We express our appreciation to the many colleagues and students who have assisted with our work and helped us appreciate more fully the nature of the urogenital flora.

References

Bruce, A. W., Chan, R. C. Y., Pinkerton, D., Morales, A. & Chadwick, P. (1983). Adherence of Gram negative uropathogens to human uroepithelial cells. *Journal of Urology*, **130**, 293–8.

Bruce, A. W. & Reid, G. (1988). Intravaginal instillation of lactobacilli for prevention of recurrent urinary tract infections. *Canadian Journal of Microbiology*, **34**, 337–43.

Bruce, A. W., Reid, G., McGroarty, J. A., Taylor, M. & Preston, C. (1992). Preliminary study on the prevention of recurrent urinary tract infections in ten adult women using intravaginal lactobacilli. *International Urogynecology Journal*, **3**, 22–5.

Chan, R. C. Y., Bruce, A. W. & Reid, G. (1984). Adherence of cervical, vaginal and distal urethral normal microbial flora to human uroepithelial cells and the inhibition of adherence of Gram negative uropathogens by competitive exclusion. *Journal of Urology*, **131**, 596–601.

Chan, R. C. Y., Reid, G., Irvin, R. T., Bruce, A. W. & Costerton, J. W. (1985). Competitive exclusion of uropathogens from uroepithelial cells by *Lactobacillus* whole cells and cell wall fragments. *Infection and Immunity*, **47**, 84–9.

Cook, R. L., Reid, G., Pond, D. G., Schmitt, C.A. & Sobel, J. D. (1989). Clue cells and bacterial vaginosis: immunofluorescent identification of the adherent Gram negative bacteria as *Gardnerella vaginalis*. *Journal of Infectious Diseases*, **160**, 490–6.

Cuperus, P. L., van der Mei, H. C., Reid, G. *et al.* (1993a). The effect of serial passaging of lactobacilli in liquid medium on their physico-chemical and structural surface characteristics. *Cells and Materials*, **2** (4), 271–80.

Cuperus, P. L., van der Mei, H. C., Reid, G. *et al.* (1993b). Physico-chemical surface characteristics of urogenital and poultry lactobacilli. *Journal of Colloids Interface Science*, **156**, 319–24.

Hagberg, L., Bruce, A. W., Reid, G., Svanborg Edén, C., Lincoln, K. & Lidin-Janson, G. (1989). Colonization of the urinary tract with live bacteria from the normal fecal and urethral flora in patients with recurrent symptomatic urinary tract infections. In *Host–parasite interactions in urinary tract infections*, ed. E. H. Kass & C. Svanborg Edén, pp. 194–7. Chicago: University of Chicago Press.

Hedges, S., Anderson, P., Lidin-Janson, G., de Man, P. & Svanborg Edén, C. (1991). Interleukin-6 response to deliberate colonization of the human urinary tract with Gram-negative bacteria. *Infection and Immunity*, **59**, 421–7.

Hertelius, M., Gorbach, S. L., Mollby, R., Nord, C. E., Pettersson, L. & Winberg, J.(1989). Elimination of vaginal colonization with *Escherichia coli* by administration of indigenous flora. *Infection and Immunity*, **57**, 2447–51.

Hughes, V. L. & Hilliar, S. L. (1990). Microbiologic characteristics of *Lactobacillus* products used for colonization of the vagina. *Obstetrics and Gynecology*, **75**, 244–8.

Kraft, J. K. & Stamey, T. A. (1977). The natural history of symptomatic recurrent bacteriuria in women. *Medicine*, **56**, 55–60.

Marrie, T. J., Swantee, C. A & Hartlen, M. (1980). Aerobic and anaerobic urethral flora of healthy females in various physiological age groups and of females with urinary tract infections. *Journal of Clinical Microbiology*, **11**, 650–4.

McGroarty, J. A., Chong, S., Reid, G. & Bruce, A. W. (1990). Influence of the spermicidal compound nonoxynol-9 on the growth and adhesion of urogenital bacteria *in vitro*. *Current Microbiology*, **21**, 219–23.

McGroarty, J. A., Lee, V., Reid, G. & Bruce, A. W. (1992). Modulation of adhesion of uropathogenic *Enterococcus faecalis* to human epithelial cells *in vitro* by *Lactobacillus* species. *Microbial Ecology in Health and Disease*, **5**, 309–14.

McGroarty, J. A. & Reid, G. (1988a). Detection of a lactobacillus substance which inhibits *Escherichia coli*. *Canadian Journal of Microbiology*, **34**, 974–8.

McGroarty, J. A. & Reid, G. (1988b). Inhibition of enterococci by *Lactobacillus* species *in vitro*. *Microbial Ecology in Health and Disease*, **1**, 215–19.

Pfau, A. & Sacks, T. (1981). The bacterial flora of the vaginal vestibule, urethra and vagina in premenopausal women with recurrent urinary tract infections. *Journal of Clinical Microbiology*, **126**, 630–4.

Reeves, P. (1965). The bacteriocins. *Bacteriological Reviews*, **29**, 24–45.

Reid, G. & Brooks, H. J. L. (1985). A fluorescent antibody staining technique to detect bacterial adherence to urinary tract epithelial cells. *Stain Technology*, **60**, 211–17.

Reid, G., Bruce, A. W. & Cook, R. L. (1987). Examination of strains of lactobacilli for properties that may influence bacterial interference in the urinary tract. *Journal of Urology*, **138**, 330–5.

Reid, G., Bruce, A. W., Cook, R. L. & Llano, M. (1990a). Effect on the urogenital flora of antibiotic therapy for urinary tract infection. *Scandinavian Journal of Infectious Diseases*, **22**, 43–7.

Reid, G., Bruce, A.W., McGroarty, J. A., Cheng, K.-J. & Costerton, J. W. (1990b). Is there a role for lactobacilli in prevention of urogenital and intestinal infections? *Clinical Microbiology Reviews*, **3**, 335–44.

Reid, G., Bruce, A. W. & Taylor, M. (1992). Influence of three day antimicrobial therapy and lactobacillus suppositories on recurrence of urinary tract infection. *Clinical Therapeutics*, **14**, 11–16.

Reid, G., Chan, R. C. Y., Bruce, A. W. & Costerton, J. W. (1985). Prevention of urinary tract infection in rats with an indigenous *Lactobacillus casei* strain. *Infection and Immunity*, **49**, 320–4.

Reid, G., Cook, R. L., Hagberg, L. & Bruce, A. W. (1989). Lactobacilli as competitive colonizers of the urinary tract. In *Host–parasite interactions in urinary tract infections*, ed. E. H. Kass & C. Svanborg Edén, pp. 390–6. Chicago: University of Chicago Press.

Reid, G., McGroarty, J. A., Angotti, R. & Cook, R. L. (1988). Lactobacillus inhibitor production against *E. coli* and coaggregation ability with uropathogens. *Canadian Journal of Microbiology*, **34**, 344–51.

Reid, G., McGroarty, J. A., Domingue, P. A. G. *et al.* (1990c). Coaggregation of urogenital bacteria *in vitro* and *in vivo*. *Current Microbiology*, **20**, 47–52.

Reid, G., Millsap, K. & Bruce, A. W. (1994). Implantation of lactobacilli in the vagina: survival, colonization and genetic stability. *Lancet* **244**, 1229.

Reid, G. & Sobel, J. D. (1987). Bacterial adherence in the pathogenesis of urinary tract infection, a review. *Review of Infectious Diseases*, **9**, 470–87.

Sadhu, K., Domingue, P. A. G., Chow, A. W., Nelligan, J., Bartlett, K. & Costerton, J. W. (1989). A morphological study of the *in situ* tissue-associated autochthonous microflora of the human vagina. *Microbial Ecology in Health and Disease*, **2**, 99–106.

Schachter, J. (1978). Chlamydial infections. *New England Journal of Medicine*, **298**, 428.

Seddon, J., Bruce, A. W., Chadwick, P. & Carter, D. (1976). Introital bacterial flora – effect of increased frequency of micturition. *British Journal of Urology*, **48**, 211–18.

Skarin, A. & Sylwan, J. (1986). Vaginal lactobacilli inhibiting growth of *Gardnerella vaginalis*, *Mobiluncus* and other bacterial species cultured from vaginal content of women with bacterial vaginosis. *Acta Pathologica, Microbiologica et Immunologica Scandinavica*, Section B, **94**, 399–403.

Sobel, J. D. (1989). Bacterial vaginosis, an ecological mystery. *Annals of Internal Medicine*, **111**, 551–2.

Svanborg Edén, C., Hanson, L. A., Jodal, U., Lindberg, U. & Sohl Akerlund, A. (1976). Variable adherence to normal human urinary-tract epithelial cells of *Escherichia coli* strains associated with various forms of urinary-tract infection. *Lancet*, **ii**, 490–2.

Tagg, J. R., Dajani, A. S. & Wannamaker, L. W. (1976). Bacteriocins of Gram positive bacteria. *Bacteriology Reviews*, **40**, 722–56.

18

Dental Plaque

Philip D. Marsh

Introduction

Dental plaque was probably the first biofilm to have been studied in terms of either its microbial composition or its sensitivity to antimicrobial agents. In the seventeenth century, Anton van Leeuwenhoek pioneered the approach of studying biofilms by direct microscopic observation when he reported on the diversity and high numbers of 'animalcules' present in scrapings taken from around human teeth. He also conducted early studies on biocides when he established the resistance of these 'sticky animalcules' to salt and vinegar.

Following these pioneering observations and until the 1960s, there were relatively few studies of the microbiology of dental plaque. In the past three decades, however, there has been an enormous expansion of knowledge of the biochemistry and bacteriology of the plaque microflora. Impetus for this expansion stemmed from the exploitation of gnotobiotic animal technology in the 1950s and 1960s which established the role of particular bacterial species in the aetiology of two of the commonest diseases to affect humans in industrialised societies, namely, dental caries and periodontal (gum) diseases. The aim of this chapter will be to review our current knowledge of the microbiology of dental plaque, with particular emphasis on properties that relate to its biofilm structure.

Definition of dental plaque

The mouth is unique in the human body in that it provides non-shedding surfaces (teeth) for microbial colonization. Because of this, large masses of bacteria (and their products) are able to accumulate, especially at stagnant sites between teeth (approximal surfaces), in the pits and fissures on occlusal surfaces of premolars and molars, and in the gingival crevice (Fig. 18.1). In contrast, elsewhere in the body, desquamation ensures that the bacterial load is light on mucosal surfaces. Dental plaque forms naturally on teeth and, indeed, forms part of the host's defences by helping to prevent colonization by exogenous (and often pathogenic) microorganisms (Marsh 1989). Dental plaque has been

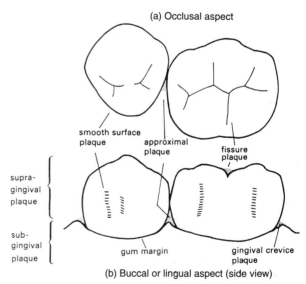

(a) Occlusal aspect

smooth surface plaque

approximal plaque

fissure plaque

supra-gingival plaque

sub-gingival plaque

gum margin

gingival crevice plaque

(b) Buccal or lingual aspect (side view)

Fig. 18.1. Diagram illustrating different surfaces of a tooth, and the range of sites from which dental plaque can be recovered. (From Marsh & Martin 1992, with permission.)

Table 18.1. Bacterial genera found in dental plaque

Gram positive	Gram negative
Cocci	Cocci
Streptococcus	*Neisseria*
Peptostreptococcus	*Veillonella*
Rods	Rods
Actinomyces	*Campylobacter*
Bifidobacterium	*Capnocytophaga*
Corynebacterium	*Eikenella*
Eubacterium	*Fusobacterium*
Lactobacillus	*Haemophilus*
Propionibacterium	*Leptotrichia*
Rothia	*Prevotella*
	Porphyromonas
	Selenomonas
	Treponema

Most genera contain more than one species. Some of these genera are found rarely and in low numbers at healthy sites. Other genera can be isolated occasionally as transient members of the plaque microflora (e.g. *Staphylococcus*).

Table 18.2. Components of saliva and gingival crevicular fluid (GCF) that can influence the growth of plaque bacteria

Host defences	Nutrients
Saliva	
s-IgA (IgG, IgM)	Glycoproteins (mucins)
Lysozyme	Proteins (amylase)
Lactoferrin/apo-lactoferrin	Peptides
Salivary peroxidase	Amino acids
Histidine-rich peptides	
Glycoproteins (agglutinins)	
Gingival crevicular fluid	
IgG (IgA, IgM)	Immunoglobulins
Complement	Proteins (albumin)
Neutrophils	α-2–globulin
Macrophages	Transferrin
B and T lymphocytes	Hormones
	Haemopexin
	Haptoglobin
	Haemoglobin

defined as the diverse microbial community found on the tooth surface embedded in a matrix of polymers of bacterial and salivary origin (Marsh & Martin 1992). Plaque that becomes calcified is referred to as calculus or tartar.

Factors affecting the growth of plaque bacteria

The resident microflora of dental plaque is extremely diverse, and consists of Gram positive and Gram negative bacteria (Table 18.1), including facultatively anaerobic and obligately anaerobic species (Marsh & Martin 1992). At present, not all of the bacteria that are seen in plaque by microscopy can be cultured in the laboratory, but the persistence of such fastidious organisms in dental plaque is evidence that all of the nutritional and environmental requirements of this diverse array of species are being met.

Nutrients

There are three main sources of nutrients for plaque bacteria. The diversity of the plaque microflora is believed to be due principally to the endogenous nutrients supplied by the host rather than by exogenous factors in the diet. Saliva pro-vides amino acids, peptides, proteins and glyco-proteins (such as mucins) for microbial growth (Table 18.2). Gingival crevicular fluid (GCF) is a serum-like exudate that bathes the gingival crevice, the flow of which is markedly increased during inflammation in periodontal disease. In addition to providing components of the host's defences, GCF also supplies peptides, proteins and glycoproteins, as well as iron- and haem-containing molecules such as transferrin, haemopexin and haptoglobin (Cimasoni 1983; Mukherjee 1985) which can be utilized by the resident flora of the gingival crevice (Table 18.2). The catabolism of complex substrates such as proteins and glycoproteins requires the concerted metabolic action of several species (ter Steeg *et al.* 1987; van der Hoeven & Camp 1991, 1993; Homer & Beighton 1992), and such metabolic cooperation is one way in which the stability of the resident plaque microflora is maintained.

Evidence for the importance of endogenous nutrients to the persistence of plaque bacteria has come from two sources. It has been found that plaque forms, and a relatively diverse microflora is maintained, in humans fed by stomach tube (Littleton *et al.* 1967), while the plaque flora of animals with dietary habits, ranging from insectivores and herbivores to carnivores, was broadly similar (Dent 1979).

Superimposed upon these endogenous nutrients is the complex array of foodstuffs ingested periodically in the diet. Fermentable carbohydrates are the only component that has been found to influence significantly the ecology of dental plaque. These substrates can be catabolized in a number of ways either to acidic fermentation products (Hamilton 1987) or to a range of extracellular or intracellular polysaccharides (Walker & Jacques 1987). Acids such as lactate transiently lower the pH of plaque, leading to the demineralization of tooth enamel and, if repeated often enough, eventually to dental caries. The repeated lowering of plaque pH can also disrupt the balance of the flora, inhibiting many of the acid sensitive species that are associated with sound enamel (de Stoppelaar *et al.* 1970). Sucrose can also be converted by glucosyl- and fructosyltransferases into a range of soluble and insoluble glucans and fructans, respectively (Walker & Jacques 1987). The soluble polymers can act as extracellular storage compounds that, subsequently, can be broken down by other plaque bacteria with exo- and endoglucanases and fructanases, while the insoluble glucans are involved in the consolidation of the attachment of certain species to the tooth surface.

Other dietary components that can influence the ecology of dental plaque, albeit to a lesser degree, include dairy products and sugar substitutes. Milk proteins and polypeptides can adsorb to enamel and reduce bacterial adhesion (Olsson *et al.* 1980; Reynolds & Wong 1983); they can also sequester calcium phosphate and encourage remineralization of enamel. Some of the sugar substitutes that are currently being used as a replacement for sucrose in confectionery and beverages have an antimicrobial action (Grenby & Saldanha 1986; Mäkinen 1989); the long-term use of xylitol, for example, can reduce the levels in plaque of bacteria implicated in dental caries (mutans streptococci: Mäkinen 1989).

Environmental factors

The temperature of the human mouth is maintained at approximately 35–36 °C, which is optimal for the growth of a wide range of microorganisms. In spite of the fact that air flows over the surfaces of the teeth, dental plaque contains few, if any, truly aerobic species; the majority of plaque bacteria are facultatively or obligately anaerobic (Marsh & Martin 1992). The redox potential (E_h) falls rapidly during the formation of dental plaque, from initial values of $> +200$ mV to -30 mV after 2 days and <-150 mV after 7 days (Kenney & Ash 1969). The E_h of the gingival crevice is usually lower than that of other sites around a healthy tooth, and this is the site with the greatest proportion of obligately anaerobic species.

The pH of plaque is normally around neutrality, due mainly to the buffering action of saliva. The pH in plaque can fall rapidly to <5.0 following the catabolism of dietary carbohydrates, before returning to resting values following the clearance of the sugar substrate (Jensen & Schachtele 1983). In contrast, the pH can rise to between 7.0 and 8.0 in the gingival crevice during an inflammatory host response (Eggert *et al.* 1991). Laboratory studies have shown that such changes in the local pH can lead to shifts in the balance of the plaque microflora (Bradshaw *et al.* 1989; McDermid *et al.* 1990). Saliva and GCF also contain a number of components of the host defences, including both non-specific (innate) and specific factors (Table 18.2), that could modify the composition or metabolic activity of dental plaque. The flow rate of saliva and GCF will detach organisms that are not firmly attached to oral surfaces. The rate of swallowing ensures that bacteria cannot be maintained in saliva by growth. Therefore, in order to be retained on the tooth surface, plaque bacteria have developed specific mechanisms of attachment, and these will be discussed in a later section.

Development and structure of dental plaque

Studies of the structure and development of dental plaque with time have been carried in human volunteers using removable artificial surfaces such as Mylar strips, epoxy resin crowns, plastic films or hydroxyapatite blocks. These surfaces can be worn at different sites in the mouth, and removed at varying time intervals for bacteriological analysis or for ultrastructural studies.

Immediately after the cleaning of a tooth, a proteinaceous conditioning film is adsorbed on to the exposed surfaces. This film, termed the acquired pellicle, is composed of molecules such as albumin, lysozyme, glycoproteins (including mucins and immunoglobulins), phosphoproteins and lipids derived from saliva and GCF (Al-

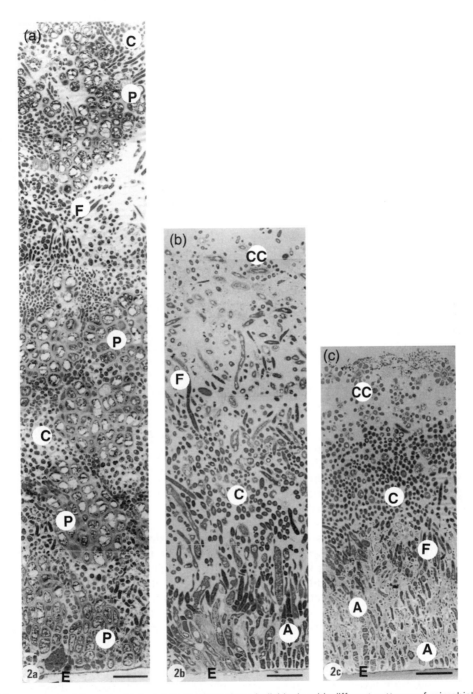

Fig. 18.2. Ultrastructure of 2–week old dental plaque from three individuals with different patterns of microbial colonisation (a–c). The thickness, structure and microbial composition show marked differences: A, thick-walled Gram positive pleomorphic bacteria; C, Gram positive coccal bacteria; F, Gram negative filamentous bacteria; CC, 'corn cob' formations; P, large polygonal bacteria; E, space remaining after demineralization of enamel. Bar = 5 μm. (From Nyvad & Fejerskov 1989, with permission.)

Hashimi & Levine 1989). In addition, there are some bacterial products, including biologically active glucosyltransferases (GTF: Scheie *et al.* 1987; Schilling & Bowen 1988), and possibly some dairy products derived from the diet (Pearce & Bibby 1966). After 2–4 h, single cells of mainly Gram positive coccoid bacteria can be seen on the pellicle coated surfaces, together with a few rod shaped organisms. Cultural studies have shown that these pioneer species are predominantly streptococci (61–78% of the total microflora) and actinomyces (4–30%), with some haemophili (Nyvad & Kilian 1987). Perhaps significantly, the early streptococcal colonizers (*Streptococcus sanguis*, *S. oralis* and *S. mitis*) generally produce IgA$_1$ proteases (Nyvad & Kilian 1990), presumably to enable them to evade the adherence inhibiting effects of s-IgA (Reinholdt & Kilian 1987). Following this, the attached cells divide rapidly to form microcolonies in the first instance, which eventually coalesce to form a confluent film of varying thickness (Nyvad & Fejerskov 1989). The fastest cell doubling times (t_d) are registered during these early phases of plaque formation. In animal studies, t_d was determined experimentally for *Actinomyces* spp. to be 2.7–2.9 h, and for *S. sanguis* and *S. mutans* to be 1.4–2.1 h, after 2–4 h of plaque development (Beckers & van der Hoeven 1982; Beighton *et al.* 1986).

After 1–2 days, Gram positive rods and filaments can be observed extending outwards from microcolonies of mainly coccoid cells. After several days of development, the morphological and cultural diversity of the microflora increases, and there is a shift from a streptococcal dominated flora to higher numbers of *Actinomyces* spp. The depth of the biofilm increases, and there is believed to be a lowering of the oxygen tension and of the redox potential, resulting in a marked increase in the proportions of obligately anaerobic species (Nyvad 1993). The structure of dental plaque becomes more varied during this period (Fig. 18.2).There is a layer of densely packed cells (3–20 cells deep) next to the tooth surface, many of which have thickened cell walls. This is termed the 'condensed layer', above which there lies the 'bulk layer' with a more variable structure and a greater morphological diversity (Fig. 18.2). Layering can be seen, the existence of which has been attributed to bacterial succession. Columns (palisades) of morpho-

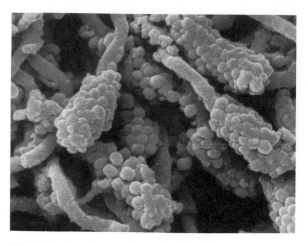

Fig. 18.3. Scanning electron micrograph showing 'corn cob' formation. Magnification *c.* × 600. (Courtesy of Dr C. A. Saxton.)

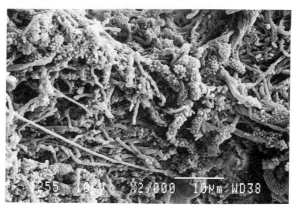

Fig. 18.4. Scanning electron micrograph of dental plaque showing morphological diversity of the resident microflora. Bar = 10 μm. (Courtesy of Dr K. M. Pang.)

logically similar bacteria can be observed throughout the entire thickness of a section of plaque while, on other occasions, microcolonies lying parallel to the tooth surface have been seen. Electron microscopy has shown the presence of an interbacterial matrix of polysaccharide. The superficial layers of dental plaque often exhibit a high species diversity, with some unusual combinations of bacteria such as 'corn-cobs' (Gram positive filaments covered by Gram positive cocci), 'rosettes' (coccal bacteria covered by small Gram positive curved rods), or 'bristle brushes' (large filaments surrounded by Gram negative rods or short filaments; Figs. 18.2–18.4; Listgarten 1976; Nyvad 1993). The generation

Fig. 18.5. Dental plaque in a fissure on the occlusal surface of a molar. Magnification *c.* × 50. (Courtesy of Dr K. M. Pang.)

time of bacteria during this phase of plaque development has been estimated crudely to be in the range 3–7 h (Brecx *et al.* 1983).

If plaque is left to develop undisturbed on exposed enamel surfaces for 2–3 weeks a climax community will establish, and the bacterial composition will become relatively constant with time. The depth of the biofilm reaches approximately 50–100 μm thick, and is probably restricted on exposed surfaces by the shear forces imposed by saliva flow and mastication (Nyvad 1993). Plaque can develop to greater depths at stagnant or protected sites, such as in the fissures (Fig. 18.5) on occlusal surfaces, between the teeth at approximal sites, and in the gingival crevice (Fig. 18.1). The climax community that develops at such anatomically distinct sites varies quite markedly, and will be described in the following section.

Climax community of dental plaque

Environmental conditions on a tooth are not uniform. Differences exist in the degree of protection from shear forces, and in the gradients of many biological and chemical factors that can influence the growth of the resident microflora. These differences are reflected in the composition of the resident microflora of plaque from sites as distinct as the gingival crevice, approximal sites, and fissures (Fig. 18.1).

Fissure plaque

The microflora of fissure plaque has been determined using either artificial fissures that are implanted in occlusal surfaces of pre-existing restorations, or by sampling natural fissures with either fine probes, blunt hypodermic needles, or toothpicks. The microflora is mainly Gram positive, and is dominated by streptococci, especially mutans streptococci (Theilade *et al.* 1974, 1982). With the exception of *Veillonella* spp., obligately anaerobic species are recovered only occasionally; *Propionibacterium* and *Eubacterium* spp. have been isolated in low numbers (Theilade *et al.* 1982). The total number of viable bacteria that could be recovered from one study of 10 fissures ranged from 0.4 to 9.7×10^6 CFU (colony forming units). The microflora from fissures is less diverse than that found at other sites on the tooth surface. The median number of species found in the above 10 fissures was eight and ranged from two to 11 (Theilade *et al.* 1982). The distribution of bacteria within a fissure has not been studied in detail, but it is probable that the environment at the base will be markedly different from that near the neck of the fissure, where saliva will be able to penetrate more easily (Fig. 18.5). For example, it has been suggested that *Actinomyces* cells appear as short rods or coccoid forms in the depth of fissures and as long rods or filaments in the orifice (Theilade *et al.* 1976; Theilade *et al.* 1978).

Approximal surface plaque

In contrast to fissures, the predominant bacteria at sites between teeth are Gram positive rods such as *Actinomyces*; there are much higher levels of obligately anaerobic species, including *Veillonella*, *Fusobacterium* and *Prevotella* spp. (Bowden *et al.* 1975; Marsh *et al.* 1989). This reflects the more reduced environment at this site. Variations in the distribution of bacteria at approximal sites have been reported depending on whether the site sampled was below or to the side of the contact area. This is the site where adjacent teeth touch and which is, therefore, devoid of plaque (Ahmady *et al.* 1993). Generally, the microflora of approximal plaque is more diverse than that of fissures.

Gingival crevice plaque

The gingival crevice provides a distinct habitat, and the ecology of the site is influenced by the flow of, and nutrients provided by, GCF. Compared to fissures and approximal sites, there are higher levels of obligately anaerobic species belonging to genera such as *Peptostreptococcus*, *Actinomyces*, *Propionibacterium*, 'Bacteroides', *Fusobacterium*, *Prevotella*, *Selenomonas* and *Veillonella* (van Palenstein Helderman 1975; Slots 1977). Some of these species are asaccharolytic but proteolytic, and derive their energy from the catabolism of host proteins and glycoproteins. Many bacterial associations have been observed in samples of plaque from the gingival crevice, including 'corn-cobs' between cocci and long rods or filaments (Fig. 18.4).

Mechanisms of dental plaque formation

Over recent years, there has been great interest in determining the mechanisms involved in the development of dental plaque because it is hoped that such knowledge could be exploited to help control the formation of dental plaque, and thereby reduce the incidence of caries and periodontal diseases.

The development of dental plaque can be subdivided into several arbitrary phases, although as plaque formation is a dynamic and continuous process, these phases will inevitably overlap. These phases include:

1. A non-specific reversible stage, involving van der Waals forces of attraction and electrostatic repulsion,
2. short range specific molecular interactions between bacterial cell surface adhesins on primary colonizers and host receptors in the acquired pellicle,
3. the attachment of secondary colonizers to primary colonizers (coaggregation),
4. extracellular polymer synthesis and growth.

Microorganisms are transported to the tooth surface by the flow of saliva. As a cell approaches the pellicle-coated surface, long range physicochemical forces operate. In brief, although the surface of oral microorganisms and of the acquired pellicle are both negatively charged, a weak van der Waals attraction is induced by the fluctuating dipoles within the molecules of the two approaching surfaces (Rutter & Vincent 1980; Busscher *et al.* 1992). Ultimately, repulsive electrostatic forces will prevent further progress towards the surface. At separation distances of between 10 and 20 nm from the surface, therefore, organisms may be held reversibly in a weak area of attraction (the secondary minimum, as described by the DLVO theory: see Korber *et al.*, Chapter 1). Then, hydrophobic cell surface components may play a role in removing water from between the interacting surfaces to bring them closer together so that short range stereochemical interactions may occur between complementary molecules on the tooth and bacterial surfaces.

Numerous examples of these molecular interactions have now been described. Some of the salivary components that can be found in the acquired pellicle have been purified and tested for their ability to bind to particular bacteria. For example, different regions of the acidic proline rich proteins can bind to *Streptococcus gordonii* and to *Actinomyces naeslundii* (Gibbons *et al.* 1988; Gibbons *et al.* 1991). Likewise, *Streptococcus gordonii* can bind to α-amylase (Scannapieco *et al.* 1989; Douglas *et al.* 1990) while *A. naeslundii* and *Fusobacterium nucleatum* interact with statherin (Strömberg *et al.* 1992; Kolenbrander & London 1993). Other mechanisms involve lectin-like bacterial proteins interacting with carbohydrates or oligosaccharides in pellicle-associated glycoproteins. Thus, *S. sanguis* can bind to terminal sialic acid residues in adsorbed salivary glycoproteins (McBride & Gislow 1977), while *S. oralis* expresses either a galactose binding lectin (Shibata *et al.* 1980) or a lectin that interacts with a trisaccharide structure containing sialic acid, galactose and *N*-acetylgalactosamine (Murray *et al.* 1982, 1986). Other examples include the interactions of bacterial lipoteichoic acid or GTF and host blood group reactive substances (Hogg & Embery 1982), and between bacterial antigens and adsorbed antibodies.

The bacterial adhesins are also being characterized. *Actinomyces* spp. have two antigenically and functionally distinct types of fimbriae (Cisar *et al.* 1984; Cisar 1986). Type 1 fimbriae mediate bacterial adherence to proline-rich peptides and statherin (i.e. a protein–protein interaction), whereas type 2 fimbriae are associated with a lactose sensitive lectin-like activity (that is, a protein–carbohydrate interaction) that is involved with the adherence of cells to streptococci (coaggregation; see later in this section). More recently, two slightly distinct receptor specificities of type 2 fimbrial lectins have been found which might influence the ability of *Actinomyces* strains to colonize either tooth or mucosal surfaces (Strömberg & Borén 1992).

A number of proteins and lipoproteins in the cell walls of streptococci have been identified as adhesins. For example, a high molecular weight protein from *S. mutans*, known by various names including antigen B, I/II, P1 or Pac, interacts with salivary agglutinins. Antibody directed against this molecule blocked adhesion of *S. mutans* to saliva coated surfaces (Douglas & Russell 1984). Similar proteins are found in other streptococci, with one segment being highly conserved (Ma *et al.* 1991), although the related proteins from *S. mutans* and *S. sanguis* recognize different binding sites on salivary agglutinins (Demuth *et al.* 1990). The large size of some of the wall proteins means that they may be involved with more than one function. For example, a protein of *S. gordonii* can interact with salivary proteins and it can coaggregate with *Actinomyces naeslundii* (Jenkinson *et al.* 1993). Cloning and sequencing has shown some adhesins from *S. sanguis* (Ganeshkumar *et al.* 1993) and *S. gordonii* (Jenkinson 1992) to be lipoproteins.

A critical factor in plaque formation centres around the nature of these specific interactions between bacterial adhesins and host receptors. These receptors lie on molecules that are not only adsorbed to the tooth surface but which are also freely in suspension in saliva. Some of these molecules are designed for aggregating bacteria, thereby facilitating their removal from the mouth by swallowing. Obviously, it is essential if plaque formation is to proceed that all bacteria are not aggregated before they reach the tooth surface. Recent work has shown that a highly selective mechanism may function to overcome this problem. It was found that although *A. naeslundii* could bind to the acidic proline rich peptides

when the latter were bound to a surface, it did not interact with this protein in solution. It has been proposed that hidden molecular segments of this molecule become exposed as a result of conformational changes when it is adsorbed to a surface (Gibbons 1989; Gibbons *et al.* 1990). Such hidden receptors for bacterial adhesins have been termed 'cryptitopes'. In this way, a controlled mechanism for facilitating plaque formation has evolved by which the host can promote the attachment of specific bacteria without compromising this selective process in the planktonic phase. Another example of a cryptitope involved in plaque formation can arise following the enzymic modification of host molecules by oral bacteria. Neuraminidase producing bacteria, such as *A. naeslundii*, *S. oralis* and *S. mitis*, can cleave the terminal sialic acid residue from oligosaccharide side chains of glycoproteins to expose the penultimate galactosyl sugar. Many bacteria possess galactosyl binding lectins, including *A. naeslundii*, *Leptotrichia buccalis*, *Fusobacterium nucleatum*, *Eikenella corrodens* and *Prevotella intermedia*, and could therefore exploit such a cryptitope (Gibbons 1989).

An important mechanism that aids plaque buildup and encourages species diversity is the phenomenon of coaggregation between bacterial cells. Coaggregation, or cell-to-cell recognition of genetically distinct partner cell types (Kolenbrander & London 1993), has been observed with isolates representing 18 bacterial genera from the mouth (Fig. 18.6). In dental plaque, the ability of cells to attach to bacteria already anchored to a surface may provide secondary (or late) colonizers with the same advantage as primary (or early) colonizers. To date, there is little or no evidence for coaggregation among resident bacteria from other ecosystems, whereas coaggregation among plaque bacteria can be intra-, inter-, or even multi-generic (Fig. 18.6; Kolenbrander & London 1993). In coaggregation, secondary colonizers generally synthesize protein adhesins that recognize receptors on streptococci and actinomycetes.

Initial plaque accumulation will be enhanced by intrageneric and intergeneric coaggregation among the early colonizers such as the streptococci and the actinomycetes. Over 90% of more than 300 isolates of *A. naeslundii*, *S. gordonii*, *S. mitis*, *S. oralis* and *S. sanguis* tested in pair-wise intergeneric coaggregations were found to

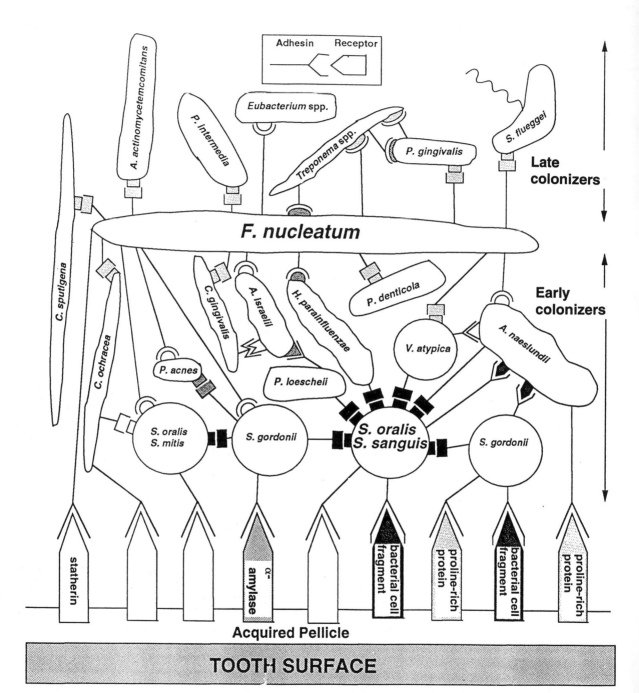

Fig. 18.6. Schematic representation of patterns of coaggregation in human dental plaque. Early colonizers bind to receptors in the acquired pellicle; subsequently, other early and late colonizers bind to these already attached cells (coaggregation). Adhesins (symbols with stems) are cell components that are heat or protease sensitive; the receptor (complementary symbol) is insensitive to either treatment. Identical symbols are not necessarily identical molecules, although they are likely to be related functionally. The symbols with rectangular shapes represent lactose inhibitable coaggregations (From Kolenbrander & London 1993, with permission.)

co-aggregate (Kolenbrander & London 1992). These *Actinomyces–Streptococcus* interactions are not random but are highly specific, and can be reversed by treatment with sugars (e.g. lactose), or by protease (or heat) treatment of either or both cell surfaces. Some bacteria can act as coaggregation bridges between otherwise non-coaggregating species (Fig. 18.6; Kolenbrander & London 1993).

Fusobacteria have been found to coaggregate with the widest range of bacterial genera but, curiously, they do not coaggregate with each other. Early colonizers of plaque coaggregate extensively with *F. nucleatum*, while late colonizers such as *Selenomonas flueggei* do not coaggregate with early colonizers, but do usually coaggregate with *F. nucleatum* (Kolenbrander *et al.* 1989). Other obligately anaerobic species, such as *Eubacterium* spp., can also coaggregate with fusobacteria (George & Falkler 1992). Thus, it has been proposed that fusobacteria act as a bridge between early and late colonizing bacteria (Kolenbrander & London 1993). In addition to enabling bacteria to anchor to a surface, coaggregation may also offer metabolic benefits, perhaps by facilitating cross-feeding or cooperation in the breakdown of complex host molecules.

Another factor in the development of plaque will be the synthesis of extracellular polysaccharides (EPS) from sucrose by adherent bacteria (Walker & Jacques 1987). These polymers can include soluble glucans and fructans (which can be metabolized by other plaque bacteria), and insoluble glucans, and these polymers comprise about 20% of the dry weight of dental plaque. Originally, it was believed that polysaccharide production was important in the initial stages of attachment. Although preformed polymer may react with receptors in these pellicle, these insoluble molecules are now considered to make a more important contribution to the structural integrity and diffusion properties of plaque. Perhaps surprisingly at first sight, neither acid nor sugar diffusion is much affected by the EPS content of plaque (McNee *et al.* 1982; Dibdin *et al.* 1983). However, EPS may lead to an increased cariogenic challenge at the tooth surface because, in a thick plaque, EPS will enable sugars to penetrate deeper into the biofilm, while the significant buffering effect of bacteria will be reduced, thereby producing a more pronounced pH fall at the plaque–enamel interface (Dibdin & Shellis 1988). In support of this, studies comparing the effects of differing layers of *S. mutans* and EPS found that demineralization of experimental enamel slices was maximal when the artificial plaque consisted of 95% EPS and only 5% bacteria (Zero *et al.* 1986).

Microbial interactions in dental plaque

In a biofilm such as plaque, microorganisms are in close proximity with one another and interact as a consequence. Such interactions can be beneficial to one or other of the interacting populations, while others can be antagonistic (Table 18.3). Microbial metabolism within plaque will produce gradients in factors such as the depletion of essential nutrients and the accumulation of inhibitory by-products, and this will affect the growth and distribution of other species. Gradients will lead to the development of vertical and horizontal stratifications within the plaque biofilm producing a mosaic of microenvironments (Marsh & Martin 1992). This enables organisms with widely differing requirements to grow, and ensures the coexistence of species that would otherwise be incompatible with one another in a homogeneous habitat.

With very few exceptions, most studies have characterized microbial interactions in the laboratory using conventional liquid culture technology, with the assumption that such interactions will operate similarly in a biofilm in the mouth. A major goal for the future will be to validate these findings using more relevant models that incorporate the biofilm characteristics of dental plaque (Tatevossian 1991). Nevertheless, competition for nutrients will inevitably be one of the primary ecological determinants in dictating the prevalence of a particular species in dental plaque (van der Hoeven *et al.* 1985). As discussed earlier, the main source of nutrients for the resident microflora is the endogenous supply of proteins and glycoproteins by host secretions. The plaque microflora produces a wide range of glycosidases and proteases, but no single species possesses the complete range to degrade in full these complex molecules. Recent studies have shown that the concerted action of several species with complementary patterns of enzyme activity is necessary

Table 18.3. Types of microbial interaction in dental plaque

Beneficial	Antagonistic
Enzyme complementation	Nutrient competition
Food chains	Hydrogen peroxide
Coaggregation	Organic acids
'Neutralization' of inhibitory molecules	Low pH
	Bacteriocins
Inactivation of host defences	Enzymes

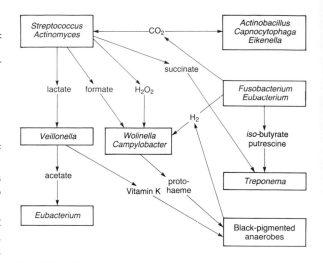

Fig. 18.7. Some potential nutritional interactions (food chains) between plaque bacteria.. (From Marsh & Martin 1992, with permission.)

for the complete breakdown of host molecules (ter Steeg *et al.* 1987; van der Hoeven & Camp 1991, 1993; Homer & Beighton 1992).

Other examples of nutritional interactions that have been identified among plaque bacteria include secondary feeding, whereby the products of metabolism of one organism (primary feeder) provide the main source of nutrients for another (secondary feeder). The best described interaction of this type is the utilization by *Veillonella* spp. of lactate produced from the metabolism of dietary carbohydrates by streptococci or actinomycetes (Mikx & van der Hoeven 1975), an interaction that may help protect enamel from demineralization by lactic acid (Mikx *et al.* 1972). Recently, a mutually beneficial interaction between *S. sanguis* and the obligate anaerobe, *Campylobacter rectus*, has been described whereby the anaerobe scavenges inhibitory oxygen, or even hydrogen peroxide produced by the streptococci, while *S. sanguis* provides *C. rectus* with formate following the fermentation of glucose under carbohydrate limiting conditions (Ohta *et al.* 1990). *C. rectus* is also able to produce protohaem for the growth of black-pigmenting, obligately anaerobic *Prevotella* and *Porphyromonas* species (Grenier & Mayrand 1986). Accumulatively, these individual food chains develop into a complex food web; the nutritional interdependence of species may be one of the reasons why it is not possible to grow in pure culture in the laboratory all of the species seen in plaque (Fig. 18.7).

Antagonism is another major contributory factor in determining the final microbial composition of dental plaque. Many genera (and especially streptococci) produce bacteriocins or bacteriocin-like substances (Marsh 1989). Studies with gnotobiotic rodents in which strains of *S. mutans* with various degrees of bacteriocinogenic activity were implanted, with or without a

bacteriocin sensitive strain of *Actinomyces* sp., showed that bacteriocin production conferred an advantage to *S. mutans* during colonization of the teeth (Rogers *et al.* 1978, 1979; van der Hoeven & Rogers 1979). Other inhibitory factors produced by plaque bacteria include organic acids, hydrogen peroxide, and certain enzymes (see Marsh 1989). The production of hydrogen peroxide by some streptococci has been proposed as a mechanism whereby the growth of some of the Gram negative bacteria that are implicated in periodontal diseases are suppressed at such low levels that they are incapable of initiating disease (Hillman *et al.* 1985). The low pH generated from carbohydrate catabolism is also inhibitory to the growth of many plaque species. Studies with both a laboratory biofilm model (Perrons & Donoghue 1990) and with gnotobiotic rats (van der Hoeven & Rogers 1979) have shown that members of the resident plaque microflora, when established, can exhibit colonization resistance against the subsequent challenge of *S. mutans*, even with strains of the latter that produce a bacteriocin (van der Hoeven & Rogers 1979). Bacterial antagonism may also be a mechanism whereby exogenous species are prevented from colonizing dental plaque. However, the production of inhibitory factors will not necessarily lead to the complete exclusion of sensitive bacteria from plaque. The presence of distinct microhabitats within a biofilm such as dental plaque enables the survival of organisms that would be incompatible with one another in a homogeneous environment.

The expression of a surface associated phenotype by oral bacteria has not been studied extensively in the laboratory. In one study, the effect of surfaces on acid production by oral streptococci was investigated (Berry & Henry 1977). Lactate production by all nine strains of mutans streptococci, but by only two out of five strains of *S. sanguis* tested, displayed an enhanced glycolytic activity following the addition of hydroxyapatite beads; acid production by the other strains of *S. sanguis* and a strain of *S. mitis* and of *S. salivarius* were reduced in the presence of the beads. Likewise, the expression of genes that code for certain GTFs (*gtf* B/C) was increased in cells of *S. mutans* bound to a surface (Hudson & Curtiss 1990). It was proposed that initial binding might signal *S. mutans* to produce insoluble glucan to enhance attachment. In contrast, fructosyltransferase activity was unaffected by surface binding (Hudson & Curtiss 1990).

Dental plaque and disease

Although dental plaque forms naturally on teeth, in the absence of adequate oral hygiene it can accumulate at stagnant sites beyond levels compatible with oral health and, at a susceptible site, dental caries or periodontal disease can occur. As plaque mass increases, the beneficial buffering and antimicrobial properties of saliva are less able to penetrate and protect enamel, and there is a shift in the balance of the predominant bacteria away from those associated with health.

Dental caries

Dental caries is the localized dissolution of the enamel (demineralization) by acids produced primarily from the catabolism of dietary carbohydrates by bacteria in dental plaque. The early stages of caries are reversible, and remineralization can occur, especially in the presence of fluoride.

The resting pH of plaque is around neutrality due to the buffering of saliva. The consumption of dietary carbohydrates leads to the rapid production of acid (predominantly lactic acid) in plaque by glycolysis, so that the local pH in plaque can fall below 5.0 within a few minutes, and remain at such values for some time, causing enamel to demineralize (Jensen & Schachtele 1983). As the pH gradually returns to resting values following the clearance of the sugars, enamel begins to remineralize. However, the frequent consumption of dietary sugars leads to more regular and prolonged conditions of low pH in plaque. Laboratory studies of complex mixed cultures of oral bacteria have shown that this favours the growth of acid tolerating bacteria such as mutans streptococci and lactobacilli, and inhibits many of the acid sensitive species (e.g. *S. sanguis*, *S. oralis*, Gram negative spp.) associated with sound enamel (Bradshaw *et al.* 1989). The regular exposure of enamel to low pH also tips the physicochemical equilibrium more towards demineralization, and eventually caries.

There have been numerous studies examining the composition of the plaque microflora associated with caries; many of these have been reviewed by Loesche (1986) and Bowden (1991). The accumulative evidence has shown a strong correlation between the presence of higher levels of mutans streptococci (especially *S. mutans* and *S. sobrinus*) and lactobacilli at sites with caries compared with plaque overlying sound enamel (Loesche 1986). Longitudinal studies suggest that lactobacilli, and possibly *Actinomyces odontolyticus*, may be involved more with progression of lesions and advanced stages of cavity formation (Boyar & Bowden 1985) rather than with caries initiation. The pathogenic traits of mutans streptococci have been investigated in detail (Loesche 1986; Bowden 1991; Marsh & Martin 1992), and are listed in Table 18.4; they include the ability to metabolize sugars rapidly and to grow well in acidic environments. In older age, recession of the gums can expose the root surface to caries attack. The microflora implicated with root surface caries is also associated with raised levels of mutans streptococci and lactobacilli, as well as with *Actinomyces* spp. (Bowden 1990).

Periodontal diseases

The term periodontal diseases embraces a number of conditions in which the supporting tissues of the teeth are attacked. The junctional epithelium at the base of the gingival crevice migrates

Table 18.4. Pathogenic determinants of bacteria implicated in dental caries

Property	Comment
Sugar transport	High and low affinity transport systems required to operate over wide range of conditions, especially at low pH.
Acid production	Rapid glycolytic rates, especially at low pH.
Aciduricity	Ability to survive, grow and metabolise in low pH environments.
Extracellular polysaccharide production	Contributes to plaque matrix and consolidates cell attachment. Aids sugar penetration and enhances acid production.
Intracellular polysaccharide production	Catabolism enables significant acid production to continue in the absence of dietary sugars.

down the root of the tooth to form a periodontal pocket. In advanced stages of disease, attachment fibres and alveolar bone are also lost; ultimately, teeth may become mobile in their sockets and require extraction. Generally, the predominant bacteria in periodontal pockets are obligately anaerobic or CO_2-requiring (capnophilic) Gram negative rods, filaments or spiral shaped bacteria, many of which are nutritionally fastidious and difficult (or, at present, impossible) to grow in the laboratory. Recent studies using sophisticated cultural techniques or gene probe technology have identified several new species or genera unique to this habitat. The plaque microflora associated with the main forms of periodontal diseases will be described briefly below.

Gingivitis is a non-specific inflammatory response to dental plaque involving the gums (gingival margins). If good oral hygiene is restored, gingivitis is usually eradicated and the gingival tissue becomes clinically normal again. Probably the whole of the dentate population is affected by gingivitis at some stage in their life. Gingivitis is associated with an increase in plaque mass around the gingival margin. This leads to a shift in the composition of plaque away from a streptococci-dominated microflora (Slots 1977) towards higher levels of *Actinomyces* spp. and an

increase in the isolation of capnophilic and obligately anaerobic Gram negative bacteria (Savitt & Socransky 1984; Moore *et al.* 1987). The microflora increases in diversity during the development of gingivitis, but no particular group of bacteria is uniquely associated with disease. Increased proportions of *F. nucleatum*, *Prevotella intermedia*, *Capnocytophaga* spp., *Eubacterium* spp., and spirochaetes have been reported in gingivitis (Savitt & Socransky 1984; Moore *et al.* 1987). It is still not clear whether gingivitis is a prerequisite for the development of more advanced forms of periodontal disease but, perhaps significantly, it has been reported that some of the species that predominate in chronic periodontitis, but which are not detectable in the healthy gingival crevice, are also found as a small percentage of the flora in gingivitis (Moore *et al.* 1987).

Chronic adult periodontitis is the most common form of advanced periodontal disease affecting the general population. Rather than pocket formation progressing at a slow but consistent rate, it is now believed that attachment loss occurs during relatively short bursts of activity, which are followed by periods of quiescence or even repair (Socransky *et al.* 1984). The microflora in adult periodontitis is extremely diverse; in one comprehensive study, 136 distinct microbial taxa were isolated from 38 samples from 22 adults (Moore *et al.* 1983). The predominant flora is highly variable both between subjects and even between sites in the same subject. It is difficult technically (i) to sample the advancing front of the lesion without also including within the sample bacteria from elsewhere in the periodontal pocket and which may have no role in disease, and (ii) to be certain that plaque is being sampled only during periods of disease activity (Socransky *et al.* 1987). There is a progressive change in the composition of the microflora from health and gingivitis to chronic periodontitis. This change involves not only the emergence of apparently previously undetected species, but also modifications to the numbers or proportions of a variety of species already present. Some of the bacterial species that have been implicated in chronic periodontitis in humans are listed in Table 18.5.

There are some acute or exaggerated forms of periodontal diseases due to a variety of predisposing conditions. The main forms include a more

Table 18.5. Some bacterial species that have been commonly implicated in chronic periodontitis in adult humans

Gram positive	Gram negative
Peptostreptococcus micros	*'Bacteroides forsythus '*[a]
Eubacterium brachy	*Prevotella intermedia*
Eubacterium nodatum	*Porphyromonas gingivalis*
Eubacterium timidium	*Fusobacterium nucleatum*
	Campylobacter rectus
	Eikenella corrodens
	Treponema spp.

[a]This species is not a true *Bacteroides*, but its taxonomic status has yet to be resolved.

Table 18.6. Pathogenic determinants of bacteria implicated in periodontal diseases

Stage of disease		Bacterial factor
Attachment to host tissues		Adhesins (proteins) Fimbriae
Evasion of host defences		Capsules and slimes Polymorph-receptor blockers Leukotoxin Immunoglobulin- and complement-degrading proteases Suppressor T cell induction
Tissue damage	(a) direct	Enzymes: proteases, collagenase, hyaluronidase, chondroitin sulphatase Bone resorbing factors: capsule, LPS, lipoteichoic acid Cytotoxins: butyric and propionic acids, indole, ammonia, amines, volatile sulphur compounds
	(b) indirect	Inflammatory response to plaque antigens Interleukin-1 production and proteinase synthesis in response to plaque antigens

aggressive form of gingivitis and periodontitis in HIV patients, presumably due to the impaired immune status of these individuals. The predominant species, however, appear to be similar to those found in conventional cases of adult periodontitis except that some opportunistic pathogens (*Candida* spp., enteric Gram negative rods) can be present (Zambon *et al.* 1990; Rams *et al.* 1991). Acute necrotizing ulcerative gingivitis is a painful condition of the gums, and microorganisms invade the gingival tissues. The predominant bacteria seen by microscopy and obtained by culture are spirochaetes, fusobacteria and *P. intermedia* (Loesche *et al.* 1982). Juvenile periodontitis is a relatively rare and aggressive condition which usually occurs in adolescents. It may be associated with neutrophil dysfunction. There is a distinct pattern of extremely rapid bone loss which is localized characteristically (for as yet unknown reasons) to the first permanent molars and the incisor teeth. Surprisingly, in view of the aggressive nature of this disease, the plaque from infected sites is relatively sparse; the disease is associated with the capnophilic organism, *Actinobacillus actinomycetemcomitans*, many strains of which produce a leukotoxin (Zambon 1985).

Bacteria that are associated with periodontal diseases are often proteolytic, and produce enzymes that can damage tissue directly and/or interfere with the activity of the host defences (Slots & Genco 1984). They also produce metabolites that can be cytotoxic (acids, ammonia and sulphur containing compounds) (Table 18.6). As no single species produces all of these factors, this is taken as further evidence that periodontal diseases are an example of a polymicrobial infection, perhaps involving pathogenic synergism (Marsh & Martin 1992). In this way, organisms that are individually unable to cause disease combine forces to do so. The host mounts an immunological response to the bacterial challenge, and the resulting inflammation can also lead to tissue damage due to the release of proteolytic enzymes from phagocytic and other host cells (Genco & Slots 1984; Meikle *et al.* 1986).

Approaches to control dental plaque and prevent disease

The mechanical removal of plaque by efficient oral hygiene (brushing and flossing) can almost completely prevent plaque mediated dental diseases, especially when this is combined with a reduced frequency of sugar intake. However, as it is difficult to alter established eating patterns and to maintain a high degree of motivation for effective oral hygiene, alternative preventive measures are being developed either to control dental plaque or to increase the resistance of the host.

Fissure sealants can be applied to newly erupted teeth to protect physically this most caries-prone surface from plaque acids. Fluoride, delivered from drinking water or from toothpastes and other sources, is incorporated into enamel to form fluorapatite, which is thermodynamically more stable than hydroxyapatite, and hence is more resistant to acid attack. Fluoride may also interfere with the metabolism of plaque bacteria by inhibiting glycolysis, especially at low pH (Hamilton & Bowden 1988). The frequency of the acid challenge to teeth can also be reduced by replacing fermentable substrates in the diet with sweetening agents that are only weakly metabolized, or not metabolized at all, by plaque bacteria. These sugar substitutes include intense sweeteners such as cyclamate, aspartame and saccharin, and bulk agents such as the sugar alcohols, sorbitol and xylitol. Some of these alternative sweeteners also possess weak antimicrobial activity (Grenby & Saldanha 1986; Mäkinen 1989). As saliva and GCF provide the mouth with all of the components necessary to mount an effective immune response, vaccines have been developed against the major group of implicated bacteria, the mutans streptococci (Russell & Johnson 1987; Krasse et al. 1987). These vaccines have been shown to be effective in reducing caries in rodent and primate models, but no human trials have yet been commissioned.

The addition of antiplaque or antimicrobial agents to toothpastes and mouthwashes is now being used widely as an adjunct to conventional mechanical oral hygiene procedures. These agents include plant extracts (sanguinarine), metal salts (stannous, zinc), enzymes (glucan hydrolases), quaternary ammonium compounds (cetylpyridinium chloride), bisbiguanides (chlorhexidine), and phenols (Triclosan) (Scheie 1989; Marsh 1993). Recent studies support data obtained with microorganisms from other ecosystems in showing that oral bacteria growing on a surface are more resistant to antimicrobial agents. This is of particular relevance to the use of such agents for plaque control. Mature biofilms of *S. sanguis* were more tolerant of chlorhexidine than cells in suspension (Millward & Wilson 1989), while mixed culture biofilms were also found to be less sensitive to a range of inhibitors used in toothpastes than the planktonic culture (Marsh & Bradshaw 1993). Although these agents have been shown in clinical studies to reduce plaque and gingivitis, their prolonged use must not disrupt the natural ecology of plaque nor lead to overgrowth by exogenous microorganisms. This apparent paradoxical requirement is achieved by the fact that, although most plaque control agents are described as having a broad spectrum of antimicrobial activity from conventional minimum inhibitory concentration tests, they operate in the mouth over short contact times and with rapid salivary clearance rates, and under these conditions their mode of action is more selective (Marsh 1993). Laboratory studies using defined mixed cultures in either a conventional planktonic continuous culture mode or on hydroxyapatite discs as a biofilm have shown that short pulses of the broad spectrum agent, Triclosan, selectively inhibit the Gram negative anaerobic species associated with gingivitis and other periodontal diseases, while leaving relatively unaffected the Gram positive organisms associated more with sound enamel (Bradshaw et al. 1993; Marsh & Bradshaw 1993).

Dental plaque is a complex biofilm which, because of its location, is readily accessible for study. Knowledge derived from fundamental studies of the physiology and ecology of dental plaque is not merely of relevance to the oral cavity but should also be applicable to other biofilm problems. In recent years, much has been learnt about dental plaque formation, and the role of this biofilm in health and disease. The current challenge is to convert this information into new strategies to control dental plaque and reduce its burden on public health resources while, at the same time, preserving the beneficial properties to the host of the resident microflora of plaque.

Acknowledgements

I would like to thank Dr M. Pang, University of Hong Kong, Dr Allan Saxton, Unilever Dental Research, and Dr Bente Nyvad, University of Aarhus, for generously providing the electron micrographs.

References

Ahmady, K., Marsh, P. D., Newman, H. N. & Bulman, J. S. (1993). Distribution of *Streptococcus mutans* and *Streptococcus sobrinus* at sub-sites in human approximal dental plaque. *Caries Research*, 27, 135–9.

Al-Hashimi, I. & Levine, M. J. (1989). Characterization of *in vivo* salivary-derived enamel pellicle. *Archives of Oral Biology*, **34**, 289–95.

Beckers, H. J. A. & van der Hoeven, J. S. (1982). Growth rates of *Actinomyces viscosus* and *Streptococcus mutans* during early colonization of tooth surfaces in gnotobiotic rats. *Infection and Immunity*, **35**, 583–7.

Beighton, D., Smith, K. & Hayday, H. (1986). The growth of bacteria and the production of exoglycosidic enzymes in the dental plaque of macaque monkeys. *Archives of Oral Biology*, **31**, 829–35.

Berry, C. W. & Henry, C. A. (1977). The effect of adsorption on the acid production of caries and non-caries-producing streptococci. *Journal of Dental Research*, **56**, 1193–1200.

Bowden, G. H. W. (1990). Microbiology of root surface caries in humans. *Journal of Dental Research*, **69**, 1205–10.

Bowden, G. H. W. (1991). Which bacteria are cariogenic in humans? In *Dental caries: markers of high and low risk groups and individuals*, ed. N. W. Johnson, pp. 266–86. Cambridge: Cambridge University Press.

Bowden, G. H., Hardie, J. M. & Slack, G. L. (1975). Microbial variations in approximal dental plaque. *Caries Research*, **9**, 253–77.

Boyar, R. M. & Bowden, G. H. (1985). The microflora associated with the progression of incipient lesions in teeth of children living in a water fluoridated area. *Caries Research*, **19**, 298–306.

Bradshaw, D. J., Marsh, P. D., Watson, G. K. & Cummins, D. (1993). The effects of Triclosan and zinc citrate, alone and in combination, on a community of oral bacteria grown *in vitro*. *Journal of Dental Research*, **72**, 25–30.

Bradshaw, D. J., McKee, A. S. & Marsh, P. D. (1989). Effect of carbohydrate pulses and pH on population shifts within oral microbial communities *in vitro*. *Journal of Dental Research*, **68**, 1298–1302.

Brecx, M., Theilade, J. & Attström, R. (1983). An ultrastructural quantitative study of the significance of microbial multiplication during early plaque growth. *Journal of Periodontal Research*, **18**, 177–86.

Busscher, H. J., Cowan, M. M. & van der Mei, H. C. (1992). On the relative importance of specific and non-specific approaches to oral microbial adhesion. *FEMS Microbiology Reviews*, **88**, 199–210.

Cimasoni, G. (1983). *The crevicular fluid updated*, ed. H. M. Myers. Basel: Karger.

Cisar, J. O. (1986). Fimbrial lectins of the oral *Actinomyces*. In *Microbial lectins and agglutinins; properties and biological activity*, ed. D. Mirelman, pp. 183–96. New York: Wiley.

Cisar, J. O., Sandberg, A. L. & Mergenhagen, S. E. (1984). The function and distribution of different fimbriae on the strains of *Actinomyces viscosus* and *Actinomyces naeslundii*. *Journal of Dental Research*, **63**, 393–6.

Demuth, D. R., Lammey, M. S., Huck, M., Lally, E. & Malamud, D. (1990). Comparison of *Streptococcus mutans* and *Streptococcus sanguis* receptors for human salivary agglutinin. *Microbial Pathogenesis*, **9**, 199–211.

Dent, V. (1979). The bacteriology of dental plaque from a variety of zoo-maintained mammalian species. *Archives of Oral Biology*, **24**, 277–82.

de Stoppelaar, J. D., van Houte, J. & Backer Dirks, O. (1970). The effect of carbohydrate restriction on the presence of *Streptococcus mutans*, *Streptococcus sanguis* and iodophilic polysaccharide-producing bacteria in human dental plaque. *Caries Research*, **4**, 114–23.

Dibdin, G. H. & Shellis, R. P. (1988). Physical and biochemical studies of *Streptococcus mutans* sediments suggest new factors linking the cariogenicity of plaque with its extracellular polysaccharide content. *Journal of Dental Research*, **67**, 890–5.

Dibdin, G. H., Wilson, C. M. & Shellis, R. P. (1983). Effect of packing density and polysaccharide to protein ratio of plaque samples cultured *in vitro* upon their permeability. *Caries Research*, **17**, 52–8.

Douglas, C. W. I., Pease, A. A. & Whiley, R. A. (1990). Amylase-binding as a discriminator among oral streptococci. *FEMS Microbiology Letters*, **66**, 193–8.

Douglas, C. W. I. & Russell, R. R. B. (1984). Effect of specific antisera upon *Streptococcus mutans* adherence to saliva-coated hydroxyapatite. *FEMS Microbiology Letters*, **25**, 211–14.

Eggert, F. M., Drewell, L., Bigelow, J. A., Speck, J. E. & Goldner, M. (1991). The pH of gingival crevices and periodontal pockets in children, teenagers and adults. *Archives of Oral Biology*, **36**, 233–8.

Ganeshkumar, N., Arora, N. & Kolenbrander, P. E. (1993). Saliva-binding protein (SsaB) from *Streptococcus sanguis* 12 is a lipoprotein. *Journal of Bacteriology*, **175**, 572–4.

Genco, R. J. & Slots, J. (1984). Host responses in periodontal diseases. *Journal of Dental Research*, **63**, 441–51.

George, K. S. & Falkler, W. A. Jr. (1992). Coaggregation studies of the *Eubacterium* species. *Oral Microbiology and Immunology*, **7**, 285–90.

Gibbons, R. J. (1989). Bacterial adhesion to oral tissues: a model for infectious diseases. *Journal of Dental Research*, **68**, 750–60.

Gibbons, R. J., Hay, D. I, Cisar, J. O. & Clark, W. B. (1988). Adsorbed salivary proline-rich protein-1 and statherin: receptors for type 1 fimbriae of *Actinomyces viscosus* T14V-J1 on apatitic surfaces. *Infection and Immunity*, **56**, 2990–3.

Gibbons, R. J., Hay, D. I., Childs, W. C. III & Davis, G. (1990). Role of cryptic receptors (cryptitopes) in bacterial adhesion to oral surfaces. *Archives of Oral Biology*, **35** (Suppl.), 107S-114S.

Gibbons, R. J., Hay, D. I. & Schlesinger, D. H. (1991). Delineation of a segment of adsorbed salivary acidic proline-rich proteins which promotes adhesion of *Streptococcus gordonii* to apatitic surfaces. *Infection and Immunity*, **59**, 2948-54.

Grenby, T. H. & Saldanha, M. G. (1986). Studies of the inhibitory action of intense sweeteners on oral microorganisms relating to dental health. *Caries Research*, **20**, 7-16.

Grenier, D. & Mayrand, D. (1986). Nutritional relationships between oral bacteria. *Infection and Immunity*, **53**, 616-20.

Hamilton, I. R. (1987). Effects of changing environment on sugar transport and metabolism by oral bacteria. In *Sugar transport and metabolism in Gram positive bacteria*, ed. A. Reizer & A. Peterkofsky, pp. 94-133. Chichester: Ellis Horwood.

Hamilton, I. R. & Bowden, G. H. (1988). Effect of fluoride on oral microorganisms. In *Fluoride in dentistry*, ed. J. Ekstrand, O. Fejerskov & L. M. Silverstone, pp. 77-103. Copenhagen: Munksgaard.

Hillman, J. D., Socransky, S. S. & Shivers, M. (1985). The relationships between streptococcal species and periodontopathic bacteria in human dental plaque. *Archives of Oral Biology*, **30**, 791-5.

Hogg, S. D. & Embery, G. (1982). Blood-group reactive glycoproteins from human saliva interact with lipoteichoic acid on the surface of *Streptococcus sanguis* cells. *Archives of Oral Biology*, **27**, 261-8.

Homer, K. A. & Beighton, D. (1992). Synergistic degradation of bovine serum albumin by mutans streptococci and other dental plaque bacteria. *FEMS Microbiology Letters*, **90**, 259-62.

Hudson, M. C. & Curtiss, R. III (1990). Regulation of expression of *Streptococcus mutans* genes important to virulence. *Infection and Immunity*, **58**, 464-70.

Jenkinson, H. F. (1992). Adherence, coaggregation, and hydrophobicity of *Streptococcus gordonii* associated with expression of cell surface lipoproteins. *Infection and Immunity*, **60**, 1225-8.

Jenkinson, H. F., Terry, S. D., McNab, R. & Tannock, G. W. (1993). Inactivation of the gene encoding surface protein SspA in *Streptococcus gordonii* DL1 affects cell interactions with human salivary agglutinin and oral *Actinomyces*. *Infection and Immunity*, **61**, 3199-208.

Jensen, M. E. & Schachtele, C. F. (1983). The acidogenic potential of reference foods and snacks at interproximal sites. *Journal of Dental Research*, **62**, 889-92.

Kenney, E. B. & Ash, M. (1969). Oxidation-reduction potential of developing plaque, periodontal pockets and gingival sulci. *Journal of Periodontology*, **40**, 630-3.

Kolenbrander, P. E., Andersen, R. N. & Moore, L. V. H. (1989). Coaggregation of *Fusobacterium nucleatum*, *Selenomonas flueggei*, *Selenomonas infelix*, *Selenomonas noxia*, and *Selenomonas sputigena* with strains from 11 genera of oral bacteria. *Infection and Immunity*, **57**, 3194-203.

Kolenbrander, P. E. & London, J. (1992). Ecological significance of coaggregation among oral bacteria. *Advances in Microbial Ecology*, **12**, 183-217.

Kolenbrander, P. E. & London, J. (1993). Adhere today, here tomorrow: oral bacterial adherence. *Journal of Bacteriology*, **175**, 3247-52.

Krasse, B., Emilson, C. & Gahnberg, L. (1987). An anticaries vaccine: report on the status of research. *Caries Research*, **21**, 255-76.

Listgarten, M. A. (1976). Structure of the microbial flora associated with periodontal health and disease in man. A light and electron microscopic study. *Journal of Periodontology*, **47**, 1-18.

Littleton, N. W., McCabe, R. M. & Carter, C. H. (1967). Studies of oral health in persons nourished by stomach tube. II. Acidogenic properties and selected bacterial components of plaque material. *Archives of Oral Biology*, **12**, 601-9.

Loesche, W. J. (1986). Role of *Streptococcus mutans* in human dental decay. *Microbiological Reviews*, **50**, 353-80.

Loesche, W. J., Syed, S. A., Laughon, B. E. & Stoll, J. (1982). The bacteriology of acute necrotizing ulcerative gingivitis. *Journal of Periodontology*, **53**, 223-30.

Ma, J. K.-C., Kelly, C. G., Munro, G., Whiley, R. A. & Lehner, T. (1991). Conservation of the gene encoding streptococcal antigen I/II in oral streptococci. *Infection and Immunity*, **59**, 2686-94.

Mäkinen, K. K. (1989). Latest dental studies on xylitol and mechanism of action of xylitol in caries limitation. In *Progress in sweeteners*, ed. T. H. Grenby, pp. 331-62. London: Elsevier.

Marsh, P. D. (1989). Host defenses and microbial homeostasis: role of microbial interactions. *Journal of Dental Research*, **68**, 1567-75.

Marsh, P. D. (1993). Microbiological aspects of the chemical control of plaque and gingivitis. *Journal of Dental Research*, **71**, 1431-8.

Marsh, P. D. & Bradshaw, D. J. (1993). Microbiological effects of new agents in dentifrices for plaque control. *International Dental Journal*, **43**, 399-406.

Marsh, P. D., Featherstone, A., McKee, A. S. *et al.* (1989). A microbiological study of early caries of approximal surfaces in schoolchildren. *Journal of Dental Research*, **68**, 1151-4.

Marsh, P. D. & Martin, M. V. (1992). *Oral microbiology*, 3rd edn. London: Chapman & Hall.

McBride, B. C. & Gislow, M. T. (1977). Role of sialic acid in saliva induced aggregation of *Streptococcus sanguis*. *Infection and Immunity*, **18**, 35–40.

McDermid, A. S., McKee, A. S. & Marsh, P. D. (1990). Interactions and pH optima for growth of three black-pigmented *Bacteroides* species. *Journal of Dental Research*, **69**, 999 (abstract).

McNee, S. G., Geddes, D. A. M., Weetman, D. A., Sweeney, D. & Beeley, J. A. (1982). Effect of extracellular polysaccharides on diffusion of NaF and ^{14}C-sucrose in human dental plaque and in sediments of the bacterium *Streptococcus sanguis* 804 (NCTC 10904). *Archives of Oral Biology*, **27**, 981–6.

Meikle, M. C., Heath, J. K. & Reynolds, J. J. (1986). Advances in understanding cell interactions in tissue resorption. Relevance to the pathogenesis of periodontal diseases and a new hypothesis. *Journal of Oral Pathology*, **15**, 239–50.

Mikx, F. H. M. & van der Hoeven, J. S. (1975). Symbiosis of *Streptococcus mutans* and *Veillonella alcalescens* in mixed continuous culture. *Archives of Oral Biology*, **20**, 407–10.

Mikx, F. H. M., van der Hoeven, J. S., König, K. G., Plasschaert, A. J. M. & Guggenheim, B. (1972). Establishment of defined microbial ecosystems in germ-free rats. I. The effect of the interaction of *Streptococcus mutans* or *Streptococcus sanguis* with *Veillonella alcalescens* on plaque formation and caries activity. *Caries Research*, **6**, 211–23.

Millward, T. A. & Wilson, M. (1989). The effect of chlorhexidine on *Streptococcus sanguis* biofilms. *Microbios*, **58**, 155–64.

Moore, W. E. C., Holdeman, L. V., Cato, E. P., Smibert, R. M., Burmeister, J. A. & Ranney, R. R. (1983). Bacteriology of moderate (chronic) periodontitis in mature adult humans. *Infection and Immunity*, **42**, 510–15.

Moore, L. V. H., Moore, W. E. C., Cato, E. P. *et al.* (1987). Bacteriology of human gingivitis. *Journal of Dental Research*, **66**, 989–95.

Mukherjee, S. (1985). The role of crevicular iron in periodontal disease. *Journal of Periodontology*, **56** (Suppl.), 22–7.

Murray, P. A., Levine, M. J., Reddy, M. S., Tabak, L. A. & Bergey, E. J. (1986). Preparation of a sialic acid-binding protein from *Streptococcus mitis* KS32AR. *Infection and Immunity*, **53**, 359–65.

Murray, P. A., Levine, M. J., Tabak, L. A. & Reddy, M. S. (1982). Specificity of salivary-bacterial interactions. II. Evidence for a lectin on *Streptococcus sanguis* with specificity for a NeuAc(α)2, 3Gal(β)1, 3GalNAc sequence. *Biochemistry and Biophysical Research Communications*, **106**, 390–6.

Nyvad, B. (1993). Microbial colonization of human tooth surfaces. *Acta Pathologica, Microbiologica et Immunologica Scandinavica*, **101** (Suppl. 32), 7–45.

Nyvad, B. & Fejerskov, O. (1989). Structure of dental plaque and the plaque–enamel interface in human experimental caries. *Caries Research*, **23**, 151–8.

Nyvad, B. & Kilian, M. (1987). Microbiology of the early colonization of human enamel and root surfaces *in vivo*. *Scandinavian Journal of Dental Research*, **95**, 369–80.

Nyvad, B. & Kilian, M. (1990). Comparison of the initial streptococcal microflora on dental enamel in caries-active and in caries-inactive individuals. *Caries Research*, **24**, 267–72.

Ohta, H., Gottschal, J. C., Fukui, K. & Kato, K. (1990). Interrelationships between *Wolinella recta* and *Streptococcus sanguis* in mixed continuous cultures. *Microbial Ecology in Health and Disease*, **3**, 237–44.

Olsson, J., Jontell, M. & Krasse, B. (1980). Effect of macromolecules on adherence of *Streptococcus mutans*. *Scandinavian Journal of Infectious Disease*, **24** (Suppl.), 173–8.

Pearce, E. I. F. & Bibby, B. G. (1966). Protein adsorption on bovine enamel. *Archives of Oral Biology*, **11**, 329–36.

Perrons, C. J. & Donoghue, H. D. (1990). Colonization resistance of defined bacterial plaques to *Streptococcus mutans* implantation on teeth in a model mouth. *Journal of Dental Research*, **69**, 483–8.

Rams, T. E., Andriolo, M. Jr, Feik, D., Abel, S. N., McGivern, T. M. & Slots, J. (1991). Microbiological study of HIV-related periodontitis. *Journal of Periodontology*, **62**, 74–81.

Reinholdt, J. & Kilian, M. (1987). Interference of IgA protease with the effect of secretory IgA on adherence of oral streptococci to saliva-coated hydroxyapatite. *Journal of Dental Research*, **66**, 492–7.

Reynolds, E. C. & Wong, A. (1983). Effect of adsorbed protein on hydroxyapatite zeta potential and *Streptococcus mutans* adherence. *Infection and Immunity*, **39**, 1285–90.

Rogers, A. H., van der Hoeven, J. S. & Mikx, F. H. M. (1978). Inhibition of *Actinomyces viscosus* by bacteriocin-producing strains of *Streptococcus mutans* in the dental plaque of gnotobiotic rats. *Archives of Oral Biology*, **23**, 477–83.

Rogers, A. H., van der Hoeven, J. S. & Mikx, F. H. M. (1979). Effect of bacteriocin production by *Streptococcus mutans* on the plaque of gnotobiotic rats. *Infection and Immunity*, **23**, 571–6.

Russell, R. R. B. & Johnson, N. W. (1987). The prospects for vaccination against dental caries. *British Dental Journal*, **162**, 29–34.

Rutter, P. R. & Vincent, B. (1980). The adhesion of microorganisms to surfaces: physico-chemical aspects. In *Microbial adhesion to surfaces*, ed. R. C. W. Berkeley, J. M. Lynch, J. Melling, P. R. Rutter & B. Vincent, pp. 79–93. Chichester: Ellis Horwood.

Savitt, E. D. & Socransky, S. S. (1984). Distribution of certain subgingival microbial species in selected periodontal conditions. *Journal of Periodontal Research*, **19**, 111–23.

Scannapieco, F. A., Bergey, E. J., Reddy, M. S. & Levine, M. J. (1989). Characterization of salivary α-amylase binding to *Streptococcus sanguis*. *Infection and Immunity*, **57**, 2853–63.

Scheie, A. Aa. (1989). Modes of action of currently known chemical anti-plaque agents other than chlorhexidine. *Journal of Dental Research*, **68**, 1609–16.

Scheie, A. Aa., Eggen, K. H. & Rölla, G. (1987). Glucosyltransferase activity in human *in vivo* formed enamel pellicle and in whole saliva. *Scandinavian Journal of Dental Research*, **95**, 212–15.

Schilling, K. M. & Bowen, W. H. (1988). The activity of glucosyltransferase adsorbed onto saliva-coated hydroxyapatite. *Journal of Dental Research*, **67**, 2–8.

Shibata, S., Nagata, K., Nakamura, R., Tsunemitsu, A. & Misaki, A. (1980). Interaction of parotid saliva basic glycoprotein with *Streptococcus sanguis* ATCC 10557. *Journal of Periodontology*, **51**, 499–504.

Slots, J. (1977). Microflora in the healthy gingival sulcus in man. *Scandinavian Journal of Dental Research*, **85**, 247–54.

Slots, J. & Genco, R. J. (1984). Black-pigmented *Bacteroides* species, *Capnocytophaga* species, and *Actinobacillus actinomycetemcomitans* in human periodontal disease: virulence factors in colonization, survival, and tissue destruction. *Journal of Dental Research*, **63**, 412–21.

Socransky, S. S., Haffajee, A. D., Goodson, J. M. & Lindhe, J. (1984). New concepts of destructive periodontal disease. *Journal of Clinical Periodontology*, **11**, 21–31.

Socransky, S. S., Haffajee, A. D., Smith, G. L. F. & Dzink, J. L. (1987). Difficulties encountered in the search for the etiologic agents of destructive periodontal diseases. *Journal of Clinical Periodontology*, **14**, 588–93.

Strömberg, N. & Borén, T. (1992). *Actinomyces* tissue specificity may depend on differences in receptor specificity for GalNAcß-containing glycoconjugates. *Infection and Immunity*, **60**, 3268–77.

Strömberg, N., Borén, T., Carlén, A. & Olsson, J. (1992). Salivary receptors for GalNAcß-sensitive adherence of *Actinomyces* spp.: Evidence for heterogeneous GalNAcß and proline-rich receptor properties. *Infection and Immunity*, **60**, 3278–86.

Tatevossian, A. (1991). Film fermenters in dental research. In *Handbook of laboratory model systems for microbial ecosystems*, Vol. 1, ed. J. W. T. Wimpenny, pp. 197–227. Boca Raton, Fla: CRC Press.

ter Steeg, P. F., van der Hoeven, J. S., de Jong, M. H., van Munster, P. J. J. & Jansen, M. J. H. (1987). Enrichment of subgingival microflora on human serum leading to accumulation of *Bacteroides* species, *Peptostreptococci* and *Fusobacteria*. *Antonie van Leeuwenhoek*, **53**, 261–71.

Theilade, J., Fejerskov, O. & Hörsted, M. (1976). A transmission electron microscopic study of 7–day old bacterial plaque in human tooth fissures. *Archives of Oral Biology*, **21**, 587–98.

Theilade, E., Fejerskov, O., Karring, T. & Theilade, J. (1978). A microbiological study of old plaque in occlusal fissures of human teeth. *Caries Research*, **12**, 313–19.

Theilade, E., Fejerskov, O., Karring, T. & Theilade, J. (1982). Predominant cultivable microflora of human dental fissure plaque. *Infection and Immunity*, **36**, 977–82.

Theilade, E., Fejerskov, O., Prachyabrued, W. & Kilian, M. (1974). Microbiologic study on developing plaque in human fissures. *Scandinavian Journal of Dental Research*, **82**, 420–7.

van der Hoeven, J. S. & Camp, P. J. M. (1991). Synergistic degradation of mucin by *Streptococcus oralis* and *Streptococcus sanguis* in mixed chemostat cultures. *Journal of Dental Research*, **68**, 1041–4.

van der Hoeven, J. S. & Camp, P. J. M. (1993). Mixed continuous cultures of *Streptococcus mutans* with *Streptococcus sanguis* or with *Streptococcus oralis* as a model to study ecological effects of the lactoperoxidase system. *Caries Research*, **27**, 26–30.

van der Hoeven, J. S., de Jong, M. H. & Rogers, A. H. (1985). Effect of utilization of substrates on the composition of dental plaque. *FEMS Microbiology Ecology*, **31**, 129–33.

van der Hoeven, J. S. & Rogers, A. H. (1979). Stability of the resident microflora and bacteriocinogeny of *Streptococcus mutans* as factors affecting its establishment in specific pathogen free rats. *Infection and Immunity*, **23**, 206–12.

van Palenstein Helderman, W. H. (1975). Total viable count and differential count of *Vibrio* (*Campylobacter*) *sputorum*, *Fusobacterium nucleatum*, *Selenomonas sputigena*, *Bacteroides ochraceus* and *Veillonella* in the inflamed and non inflamed gingival crevice. *Journal of Periodontal Research*, **10**, 230–41.

Walker, G. J. & Jacques, N. A. (1987). Polysaccharides of oral streptococci. In *Sugar transport and metabolism in Gram positive bacteria*, ed. A. Reizer & A. Peterkofsky, pp. 39–68. Chichester: Ellis Horwood.

Zambon, J. J. (1985). *Actinobacillus actinomycetemcomitans* in human periodontal disease. *Journal of Clinical Periodontology*, **12**, 1–20.

Zambon, J. J., Reynolds, H. S. & Genco, R. J. (1990). Studies of the sub-gingival microflora in patients with acquired immunodeficiency syndrome. *Journal of Periodontology*, **61**, 699–704.

Zero, D. T., van Houte, J. & Russo, J. (1986). The intraoral effect on enamel demineralization of extracellular matrix material synthesized from sucrose by *Streptococcus mutans*. *Journal of Dental Research*, **65**, 918–23.

Index